ASPECTS OF TOPOLOGY

MONOGRAPHS AND TEXTBOOKS IN
PURE AND APPLIED MATHEMATICS

ASPECTS OF TOPOLOGY

CHARLES O. CHRISTENSON
WILLIAM L. VOXMAN

Department of Mathematics
College of Letters and Science
University of Idaho
Moscow, Idaho

MARCEL DEKKER, INC.   New York and Basel

QA
611
C436

Library of Congress Cataloging in Publication Data

Christenson, Charles O
   Aspects of topology.

   (Pure and applied mathematics ; 39)
   Bibliography: p.
   Includes index.
   1. Topology. I. Voxman, William L., joint
author. II. Title.
QA611.C436   514   75-40764
ISBN 0-8247-6331-9

MARCEL DEKKER, INC.
270 Madison Avenue, New York, New York 10016

Current printing (last digit):
10 9 8 7 6 5 4 3 2 1

PRINTED IN THE UNITED STATES OF AMERICA

The authors dedicate this work to their former mentors,
S. Armentrout, J. Giever, A. Kruse, and T. Price.

I had a feeling about Mathematics—that I saw it all. Depth beyond Depth was revealed to me—the Byss and the Abyss. I saw—as one might see the transit of Venus or even the Lord Mayor's Show—a quantity passing through infinity and changing its sign from plus to minus. I saw exactly how it happened and why the tergiversation was inevitable—but it was after dinner and I let it go.

Winston S. Churchill

PREFACE

In this book we have tried to present a blend of classic and contemporary topology. Thus, in addition to that material usually found in a general topology text, we have included fairly detailed accounts of homotopy and simplicial theory, a solid introduction to the study of manifolds, as well as a variety of other somewhat more esoteric topics. Although the book is primarily designed as a text, it is our intent that it should also be of use as a general reference work. We point out, however, that whenever there seemed to be a conflict between generality and clarity, the latter virtue was favored.

We abscribe wholeheartedly to the idea that learning mathematics is synonymous with doing mathematics. Exercises have been interspersed throughout the text and made an integral part of the exposition. The generous instructor may wish to present solutions to these exercises as a part of the lecture content, but in any case, the reader is advised to work as many of them as possible. A number of additional problems are given at the end of each chapter. Although the continuity of the text does not depend on the results of these problems, many of them are of interest in their own right. The problems that we think may cause the student exceptional difficulty are marked with an asterisk.

Prerequisites for a successful reading of the text are minimal. Previous contact with courses in advanced calculus and set theory would be advantageous, but such a background is not essential. However, starting with Chapter 12, we do use some elementary group theory commonly included in a first course in modern algebra. The illusive quality known as mathematical maturity should not be considered as a handicap.

The quantity of material in this book permits the instructor to tailor a year's course to suit his interests. In fact, there is probably sufficient material for a leisurely three semester course. Most of the standard topics in general topology are treated in the first twelve chapters; however, considerable additional material is also presented here. It is our feeling, that the remaining chapters are more in keeping with the trends in modern topology, and in a one year course, the instructor should endeavor to pursue as much of this material as time and inclination dictate. Although the topics in Chapters 13 through 18 are not totally independent of each other, if one is willing to accept an occasional result from previous chapters, then it is possible to read the chapters in any order. For instance, the theorem concerning Invariance of Domain proven in Chapter 15 is this chapter's only connection with Chapter 17.

At the University of Idaho approximately four-fifths of a preliminary version of the text was completed during a two semester course in a class that included one junior, two seniors, and various graduate students. The class met in a lecture setting for three hours a week.

This book, like most texts, is often eclectic in nature; few of the results or proofs are original. The outline of Chapters 16 and 17 follows notes developed by one of the authors and R. P. Osborne in 1965. A set of mimeographed notes of T. Price has greatly influenced our presentation of the material in Chapter 5.

The bibliography, while extensive, is by no means complete. The ultimate origin of many results are unknown to us and furthermore we have not (for pedagogical reasons) given references to the majority of the problems.

The authors would like to thank Edwin Hewitt for his kind comments and good advice. We also thank John Cobb for his insights into various problems and Leo Boron for his generous help in reading the manuscript. In addition, we are grateful to our students, C. Blais, D. Boley, W. Cordwell, D. Grey, J. Olson, L. Showalter, J. Taylor, D. Anderson, S. Cambareri, and B. Smith for vociferously calling our attention to many errors in preliminary versions of the text (any remaining errors may be viewed as learning aids). Finally, one typist, Celeste Cummins, performed unstintingly throughout several revisions of the manuscript.

Charles O. Christenson
William L. Voxman

CONTENTS

Contents

ASPECTS OF
TOPOLOGY

Chapter 0

PRELIMINARIES

In this chapter, we introduce much of the notation to be employed throughout the book. In addition, we present a sketch of the elements of set theory that will be needed in ensuing chapters. No attempt has been made to give a full presentation of this material; for more complete expositions of set theory, the books by Monk [1969] and Halmos [1960] are recommended.

Proofs of most of the propositions found in this chapter are left to the reader, presumably as review exercises.

A. BASIC NOTATION

The following symbols will be frequently used:

$\subset$	is a subset of
$\subsetneq$	is a proper subset of
$\in$	is an element of
$\notin$	is not an element of
$\mathbf{Z}$	the set of integers (also considered as an additive group)
$\mathbf{Z}^+$	the set of positive integers
$\mathbf{N}$	the set of natural numbers
$\mathbf{R}^1$	the set of real numbers
$\mathbf{Q}$	the set of rational numbers
$\mathbf{I}$	the interval $\{x \in \mathbf{R}^1 \mid 0 \leq x \leq 1\}$
$\mathbf{R}^n$	the set $\{(x_1, x_2, \ldots, x_n) \mid x_i \in \mathbf{R}^1\}$

$\varnothing$ the empty set
$\mathscr{P}(X)$ the set of subsets of X

Standard set notation is employed. The words "set," "family," and "collection" are used interchangeably throughout the book.

(0.A.1) Definition. Suppose that $\mathscr{A} = \{A_\alpha \mid \alpha \in \Lambda\}$ is a family of sets indexed by Λ. Then the *union* of the A_α is defined to be $\{x \mid x \in A_\alpha \text{ for some } \alpha\}$ and is denoted either by $\bigcup \{A_\alpha \mid \alpha \in \Lambda\}$ or $\bigcup\limits_{\alpha\in\Lambda} A_\alpha$.

The *intersection* of the A_α is defined to be $\{x \mid x \in A_\alpha \text{ for each } \alpha\}$ and is denoted by $\bigcap \{A_\alpha \mid \alpha \in \Lambda\}$ or $\bigcap\limits_{\alpha\in\Lambda} A_\alpha$.

(0.A.2) Definition. If X is a set and A is a subset of X, then the *complement of A in X* is defined to be $\{x \in X \mid x \notin A\}$ and is denoted by $X \setminus A$.

(0.A.3) Definition. A set X is a *singleton* if and only if X has just one element.

Two essential and easily established results involving set operations are given in the following proposition. They are commonly referred to as *De Morgan's rules.*

(0.A.4) Proposition. Suppose that $\{A_\alpha \mid \alpha \in \Lambda\}$ is a family of subsets of a set X. Then

(i) $X \setminus (\bigcup \{A_\alpha \mid \alpha \in \Lambda\}) = \bigcap \{X \setminus A_\alpha \mid \alpha \in \Lambda\}$, and
(ii) $X \setminus (\bigcap \{A_\alpha \mid \alpha \in \Lambda\}) = \bigcup \{X \setminus A_\alpha \mid \alpha \in \Lambda\}$.

(0.A.5) Proposition. Suppose that $\{A_\alpha \mid \alpha \in \Lambda\}$ is a family of subsets of a set X and that $A \subset X$. Then

(i) $A \cap (\bigcup \{A_\alpha \mid \alpha \in \Lambda\}) = \bigcup \{A \cap A_\alpha \mid \alpha \in \Lambda\}$, and
(ii) $A \cup (\bigcap \{A_\alpha \mid \alpha \in \Lambda\}) = \bigcap \{A \cup A_\alpha \mid \alpha \in \Lambda\}$.

(0.A.6) Definition. A *subset A of $\mathbf{R}^1$* is an interval if and only if whenever $x,y \in A$ and $x < z < y$, then $z \in A$. We denote the various kinds of intervals as follows:

$$(a,b) = \{x \in \mathbf{R}^1 \mid a < x < b\}$$
$$[a,b] = \{x \in \mathbf{R}^1 \mid a \leq x \leq b\}$$
$$[a,b) = \{x \in \mathbf{R}^1 \mid a \leq x < b\}$$
$$(a,b] = \{x \in \mathbf{R}^1 \mid a < x \leq b\}$$

$$(-\infty,a) = \{x \in \mathbf{R}^1 \mid x < a\}$$
$$(-\infty,a] = \{x \in \mathbf{R}^1 \mid x \le a\}$$
$$(b,\infty) = \{x \in \mathbf{R}^1 \mid x > b\}$$
$$[b,\infty) = \{x \in \mathbf{R}^1 \mid x \ge b\}$$
$$(-\infty,\infty) = \mathbf{R}^1$$

(0.A.7) Definition. The *dyadic rationals* are defined to be all real numbers of the form $m/2^n$ where $m,n \in \mathbf{Z}$.

B. CARTESIAN PRODUCTS AND RELATIONS

(0.B.1) Definition. If X and Y are sets, then the *Cartesian product of X and Y*, denoted by $X \times Y$, consists of all ordered pairs (x,y) where $x \in X$ and $y \in Y$. That is, $X \times Y = \{(x,y) \mid x \in X$ and $y \in Y\}$.

The Cartesian product of n spaces $X_1, X_2, \ldots, X_n$ is defined by $X_1 \times X_2 \times \cdots \times X_n = \{(x_1, x_2, \ldots, x_n) \mid x_i \in X_i$ for each $i\}$.

Cartesian products are used to define relations in a set X.

(0.B.2) Definition. A *relation R in a set X* is any subset of $X \times X$. If $(x_1,x_2) \in R$, we write $x_1 R x_2$.

(0.B.3) Definition. If R is a relation on a set X, then

   (i) R is *reflexive* on X if and only if xRx for each $x \in X$,
   (ii) R is *symmetric* if and only if yRx whenever xRy, and
   (iii) R is *transitive* if and only if xRy and yRz implies that xRz.

A relation R is an *equivalence relation* on X if and only if it is reflexive on X, symmetric, and transitive.

Suppose that R is an equivalence relation in X and $x \in X$. Then the *equivalence class of x* is $\{y \in X \mid xRy\}$ and is denoted by $[x]$ or $[x]_R$.

(0.B.4) Proposition. Suppose that R is an equivalence relation in a set X. Then $\{[x] \mid x \in X\}$ forms a partition of X, i.e., $\bigcup \{[x] \mid x \in X\} = X$, and if $x,y \in X$, then either $[x] \cap [y] = \varnothing$ or $[x] = [y]$.

Relations are used to define various kinds of orderings.

(0.B.5) Definition. Suppose that R is a relation in a set X. Then R is a *partial ordering* for X if and only if

(i) R is reflexive and transitive, and
(ii) if xRy and yRx, then $x = y$.

In this context R is frequently replaced by the symbol $\leq$. Thus, xRy if and only if $x \leq y$. We write $x < y$ if and only if $x \leq y$ and $x \neq y$. If $\leq$ is a partial ordering for a set X, we write $(X, \leq)$.

(0.B.6) Definition.   A partial ordering $\leq$ for a set X is a *linear ordering* if and only if for each $x,y \in X$ either $x \leq y$ or $y \leq x$.

(0.B.7) Definition.   A linear ordering $\leq$ for a set X is a *well ordering* if and only if whenever $A \neq \emptyset$ and $A \subset X$, there is an $a \in A$ such that $a \leq x$, for each $x \in A$.

(0.B.8) Definition.   Suppose that $\leq$ is a partial ordering for a set X and $A \subset X$. Then $u \in X$ is an *upper bound* for A if and only if $a \leq u$ for each $a \in A$; u is a *least upper bound* for A if and only if u is an upper bound for A and if v is also an upper bound for A, then $u \leq v$. The least upper bound u is denoted by either lub A or sup A.

(0.B.9) Definition.   Suppose that $\leq$ is a partial ordering for a set X and $A \subset X$. Then $b \in X$ is a *lower bound* for A if and only if $b \leq a$ for each A; b is a *greatest lower bound* for A if and only if b is a lower bound, and whenever c is a lower bound for A, then $c \leq b$. The greatest lower bound b is denoted either by glb A or inf A.

(0.B.10) Definition.   Suppose that $\leq$ is a partial ordering for X and that $A \subset X$. Then $m \in A$ is a *maximal element* in A if and only if whenever $a \in A$ and $m \leq a$, then $m = a$; $n \in A$ is a *minimal element* in A if and only if whenever $a \in A$ and $a \leq n$, then $a = n$.

C. FUNCTIONS

(0.C.1) Definition.   Suppose that X and Y are sets. Then a *function from X into Y* is a subset f of $X \times Y$, denoted by $f : X \to Y$, with the property that for each $x \in X$, there is a unique $y \in Y$ such that $(x,y) \in f$. This is usually written $f(x) = y$. The set X is called the *domain* of f and $\{y \in Y \mid f(x) = y$ for some $x \in X\}$ is the *image* of f; the image of f will often be denoted by Im f.

(0.C.2) Definition.   Suppose that $f : X \to Y$ and $A \subset X$. Then $f(A)$ is defined to be the set $\{f(x) \mid x \in A\}$.

(0.C.3) **Definition.** Suppose that $f : X \to Y$ is a function. Then f is *onto* (or is a *surjection*) if and only if $f(X) = Y$.

(0.C.4) **Definition.** Suppose that $f : X \to Y$ is a function. Then f is *one to one* (or an *injection*), denoted by 1–1, if and only if whenever $x_1, x_2 \in X$ and $x_1 \neq x_2$, then $f(x_1) \neq f(x_2)$. If f is both 1–1 and onto, f is called a *bijection*.

(0.C.5) **Definition.** Suppose that $f : X \to Y$ is a function and $A \subset X$. Then the *restriction* of f to A is the function $f_{|A} = \{(x,y) \in f \mid x \in A\}$.

(0.C.6) **Definition.** Suppose $f : X \to Y$ is a function and $B \subset Y$. Then the *inverse of B under f*, denoted by $f^{-1}(B)$, is defined to be $\{x \in X \mid f(x) \in B\}$.

(0.C.7) **Proposition.** Suppose that $\{A_\alpha \mid \alpha \in \Lambda\}$ is a collection of subsets of a set X and $\{B_\beta \mid \beta \in \Psi\}$ is a collection of subsets of a set Y. If $f : X \to Y$ is a function, then

(i) $f(\bigcap_{\alpha \in \Lambda} A_\alpha) \subset \bigcap_{\alpha \in \Lambda} f(A_\alpha)$,

(ii) $f(\bigcup_{\alpha \in \Lambda} A_\alpha) = \bigcup_{\alpha \in \Lambda} f(A_\alpha)$,

(iii) $f^{-1}(\bigcap_{\beta \in \Psi} B_\beta) = \bigcap_{\beta \in \Psi} f^{-1}(B_\beta)$,

(iv) $f^{-1}(\bigcup_{\beta \in \Psi} B_\beta) = \bigcup_{\beta \in \Psi} f^{-1}(B_\beta)$,

(v) $f^{-1}(Y \setminus B_\beta) = X \setminus f^{-1}(B_\beta)$,

(vi)   if f is onto, then $Y \setminus f(A_\alpha) \subset f(X \setminus A_\alpha)$,

(vii)   if f is 1–1, then $f(X \setminus A_\alpha) \subset Y \setminus f(A_\alpha)$, and

(viii) $f(f^{-1}(B) \cap A) = B \cap f(A)$.

Suppose that $f : X \to Y$ is a bijection, and that for each $y \in Y$, $f^{-1}(y)$ consists of a single point in X. A function $g : Y \to X$ may be defined by setting $g(y) = x$, where $f(x) = y$. In this context, g is usually denoted by f^{-1}, and is called the *inverse* of f.

(0.C.8) **Definition.** Suppose that $f : X \to Y$ and $g : Y \to Z$ are functions. Then the *composite function* $gf : X \to Z$ is defined by $gf(x) = g(f(x))$.

(0.C.9) **Proposition.** If $f : X \to Y$ and $g : Y \to Z$, then $(gf)^{-1}(B) = f^{-1}(g^{-1}(B))$ for each $B \subset Z$.

(0.C.10) **Definition.** If X is a set, then the *identity function* $id_X : X \to X$ is defined by $id_X(x) = x$, for each $x \in X$. When the context is clear, we write *id* instead of id_X.

(0.C.11) *Proposition.* If $f : X \to Y$ and $g : Y \to X$ are functions such that $gf = id_X$, then g is onto and f is 1–1. If in addition $fg = id_Y$, then f and g are bijections.

(0.C.12) *Definition.* Let W, X, Y, and Z be sets. Suppose that $f : X \to Y$, $g : Y \to Z$, $h : X \to Z$, are functions. Then the diagram

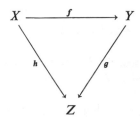

is *commutative* if and only if $gf = h$. Suppose that $f : X \to Y$, $g : Y \to Z$, $p : X \to W$, and $q : W \to Z$ are functions. The diagram

is *commutative* if and only if $gf = qp$. Similar definitions hold for other configurations.

Functions are used to define infinite Cartesian products.

(0.C.13) *Definition.* Suppose that $\{X_\alpha \mid \alpha \in \Lambda\}$ is a family of sets. Then the *Cartesian product of the X_α's*, denoted by $\prod_{\alpha \in \Lambda} X_\alpha$, is defined to be $\{f \mid f : \Lambda \to \bigcup_{\alpha \in \Lambda} X_\alpha$ such that $f(\alpha) \in X_\alpha$ for each $\alpha \in \Lambda\}$. An element $f \in \prod_{\alpha \in \Lambda} X_\alpha$ is usually denoted by $\{x_\alpha\}$ or $\{x_\alpha\}_{\alpha \in \Lambda}$ where it is understood that $f(\alpha) = x_\alpha$ for each $\alpha \in \Lambda$.

(0.C.14) *Notation.* Suppose that $\{X_\alpha \mid \alpha \in \Lambda\}$ and $\{Y_\alpha \mid \alpha \in \Lambda\}$ are families of sets indexed by the same set Λ, and that for each $\alpha \in \Lambda, f_\alpha : X_\alpha \to Y_\alpha$. Then the function $\prod f_\alpha : \prod_{\alpha \in \Lambda} X_\alpha \to \prod_{\alpha \in \Lambda} Y_\alpha$ is defined by $\prod f_\alpha(\{x_\alpha\}) = \{f(x_\alpha)\}_{\alpha \in \Lambda}$.

D. THE AXIOM OF CHOICE AND SOME OF ITS VARIANTS

(0.D.1)   The Axiom of Choice. Suppose that $\{X_\alpha \mid \alpha \in \Lambda\}$ is a family of nonempty sets. Then there is a function $f : \Lambda \to \bigcup_{\alpha \in \Lambda} X_\alpha$ such that $f(\alpha) \in X_\alpha$ for each $\alpha \in \Lambda$.

Note that the "choice" function f selects a point x_α from each X_α. Thus it follows from (0.D.1) that the Cartesian product (0.C.13) is not empty. There are a number of propositions which are equivalent to the axiom of choice. We list a few of these after the following definition.

(0.D.2)   Definition. Suppose that A is a subset of a partially ordered set $(X, \leq)$. Then A is a *nest* in X if and only if for each $a, b \in A$, one of the relations $a \leq b$ or $b \leq a$ holds, i.e., A is a linearly ordered subset of the partially ordered set X.

A proof of the next proposition may be found in most books on set theory.

(0.D.3)   Proposition (Zorn's Lemma). If X is a partially ordered set such that every nest in X has an upper bound (in X), then X contains a maximal element.

(0.D.4)   Proposition (Kuratowski's Lemma). Suppose that $(X, \leq)$ is a partially ordered set. Then each nest A in X is contained in a maximal nest M, i.e., if M is a nest, $A \subset M$, and if $x \in X \setminus M$, then $M \cup \{x\}$ is not a nest.

Proof. Suppose that N is a nest in $(X, \leq)$. Let $\mathcal{M} = \{M \subset \mathcal{P}(X) \mid M$ is a nest in $(X, \leq)$ and $N \subset M\}$. We apply Zorn's lemma to find a maximal element in $\mathcal{M}$. Partially order $\mathcal{M}$ by declaring $M_1 \leq M_2$ if and only if $M_1 \subset M_2$. Let $\mathscr{C} = \{M_\alpha \mid \alpha \in \Lambda\}$ be a nest in $\mathcal{M}$. We show that $D = \bigcup \{M_\alpha \mid \alpha \in \Lambda\}$ is an upper bound (in $\mathcal{M}$) for $\mathscr{C}$.

Certainly $N \subset D$ and $M_\alpha \subset D$; hence it remains to establish that D is a nest in $(X, \leq)$. Suppose that $A, B \in D$, $A \in M_{\alpha_1} \in \mathscr{C}$, and $B \in M_{\alpha_2} \in \mathscr{C}$. Then since either $M_{\alpha_1} \subset M_{\alpha_2}$ or $M_{\alpha_2} \subset M_{\alpha_1}$, we have that either $A, B \in M_{\alpha_1}$ or $A, B \in M_{\alpha_2}$. Since M_{α_1} and M_{α_2} are nests, it follows that $A \leq B$ or $B \leq A$ and, consequently D is a nest in $(X, \leq)$.

Thus, D is an upper bound for $\mathscr{C}$, and therefore, by Zorn's lemma, $\mathcal{M}$ has a maximal element.

(0.D.5)   Remark. Kuratowski's lemma is usually applied in the following context. Suppose that X is a set and $\mathscr{W}$ is a subset of $\mathcal{P}(X)$. Partially order $\mathcal{P}(X)$ by inclusion: if $A, B \in \mathscr{W}$, then $A \leq B$ if and only if $A \subset B$. A nest in $\mathscr{W}$ is a collection $\mathscr{C}$ of sets in $\mathscr{W}$ such that if $A, B \in \mathscr{C}$, then either $A \subset B$ or $B \subset A$.

We outline a proof of the following proposition.

(0.D.6)   Proposition (Zermelo's Theorem).   Every set can be well ordered.

Proof (sketch).   Suppose that X is a set. Let $\mathscr{C}$ be the set of all ordered pairs of the form $(A, \leq_A)$ where $A \subset X$ and $\leq_A$ is a well ordering for A. Partially order $\mathscr{C}$ by setting $(A, \leq_A) \leq (B, \leq_B)$, provided that

    (i) $A \subset B$,
    (ii) if $x, y \in A \subset B$ then $x \leq_B y$ if and only if $x \leq_A y$, and
    (iii) if $y \in B \setminus A$ and $x \in A$, then $x \leq_B y$.

Zorn's lemma may be used to show that $(\mathscr{C}, \leq)$ has a maximum element $(Y, \leq_Y)$, and it is then easy to establish that $Y = X$.

(0.D.7)   Definition.   Suppose that $(X, \leq)$ is a well ordered set and $x \in X$. Then the *initial segment determined by* x, denoted by X_x, is defined to be $\{y \in X \mid y < x\}$.

(0.D.8)   Proposition (Transfinite Induction).   Suppose that X is a well ordered set and that E is a nonempty subset of X. Suppose further that for each $x \in X$, if $X_x \subset E$, then $x \in E$. Then E is all of X.

Proof.   If $E \neq X$, then $X \setminus E \neq \varnothing$. Let x be the first element of $X \setminus E$. Since every element of X_x is an element of E, we have $X_x \subset E$. Thus, by hypothesis, $x \in E$, which contradicts that $x \in X \setminus E$.

E.   CARDINAL NUMBERS AND ORDINAL NUMBERS

It is beyond the scope and intent of this book to provide a systematic treatment of either cardinal or ordinal number theory. For our purposes, it suffices to postulate (and accept) the existence of certain sets to be described below.

(0.E.1)   Definition.   Suppose that X and Y are sets. Then X and Y are *equipotent* if and only if there exists a bijection $f : X \to Y$.

We postulate the existence of a class $\mathscr{C}$ of sets with the property that every set X is equipotent with precisely one of the sets in $\mathscr{C}$. If X is equipotent with $C \in \mathscr{C}$, we write $C = \operatorname{card} X$.

(0.E.2)   Definition.   A set X is *countable* if and only if it is equipotent with $\mathbf{Z}^+$ or with a finite set $\{1, 2, \ldots, n\}$. Otherwise X is said to be *uncountable*.

(0.E.3) *Proposition.*

   (i)   The set of rational numbers **Q** is countable.

   (ii) The union of countably many countable sets is countable.

 (iii) The finite product of countable sets is countable.

We dispose of ordinal numbers (up to and including the first uncountable ordinal) by postulating the existence of an uncountable well-ordered set $(\Theta, \leq)$, containing a largest element Ω, and with the property that if $\alpha \in \Theta$ and $\alpha < \Omega$, then $\{\beta \in \Theta \mid \beta \leq \alpha\}$ is countable. The element Ω will be referred to as the *first uncountable ordinal.*

Since $(\Theta, \leq)$ is well ordered, there is a first element ω with the property that $\{\alpha \in \Theta \mid \alpha < \omega\}$ is countable but not finite. We call ω the *first infinite ordinal.*

The smallest ordinal in Θ will be denoted by 0, the second smallest by 1, etc. Thus the ordinals may be thought of as being strung out in the following manner: $0, 1, 2, \ldots, \omega, \omega + 1, \omega + 2, \ldots, 2\omega, 2\omega + 1, \ldots, \Omega$.

A key property of the ordinals is the following.

(0.E.4) *Proposition.* If $A \subset \Theta$ is countable and $\Omega \notin A$, then sup $A < \Omega$.

Proof. It follows from the properties of Θ that for each $\alpha \in A$, the set $\{\beta \in \Theta \mid \beta \leq \alpha\}$ is countable. Let $X = \{\beta \mid \beta \leq \alpha \text{ for some } \alpha \in A\}$. Then by (0.E.3), X is countable (it is the countable union of countable sets). Since Θ is well ordered, there is a smallest element γ in Θ that is not contained in X. Clearly γ is an upper bound for X, and furthermore, $\beta < \gamma$ if and only if $\beta \in X$. Hence, $\{\delta \in \Theta \mid \delta \leq \gamma\}$ is countable, and consequently γ cannot be Ω. Since sup $A \leq \gamma$, the proof is complete.

F.   MINKOWSKI'S INEQUALITY

The following inequality (Minkowski's inequality) will be frequently useful:

$$\sqrt{\sum_{i=1}^{n} (a_i + b_i)^2} \leq \sqrt{\sum_{i=1}^{n} a_i^2} + \sqrt{\sum_{i=1}^{n} b_i^2}$$

where a_i and b_i are arbitrary real numbers and $n \in \mathbf{Z}^+$.

This inequality follows readily from another famous inequality, that of Cauchy and Schwarz.

(0.F.1) *Proposition* (*Cauchy-Schwarz Inequality*). Suppose that $a_1, a_2, \ldots,$

$a_n, b_1, b_2, \ldots, b_n$ are real numbers. Then we have

$$\sum_{i=1}^n a_i b_i \leq \sqrt{\sum_{i=1}^n a_i^2} \sqrt{\sum_{i=1}^n b_i^2}$$

Proof. It suffices to show that

$$\left(\sum_{i=1}^n a_i b_i\right)^2 \leq \left(\sum_{i=1}^n a_i^2\right)\left(\sum_{i=1}^n b_i^2\right)$$

If $b_i = 0$ for each i, then clearly the above inequality holds. Suppose then that not all the b_i's are 0. If α is any real number, then

$$0 \leq \sum_{i=1}^n (a_i + \alpha b_i)^2 = \sum_{i=1}^n a_i^2 + 2\alpha \sum_{i=1}^n a_i b_i + \alpha^2 \sum_{i=1}^n b_i^2$$

Let

$$\alpha = -\frac{\sum_{i=1}^n a_i b_i}{\sum_{i=1}^n b_i^2}$$

Then we have

$$0 \leq \sum_{i=1}^n a_i^2 - 2\frac{(\sum_{i=1}^n a_i b_i)^2}{\sum_{i=1}^n b_i^2} + \frac{(\sum_{i=1}^n a_i b_i)^2}{\sum_{i=1}^n b_i^2} = \sum_{i=1}^n a_i^2 - \frac{(\sum_{i=1}^n a_i b_i)^2}{\sum_{i=1}^n b_i^2}$$

or

$$\left(\sum_{i=1}^n a_i b_i\right)^2 \leq \left(\sum_{i=1}^n a_i^2\right)\left(\sum_{i=1}^n b_i^2\right)$$

(0.F.2)   **Proposition (*Minkowski's Inequality*).**   Suppose that $a_1, a_2, \ldots,$ $a_n, b_1, b_2, \ldots, b_n$ are real numbers. Then

$$\sqrt{\sum_{i=1}^n (a_i + b_i)^2} \leq \sqrt{\sum_{i=1}^n a_i^2} + \sqrt{\sum_{i=1}^n b_i^2}$$

Proof.

$$\sum_{i=1}^n (a_i + b_i)^2 = \sum_{i=1}^n (a_i^2 + 2a_i b_i + b_i^2)$$

$$= \sum_{i=1}^n a_i^2 + 2\sum_{i=1}^n a_i b_i + \sum_{i=1}^n b_i^2$$

$$\leq \sum_{i=1}^n a_i^2 + 2\sqrt{\sum_{i=1}^n a_i^2}\sqrt{\sum_{i=1}^n b_i^2} + \sum_{i=1}^n b_i^2$$

$$= \left(\sqrt{\sum_{i=1}^n a_i^2} + \sqrt{\sum_{i=1}^n b_i^2}\right)^2$$

The desired inequality is now obtained by taking square roots of both sides.

Minkowski's inequality can be extended to infinite sequences.

(0.F.3) Proposition. Suppose that $a_1, a_2, a_3, \ldots$ and $b_1, b_2, b_3, \ldots$ are sequences of real numbers such that

$$\sum_{i=1}^{\infty} a_i^2 \quad \text{and} \quad \sum_{i=1}^{\infty} b_i^2$$

are convergent. Then

$$\sqrt{\sum_{i=1}^{\infty} (a_i + b_i)^2} \leq \sqrt{\sum_{i=1}^{\infty} a_i^2} + \sqrt{\sum_{i=1}^{\infty} b_i^2}$$

Of special interest will be the following corollary to (0.F.2) and (0.F.3).

(0.F.4) Proposition. Suppose that $x_1, x_2, \ldots$ and $y_1, y_2, \ldots$ and $z_1, z_2, \ldots$ are sequences of real numbers. Then for each $n \in \mathbf{Z}^+$,

$$\sqrt{\sum_{i=1}^{n} (x_i - z_i)^2} \leq \sqrt{\sum_{i=1}^{n} (x_i - y_i)^2} + \sqrt{\sum_{i=1}^{n} (y_i - z_i)^2}$$

Furthermore, if $\sum_{i=1}^{\infty} x_i^2$, $\sum_{i=1}^{\infty} y_i^2$, and $\sum_{i=1}^{\infty} z_i^2$ converge, then

$$\sqrt{\sum_{i=1}^{\infty} (x_i - z_i)^2} \leq \sqrt{\sum_{i=1}^{\infty} (x_i - y_i)^2} + \sqrt{\sum_{i=1}^{\infty} (y_i - z_i)^2}$$

Proof. Let $a_i = x_i - y_i$, $b_i = y_i - z_i$, and apply the previous two propositions.

Chapter 1

THE BASIC CONSTRUCTS

A. TOPOLOGICAL SPACES AND CONTINUITY

To a significant extent, topology may be viewed as the study of the meta-morphosis that sets or spaces undergo when they fall under the influence of continuous functions. The idea of continuity should be familiar to the student from his experience with calculus, but the notion of space, or more accurately topological space, perhaps remains a rather vague concept. The reader is certainly aware of some special sets such as the real line $\mathbf{R}^1$, the plane $\mathbf{R}^2$, closed intervals, spheres, balls, doughnuts (tori), etc. Our first task is to give each of these sets a precise structure that will enable us to define the continuity of functions between any two of them.

The reader undoubtedly remembers the $\varepsilon - \delta$ concept of continuity employed in calculus, wherein a function $f : \mathbf{R}^1 \to \mathbf{R}^1$ is said to be continuous at a point $c \in \mathbf{R}^1$ if and only if for each positive number ε, there is a positive number δ such that $|f(x) - f(c)| < \varepsilon$ whenever $|x - c| < \delta$. This definition can be stated in more geometric terms as follows: a function f is continuous at a point c if and only if for each open interval (w,z) containing $f(c)$, there is an open interval (a,b) containing c such that $f((a,b)) \subset (w,z)$.

The reader should verify that the two definitions are indeed equivalent. The advantage of the latter definition is that if we could establish the notion of "open interval" or, more generally, open subset of a given set, then our definition could be generalized to functions between arbitrary sets.

We begin by defining an open set in $\mathbf{R}^1$ to be any set that can be written as the union of a family of open intervals (in particular, this will make the empty set open, since it is the union over an empty family of open intervals). It is easy to verify that

  (i)   the union of any collection of open sets is open, and
  (ii) the intersection of any finite number of open sets is open.

We now say that a function $f : \mathbf{R}^1 \to \mathbf{R}^1$ is continuous at a point c if and only if for each open set U containing $f(c)$, there is an open set V containing c such that $f(V) \subset U$. The reader should check that this definition of continuity is compatible with the previous ones. The appropriate designation of open subsets of a set X determines what is called a topological structure for X. Much of the beauty of topology stems from the fact that properties (i) and (ii) given above are all that is needed to determine whether or not a subfamily $\mathcal{U}$ of $\mathcal{P}(X)$ (the collection of all subsets of X) is a topological structure, and that the idea of continuity is so easily defined using this structure. Formally, then, we have the following definition.

(1.A.1) Definition. Suppose that X is a set and that $\mathcal{U}$ is a collection of subsets of X. The pair $(X, \mathcal{U})$ is a *topological space* if and only if the following conditions are satisfied:

  (i)   if $\{U_\alpha \mid \alpha \in \Lambda\}$ is a collection of members of $\mathcal{U}$, then $(\bigcup_{\alpha \in \Lambda} U_\alpha) \in \mathcal{U}$;
  (ii) if $\{U_\alpha \mid \alpha \in K\}$ is a finite subcollection of members of $\mathcal{U}$, then $(\bigcap_{\alpha \in K} U_\alpha) \in \mathcal{U}$;
  (iii) $\varnothing \in \mathcal{U}$ and $X \in \mathcal{U}$.

The family $\mathcal{U}$ is called a *topology* for X, and the elements of $\mathcal{U}$ are called the *open sets* of $(X, \mathcal{U})$.

Note that if each point in a set U is contained in an open set $V \subset U$, then U is open.

The reader should observe that the family of unions of open intervals in $\mathbf{R}^1$ yields a topology for $\mathbf{R}^1$.

(1.A.2) Definition. If $\mathcal{U}$ and $\mathcal{V}$ are topologies for a set X, then $\mathcal{U}$ is *smaller* than $\mathcal{V}$ (or $\mathcal{V}$ is *larger* than $\mathcal{U}$) if and only if $\mathcal{U} \subset \mathcal{V}$.

(1.A.3) *Definition.* Suppose that $(X,\mathcal{U})$ and $(Y,\mathcal{V})$ are topological spaces and that $f : X \to Y$. Then f is *continuous at a point* $c \in X$ if and only if whenever $f(c) \in V \in \mathcal{V}$, there is a $U \in \mathcal{U}$ containing c such that $f(U) \subset V$. If f is continuous at each point $c \in X$, we say that f is *continuous*.

It is often convenient to express continuity in the following terms.

(1.A.4) *Theorem.* Suppose that $(X,\mathcal{U})$ and $(Y,\mathcal{V})$ are topological spaces and that $f : X \to Y$. Then f is continuous if and only if for each $V \in \mathcal{V}$, $f^{-1}(V) \in \mathcal{U}$.

Proof. Suppose that f is continuous and that $V \in \mathcal{V}$. Let $x \in f^{-1}(V)$. Then $f(x) \in V$, and since f is continuous, there is a set $U_x \in \mathcal{U}$ such that $x \in U_x$ and $f(U_x) \subset V$. Hence we have that $U_x \subset f^{-1}(V)$. It follows from (i) of (1.A.1) that $f^{-1}(V)$ is open, since $f^{-1}(V) = \bigcup \{U_x \mid x \in f^{-1}(V)\}$.

Conversely, suppose that $f^{-1}(V) \in \mathcal{U}$ for each $V \in \mathcal{V}$. Let $x \in X$ and suppose that $f(x) \in V \in \mathcal{V}$. Then we have that $f^{-1}(V) \in \mathcal{U}$, $x \in f^{-1}(V)$, and $f(f^{-1}(V)) \subset V$. Consequently, f is continuous at x, and since x was arbitrary, we conclude that f is continuous.

Frequently, the topology $\mathcal{U}$ for a set X proves to be rather unwieldy; this situation is rendered more tractable by the introduction of the notion of a basis.

(1.A.5) *Definition.* If $(X,\mathcal{U})$ is a topological space, then a *basis* for $\mathcal{U}$ is a subcollection **B** of $\mathcal{U}$ with the property that if $x \in X$ and U is an open set containing x, then there is a $V \in \mathbf{B}$ such that $x \in V \subset U$. In other words, each $U \in \mathcal{U}$ can be written as a union of sets in **B**.

For example, the open intervals of $\mathbf{R}^1$ constitute a basis for the topology determined by the open sets in $\mathbf{R}^1$ discussed earlier. This topology for $\mathbf{R}^1$ will henceforth be referred to as the *usual* topology for $\mathbf{R}^1$ (other topological structures for the set of real numbers are to be introduced shortly).

Let us next define a reasonable topology for the plane $\mathbf{R}^2$. Here it is particularly advantageous to define a topology in terms of a basis. For each point $z = (x,y) \in \mathbf{R}^2$ and each $\varepsilon > 0$, let

$$S_\varepsilon(z) = \{(u,v) \in \mathbf{R}^2 \mid \sqrt{(x - u)^2 + (y - v)^2} < \varepsilon\}$$

Then $\mathbf{B} = \{S_\varepsilon(z) \mid z \in \mathbf{R}^2 \text{ and } \varepsilon > 0\}$ is a basis for a topology $\mathcal{U}$ for $\mathbf{R}^2$, where $\mathcal{U}$ consists of all possible unions of members of **B**. Typical elements from **B** and $\mathcal{U}$ are indicated below. The sets $S_\varepsilon(x)$ are called *open disks in* $\mathbf{R}^2$.

That **B** is a basis for a unique topology for $\mathbf{R}^2$ is a consequence of the following theorem and exercise.

(1.A.6)   Theorem.   Suppose that **B** is a collection of subsets of a set X. Then **B** is a basis for some topology for X if and only if $X = \bigcup \{B \mid B \in \mathbf{B}\}$, and whenever $B_1, B_2 \in \mathbf{B}$ and $x \in B_1 \cap B_2$, there is a $B \in \mathbf{B}$ such that $x \in B \subset B_1 \cap B_2$.

Proof. Suppose that **B** is a basis for a topology $\mathscr{U}$ on X. If $x \in X$, there is a $U \in \mathscr{U}$ such that $x \in U$. Since **B** is a basis, there is a $B \in \mathbf{B}$ such that $x \in B \subset U$, and hence $X = \bigcup \{B \mid B \in \mathbf{B}\}$. If $B_1, B_2 \in \mathbf{B}$ and $x \in B_1 \cap B_2$, then since B_1 and B_2 also belong to $\mathscr{U}$, we have that $B_1 \cap B_2 \in \mathscr{U}$. Consequently, there is a $B \in \mathbf{B}$ such that $x \in B \subset B_1 \cap B_2$.

Conversely, let $\mathscr{U} = \{U \subset X \mid U$ is a union of members of $\mathbf{B}\}$. Clearly, $X, \varnothing \in \mathscr{U}$ and arbitrary unions of members of $\mathscr{U}$ are in $\mathscr{U}$. If $U, V \in \mathscr{U}$ and $x \in U \cap V$, then there are sets $B, B_1, B_2 \in \mathbf{B}$ such that $x \in B \subset B_1 \cap B_2 \subset U \cap V$, and hence $U \cap V$ is a union of members of **B**.

(1.A.7)   Exercise.   Suppose that **B** is a basis for two topologies $\mathscr{U}$ and $\mathscr{U}'$ on a set X. Show that $\mathscr{U} = \mathscr{U}'$, and hence that a basis determines a unique topology.

A given topology, however, may have distinct bases. For example, for each $z = (x,y) \in \mathbf{R}^2$ and each $\varepsilon > 0$, let $T_\varepsilon(z) = \{(u,v) \in \mathbf{R}^2 \mid |x - u| < \varepsilon$ and $|y - v| < \varepsilon\}$. Then $\mathbf{B}' = \{T_\varepsilon(z) \mid z \in \mathbf{R}^2$ and $\varepsilon > 0\}$ is a basis for precisely the same topology generated by $\mathbf{B} = \{S_\varepsilon(z) \mid z \in \mathbf{R}^2$ and $\varepsilon > 0\}$. This follows immediately from the next exercise.

(1.A.8)   Exercise.   Show that two bases **B** and **B**′ generate the same topology on a set X if and only if whenever $x \in B \in \mathbf{B}$, there is a $B' \in \mathbf{B}'$ such that $x \in B' \subset B$, and whenever $x \in B' \in \mathbf{B}'$, there is a $B \in \mathbf{B}$ such that $x \in B \subset B'$.

(1.A.9)   Definition.   Bases that generate the same topology on a set X are said to be *equivalent*.

It follows from the previous exercise that the families of open intervals $\mathbf{B} = \{(x,y) \mid x,y \text{ are rational}\}$ and $\mathbf{B}' = \{(x,y) \mid x,y \text{ are irrational}\}$ are equivalent bases for the usual topology for $\mathbf{R}^1$.

Continuity may also be expressed in terms of basis elements.

(1.A.10) Theorem. Suppose that $(X,\mathcal{U})$ and $(Y,\mathcal{V})$ are topological spaces and $\mathbf{B}$ is a basis for $\mathcal{U}$ and $\mathbf{B}'$ is a basis for $\mathcal{V}$. Then a function $f : X \to Y$ is continuous at a point c if and only if whenever $f(c) \in B' \in \mathbf{B}'$, there is a $B \in \mathbf{B}$ such that $c \in B$ and $f(B) \subset B'$.

Proof. Suppose that $f(c) \in B' \in \mathbf{B}'$. Since B' is also in $\mathcal{V}$ and f is continuous, there are sets $U \in \mathcal{U}$ and $B \in \mathbf{B}$ such that $c \in B \subset U$ and $f(c) \in f(B) \subset f(U) \subset B'$.

Conversely, suppose that $f(c) \in V \in \mathcal{V}$. Then there is a $B' \in \mathbf{B}'$ such that $f(c) \in B' \subset V$, and consequently, there exists $B \in \mathbf{B} \subset \mathcal{U}$ such that $c \in B$ and $f(B) \subset B' \subset V$.

Thus, it follows from (1.A.10) that if one wants to determine whether or not a function mapping $\mathbf{R}^2$ into itself is continuous, it suffices to check that f carries appropriate open disks into given open disks.

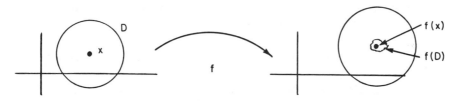

B. ADDITIONAL EXAMPLES OF TOPOLOGICAL SPACES

1. *Discrete topology.* Let X be arbitrary and $\mathcal{U} = \mathscr{P}(X)$.
2. *Indiscrete topology.* Let X be arbitrary and $\mathcal{U} = \{\varnothing, X\}$.
3. *Half-open interval topology.* Let X be $\mathbf{R}^1$. The sets of the form $[a,b)$ form a basis for a topology $\mathcal{U}$, which is distinct from the usual topology for $\mathbf{R}^1$.
4. *Open ray topology.* Let X be $\mathbf{R}^1$ and $\mathcal{U}$ be the topology determined by a basis consisting of sets of the form (a,∞) for $a \in \mathbf{R}^1$.
5. *Usual topology for $\mathbf{R}^n$.* For $z = (x_1, x_2, \ldots, x_n) \in \mathbf{R}^n$ and $\varepsilon > 0$, let

$$S_\varepsilon(z) = \{(u_1, u_2, \ldots, u_n) \in \mathbf{R}^n \mid \sqrt{\sum_{i=1}^{n} (x_i - u_i)^2} < \varepsilon\}.$$ Then $\mathbf{B} = \{S_\varepsilon(z) \mid z \in \mathbf{R}^n$ and $\varepsilon > 0\}$ is a basis for the usual topology for $\mathbf{R}^n$.

6. *Order topology.* Let $(X, \leq)$ be a linearly ordered set. For each $x, y \in X$, with $x < y$, let $(x,y) = \{c \in X \mid x < c < y\}$. The sets of the form (x,y) together with sets of the form $\{c \in X \mid c < x\}$ and sets of the form $\{c \in X \mid c > x\}$ constitute a basis for a topology for X. Note that the usual topology for $\mathbf{R}^1$ is an order topology.

7. *Finite complement topology.* Let X be an infinite set. Define a topology $\mathcal{U}$ to be the collection that consists of the empty set together with all subsets A of X with the property that $X \setminus A$ is finite.

8. Let X be any uncountable set and p be a particular point of X. Then $\mathcal{U} = \{U \subset X \mid X \setminus U$ is countable or $p \in X \setminus U\}$ is a topology for X.

9. Let X be an arbitrary set and p be a particular point in X. Then $\mathcal{U} = \{U \subset X \mid U = X$ or $p \notin U\}$ is a topology for X.

(1.B.1) Notation. Henceforth, we shall let $\mathscr{E}^n$ denote $\mathbf{R}^n$ with the usual topology.

We conclude this section with a brief introduction to a few of the more common properties that a topological space may satisfy.

(1.B.2) Definition. A topological space is a *Hausdorff* (or T_2) space if and only if for each pair of distinct points a and b in X, there are disjoint open sets U and V that contain a and b respectively.

(1.B.3) Exercise. Determine which of the above spaces are Hausdorff.

(1.B.4) Definition. A topological space $(X, \mathcal{U})$ is *second countable* if and only if $\mathcal{U}$ has a basis consisting of a countable number of sets.

For instance, $\mathscr{E}^1$ is second countable, since the family of open intervals with rational endpoints constitutes a countable basis for the usual topology for $\mathbf{R}^1$. An uncountable discrete space is obviously not second countable.

(1.B.5) Exercise. Determine which of the foregoing spaces are second countable.

(1.B.6) Definition. If $(X, \mathcal{U})$ is a topological space and $x \in X$, then a *basis for $\mathcal{U}$ at x* (often called a *neighborhood basis for $\mathcal{U}$ at x*) is a subcollection $\mathbf{B}_x \subset \mathcal{U}$ with the property that whenever $x \in U \in \mathcal{U}$, there exists $V \in \mathbf{B}_x$ such that $x \in V \subset U$.

(1.B.7) Definition. A topological space $(X, \mathcal{U})$ is *first countable* if and only if there is a countable neighborhood basis at each point $x \in X$.

Thus $\mathscr{E}^n$ is first countable, since for each $z \in \mathscr{E}^n$, $\{S_{1/i}(z) \mid i \in \mathbf{Z}^+\}$ forms a countable neighborhood basis at z.

(1.B.8) *Exercise.* Find a space that is first countable but is not second countable.

C. METRIC SPACES: A PREVIEW

A very important class of topological spaces is that in which the concept of distance between two points is defined. These spaces are known as metric spaces, and they will play a fundamental role throughout the book.

(1.C.1) *Definition.* A *metric space* is a pair (X,d), where X is a set and $d : X \times X \to [0,\infty)$ is a function with the property that for all $x,y,z \in X$:
   (i) $d(x,y) = 0$ if and only if $x = y$ *(reflexive property)*;
   (ii) $d(x,y) = d(y,x)$ *(symmetric property)*;
   (iii) $d(x,z) \leq d(x,y) + d(y,z)$ *(triangle inequality)*.
The function d is called the *distance function* or the *metric* for the metric space (X,d).

The first condition implies that the distance from a point to itself is 0 and that the distance between distinct points is positive. Condition (ii) states that the distance from a point x to a point y is the same as the distance from y to x. The third condition can be conceived of as a generalization of the idea that the shortest distance between two points is a straight line.

(1.C.2) *Notation.* Suppose that (X,d) is a metric space, $x \in X$, and $\varepsilon > 0$. We denote the set $\{y \in X \mid d(x,y) < \varepsilon\}$ by $S_\varepsilon^d(x)$. Whenever the metric is clear from the context, "d" is omitted and we simply write $S_\varepsilon(x)$.

The distance function d is used to define a topology for X. In the next theorem, it is shown that if (X,d) is a metric space, then $\mathbf{B} = \{S_\varepsilon^d(x) \mid x \in X, \varepsilon > 0\}$ is a basis for a (unique) topology for X. This topology is called the *metric topology* or the *topology induced by the metric d*.

(1.C.3) *Theorem.* If (X,d) is a metric space, then $\mathbf{B} = \{S_\varepsilon^d(x) \mid x \in X, \varepsilon > 0\}$ is a basis for a topology for X.

Proof. By (1.A.6) it suffices to show that if $p \in S_\varepsilon^d(x) \cap S_{\varepsilon'}^d(y)$, then there is a positive number ε'' such that $S_{\varepsilon''}^d(p) \subset S_\varepsilon^d(x) \cap S_{\varepsilon'}^d(y)$. Let $\lambda_1 = d(x,p)$ and $\lambda_2 = d(y,p)$. Then if ε'' is any positive number less than both $\varepsilon - \lambda_1$ and $\varepsilon' - \lambda_2$, it is easy to see that $S_{\varepsilon''}^d(p)$ is the required basis element containing p.

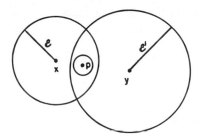

We list some examples of metric spaces. The reader should verify that each example does satisfy properties (i), (ii), and (iii) of (1.C.1).

(1.C.4) Examples.
1. Let X be $\mathbf{R}^1$ and define $d(x,y) = |x - y|$ *(usual metric for $\mathbf{R}^1$).*
2. Let X be $\mathbf{R}^2$ and define $d(x,y) = \sqrt{(x_1 - y_1)^2 + (x_2 - y_2)^2}$ where $x = (x_1,x_2)$ and $y = (y_1,y_2)$ *(usual metric for $\mathbf{R}^2$).* Use the Minkowski inequality to show that d satisfies the triangle inequality.
3. Let X be $\mathbf{R}^2$ and define $d(x,y) = |x_1 - x_2| + |y_1 - y_2|$ *(the taxicab metric).*
4. Let X be $\mathbf{R}^n$ and define $d(x,y) = \sqrt{\sum_{i=1}^{n} (x_i - y_i)^2}$, where $x = (x_1, x_2, \ldots, x_n)$ and $y = (y_1, y_2, \ldots, y_n)$ *(usual metric for $\mathbf{R}^n$).* Use the Minkowski inequality to show that d satisfies the triangle inequality.
5. Let X be arbitrary and define $d(x,y) = 1$ if $x \neq y$ and $d(x,y) = 0$ if $x = y$ *(discrete metric).*
6. Let X be $\{f | f : y \to \mathbf{R}^1$ and f is bounded$\}$ and define $d(f,g) = \sup\{|f(y) - g(y)| \, | \, y \in Y\}$. Here Y may be any set. (A function $f : Y \to \mathbf{R}^1$ is *bounded* if and only if there is a positive number M such that $|f(y)| \leq M$ for each $y \in Y$.)

Note that the metrics in examples 2 and 3 induce the same topology; that is, if $\mathcal{U}$ is the topology generated by the usual metric for $\mathbf{R}^2$ and $\mathcal{U}'$ is the topology arising from the taxicab metric, then $\mathcal{U} = \mathcal{U}'$. To see this, suppose that d is the usual metric and d' is the metric described in example 3. Let $x \in X$, and let $B' = \{y \in X \, | \, d'(x,y) < \varepsilon\}$ be a typical basis element for $\mathcal{U}'$. Suppose that $d(x,y) < \varepsilon/2$. Then $d'(x,y) = |x_1 - x_2| + |y_1 - y_2| \leq 2\sqrt{(x_1 - x_2)^2 + (y_1 - y_2)^2} = 2d(x,y) < \varepsilon$. Hence, $x \in S_{\varepsilon/2}(x) \subset B'$, and it follows that $\mathcal{U}' \subset \mathcal{U}$. On the other hand, if $B = S_{\varepsilon}^d(x)$ is a member of the basis for $\mathcal{U}$, then since $d(x,y) \leq d'(x,y)$ for all $x,y \in X$, it is clear that $x \in S_{\varepsilon}^{d'}(x) \subset S_{\varepsilon}^d(x)$ and consequently $\mathcal{U} \subset \mathcal{U}'$.

(1.C.5) Exercise. Prove or find a counterexample: (i) all metric spaces are second countable; (ii) all metric spaces are first countable.

(1.C.6) Exercise. Show that all metric spaces are Hausdorff.

We have just observed that a topology can be derived in a natural way from a metric space. An interesting and actually quite important question (which will be treated in some detail in Chapter 10) concerns the reverse implication—given a topological space $(X,\mathscr{U})$, when does there exist a metric for X that induces $\mathscr{U}$? A topological space $(X,\mathscr{U})$ is *metrizable* if and only if it is possible to define a metric d on X such that the metric topology induced by d coincides with $\mathscr{U}$. Note that example 5 above guarantees the existence of at least one metric for any set, but clearly the topology associated with this metric is discrete. There is a great advantage in having a metric function available, although fairly stringent conditions are necessary to guarantee the existence of one.

In the following exercises, determine whether or not there is a metric function whose induced topology coincides with the given topology, and if there is, find one.

(1.C.7) Exercise. Let X be an arbitrary set with the indiscrete topology. (Assume X has more than one point.)

(1.C.8) Exercise. Let X be an arbitrary set with the discrete topology.

(1.C.9) Exercise. Let X be $\mathbf{R}^1$ with the open ray topology.

It is possible that two distinct metrics d and d' on a set X will induce or generate the same topology. In this case, d and d' are said to be *equivalent metrics*.

(1.C.10) Example. Let (X,d) be any metric space. Define $d': X \times X \to [0,\infty)$ by setting $d'(x,y) = \min\{d(x,y), 1\}$. It is easy to show that d' is a metric; that d and d' are equivalent follows immediately from the next theorem.

(1.C.11) Theorem. Suppose that d and d' are metrics for a set X. Then d and d' are equivalent if and only if for each $x \in X$ and each $r > 0$, there exist numbers $s > 0$ and $s' > 0$ such that $S_s^d(x) \subset S_r^{d'}(x)$ and $S_{s'}^{d'}(x) \subset S_r^d(x)$.

Proof. Suppose that d and d' are equivalent, $x \in X$, and $r > 0$. Since $S_r^{d'}(x)$ is an open set containing x, there is a basis element $S_s^d(x)$ in the topology generated by d such that $S_s^d(x) \subset S_r^{d'}(x)$. Similarly, an $s' > 0$ can be found so that $S_{s'}^{d'}(x) \subset S_r^d(x)$.

To prove the converse, let $\mathcal{U}$ be the topology induced by d and $\mathcal{U}'$ be the topology induced by d'. Suppose that $U \in \mathcal{U}$ and that x is an arbitrary point of U. Then there is $r > 0$ such that $S_r^d(x) \subset U$. By the hypothesis there exists an $s' > 0$ such that $x \in S_{s'}^{d'}(x) \subset S_r^d(x) \subset U$; thus, U is a union of open sets in $\mathcal{U}'$, and consequently $\mathcal{U} \subset \mathcal{U}'$. An analogous argument shows that $\mathcal{U}' \subset \mathcal{U}$.

(1.C.12) Definition. A metric space (X,d) is *bounded* if and only if $\sup\{d(x,y) \mid x,y \in X\}$ is finite.

(1.C.13) Corollary. Every metric space is equivalent to a bounded metric space.

Proof. The proof is immediate from (1.C.10) and (1.C.11).

D. BUILDING NEW TOPOLOGICAL SPACES FROM OLD ONES

The Relative Topology

Frequently we will wish to consider a subset A of a space $(X,\mathcal{U})$ as a topological space in its own right. This is possible once we have determined which subsets of A are to be designated as open (in A). The next definition handles this in a natural manner.

(1.D.1) Definition. Suppose that $(X,\mathcal{U})$ is a topological space and that A is a subset of X. Declare a set $V \subset A$ to be *open in A* (or *A-open*) if and only if $V = U \cap A$ for some $U \in \mathcal{U}$. The collection of all A-open sets forms a topology for A, called the *relative topology*. Henceforth, whenever a subset A of a space $(X,\mathcal{U})$ is considered as a topological space, it is assumed that A has the relative topology. In this context, A is often called a *subspace* of $(X,\mathcal{U})$.

(1.D.2) Exercise. Show that the relative topology is in fact a topology.

It should be noted that A-open sets need not be open in the original space X. For instance, if $X = \mathcal{E}^1$ and $A = [0,1)$, then $[0,\frac{1}{2})$ is open in A but not in X. In fact, if $A = [0,1]$, then A is an open subset of itself, but not open in $\mathcal{E}^1$. However, whenever A is itself open in X, then A-open sets are also open in X.

(1.D.3) **Theorem.** Suppose that $U \subset A \subset X$ where A is open in X and U is open in A. Then U is open in X.

Proof. Since U is open in A, there is an open set V in X such that $V \cap A = U$. However, since A is open in X, we have that $A \cap V$ is open in X, and the theorem follows.

A second trivial but useful result is the following.

(1.D.4) **Theorem.** Suppose that W and Z are subspaces of a topological space X, and that $U \subset W \cap Z$ is open in both W and Z. Then U is open in $W \cup Z$.

Proof. Since U is open in both W and Z, there are open sets (in X), W' and Z', such that $W' \cap W = U$ and $Z' \cap Z = U$. Then $(W' \cap Z') \cap (W \cup Z) = U$, which implies that U is an open set in the relative topology for $W \cup Z$.

(1.D.5) **Exercise.** Show that if X and Y are topological spaces, $f : X \to Y$ is continuous, and $A \subset X$, then $f_{|A}$ is continuous.

(1.D.6) **Exercise.** Show that if (X,d) is a metric space and $A \subset X$, then the relative topology for A is metrizable. In fact, $d_{|A \times A}$ is a metric that induces the relative topology.

The (Finite) Product Topology

The point set $\mathbf{R}^n$ is the Cartesian product of $\mathbf{R}^1$ with itself n times. Thus, it is not unreasonable to expect that the usual topology defined for $\mathbf{R}^n$ might be related in some way to the usual topology for $\mathbf{R}^1$. This relationship is perhaps best seen if we generalize somewhat and consider the notion of a finite Cartesian product of arbitrary spaces. In a later chapter, infinite Cartesian products will be treated.

(1.D.7) **Definition.** Suppose that $(X_1, \mathcal{U}_1)$, $(X_2, \mathcal{U}_2), \ldots$, $(X_n, \mathcal{U}_n)$ are topological spaces and $\prod_{i=1}^{n} X_i$ is the Cartesian product of the sets $X_1, X_2, \ldots$, X_n. The *product topology* for $\prod_{i=1}^{n} X_i$ is the topology which has a basis consisting of sets of the form $U = U_1 \times U_2 \times \cdots \times U_n$, where $U_i \in \mathcal{U}_i$ for $i = 1, 2, \ldots, n$.

(1.D.8) Remark. If $(X_1, \mathcal{U}_1)$, $(X_2, \mathcal{U}_2)$, ..., $(X_n, \mathcal{U}_n)$ are topological spaces, then, unless otherwise stated, $\prod_{i=1}^{n} X_i$ will be assumed to have the product topology.

(1.D.9) Exercise. Show that the product topology is in fact a topology.

(1.D.10) Theorem. If $\mathcal{U}$ is the usual topology for $\mathbf{R}^n$ and $\mathcal{V}$ is the product topology for $\mathscr{E}^1 \times \mathscr{E}^1 \times \cdots \times \mathscr{E}^1$, then $\mathcal{U} = \mathcal{V}$.

Proof. To see that $\mathcal{U} \subset \mathcal{V}$, let $U = S_\varepsilon^d(x)$ be a member of the usual basis for $\mathcal{U}$, where $x = (x_1, x_2, \ldots, x_n) \in \mathbf{R}^n$ and $\varepsilon > 0$. For each x_i, let $U_i = \{y \in \mathscr{E}^1 \mid |x_i - y| < \varepsilon/\sqrt{n}\}$. Then $x \in U_1 \times U_2 \times \cdots \times U_n$, and if $u = (u_1, u_2, \ldots, u_n) \in U_1 \times U_2 \times \cdots \times U_n$, we have that

$$d(u,x) = \sqrt{\sum_{i=1}^{n} (x_i - u_i)^2} < \sqrt{\frac{n\varepsilon^2}{n}} = \varepsilon$$

Hence, $U \in \mathcal{V}$.

Now suppose that $U_1 \times U_2 \times \cdots \times U_n$ is a typical member of the basis for the product topology, and let

$$x = (x_1, x_2, \ldots, x_n) \in U_1 \times U_2 \times \cdots \times U_n.$$

Then for each x_i, there is an $\varepsilon_i > 0$ such that $\{y \in \mathscr{E}^1 \mid |x_i - y| < \varepsilon_i\} \subset U_i$. Let $\varepsilon = \min\{\varepsilon_i \mid i = 1, 2, \ldots, n\}$. We show that $S_\varepsilon^d(x) \subset U_1 \times U_2 \times \cdots \times U_n$. If $z = (z_1, z_2, \ldots, z_n) \in S_\varepsilon^d(x)$, then

$$|x_i - z_i| \le \sqrt{\sum_{i=1}^{n} (x_i - z_i)^2} < \varepsilon \le \varepsilon_i$$

and therefore $z_i \in U_i$ for each i, which completes the proof.

(1.D.11) Definition. For $i = 1, 2, \ldots, n$, the function $p_i : (X_1 \times \cdots \times X_i \times \cdots \times X_n) \to X_i$ defined by $p_i(x_1, \ldots, x_i, \ldots, x_n) = x_i$ is called the ith *projection map*.

(1.D.12) Theorem. Suppose that $(X_1, \mathcal{U}_1)$, $(X_2, \mathcal{U}_2)$, ..., $(X_n, \mathcal{U}_n)$ are topological spaces and that $\prod_{i=1}^{n} X_i$ is the associated product space. Then the projection maps p_i are continuous. Furthermore, the product topology is the smallest topology for which all the p_i are continuous, i.e., if $\mathcal{U}$ denotes the product topology and $\mathcal{V} \subset \mathcal{U}$ is any other topology for the set $\prod_{i=1}^{n} X_i$ for which each p_i is continuous, then $\mathcal{V} = \mathcal{U}$.

Proof. Suppose that U_i is an open set in X_i. Then $p_i^{-1}(U_i) = X_1 \times X_2 \times \cdots \times U_i \times X_{i+1} \times \cdots \times X_n$ is clearly a member of $\mathcal{U}$, and therefore p_i is continuous (1.A.4).

Suppose now that $\mathcal{V} \subset \mathcal{U}$ and that each p_i is continuous with respect to $\mathcal{V}$. To demonstrate that $\mathcal{V} = \mathcal{U}$, it suffices to show that each basis set $U = U_1' \times U_2 \times \cdots \times U_n$ in $\mathcal{U}$ belongs to $\mathcal{V}$ (why?). Since for each i, p_i is continuous, it follows that $p_i^{-1}(U_i)$ is open with respect to $\mathcal{V}$. But $U = U_1 \times U_2 \times \cdots \times U_n = \bigcap_{i=1}^{n} p_i^{-1}(U_i)$, and hence $U \in \mathcal{V}$, which concludes the proof.

(1.D.13) Exercise. Suppose that for $i = 1, 2, \ldots, n$, $f_i : X_i \to Y_i$ is continuous. Show that the map $\prod f_i : \prod_{i=1}^{n} X_i \to \prod_{i=1}^{n} Y_i$ defined by $\prod f_i(x_1, x_2, \ldots, x_n) = (f_1(x_1), f_2(x_2), \ldots, f_n(x_n))$ is continuous.

The Disjoint Union Topology

We now give one further example of how new topologies may be generated from old ones. Suppose that $\{(X_\alpha, \mathcal{U}_\alpha) \mid \alpha \in \Lambda\}$ is a collection of pairwise disjoint topological spaces, i.e., if $\alpha, \beta \in \Lambda$, then $X_\alpha \cap X_\beta = \varnothing$. Let $X = \bigcup_{\alpha \in \Lambda} X_\alpha$. Define a topology $\mathcal{U}$ for X by declaring a set $U \subset X$ to be open if and only if $U \cap X_\alpha$ is open in X_α for each $\alpha \in \Lambda$.

(1.D.14) Exercise. Show that the family $\mathcal{U}$ defined above forms a topology.

The topological space obtained in this matter is customarily called the *free union* of the spaces X_α, and the corresponding topology is frequently referred to as the *disjoint union topology*. Requiring that the spaces in question be pairwise disjoint can be avoided, since any collection of sets may be replaced by a disjoint collection of sets as follows. Suppose that $\{X_\alpha \mid \alpha \in \Lambda\}$ is a family of sets. For each $\alpha \in \Lambda$, define $\hat{X}_\alpha = X_\alpha \times \{\alpha\}$. Then clearly for $\alpha \neq \beta$, $\hat{X}_\alpha \cap \hat{X}_\beta = \varnothing$. Now suppose that $\{(X_\alpha, \mathcal{U}_\alpha) \mid \alpha \in \Lambda\}$ is a collection of topological spaces. If $\alpha \in \Lambda$, let $\hat{X}_\alpha = X_\alpha \times \{\alpha\}$ and define a topology $\hat{\mathcal{U}}_\alpha$ for $\hat{X}_\alpha$ by declaring a subset $U_\alpha \times \{\alpha\} \subset X_\alpha \times \{\alpha\}$ open if and only if U_α is open in X_α. Then the *free union* of the X_α is defined to be the free union of the disjoint spaces $(\hat{X}_\alpha, \hat{\mathcal{U}}_\alpha)$ described previously.

(1.D.15) Exercise. Suppose that $\{A_\alpha \mid \alpha \in \Lambda\}$ is a collection of disjoint subsets of a topological space X. Let $\mathcal{U}$ denote the relative topology for

$\bigcup \{A_\alpha \mid \alpha \in \Lambda\}$ and let $\mathscr{V}$ denote the disjoint union topology for the same set, where each A_α is endowed with the relative topology. Discuss the relationship between $\mathscr{U}$ and $\mathscr{V}$.

E. A POTPOURRI OF FUNDAMENTAL CONCEPTS

Although the concepts to be introduced in this section are relatively simple, the reader should give them careful consideration. A firm grasp of these ideas is needed before one can successfully proceed to the more intriguing aspects of topology.

Closed Sets

(1.E.1) Definition. If $(X,\mathscr{U})$ is a topological space, then a subset A of X is *closed* in X if and only if $X \setminus A$ is open.

Note that if U is open in X, then $X \setminus A$ is closed. It follows easily from De Morgan's rules and (1.A.1.) that both finite unions and arbitrary intersections of closed sets are closed. Of course, subsets of a space may be neither open nor closed, e.g., $[0,1) \subset \mathscr{E}^1$. Furthermore, infinite unions of closed subsets may fail to be closed, e.g., the set of rational numbers is a countable union of singletons (each of which is closed in $\mathscr{E}^1$), and yet the rationals are not closed in $\mathscr{E}^1$.

(1.E.2) Exercise. Suppose that A is a subspace of a topological space X and that $F \subset A$. Show that F is closed in A if and only if there is a closed set F' in X such that $F = F' \cap A$.

(1.E.3) Exercise. Show that if F is closed in G and G is closed in X, then F is closed in X.

All subsets of a topological space may be "closed off" in the following way.

(1.E.4) Definition. Let $(X,\mathscr{U})$ be a topological space and suppose that $A \subset X$. Then the *closure of A* (in X) is defined to be $\bigcap \{F \mid F$ is closed in X and $A \subset F\}$, and is denoted by $\bar{A}^X$ (or when the context is clear, simply by $\bar{A}$).

(1.E.5) Theorem. If A and B are subsets of a topological space X, then

    (i) $A \subset \bar{A}$,
    (ii) $\bar{A}$ is closed,

  (iii) if A is closed, then $A = \bar{A}$,

  (iv) $\bar{A}$ is the smallest closed subset of X containing A, in the sense that if $A \subset B \subset \bar{A}$ and B is closed, then $B = \bar{A}$,

  (v) if $A \subset B$, then $\bar{A} \subset \bar{B}$,

  (vi) $\overline{A \cup B} = \bar{A} \cup \bar{B}$, and

  (vii) $\overline{A \cap B} \subset \bar{A} \cap \bar{B}$.

Proof. (i) Trivial. (ii) Recall that the intersection of closed sets is closed. (iii) Since A is closed, $\bar{A}$ is contained in A; on the other hand, A is always contained in $\bar{A}$. (iv) Since B is closed, we have that $\bar{A} \subset B$. (v) Trivial. (vi) Since the finite union of closed sets is closed, we have by (iv) that $\overline{A \cup B} \subset \bar{A} \cup \bar{B}$. On the other hand, $A \subset (A \cup B)$, and hence by (v), $\bar{A} \subset \overline{A \cup B}$. Similarly, $\bar{B}$ is contained in $\overline{A \cup B}$ and therefore $\bar{A} \cup \bar{B} \subset \overline{A \cup B}$. (vii) Since $A \cap B \subset A$ and $A \cap B \subset B$, it follows from (v) that $\overline{A \cap B} \subset \bar{A} \cap \bar{B}$.

(1.E.6) *Exercise.* Find an example to show that $\overline{A \cap B}$ is not necessarily equal to $\bar{A} \cap \bar{B}$.

(1.E.7) *Theorem.* Let $X_1, X_2, \ldots, X_n$ be topological spaces and suppose that for $i = 1, 2, \ldots, n$, $A_i \subset X_i$. Then $\overline{A_1 \times \cdots \times A_n} = \bar{A}_1 \times \cdots \times \bar{A}_n$.

Proof. We prove the theorem for $n = 2$; an easy induction argument yields the more general result. First, observe that if F and G are closed subsets of spaces X and Y respectively, then $F \times G$ is closed in $X \times Y$. To see this, suppose that $(x,y) \notin F \times G$, and without loss of generality assume that $x \notin F$. Then $(X \setminus F) \times Y$ is an open set in $X \times Y$ that contains (x,y) and lies in the complement of $F \times G$. Consequently, $(X \times Y) \setminus (F \times G)$ is open, and therefore $F \times G$ is closed. It now follows immediately from part (iv) of the previous theorem that $\overline{A_1 \times A_2} \subset \bar{A}_1 \times \bar{A}_2$.

To establish the converse we show that $X \setminus \overline{(A \times B)} \subset X \setminus (\bar{A} \times \bar{B})$. Let $(x,y) \in X \setminus \overline{(A \times B)}$. Note that $X \setminus \overline{(A \times B)}$ is an open set and clearly $(X \setminus \overline{(A \times B)}) \cap (A \times B) = \varnothing$. Hence there is a basic open set $U \times V$ such that $(x,y) \in U \times V \subset X \setminus \overline{(A \times B)}$ and thus $(U \times V) \cap (A \times B) = \varnothing$. To finish the proof it suffices to show that $U \cap A = \varnothing$ or $V \cap B = \varnothing$ (why?). But this is clear since $(U \cap A) \times (V \cap B) \subset (U \times V) \cap (A \times B) = \varnothing$.

(1.E.8) *Definition.* A topological space $(X, \mathscr{U})$ is T_1 if and only if each point in X is closed.

(1.E.9) *Exercise.* Show that a space X is T_1 if and only if for each $x, y \in X$ such that $x \neq y$, there is an open set containing x that does not contain y.

Accumulation Points

Probably no concept in either topology or analysis is of more importance than the notion of an accumulation point. It will become increasingly apparent as we proceed that the entire fabric of topology is permeated with this idea.

(1.E.10) Definition. Suppose that X is a topological space and that A is a subset of X. Then a point $x \in X$ is an *accumulation point* of A if and only if each open set containing x has nonempty intersection with $A \setminus \{x\}$.

It should be noted that in the literature, accumulation points are often referred to as limit points.

(1.E.11) Examples.
   1.   If $X = \mathscr{E}^1$ and $A = [0,1)$, then any x such that $0 \le x \le 1$ is an accumulation point of A.
   2.   If $X = \mathscr{E}^1$ and A is the set of rationals, then every point of X is an accumulation point of A.
   3.   If $X = \mathscr{E}^1$ and $A = \mathbf{Z}$, then no point of X is an accumulation point of A.

(1.E.12) Exercise. In $\mathbf{R}^1$ let $A = (0,1)$. Find the accumulation points of A with respect to each of the different topologies we have imposed on $\mathbf{R}^1$.

(1.E.13) Definition. If A is a subset of a space X, then the *derived set* of A is $\{x \mid x$ is an accumulation point of $A\}$. The derived set of A is denoted by A'.

(1.E.14) Theorem. Suppose that A is a subset of a space X. Then (i) $\bar{A} = A \cup A'$, and (ii) A is closed if and only if $A' \subset A$.

   Proof.   (i) If a point x does not lie in $A \cup A'$, then there is an open set containing x that misses A. Consequently, the complement of $A \cup A'$ is open and $A \cup A'$ is closed. Thus, we have that $\bar{A} \subset A \cup A'$. If $x \notin \bar{A}$, then since $\bar{A}$ is closed, x belongs to an open set that does not intersect A. Thus, $x \notin A \cup A'$, and hence $\bar{A} = A \cup A'$. (ii) If A is closed, then $A = \bar{A} = A \cup A'$ and therefore $A' \subset A$. Conversely, if $A' \subset A$, then $A = A \cup A' = \bar{A}$ and hence A is closed.

Interior, Exterior, and Frontier

If A is any subset of a topological space X, then X may be split in a natural way into three distinct sets as follows.

(1.E.15) Definition. Suppose that A is a subset of a space X. The *interior* of A denoted by $A°$ or int A, is the union of all open sets of X that are contained in A. The *exterior* of A, denoted by ext A, is the union of all open sets in X that are contained in $X \setminus A$. The *frontier* of A, denoted by Fr A, consists of all points $x \in X$ with the property that each open set containing x intersects both A and $X \setminus A$.

For example, if $A = (0,4] \subset \mathscr{E}^1$, then int $A = (0,4)$, ext $A = (-\infty,0) \cup (4,\infty)$, and Fr $A = \{0,4\}$.

(1.E.16) Theorem. If A is a subset of a space X, then $X = (\text{int } A) \cup (\text{ext } A) \cup (\text{Fr } A)$ and these sets are pairwise disjoint.

Proof. Suppose that $x \in X$ fails to be in either int A or ext A. Then it is clear from the foregoing definition that each open set that contains x must intersect both A and $X \setminus A$. Therefore, $x \in $ Fr A. That the three sets are pairwise disjoint is also immediate from the definition.

Although any of the three sets int A, ext A, and Fr A may be empty, it should be obvious that int A is the largest open set contained in A and ext A is the largest open set contained in $X \setminus A$.

(1.E.17) Exercise. Show that if $A \subset X$, then Fr $A = \bar{A} \cap (\overline{X \setminus A})$.

Frontiers and products are related as follows.

(1.E.18) Theorem. Suppose that $A \subset X$ and $B \subset Y$. Then $\text{Fr}(A \times B) = ((\text{Fr } A) \times \bar{B}) \cup (\bar{A} \times (\text{Fr } B))$.

Proof. Suppose that $(x,y) \in \text{Fr}(A \times B)$. Then $x \in \bar{A}$ and $y \in \bar{B}$ (why?). If $x \notin $ Fr A and $y \notin $ Fr B, then by the previous exercise $x \notin \overline{X \setminus A}$ and $y \notin \overline{Y \setminus B}$. Let $U = X \setminus (\overline{X \setminus A})$ and $V = Y \setminus (\overline{Y \setminus B})$. Then $(x,y) \in U \times V$ and $U \times V$ fails to intersect $(X \times Y) \setminus (A \times B)$. Consequently, either $x \in $ Fr A or $y \in $ Fr B, and we have that $\text{Fr}(A \times B) \subset ((\text{Fr } A) \times \bar{B}) \cup (\bar{A} \times (\text{Fr } B))$.

Suppose now that $(x,y) \in (\text{Fr } A) \times \bar{B}$ and let $U \times V$ be a basic open set such that $(x,y) \in U \times V$. We want to show that $(U \times V) \cap (A \times B) \neq \emptyset$ and $(U \times V) \cap ((X \times Y) \setminus (A \times B)) \neq \emptyset$. Since $x \in $ Fr A and $y \in \bar{B}$, we have that $U \cap A \neq \emptyset$ and $V \cap B \neq \emptyset$; hence it follows that $(U \times V) \cap (A \times B) \neq \emptyset$.

Since $x \in $ Fr A, we have that $U \cap (X \setminus A) \neq \emptyset$. Then for any $z \in U \cap (X \setminus A)$, it is clear that $\{z\} \times V \subset (U \times V) \cap ((X \times Y) \setminus (A \times B))$. Consequently, $((\text{Fr } A) \times \bar{B}) \subset $ Fr $(A \times B)$.

In a similar manner, one may show that $\bar{A} \times (\text{Fr } B) \subset $ Fr $(A \times B)$.

(1.E.19) Exercise.   Generalize (1.E.18) to finite products.

We conclude this section with an important definition.

(1.E.20) Definition.   If X is a topological space and $x \in X$, then a sub-set N of X is a *neighborhood* of x if and only if there is an open set U in X such that $x \in U \subset N$. (The reader should check to see in which of the previous definitions the words "open set containing x" may be replaced by the words "neighborhood of x" without changing the meaning of the definition).

F. CONTINUITY

Recall that we used the notion of continuity to motivate the definition of a topology. We now reexamine the concept of continuity in light of our newly developed topological ideas. We first establish some useful generalizations of presumably familiar results.

(1.F.1) Theorem.   Suppose that X, Y, and Z are topological spaces and that $f : X \to Y$ and $g : Y \to Z$ are continuous functions. Then $gf : X \to Z$ is continuous.

Proof.   We show continuity at each point $x \in X$. Let $x \in X$ and suppose that V is an open set containing $gf(x)$. Since g is continuous, there is an open set W in Y that contains $f(x)$ and such that $g(W) \subset V$. Similarly, from the continuity of f at x, we can find an open set U in X such that $x \in U$ and $f(U) \subset W$. Then $fg(U) \subset V$.

The solutions to the next exercise embody nothing more than manipulating ε's and δ's.

(1.F.2) Exercise.   Show that the following functions are continuous:
   (i) $f : \mathscr{E}^1 \to \mathscr{E}^1$, where f is defined by $f(x) = x^2$;
   (ii) $f : \mathscr{E}^1 \setminus \{0\} \to \mathscr{E}^1$, where f is defined by $f(x) = 1/x$;
   (iii) $f : \mathscr{E}^1 \to \mathscr{E}^1$, where f is defined by $f(x) = |x|$;
   (iv) $f : \mathscr{E}^1 \to \mathscr{E}^1$, where f is defined by $f(x) = ax$ where $a \in \mathbf{R}^1$.

(1.F.3) Theorem.   Suppose that f and g are continuous functions mapping a topological space X into $\mathscr{E}^1$. Let $h = f + g$, $k = f \cdot g$, and if $g(x) \neq 0$ for all x, let $j = f/g$. Then h, k, and j are continuous. (Here, of course, $(f + g)(x) = f(x) + g(x)$, $(f \cdot g)(x) = f(x)\, g(x)$, and $f/g(x) = f(x)/g(x)$.)

Proof.   Suppose that $x \in X$. We first establish the continuity of h at x. Let U be an open interval of radius ε with $h(x)$ as its center. Since f and g are continuous at x, there are open sets V and W containing x such that $f(V)$ is contained in the interval $(f(x) - \varepsilon/2, f(x) + \varepsilon/2)$ and $g(W)$ is contained in $(g(x) - \varepsilon/2, g(x) + \varepsilon/2)$. If $Y = V \cap W$, it is easily checked that $h(Y) \subset U$.

In order to show that k is continuous at x, one begins by demonstrating the continuity of $f^2 = f \cdot f$. This, however, follows immediately from the exercises above and the previous theorem, since f^2 can be written as a composition of f and the function $p(x) = x^2$. Since $f \cdot g = (1/4)((f + g)^2 - (f - g)^2)$, (1.F.2) may again be applied along with the first part of this theorem to see that k is continuous.

The function $j = f/g$ may be considered as the product of f and $1/g$. The continuity of $1/g$ is immediate from the observation that $1/g$ is merely a composition of g and $q(x) = 1/x$. Then, since j is a product of continuous functions, the proof of the theorem is complete.

The next theorem shows that continuity may be expressed equally as well in terms of closed sets as in terms of open sets.

(1.F.4) Theorem. Suppose that $(X, \mathscr{U})$ and $(Y, \mathscr{V})$ are topological spaces and that $f : X \to Y$. Then the following conditions are equivalent:

    (i) f is continuous;
    (ii) for each $V \in \mathscr{V}, f^{-1}(V) \in \mathscr{U}$;
    (iii) for each closed set $D \subset Y, f^{-1}(D)$ is closed in X.

Proof.

    (i) $\leftrightarrow$ (ii) This is just (1.A.4).

    (ii) $\to$ (iii) Suppose that D is a closed subset of Y and let $V = Y \setminus D$. Then V is open, and hence, $f^{-1}(V)$ is open in X. But $f^{-1}(D) = X \setminus f^{-1}(V)$, which proves that $f^{-1}(D)$ is closed in X.

    (iii) $\to$ (ii) Suppose that $V \in \mathscr{V}$. Then $Y \setminus V$ is closed and hence $f^{-1}(Y \setminus V)$ is closed in X. But $f^{-1}(Y \setminus V) = X \setminus f^{-1}(V)$ and therefore $f^{-1}(V)$ is open in X (why?).

We next give a criterion for continuity that is expressed in terms of the closure operator.

(1.F.5) Theorem. Suppose that $(X, \mathscr{U})$ and $(Y, \mathscr{V})$ are topological spaces and that $f : X \to Y$. Then f is continuous if and only if $f(\bar{A}) \subset \overline{f(A)}$ for each subset A of X.

Proof. Suppose that f is continuous. By part (iii) of the previous theorem, $f^{-1}(\overline{f(A)})$ is a closed set. Since $A \subset f^{-1}f(A)$, it follows that $\bar{A} \subset f^{-1}(\overline{f(A)})$, and consequently $f(\bar{A}) \subset ff^{-1}(\overline{f(A)}) \subset \overline{f(A)}$.

Conversely, suppose that $f(\bar{A}) \subset \overline{f(A)}$ for every subset A of X. First we observe that under the hypothesis, if $B \subset Y$, then $\overline{f^{-1}(B)} \subset f^{-1}(\bar{B})$. This is immediate, since if $A = f^{-1}(B)$, then $f(\bar{A}) \subset \overline{f(A)} \subset \bar{B}$, which implies that $\overline{f^{-1}(B)} = \bar{A} \subset f^{-1}(\bar{B})$. To establish the continuity of f, one now shows that inverse images of closed sets are closed. If $B \subset Y$ is closed, then $\overline{f^{-1}(B)} \subset f^{-1}(\bar{B}) = f^{-1}(B)$, and hence $f^{-1}(B)$ is closed.

The next result will see continual service during succeeding chapters.

(1.F.6) Theorem (Map Gluing Theorem). Suppose that A and B are closed subsets of a topological space X. Let Y be an arbitrary space and suppose that $f : A \to Y$ and $g : B \to Y$ are continuous functions such that $f_{|A \cap B} = g_{|A \cap B}$. Then the function $h : (A \cup B) \to Y$, defined by

$$h(x) = \begin{cases} f(x) & \text{if} \quad x \in A \\ g(x) & \text{if} \quad x \in B \end{cases}$$

is continuous.

　　Proof. We apply (1.F.4). Suppose that C is a closed subset of Y. Then $f^{-1}(C)$ is closed in A and hence also in X, since A is closed in X. Similarly, $g^{-1}(C)$ is closed in both B and X. However, $h^{-1}(C) = f^{-1}(C) \cup g^{-1}(C)$, and thus $h^{-1}(C)$ is closed in X, which completes the proof.

(1.F.7) Theorem. Suppose that X_1, X_2, and Y are topological spaces and that $f : Y \to X_1 \times X_2$, where $X_1 \times X_2$ is assumed to have the product topology. Let $f_1 = p_1 f$ and $f_2 = p_2 f$, where p_1 and p_2 are the projection maps. Then f is continuous if and only if f_1 and f_2 are continuous.

　　Proof. By (1.F.1), if f is continuous, then so are f_1 and f_2. To prove the converse, suppose that $U \times V$ is a typical basis element in the product topology. We are to show that $f^{-1}(U \times V)$ is open in Y. Note that $U \times V = (U \times X_2) \cap (X_1 \times V) = p_1^{-1}(U) \cap p_2^{-1}(V)$, and consequently we have that $f^{-1}(U \times V) = f^{-1}(p_1^{-1}(U) \cap p_2^{-1}(V)) = f^{-1} p_1^{-1}(U) \cap f^{-1} p_2^{-1}(V) = f_1^{-1}(U) \cap f_2^{-1}(V)$, which is an open set under the assumption that f_1 and f_2 are continuous.

In metric spaces, continuity may be expressed neatly in terms of sequences. We remind the reader that a *sequence* in a set X is a function $f : D \to X$, where D is a subset of N. If for each $i \in D$, $f(i) = x_i$, then the sequence f is usually denoted by $\{x_i\}$. Unless otherwise stated, it will be assumed that $D = \mathbf{Z}^+$.

(1.F.8) Definition. A sequence $\{x_i\}$ in a topological space X *converges to a point* $x \in X$ if and only if for every neighborhood U of x, there is a positive integer N_U such that $x_i \in U$ whenever $i > N_U$. In this case, x is called a *limit point of the sequence* $\{x_i\}$.

Thus, if U is any neighborhood of x, members of a sequence $\{x_i\}$ converging to x are eventually trapped in U.

(1.F.9) Theorem. Suppose that (X,d) and (Y,d') are metric spaces. A function $f : X \to Y$ is continuous at a point $x \in X$ if and only if whenever a sequence $\{x_i\}$ in X converges to x, then the sequence $\{f(x_i)\}$ converges to $f(x)$.

Proof. Suppose that f is continuous at x, and let V be any open set containing $f(x)$. Since f is continuous, there is a neighborhood U of x such that $f(U) \subset V$. The convergence of $\{x_i\}$ implies that there is some integer N such that $x_i \in U$ whenever $i > N$. Thus, for each $i > N$, we have that $f(x_i) \in V$, and therefore $\{f(x_i)\}$ converges to $f(x)$.

To prove the converse, suppose that f is not continuous at $x \in X$. Then there must be an $\varepsilon > 0$ such that no neighborhood of x is mapped by f into $S_\varepsilon(f(x))$. For each positive integer n, select a point $x_n \in S_{1/n}(x)$ such that $f(x_n) \notin S_\varepsilon(f(x))$. Clearly, the sequence $\{x_n\}$ converges to x, but the corresponding sequence $\{f(x_n)\}$ fails to converge to $f(x)$.

The preceding theorem gives rigor to the concept of continuity as often taught in elementary calculus: a function is continuous at a point $x \in \mathscr{E}^1$ if and only if whenever points in $\mathscr{E}^1$ approach x, then their images under f approach $f(x)$.

(1.F.10) Definition. Suppose that (X,d) is a metric space and that A and B are nonempty subsets of X. Then the *distance between A and B, $d(A,B)$,* is defined to be $\mathrm{glb}\{d(a,b) \mid a \in A \text{ and } b \in B\}$.

In the two exercises that conclude this section, we see that both d and a distance function closely related to d are continuous.

(1.F.11) Exercise. Suppose that (X,d) is a metric space. Show that the metric function $d : X \times X \to [0,\infty)$ is continuous.

(1.F.12) Exercise. Suppose that (X,d) is a metric space, and let $A \subset X$. Show that the function $\Psi : X \to [0,\infty)$ is continuous, where $\Psi(x) = d(\{x\},A)$.

G. HOMEOMORPHISMS

Suppose that $f : X \to Y$ is a continuous function from one topological space onto another. In spite of the continuity of f, the space X may differ considerably from Y: a flagrant example of this occurs when X is arbitrary and Y

consists of a single point. Not only may Y be a good deal simpler than X, but, surprisingly, as we shall see later, the reverse situation can happen as well. In Chapter 9, a continuous function is constructed that maps the unit interval $\mathbf{I} = [0,1]$ onto the unit square $\mathbf{I} \times \mathbf{I}$. In fact, maps can be found that carry $\mathbf{I}$ onto cubes as well as onto a host of far more exotic spaces.

In these cases, the continuous function in question will not be 1–1; nevertheless, even with this further imposition, the spaces X and Y may still have little similarity. To see this, let a set X be given the discrete topology, and let Y be any space with the same cardinality as X. Then all 1–1, onto mappings from X to Y are continuous; and hence, although X might have uncountably many open sets, Y might have only two.

Suppose now, however, that $f : X \to Y$ is a continuous bijection such that f^{-1} (which makes sense, since f is a bijection) is also continuous. The reader should verify that not only is there a 1–1 correspondence between the points of X and Y, but also between the open sets of the respective topologies. Thus, the topological structure of X has an almost exact counterpart in Y. Bijective, continuous mappings with continuous inverses are called *homeomorphisms*, and the corresponding spaces are said to be *homeomorphic* or *topologically equivalent*. An *embedding* is a 1–1 map $f : X \to Y$ such that $f^{-1} : f(X) \to X$ is continuous. At least a germinal feel for topology should arise from our discussion of homeomorphisms.

(1.G.1)   Definition.   A function $f : X \to Y$ is *open* (respectively, *closed*) if and only if f maps open (respectively, closed) sets onto open (respectively, closed) sets.

(1.G.2)   Exercise.   Show that homeomorphisms are both open and closed.

(1.G.3)   Exercise.   Show that a continuous open (or closed) bijection is a homeomorphism.

(1.G.4)   Exercise.   Show that if $X_1 \times \cdots \times X_n$ has the product topology, then the projection maps, p_i, are open.

(1.G.5)   Exercise.   Show that two metrics d and d' for a set X are equivalent if and only if the identity map $id : (X,d) \to (X,d')$ is a homeomorphism.

For a topologist, homeomorphic spaces are essentially identical. Thus, little distinction (if any) would be made between the following pair of spaces (considered as subspaces of $\mathscr{E}^2$), since one can be mapped homeomorphically onto the other.

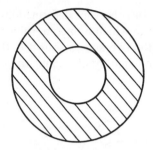 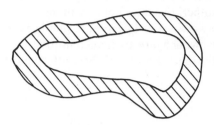

Similarly, a topologist is resigned to the fact that no distinction is to be made between the following trio of spaces.

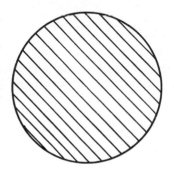 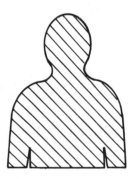 

In a like vein, since a doughnut can be transformed into a coffee cup via a homeomorphism (see figure), a topologist has been classically defined to be an otherwise enlightened individual who fails to note any difference between these two objects.

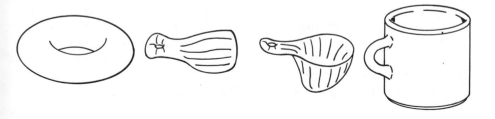

A word of caution, however. At this stage it might appear that two spaces *A* and *B* are homeomorphic if, intuitively, *A* could be molded into *B* without either *A* or *B* being torn apart. While it is true that a homeomorphism will result whenever one space is stretched, bent, or otherwise deformed into another, this does not constitute the only manner in which homeomorphisms may arise. For instance, it should be clear that the two spaces shown in the following figure are homeomorphic even though no amount of tugging (in $\mathscr{E}^3$) will transform one of them into the other.

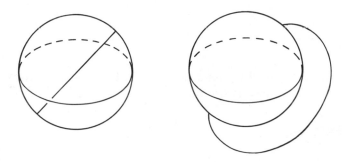

Thus, in order for two spaces to be homeomorphic, it suffices that we be able to jump immediately from one to another without any intermediate steps.

(1.G.6)   *Exercise.*   Show that $(-1,1)$ and $\mathscr{E}^1$ are homeomorphic via the map $h : \mathscr{E}^1 \to (-1,1)$ defined by $h(x) = x/(1 + |x|)$.

The problem of establishing the existence or nonexistence of a homeomorphism between two topological spaces is by no means a trivial one. The reader might enjoy discovering this for himself by attempting to show that $\mathscr{E}^1$ and $\mathscr{E}^2$ are not homeomorphic. In order to attack this problem in general, we shall investigate in the ensuing chapters certain *topological invariants*, i.e., properties of topological spaces that are preserved by homeomorphisms. If a space X possesses a certain property that is topologically invariant, and a space Y does not, then clearly X and Y cannot be homeomorphic. One obvious invariant (which does not turn out to be very useful) is the cardinality of the sets in question; another is the cardinality of their respective topologies. Note that neither of these invariants is of any help in deciding whether or not $\mathscr{E}^1$ and $\mathscr{E}^2$ are homeomorphic.

Metrizability is a prime example of topological invariance.

(1.G.7)   *Theorem.*   Suppose that $(X,\mathscr{U})$ is homeomorphic to $(Y,\mathscr{V})$ and that $(X,\mathscr{U})$ is metrizable. Then $(Y,\mathscr{V})$ is metrizable.

Proof.   Let d be a metric for X which induces the topology $\mathscr{U}$ and let

$h : X \to Y$ be a homeomorphism. Define $d' : Y \times Y \to [0,\infty)$ by $d'(y_1,y_2) = d(h^{-1}(y_1),h^{-1}(y_2))$. Since d' is clearly a metric, it remains to be shown that the topology $\mathcal{W}$ generated by d' coincides with $\mathcal{V}$. Suppose that $y \in V \in \mathcal{V}$. Then $h^{-1}(y) \in h^{-1}(V) \in \mathcal{U}$, and hence there is an $\varepsilon > 0$ such that $S_\varepsilon^d(h^{-1}(y)) \subset h^{-1}(V)$. Note that if $d'(z,y) < \varepsilon$, then we have $d(h^{-1}(z),h^{-1}(y)) < \varepsilon$ and consequently $h^{-1}(z) \in h^{-1}(V)$, so $z \in V$. Therefore, $S_\varepsilon^{d'}(y) \subset V$ and $\mathcal{V} \subset \mathcal{W}$.

Now consider $S_\varepsilon^{d'}(y)$. If $z \in S_\varepsilon^{d'}(y)$, then we have $h^{-1}(z) \in S_\varepsilon^d(h^{-1}(y))$, and therefore $h(S_\varepsilon^d(h^{-1}(y)) \subset S_\varepsilon^{d'}(y)$. Since $S_\varepsilon^d(h^{-1}(y)) \in \mathcal{U}$, it follows that $h(S_\varepsilon^d(h^{-1}(y)) \in \mathcal{V}$, and hence $\mathcal{W} \subset \mathcal{V}$.

(1.G.8) *Exercise.* Show that the properties T_2, first countability, and second countability are topological invariants. Use the topological invariance of T_2 to show that $\mathcal{E}^1$ and $\mathbf{R}^1$ with the finite complement topology are not homeomorphic.

(1.G.9) *Definition.* A subset D of a space X is *dense* in X if and only if every nonempty open set in X intersects D. A space X is *separable* if and only if X contains a countable dense subset.

The set of points in $\mathcal{E}^n$ with rational coordinates is easily seen to be a countable dense subset, and hence $\mathcal{E}^n$ is a separable metric space. Separability is obviously a topological invariant.

(1.G.10) *Theorem.* If (X,d) is a separable metric space, then X is second countable.

Proof. Let $x_1, x_2, \ldots$ be a countable dense subset of X. Then the collection $\{S_{1/n}(x_i) \mid n \in \mathbf{Z}^+, i \in \mathbf{Z}^+\}$ is a countable basis for the topology generated by d.

The following theorem has numerous applications.

(1.G.11) *Theorem.* Suppose that $f,g : X \to Y$ are continuous and that Y is a T_2 space. If $f(x) = g(x)$ for each x in a dense subset D of X, then $f = g$.

Proof. Suppose that for some $z \in X$, we have $f(z) \neq g(z)$. Then there are disjoint open subsets U and V in Y that contain $f(z)$ and $g(z)$, respectively. However, this is impossible, since $W = f^{-1}(U) \cap g^{-1}(V)$ is an open set that clearly lies in the complement of the dense set D. Therefore, f equals g.

We conclude this chapter with the definition of a class of spaces whose properties will serve as a focus for much of our attention during the remainder of the text.

(1.G.12) *Definition.* A separable metric space X is an *n-manifold* if and only if each point of X is contained in a neighborhood that is homeomorphic to $\mathscr{E}^n$.

In general, *n*-manifolds are strongly geometric and have considerable visual appeal; nevertheless, they may become exceedingly complex. Observe that a point with little wanderlust living in an exotic *n*-manifold, might well consider its domicile to be nothing more exciting than $\mathscr{E}^n$.

(1.G.13) *Examples of 1-Manifolds.*
   1. $\mathscr{E}^1$.
   2.   For each $n \in \mathbf{Z}$, let $A_n = \{(x,y) \in \mathscr{E}^2 \mid y = n\}$. Then $M = \bigcup \{A_n \mid n \in \mathbf{Z}\}$ is a 1-manifold.
   3.   The unit circle, $\mathscr{S}^1$, considered as a subspace of $\mathscr{E}^2$.

(1.G.14) *Examples of 2-Manifolds.*
   1. $\mathscr{E}^2$.
   2. $\mathscr{S}^2 = \{(x,y,z) \in \mathscr{E}^3 \mid x^2 + y^2 + z^2 = 1\}$.

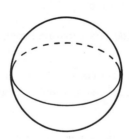

   3.   A torus (the outside of a tire or, alternatively, $\mathscr{S}^1 \times \mathscr{S}^1$).

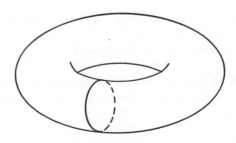

4. A Moebius strip (without the edge).

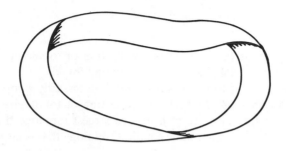

5. Any open subset of $\mathscr{E}^2$.

(1.G.15) *Examples of 3-Manifolds.*
  1. $\mathscr{E}^3$.
  2. $\mathscr{S}^3 = \{(x,y,z,w) \in \mathscr{E}^4 \mid x^2 + y^2 + z^2 + w^2 = 1\}$.
  3. The inside of a torus.
  4. The inside of a tin can.
  5. Any open subset of $\mathscr{E}^3$.

(1.G.16) *Exercise.* Determine which of the following subspaces of $\mathscr{E}^2$ are manifolds ($\mathbf{O} = (0,0)$ and $S_r(\mathbf{O}) = \{y \in \mathscr{E}^2 \mid d(y,\mathbf{O}) < r\}$):

  (i) $S_1(\mathbf{O})$;
  (ii) $\overline{S_1(\mathbf{O})}$;
  (iii) $S_2(\mathbf{O}) \setminus S_1(\mathbf{O})$;
  (iv) $S_2(\mathbf{O}) \setminus \overline{S_1(\mathbf{O})}$;
  (v) $\{(x,y) \mid -1 < x < 1, y = 0\} \cup \{(x,y) \mid x = 0, 0 \le y < 1\}$;

(1.G.17) *Exercise.*
  Let X be the union of the rays in $\mathbf{R}^2$:
  $A = \{(x,1) \mid x \ge 0\}$,
  $B = \{(x,-1) \mid x \ge 0\}$, and
  $C = \{(x,0) \mid x < 0\}$.
Points x in X other than $(0,-1)$ and $(0,1)$ are given neighborhood bases consisting of open intervals centered at x. Base neighborhoods of $(0,1)$ are of the form $\{(x,y) \mid b < x < 0$ and $y = 0$, or $0 \le x < a$ and $y = 1\}$, and base neighborhoods of $(0,-1)$ are of the form $\{(x,y) \mid b < x < 0$ and $y = 0$, or $0 \le x < a$ and $y = -1\}$. Show that X with the resulting topology is separable, T_1, first countable, and locally Euclidean, but that X is not T_2 and, hence, is not a 1-manifold.

PROBLEMS

Section A

1. (a) Suppose that $\{\mathcal{U}_\alpha \mid \alpha \in \Lambda\}$ is a collection of topologies for a set X. Show that $\bigcap \{\mathcal{U}_\alpha \mid \alpha \in \Lambda\}$ is a topology for X.
   (b) Is the union of topologies for a set X necessarily a topology?
   (c) Suppose that $\{\mathcal{U}_\alpha \mid \alpha \in \Lambda\}$ is a collection of topologies for a set X. Establish the existence of a unique largest topology that is smaller than each $\mathcal{U}_\alpha$ and a unique smallest topology that is larger than each $\mathcal{U}_\alpha$.

2. Find all possible topologies for the set $X = \{a,b,c\}$.

3. True or false? If $(X,\mathcal{U})$ is a topological space and $\mathcal{C}$ is a family of open sets in X, then $\bigcap \{C \mid C \in \mathcal{C}\}$ is open.

4. For each subset A of a set X, let $\mathcal{T}(A)$ be the topology on X whose open sets are $\varnothing$, X, and all subsets of X containing A. Assume that X has at least two elements.
   (i) Show that $A \subset B$ if and only if $\mathcal{T}(B) \subset \mathcal{T}(A)$.
   (ii) Suppose that $A_1, A_2, \ldots, A_n$ are subsets of X, and that $\mathcal{T}$ is a topology for X such that $\mathcal{T}(A_i) \subset \mathcal{T}$ for each i. Show that
   $$\mathcal{T}(\bigcap_{i=1}^{n} A_i) \subset \mathcal{T}.$$

5. Find four equivalent bases for $\mathcal{E}^2$.

6. Suppose that $\mathcal{U}_1$ and $\mathcal{U}_2$ are topologies for a set X and that $id : (X,\mathcal{U}_1) \to (X,\mathcal{U}_2)$ is the identity map. Show that id is continuous if and only if $\mathcal{U}_2 \subset \mathcal{U}_1$.

7. Show that a function $f : \mathcal{E}^1 \to \mathcal{E}^1$ is continuous if and only if $f^{-1}(-\infty,a)$ and $f^{-1}(a,\infty)$ are open for each $a \in \mathcal{E}^1$.

8. Show that a subset A of $\mathcal{E}^1$ (with the usual topology) is open if and only if A can be written as the countable union of pairwise disjoint open intervals (intervals such as $(-\infty,a)$, (b,∞), $(-\infty,\infty)$ are allowed).

9. Suppose that $(X,\mathcal{U})$ and $(Y,\mathcal{V})$ are topological spaces. A map $f : X \to Y$ is *open* if and only if $f(U) \in \mathcal{V}$ whenever $U \in \mathcal{U}$. Find an open map that is not continuous.

Section B

1. Given a topological space $(X,\mathcal{U})$, show that $\mathbf{B} \subset \mathcal{U}$ is a basis for $\mathcal{U}$ if and only if for each $x \in X$, $\mathbf{B}_x = \{B \in \mathbf{B} \mid x \in B\}$ is a neighborhood basis at x.

2. Suppose that X is a first countable space, Y is an arbitrary space, and $f : X \to Y$ is continuous and onto. Is Y necessarily first countable?

3. Determine which subsets A of $\mathscr{E}^1$ (with the usual topology) have the property that both A and $\mathscr{E}^1 \setminus A$ are open.

4. Suppose that $\mathscr{U}$ and $\mathscr{U}'$ are topologies for a set X such that $(X, \mathscr{U})$ is T_2 and $\mathscr{U} \subset \mathscr{U}'$. Show that $(X, \mathscr{U}')$ is T_2.

5. Suppose that $(X, \mathscr{U})$ is second countable. Show that any basis for $(X, \mathscr{U})$ contains a countable subcollection that is a basis for $\mathscr{U}$.

6. Show that $(X, \mathscr{U})$ is a topological space, and determine if it is T_2, first countable, or second countable where:

   (a) X is any uncountable set and $\mathscr{U}$ is $\varnothing$ together with $\{A \subset X \mid X \setminus A \text{ is countable}\}$;

   (b) X is any uncountable set and $\mathscr{U}$ consists of $\varnothing$ together with those subsets A of X such that either $X \setminus A$ is countable or $p \in (X \setminus A)$, where p is a fixed point in X;

   (c) $X = [-1,1]$, and $\mathscr{U}$ is the topology generated by a basis consisting of sets of the form $[-1,b)$, $(a,1]$, and (a,b), where $a < 0$ and $0 < b$.

7.* Let $X = \{(x,y) \mid 0 \le x \le 1, 0 \le y \le 1\}$ and define an order on X by declaring $(a,b) \le (u,v)$ if and only if $a < u$, or $a = u$ and $b \le v$. This ordering is called the *lexicographic ordering*. Show that this ordering is a linear ordering and determine if X (with the order topology) is T_2, first countable, or second countable.

8. For each $x \in \mathbf{Z}$ and each $n \in \mathbf{N}$, let $B_x^n = \{y \mid y = rn + x, r \in \mathbf{Z}\}$. Show that $\mathbf{B} = \{B_x^n \mid x \in \mathbf{Z}, n \in \mathbf{N}\}$ is a basis for a topology on $\mathbf{Z}$.

9. Suppose that X is first countable (respectively, second countable) and that $f : X \to Y$ is open and onto. Show that Y is first countable (respectively, second countable).

10. True or false? Suppose that X is second countable. Then every nest of distinct open sets in X is countable.

Section C

1. Suppose that (X,d) is a metric space and that p is a point of X. Show that for each $r > 0$, $\{x \in X \mid d(x,p) > r\}$ is open.

2. Give an $\varepsilon - \delta$ definition of continuity for a function between two metric spaces that is equivalent to (1.A.3).

3. Suppose that (X,d) is a metric space and $d' : X \times X \to [0,\infty)$ is defined by $d'(x,y) = d(x,y)/(1 + d(x,y))$. Show that d' is a metric for X and that d and d' are equivalent. [Hint: If $a \ge b \ge 0$ and $m \ge n \ge 0$, then $a/(1 + a) \ge b/(1 + b)$ and $(b + m)/(a + n) \ge b/a$.]

4. Suppose that X is a set consisting of a finite number of points. Describe

all possible topological structures that X can have in order to be metrizable.

5. *The post office metric.* Let $X = \mathbf{R}^2$ and d be the usual metric. Denote $(0,0)$ by $\mathbf{O}$. Define $\hat{d} : X \times X \to [0,\infty)$ by $\hat{d}(p,q) = d(\mathbf{O},p) + d(\mathbf{O},q)$ for $p,q \in X$ and $p \neq q$, and $d(p,p) = 0$ for all $p \in X$.
   (a) Show that $\hat{d}$ is a metric.
   (b) Show that all points other than $\mathbf{O}$ are open.
   (c) What are the neighborhoods of $\mathbf{O}$?

6. If X is a set, a function $d : X \times X \to [0,\infty)$ is called a *pseudometric* if and only if

   (i) $d(x,y) = d(y,x)$ for all $(x,y) \in X \times X$,
   (ii) $d(x,x) = 0$ for all $x \in X$, and
   (iii) $d(x,z) \leq d(x,y) + d(y,z)$ for $x,y,z \in X$.

   Let $X = \{f \mid f : \mathscr{E}^1 \to \mathscr{E}^1$ and f is integrable on $[0, 1]\}$. Define $d : X \times X \to [0,\infty)$ by $d(f,g) = \int_0^1 |f(x) - g(x)| \, dx$. Show that d is a pseudometric but not a metric. Is there a theorem analogous to (1.C.3) for pseudometric spaces?

7. Let $X = \mathbf{R}^2$, and for points $x = (x_1,x_2)$ and $y = (y_1,y_2)$ define
$$d(x,y) = \begin{cases} 1/2 & \text{if } x_1 = y_1, x_2 \neq y_2, \text{ or } x_1 \neq y_1 \text{ and } x_2 = y_2 \\ 1 & \text{if } x_1 \neq y_1 \text{ and } x_2 \neq y_2 \\ 0 & \text{otherwise} \end{cases}$$
Show that d is a metric and that the sets shown in the figure possess different "area" using d to measure the length of sides.

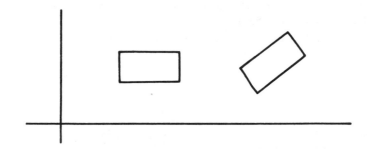

Describe the topology induced by this metric.

8. Suppose that is V a vector space over $\mathbf{R}^1$. Then a function $\phi : V \to [0,\infty)$ is a *norm* for V if and only if

   (i) $\phi(x) = 0$ if and only if $x = \mathbf{O}$,
   (ii) $\phi(x + y) \leq \phi(x) + \phi(y)$,
   (iii) $\phi(\alpha x) = |\alpha|\phi(x)$, where α is a real number.

The pair (V, ϕ) is called a *normed vector space* and $\phi(x)$ is frequently denoted by $\|x\|$.

Suppose that (V, ϕ) is a normed vector space and $d : V \times V \to [0, \infty)$ is defined by $d(x, y) = \|x - y\|$. Show that d is a metric for V.

9. *French railway metric.* Let $X = \mathbf{R}^2$. Suppose that x and y are points in X. If x and y are on a line passing through the origin, define their distance to be the usual Euclidean distance. If x and y do not lie on such a line, then define their distance as in the post office metric (problem 5). Show that the French railway metric is a metric and describe the induced topology.

Section D

1. Determine the relative topology that is induced by $\mathscr{E}^1$ for the integers.
2. Suppose that $\mathbf{B}$ is a basis for a topology on a set X and that $A \subset X$. Show that $\{B \cap A \mid B \in \mathbf{B}\}$ is a basis for the relative topology on A.
3. Suppose that $\mathbf{B}$ is a basis for a topology on a space X and $\mathbf{B}'$ is a basis for a topology on a space Y. Show that $\{B \times B' \mid B \in \mathbf{B}, B' \in \mathbf{B}'\}$ is a basis for the product topology on $X \times Y$.
4. Show that if X and Y are T_2 spaces, then $X \times Y$ is a T_2 space.
5*. Suppose that (X_1, d_1), (X_2, d_2), ..., (X_n, d_n) are metric spaces. Let $Y = X_1 \times X_2 \times \cdots \times X_n$ and define $\rho : Y \times Y \to [0, \infty)$ by setting

$$\rho((x_1, x_2, \ldots, x_n), (y_1, y_2, \ldots, y_n)) = \sqrt{\sum_{i=1}^{n} (d_i(x_i, y_i))^2}$$

Show that ρ is a metric and the topology induced by ρ coincides with the product topology on Y.

6. Consider the following diagram where the p_i's and q_i's are projection maps, and the f_i's are continuous. Show that a unique continuous function $\phi : X_1 \times X_2 \to Y_1 \times Y_2$ may be defined which yields the following commutative diagram.

$$
\begin{array}{ccccc}
X_1 & \xleftarrow{\;p_1\;} & X_1 \times X_2 & \xrightarrow{\;p_2\;} & X_2 \\
\Big\downarrow{\scriptstyle f_1} & & \Big\downarrow{\scriptstyle \phi} & & \Big\downarrow{\scriptstyle f_2} \\
Y_1 & \xleftarrow[\;q_1\;]{} & Y_1 \times Y_2 & \xrightarrow[\;q_2\;]{} & Y_2
\end{array}
$$

7. Suppose that $X_1, X_2, \ldots, X_n$ are topological spaces and that $Y = \prod_{i=1}^{n} X_i$. Let $g : Z \to Y$ be a function from a topological space Z into Y.

Show that g is continuous if and only if $p_i g$ is continuous for each projection map p_i (cf. 1.F.7.).

8. Is the product of a finite number of first countable spaces first countable? Is the product of a finite number of second countable spaces second countable?

9. Is the free union of a family of first countable spaces first countable? Is the free union of a family of second countable spaces second countable?

10.* Show that the free union of metrizable spaces is metrizable.

11.* Suppose that X and Y are topological spaces and $V \subset X \times Y$ is open. Show that $V[x] = \{y \mid (x,y) \in V\}$ and $V[y] = \{x \mid (x,y) \in V\}$ are open in Y and X respectively. Is the converse true: If $A \subset X \times Y$ and $A[x]$ and $A[y]$ are open for each $x \in X$ and $y \in Y$, then is A open in $X \times Y$?

Section E

1. Show that a subset U of a space X is open if and only if $A \cap U = \varnothing$ implies that $\bar{A} \cap U = \varnothing$ for each $A \subset X$.

2. Suppose that X is a space and that $A \subset B \subset X$. Show that $\bar{A}^B = \bar{A}^X \cap B$.

3. Find a topological space which has neither the discrete nor indiscrete topology in which subsets are open if and only if they are closed.

4. Suppose that f and g are continuous maps from a space X into a T_2 space Y. Show that $\{x \mid f(x) = g(x)\}$ is closed.

5. Show that $(A \cap B)^\circ = A^\circ \cap B^\circ$, and $(A^\circ)^\circ = A^\circ$.

6. Suppose that X is a finite T_1 space. Show that X must have the discrete topology.

7. Show that a set $U \subset X$ is open if and only if Fr $U \subset (X \setminus U)$.

8. Show that a set A is open and closed if and only if Fr $A = \varnothing$.

9. Find an example to show that if $A \subset B \subset X$, then int A in B is not necessarily the same as int A in X.

10. Show that $(A \cup B)' = A' \cup B'$.

11. Suppose that $(X,\mathcal{U})$ is a topological space, $A \subset X$, and x is an accumulation point of A. Discuss whether or not x must be an accumulation point of A with respect to topologies $\mathcal{U}'$ and $\mathcal{U}''$, where $\mathcal{U} \subset \mathcal{U}'$ and $\mathcal{U}'' \subset \mathcal{U}$.

12. Let $X = \{a,b,c\}$ and $\mathcal{U} = \{\varnothing, X, \{a\}, \{b\}, \{a,b\}\}$ be a topology for X. Find the derived sets of all subsets of X.

13. Let X be an infinite set with the finite complement topology. Find the derived set, interior, exterior, frontier, and closure for each of the subsets of X.

14.* A space X is called a *door space* if and only if each subset of X is either open or closed.
   (a) Find a door space that does not have either the discrete or the indiscrete topology.
   (b) Suppose that X is a T_2 door space. Show that X has at most one accumulation point and that nonaccumulation points are open.

15.* A space X is called a *semi-door space* if and only if for each $A \subset X$, there is an open set U such that either $U \subset A \subset \bar{U}$ or $U \subset (X \setminus A) \subset \bar{U}$.
   (a) Find a semi-door space which is not a door space.
   (b) Show that a T_2 semi-door space is a door space. [Hint: Show that there is at most one point that is not open.]

16. Find a metric space to illustrate that $\overline{S_\varepsilon(x)} \neq \{y \in X \mid d(x,y) \leq \varepsilon\}$.

17. (a) Find a topological space that is not T_1.
   (b) Find a T_1 space that is not T_2.
   (c) Determine which of the spaces listed in the examples given in Section B are T_1.

18. Show that in a T_1 space, derived sets are closed.

19. Let $(X, \mathcal{U})$ have the finite complement topology, where X is infinite. Show that if $\mathcal{U}' \subset \mathcal{U}$ and $(X, \mathcal{U}')$ is T_1, then $\mathcal{U} = \mathcal{U}'$. Furthermore, show that if $(X, \mathcal{V})$ is T_1 then $\mathcal{U} \subset \mathcal{V}$. (Hence, $\mathcal{U}$ is the smallest T_1 topology for X.) If $\mathcal{T}$ is the family of all T_1 topologies for X, show that $\mathcal{U} = \bigcap \{T \mid T \in \mathcal{T}\}$.

20. Show that an uncountable subset of a second countable space has an accumulation point.

21.* Show that there exist closed subsets of $[0,1]$ (usual topology) which are uncountable and consist only of irrational points.

22. Suppose that X is a set and $c : \mathcal{P}(X) \to \mathcal{P}(X)$ is a function with the following properties:

   (i) $c(\emptyset) = \emptyset$;
   (ii) $A \subset c(A)$ for each $A \in \mathcal{P}(X)$;
   (iii) $c(c(A)) = c(A)$ for each $A \in \mathcal{P}(X)$;
   (iv) $c(A \cup B) = c(A) \cup c(B)$ for each A and B in $\mathcal{P}(X)$.

   Let $\mathcal{U} = \{(X \setminus c(A)) \mid A \in \mathcal{P}(X)\}$. Show that $\mathcal{U}$ is a topology for X with the property that for each $A \in \mathcal{P}(X)$, $\bar{A} = c(A)$. A function with the properties of c is called a *closure operator*.

23. Let X be a set and $I : \mathcal{P}(X) \to \mathcal{P}(X)$ have the properties that:
   (i) $I(X) = X$;
   (ii) $I(I(A)) = I(A)$;
   (iii) $I(A) \subset A$;
   (iv) $I(A \cap B) = I(A) \cap I(B)$.

Let $\mathscr{U} = \{I(A) \mid A \in \mathscr{P}(X)\}$. Show that $\mathscr{U}$ is a topology for X, and that for each $A \in \mathscr{P}(X)$, $I(A) = A^\circ$. A function with the properties of I is called an *interior operator*.

24.* Formulate and prove a similar theorem for the derived operator, and for the frontier operator.

25. Prove that in a T_1 space, a point x is an accumulation point of a set A if and only if every neighborhood of x contains infinitely many points of A.

26. Suppose that A and B are disjoint closed subsets of a metric space. Show that there are disjoint open sets U and V containing A and B respectively.

27. Suppose that A is an open subset of a topological space X. Prove or disprove $\text{int}(\bar{A}) = A$.

Section F

1. Use (1.A.4) to give an alternate proof of (1.F.1).

2. Suppose that X is a set and that $A \subset X$. The *characteristic function* f_A *associated with* A is a map from X into $\{0,1\}$ which assumes the value 0 for points not in A and 1 for points in A. If X is a topological space and $A \subset X$, show that f_A is continuous if and only if A is open and closed (assume that $\{0,1\}$ has the discrete topology).

3. Show that a function $f : X \to Y$ is continuous if and only if $\overline{f^{-1}(B)} \subset f^{-1}(\bar{B})$ for each $B \subset Y$.

4. Suppose that $f, g : X \to \mathscr{E}^1$ are continuous. Show that $h : X \to \mathscr{E}^1$ defined by $h(x) = \max\{f(x), g(x)\}$ is continuous.

5.* A map $f : X \to \mathscr{E}^1$ is *upper semicontinuous* if and only if for each $b \in \mathscr{E}^1$, $\{x \mid f(x) < b\}$ is open.
   (a) Suppose that $\{f_\alpha\}_{\alpha \in \Lambda}$ is a family of continuous functions each mapping a space X into $(0,1) \subset \mathscr{E}^1$. Show that $h : X \to \mathscr{E}^1$ defined by $h(x) = \inf\{f_\alpha(x) \mid \alpha \in \Lambda\}$ is upper semicontinuous.
   (b) Suppose that $f : X \to \mathscr{E}^1$ has the property that for each rational r, $\{x \mid f(x) < r\}$ is open. Show that f is upper semicontinuous.

6. Prove that the following are equivalent:
   (a) $f : X \to Y$ is continuous;
   (b) $f(A') \subset (f(A))'$ for each $A \subset X$;
   (c) $\text{Fr} f^{-1}(B) \subset f^{-1}(\text{Fr } B)$ for each $B \subset Y$.

7. Let $f : X \to Y$ and $A \subset X$. Show that it is possible for $f_{|A}$ to be continuous even though f is not continuous at any point of A.

8. Construct a function $f : \mathscr{E}^1 \times \mathscr{E}^1 \to \mathscr{E}^1$ such that f is continuous in each variable separately, but is not continuous on $\mathscr{E}^1 \times \mathscr{E}^1$.

9. Suppose that $(X, \mathscr{U})$ and $(Y, \mathscr{V})$ are topological spaces. A function

$f: X \rightarrow Y$ is *weakly continuous at a point* $x \in X$ if and only if whenever $f(x) \in V \in \mathscr{V}$, there is a $U \in \mathscr{U}$ such that $x \in U$ and $f(U) \subset \overline{V}$. The function f is *weakly continuous* if and only if it is weakly continuous at each $x \in X$. Show that f is weakly continuous if and only if for each $V \in \mathscr{V}$, $f^{-1}(V) \subset \text{int}(f^{-1}(\overline{V}))$.

10.* Suppose that $(X,\mathscr{U})$ and $(Y,\mathscr{V})$ are topological spaces. A function $f: X \rightarrow Y$ is *feebly continuous* if and only if whenever $V \in \mathscr{V}, f^{-1}(\text{Fr } V)$ is closed in X. Show that weak continuity and feeble continuity are unrelated, but that a function f is continuous if and only if it is both feebly and weakly continuous.

11. Suppose that $(X,\mathscr{U})$ and $(Y,\mathscr{V})$ are topological spaces. A function $f: X \rightarrow Y$ is *strongly continuous* if and only if $f(\overline{A}) \subset f(A)$ for each $A \subset X$. Show that f is strongly continuous if and only if $f^{-1}(B)$ is closed for each $B \subset Y$.

12. (a) Find an example of a topological space X and a sequence $\{x_i\}$ in X such that $\{x_i\}$ has more than one limit point in X.

(b) Show that if X is Hausdorff then a sequence can have at most one limit point.

13. Find a counterexample to the following proposition: Suppose that X and Y are topological spaces and $f: X \rightarrow Y$ is a function with the property that whenever $\{x_n\}$ is a sequence in X and x is a point in X to which $\{x_n\}$ converges, the sequence $\{f(x_n)\}$ converges to $f(x)$. Then f is continuous.

Section G

1. Show that $\mathscr{E}^n$ is homeomorphic with $\{x \in \mathscr{E}^n \mid d(x,\mathbf{O}) < 1\}$.

2. A subset A of $\mathbf{R}^n$ is *convex* if and only if the line segment $\{tx + (1 - t)y \mid 0 \le t \le 1\}$ is contained in A whenever $x,y \in A$. Show that bounded convex open subsets of $\mathscr{E}^n$ are homeomorphic. Is the modifier "bounded" necessary?

3. Suppose that X and Y are topological spaces and that $f: X \rightarrow Y$ is bijective. Show that f is a homeomorphism if and only if for each $A \subset X$, $f(\overline{A}) = \overline{f(A)}$ holds.

4. Fill in the blank: Let $id : (X,\mathscr{U}) \rightarrow (X,\mathscr{U}')$. Then id is open if and only if $\mathscr{U} \underline{\qquad} \mathscr{U}'$.

5. Suppose that X and X' are homeomorphic and that Y and Y' are homeomorphic. Show that $X \times Y$ and $X' \times Y'$ are homeomorphic.

6. Find mappings that are open but not closed, closed but not open, continuous but not open or closed.

7. Let A be the set of irrationals in $\mathscr{E}^1$ with the relative topology. Is A separable?

8. Suppose that A is a dense subset of X and G is a dense open subset of X. Show that $A \cap G$ is dense in X.

9. Suppose that A is dense in X and B is dense in Y. Show that $A \times B$ is dense in $X \times Y$.

10. A map $f : X \to Y$ is a *local homeomorphism* if and only if for each point $x \in X$, there is an open set U containing x that is mapped homeomorphically by f onto an open subset of Y. Show that local homeomorphisms are continuous and open.

11.* Prove the following converse of (1.G.11). Suppose that Y has the property that for any space X, any dense subset D of X, and any two continuous functions $f,g : X \to Y$ which agree on D, then $f = g$. Then Y is T_2.

12. A metric space X is *totally bounded* if and only if for each $\varepsilon > 0$, there are points $x_1, x_2, \ldots, x_n$ such that $X = \bigcup \{S_\varepsilon(x_i) \mid i = 1, 2, \ldots, n\}$. Show that a totally bounded metric space is separable.

13.* Give an example of a nonseparable metric space that does not have the discrete topology.

14. Suppose that (X,d) is a separable metric space and that $A \subset X$. Show that $(A,d_{|A})$ is separable. In general, is it true that a subspace of a separable space is separable?

15.* Suppose that $\mathbf{R}^1$ has the half-open interval topology $\mathscr{U}$, and $A = [0,1)$ is given the relative topology (with respect to $\mathscr{U}$). Show that $\mathbf{R}^1$ and A are homeomorphic.

16. True or false? If U is an open subset of X and $h : U \to X$ is an embedding, then $h(U)$ is open in X.

17.* Assume that the statement in problem 16 is true whenever $X = \mathscr{E}^n$, and prove that it is true for n-manifolds.

18.* A subset A of a space X is *semiopen* if and only if there is an open set U in X such that $U \subset A \subset \overline{U}$.
    (a) Show that A is semiopen if and only if $\overline{A} = \overline{A \cap D}$ for each dense subset D of X.
    (b) Suppose that $D \subset X$. Show that D is dense if and only if $\overline{A} = \overline{A \cap D}$ for each semiopen set A in X.

19. (a) A topological space is *0-dimensional* if and only if whenever $x \in V$, and V is open, there is an open set U with empty frontier such that $x \in U \subset V$. Show that the rationals and the irrationals (with the relative topology) are 0-dimensional sets.
    (b) The higher dimensions are defined inductively. A space X is said to have *dimension* $\leq n$, if and only if for each point $x \in X$ and for each open set V containing x there is an open set U such that $x \in U \subset V$, and dimension Fr $U \leq n - 1$; X has *dimension* n if and only if dimension $X \leq n$, but it is false that dimension $X \leq n - 1$. Show that $\mathscr{E}^1$ has dimension 1.

(c) Show that dimension is a topological invariant.

20. Are restrictions of open (closed) mappings open (closed)? Is the restriction of an open (closed) mapping to an open (closed) subset open (closed)?

21. Find a set X and distinct topologies $\mathscr{U}_1$ and $\mathscr{U}_2$ for X such that there is a homeomorphism $h : (X,\mathscr{U}_1) \to (X,\mathscr{U}_2)$.

22. Let $(X,\mathscr{U})$ and $(Y,\mathscr{V})$ be topological spaces. Show that a map $f : (X,\mathscr{U}) \to (Y,\mathscr{V})$ is open if and only if for each $A \subset X, f(\text{int } A) \subset \text{int}(f(A))$.

Chapter 2

CONNECTEDNESS AND COMPACTNESS

The two principal topological properties studied in this chapter occupy a central position in both topology and analysis. The concepts of connectedness and compactness may be viewed as generalizations (in different ways) of two basic properties of intervals. Connectedness represents an extension of the idea that an interval is all in "one piece," while compactness may be construed as a generalization of the fact that a closed interval $[a,b]$ is both closed (as a subset of $\mathscr{E}^1$) and bounded.

A. CONNECTEDNESS: GENERAL RESULTS AND DEFINITIONS

The problem of deciding if a space X consists of just "one piece" is resolved by determining whether or not X may be broken up into disjoint open subsets.

(2.A.1) Definition. A pair (U,V) of nonempty open subsets of a space X is a *separation of X* if and only if $X = U \cup V$ and $U \cap V = \varnothing$.

(2.A.2) Remark. Suppose that X is a topological space and that A and B are disjoint nonempty closed subsets of X such that $X = A \cup B$. Then A and B effect a separation of X, since their complements are open. Thus, if X is separated by subsets A and B, then A and B are both open and closed.

(2.A.3) Definition. A topological space X is *connected* if and only if no separation of X exists. If a separation of X does exist, then X is *disconnected*. A subspace A of X is *connected* if and only if A with the relative topology is connected.

(2.A.4) Exercise. Show that a space X is connected if and only if X does not contain a proper nonempty subset that is both open and closed.

(2.A.5) Exercise. Suppose that A is a disconnected subspace of a space X. Show that there are closed subsets C_1 and C_2 of X such that $A \subset C_1 \cup C_2$, $A \cap C_1 \cap C_2 = \varnothing$, and both $A \cap C_1$ and $A \cap C_2$ are nonempty.

(2.A.6) Exercise. Show that if (U,V) is a separation of a space X and $C \subset X$ is connected, then either $C \subset U$ or $C \subset V$.

(2.A.7) Examples. (Proofs will follow from later results.)

Connected Spaces.

1. Intervals in $\mathscr{E}^1$
2. Disks (spaces homeomorphic with $\{(x,y) \in \mathscr{E}^2 \mid x^2 + y^2 \leq 1\}$)
3. $\mathscr{E}^n$
4. A sine curve
5. Any set with the indiscrete topology

Disconnected Spaces.

1. $(0,1) \cap (3,4)$ (with the relative topology)
2. $[0,1) \cap (1,4]$
3. Rational numbers (why?)
4. Any set with the discrete topology (and consisting of two or more points)

Although quite reasonable from a visual standpoint, the criterion that a space is connected if it consists of solely "one piece" is often difficult to apply in practice. For instance, while it seems apparent that $\mathscr{E}^2$ is connected, what can be said about $\mathscr{E}^2 \setminus \{(x,y) \mid x,y \in \mathbf{Q}\}$? Is this subspace still in "one piece" in spite of all the holes? In $\mathscr{E}^1$ no such complications arise, as we see from the following theorem.

(2.A.8) Theorem. A subspace B of $\mathscr{E}^1$ is connected if and only if B is a point or an interval.

Proof. Suppose that B is a subset of $\mathscr{E}^1$ that is neither an interval nor a point. Then there are points $a,b,c \in \mathscr{E}^1$ such that $a < b < c$, where $a,c \in B$ and $b \notin B$. Let $U = (-\infty,b) \cap B$ and $V = (b,\infty) \cap B$. Since U and V are clearly disjoint B-open subsets whose union is B, they form a separation of B, and consequently B is not connected.

To prove the converse, we suppose that B is not connected. We show that B cannot be an interval. It follows from (2.A.5) that there are closed subsets C and D of $\mathscr{E}^1$ such that $C \cap D \cap B = \emptyset$, $B \subset C \cap D$, and both $B \cap C$ and $B \cap D$ are nonempty. Let $c \in C \cap B$ and $d \in D \cap B$, and assume (relabel if necessary) that $c < d$. Let $s = \sup\{x \in B \mid x \in C \text{ and } c \leq x \leq d\}$. Note that $s \neq d$ (why?). Now let $t = \inf\{y \in B \mid y \in D \text{ and } s \leq y \leq d\}$. We consider two possibilities: $s = t$ and $s < t$. If $s = t$, then $s \in C \cap D$ (why?) and consequently s is not in B. If $s < t$, then $B \cap (s,t) = \emptyset$. In either case, it follows easily that B is not an interval.

(2.A.9) Exercise. Fill in the details of the foregoing proof.

(2.A.10) Theorem. Suppose that $\{A_\alpha \mid \alpha \in \Lambda\}$ is a collection of connected subspaces of a topological space X, and that for each $\alpha,\beta \in \Lambda$, $A_\alpha \cap A_\beta \neq \emptyset$. Then $\bigcup \{A_\alpha \mid \alpha \in \Lambda\}$ is connected.

Proof. Let $Y = \bigcup \{A_\alpha \mid \alpha \in \Lambda\}$ and suppose that Y is not connected. Then there are disjoint, nonempty, Y-open sets U and V whose union is Y. Since $U \neq \emptyset$, there is an $\alpha \in \Lambda$ such that $A_\alpha \cap U \neq \emptyset$, and similarly there is a $\beta \in \Lambda$ such that $A_\beta \cap V \neq \emptyset$. By (2.A.6), either $A_\alpha \subset U$ or $A_\alpha \subset V$, and since $A_\alpha \cap U \neq \emptyset$, it follows that $A_\alpha \subset U$. Similarly, we have that $A_\beta \subset V$. However, this is impossible, since $A_\alpha \cap A_\beta \neq \emptyset$.

(2.A.11) Exercise.
1. Show that $\mathscr{E}^1$ is connected.
2. Suppose that $\{A_\alpha \mid \alpha \in \Lambda\}$ is a collection of connected subspaces of a space X, and that $B \subset X$ is also connected. Show that if $B \cap A_\alpha \neq \emptyset$ for each $\alpha \in \Lambda$, then $B \cup \{A_\alpha \mid \alpha \in \Lambda\}$ is connected.

The following theorem, quite useful in spite of its trivial proof, gives a convenient characterization of connectedness. For the remainder of this chapter, the discrete space consisting of just two points, $\{0,1\}$, will be denoted by $\mathscr{S}$.

(2.A.12) Theorem. A space X is connected if and only if there is no continuous function $f : X \to \mathscr{S}$ which is onto.

Proof. If $f : X \to \mathscr{S}$ is onto, then the sets $f^{-1}(0)$ and $f^{-1}(1)$ form a separation of X. On the other hand, if X is not connected, then there is a separation (U,V) of X. Let $f : X \to \mathscr{S}$ be defined by $f(x) = 0$ if $x \in U$ and $f(x) = 1$ if $x \in V$. Then f is clearly continuous and maps X onto $\mathscr{S}$.

(2.A.13)   Theorem.   If A is a connected subspace of a topological space X and $A \subset B \subset \bar{A}$, then B is connected.

Proof.   We apply (2.A.12). Suppose that $f : B \to \mathscr{S}$ is continuous. We must show that f fails to be onto. Since $f_{|A}$ is continuous and A is connected, we may assume $f(a) = 0$ for each $a \in A$. Suppose that for some $b \in B, f(b) = 1$. Then $f^{-1}(1)$ is a B-open set that contains b but misses A. This, however, is impossible, since b is an accumulation point of A.

Connectedness is preserved by continuous functions.

(2.A.14)   Theorem.   If f is a continuous function from a connected space X onto a space Y, then Y is connected.

Proof.   If Y is not connected, then by (2.A.12) there is a continuous map g from Y onto $\mathscr{S}$. Hence, gf is a continuous function from X onto $\mathscr{S}$, which contradicts the fact that X is connected.

We have two immediate corollaries, the second of which should be familiar from elementary calculus.

(2.A.15)   Corollary.   Connectedness is a topological invariant.

(2.A.16)   Corollary (The Intermediate Value Theorem).   Let $f : \mathscr{E}^1 \to \mathscr{E}^1$ be a continuous function. Suppose that $a, b \in f(\mathscr{E}^1)$ and that $a < b$. Then if z is any number such that $a < z < b$, there is at least one point $c \in \mathscr{E}^1$ such that $f(c) = z$.

Proof.   Suppose that no such point c exists. Let $U = \{x \in \mathscr{E}^1 \mid x < z\}$ and $V = \{x \in \mathscr{E}^1 \mid x > z\}$. U and V form a separation of $f(\mathscr{E}^1)$, which contradicts (2.A.11) and (2.A.14).

One should observe that the domain of f in the foregoing corollary can be replaced by any connected space and the corresponding proposition will still hold.

B. SLIGHTLY DEEPER RESULTS CONCERNING CONNECTEDNESS

Corollary (2.A.16) of Theorem (2.A.14) proves to be a key tool in establishing certain basic theorems in calculus. However, rather than pursue these results, we give an illustration of how this corollary may be employed in a non-calculus setting.

An interesting problem that has kept many topologists and analysts occupied concerns the study of fixed point properties. A space X is said to have the *fixed point property* if and only if each continuous function mapping X into itself leaves some point alone, i.e., for each continuous map $f : X \rightarrow X$, there is an $x \in X$ such that $f(x) = x$. Theorems involving the fixed point property are not only aesthetically pleasing in themselves, but also have significant applications to other parts of mathematics. Our first fixed point result is a special case of the Brouwer fixed point theorem, which states that the space obtained by taking the product of $I = [0,1]$ with itself a finite number of times has the fixed point property.

(2.B.1)   Theorem.   The interval I has the fixed point property.

Proof.   Suppose that $f : I \rightarrow I$ is continuous. We will find a point $c \in I$ such that $f(c) = c$. If $f(0) = 0$ or $f(1) = 1$, we are finished. Hence, let us assume that $0 < f(0)$ and $f(1) < 1$. Define a continuous function g from I into $\mathscr{E}^1$ by setting $g(x) = x - f(x)$. Note that $g(0) = -f(0) < 0$ and $g(1) = 1 - f(1) < 0$. From the intermediate value theorem it follows that there is a point $c \in I$ such that $g(c) = 0$. However, this implies that $c - f(c) = 0$, and hence we have that $f(c) = c$.

Given spaces $X_1, X_2, \ldots , X_n$ all having a common topological property P, it is natural to inquire if the product space $\prod_{i=1}^{n} X_i$ also enjoys P. This proves to be the case if the property in question is connectedness.

(2.B.2)   Theorem.   If $X_1, X_2, \ldots , X_n$ are connected topological spaces, then $Y = \prod_{i=1}^{n} X_i$ is also connected.

Proof.   We prove the theorem for $n = 2$, and a trivial inductive argument will yield the more general result. If Y is not connected, then by (2.A.12) there is a map f from Y onto $\mathscr{S}$. Let $a = (a_1,a_2)$ and $b = (b_1,b_2)$ be points in Y with $f(a) = 0$ and $f(b) = 1$. Define functions $f_1 : X_1 \rightarrow \mathscr{S}$ and $f_2 : X_2 \rightarrow \mathscr{S}$, by setting $f_1(x_1) = f(x_1,b_2)$ for each $x_1 \in X_1$, and $f_2(x_2) = f(a_1,x_2)$ for each $x_2 \in X_2$. It is easily seen that f_1 and f_2 are continuous, and since X_1 and X_2 are connected, we conclude that f_1 and f_2 must be constant maps. Since $f_1(b_1) = f(b_1,b_2) = 1$, it follows that f_1 must be identically 1. Since $f_2(a_2) = f(a_1,a_2) = 0$, it follows that f_2 is identically 0. On the other hand we have that $f_1(a_1) = f_2(b_2) = f(a_1,b_2)$, and thus we may conclude that Y is connected.

(2.B.3)   Corollary.   The space $\mathscr{E}^n$ is connected.

(2.B.4)   Exercise.   Prove the converse of (2.B.2).

(2.B.5) Theorem. If B is a subset of $\mathscr{E}^n$ $(n > 1)$ and B contains a countable number of points, then $\mathscr{E}^n \setminus B$ is connected.

Proof. Let $c \in \mathscr{E}^n \setminus B$ and select a line L_c in $\mathscr{E}^n \setminus B$ containing c. (Why does such a line exist?) For each point $x \in \mathscr{E}^n \setminus B$, there is a line L_x in $\mathscr{E}^n \setminus B$ that contains x and intersects L_c. Thus, $\mathscr{E}^n \setminus B$ may be considered as the union of L_c and the L_x's, and hence by (2.A.11), $\mathscr{E}^n \setminus B$ is connected.

We have stated before that a central problem in topology is that of classifying spaces up to homeomorphism. This involves finding a method for deciding whether any arbitrary pair of topological spaces are homeomorphic. In this generality the problem cannot be solved, as mathematical logicians have shown (Markov, [1960]). However, if one reduces the problem to that of trying to determine whether a particular pair of topological spaces are homeomorphic, positive results are sometimes obtainable. As indicated in Chapter 1, a common way of attacking this problem is to find a topological invariant that is shared by one but not by both of the two spaces in question. Theorem (2.A.14) shows that connectedness is a topological (actually even a continuous) invariant. The reader is asked to show in the next exercises that the fixed point property is another such topological invariant.

(2.B.6) Exercises.
1.   Show that the fixed point property is a topological invariant, i.e., show that if $h : X \to Y$ is a homeomorphism and X has the fixed point property, then so does Y.
2.   Show that $[0,1]$ and $\mathscr{S}^1$ are not homeomorphic.

The topological invariant, connectedness, may be exploited to show that $\mathscr{E}^1$ is homeomorphically distinct from $\mathscr{E}^2$.

(2.B.7) Theorem.   The space $\mathscr{E}^1$ is not homeomorphic to $\mathscr{E}^2$.

Proof.   Suppose that a homeomorphism $h : \mathscr{E}^2 \to \mathscr{E}^1$ exists. Let $c \in \mathscr{E}^2$ and note that by (2.B.5), $\mathscr{E}^2 \setminus \{c\}$ is connected. However, $h(\mathscr{E}^2 \setminus \{c\}) = \mathscr{E}^1 \setminus \{h(c)\}$ and $\mathscr{E}^1 \setminus \{h(c)\}$ is not connected, which contradicts (2.A.14).

(2.B.8) Exercise.   Show that $(0,1)$ does not have the fixed point property, and hence $(0,1)$ is not homeomorphic with $[0,1]$.

(2.B.9) Exercise.   Using the topological invariant, connectedness, give another proof that $(0,1)$ and $[0,1]$ are not homeomorphic.

We conclude this section with a rather strange result, which will be used frequently in succeeding chapters.

(2.B.10) *Theorem.* Suppose that X is connected and that A is a connected subset of X. Suppose further that $X \setminus A = U \cup V$, where U and V are nonempty disjoint open (in $X \setminus A$) subsets of $X \setminus A$. Then $A \cup U$ is connected.

Proof. If $A \cup U$ is not connected, then $A \cup U = C_1 \cup C_2$, where C_1 and C_2 are nonempty, disjoint open and closed subsets of $A \cup U$. Since A is connected, either $A \subset C_1$ or $A \subset C_2$. If $A \subset C_1$, then we have that $C_2 \subset U$. Therefore, C_2 is open and closed in U. This, however, is impossible, since U is open and closed in $X \setminus A$, and hence it follows from (1.D.4) that C_2 is both open and closed in $(X \setminus A) \cup (A \cup U) = X$, which contradicts the connectedness of X. (Obviously, (1.D.4) also holds for closed subsets.) A similar argument holds if $A \subset C_2$.

C. PATH CONNECTEDNESS

In this section, we introduce a somewhat different concept of connectedness—one that will find repeated application in later chapters. A layman's notion of connectedness would probably center on the idea that a space X is connected if one can move from one point in X to another without ever leaving X. This concept is given mathematical import by the following two definitions.

(2.C.1) *Definition.* Suppose that X is a topological space. A *path* in X is a continuous function $f : \mathbf{I} \to X$. If $f(0) = a$ and $f(1) = b$, then f is said to be a *path from a to b*; a is the *initial point* of f, and b is the *end* or *terminal point*.

(2.C.2) *Definition.* A space X is *path connected* if and only if each pair of distinct points in X can be joined by a path, i.e., for each $x, y \in X$ there is a continuous map $f : \mathbf{I} \to X$ such that $f(0) = x$ and $f(1) = y$.

It should be obvious to the reader that $\mathscr{E}^n$ is path connected for every $n \in \mathbf{Z}^+$.

(2.C.3) *Theorem.* Suppose that a, b, and c are points in a space X and that there are paths from a to b and from b to c. Then there is a path from a to c.

Proof. Let $f : \mathbf{I} \to X$ and $g : \mathbf{I} \to X$ be paths such that $f(0) = a$, $f(1) = b$, $g(0) = b$, and $g(1) = c$. Define $h : \mathbf{I} \to X$ by setting

$$h(x) = \begin{cases} f(2t) & \text{if} \quad 0 \le t \le \tfrac{1}{2} \\ g(2t - 1) & \text{if} \quad \tfrac{1}{2} \le t \le 1 \end{cases}$$

Then by the map gluing theorem (1.F.6), (1.F.3), and (1.F.1), h is continuous. Clearly, $h(0) = a$ and $h(1) = c$.

(2.C.4) Exercise. Suppose that $\{A_\alpha \mid \alpha \in \Lambda\}$ is a collection of path connected subsets of a space X such that $\bigcap \{A_\alpha \mid \alpha \in \Lambda\} \neq \varnothing$. Show that $\bigcup \{A_\alpha \mid \alpha \in \Lambda\}$ is path connected.

(2.C.5) Exercise. Show that every path connected space is connected.

Although path connected spaces are connected, there do exist connected spaces that are not path connected, as we see in the following example.

(2.C.6) Example. Let $Y = \{(x, \sin 1/x) \in \mathscr{E}^2 \mid 0 < x < 1/\pi\}$ and let $Z = \{(x,y) \in \mathscr{E}^2 \mid x = 0 \text{ and } -1 \le y \le 1\}$. The space $X = Y \cup Z$ with the relative topology inherited from $\mathscr{E}^2$ is frequently referred to as the *topologist's sine curve.*

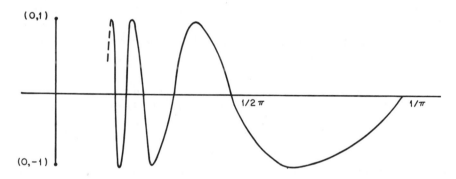

It follows immediately from (2.A.13) that X is connected. However, X is not path connected. To see this, we show that no path can stretch from the point $(1/\pi,0)$ to the point $(0,0)$. Suppose to the contrary that there is a path $f : \mathbf{I} \to X$ with $f(0) = (1/\pi,0)$ and $f(1) = (0,0)$. Since $f(\mathbf{I})$ is connected, every point on the sine curve (for $0 < x < 1/\pi$) must be included in the range of f. Thus we may select a sequence of points in $\mathbf{I}$, $x_1 < x_2 < x_3 < \cdots$, such that the sequence $\{x_i\}$ converges to 1 and the second coordinate of $f(x_i)$ is 1 if i is odd and is -1 if i is even. This, however, is absurd, since as $\{x_i\}$ converges to 1, the sequence $\{f(x_i)\}$ attempts to simultaneously approach both $(0,-1)$ and $(0,1)$ which is impossible by (1.F.9) and (1.F.8).

(2.C.7) Definition. A topological space X is *locally path connected* if and only if each point $x \in X$ has a neighborhood base of path connected open sets, i.e., if $x \in U$, where U is open, then there is a path connected open subset V of X such that $x \in V \subset U$.

(2.C.8)   *Theorem.*   If X is a connected, locally path connected topological space, then X is path connected.

Proof.   Let $x \in X$, and consider the set $A = \{z \in X \mid$ there is a path from x to $z\}$. We show that A is nonempty, open, and closed. Then, since X is connected, this will imply that $A = X$ (2.A.4). It is clear that A is nonempty, since $x \in A$. To show that A is open, suppose that $z \in A$. Since X is locally path connected, there is an open set V containing z such that every two points in V can be joined by a path lying in V. Hence, by (2.C.3), we have that $V \subset A$ and thus A must be open. To see that A is closed, we show that $X \setminus A$ is open. Suppose that $z \in X \setminus A$ and let V be a neighborhood of z that is path·connected. If $V \cap A \neq \varnothing$, then we may join z to x by a path (2.C.3), which contradicts the fact that $z \in X \setminus A$. Hence, $V \cap A = \varnothing$, and $X \setminus A$ is open.

(2.C.9)   *Corollary.*   Connected open subsets of $\mathscr{E}^n$ are path connected.

The concept of a path is perhaps less intuitive than it originally appears. For example, we have noted earlier that there are continuous functions mapping the unit interval onto the unit square. By definition, such functions are paths, even though they are probably not quite what the reader had in mind when we first described paths as a means of traveling from one point to another in a topological space. A notion undoubtedly much closer to the reader's preconception of what a path should be is that of an arc. An *arc* is an embedding of I into a space X. We shall eventually prove that in T_2 spaces, an arc is merely a path which is 1–1 (2.G.11.). As was the case with paths, if $h : I \to X$, is an arc, then $h(0)$ is referred to as the *initial point* of h and $h(1)$ is called the *terminal point*. Spaces in which any two distinct points may be connected by an arc are called *arc connected*, and *local arc connectedness* is defined in a manner analogous to local path connectedness.

(2.C.10)   *Exercise.*   Prove or disprove: If a space X is path connected, then X is always arc connected.

(2.C.11)   *Exercise.*   Prove (2.C.8) with paths replaced by arcs.

(2.C.12)   *Exercise.*   Prove that connected n-manifolds are arc connected.

D.   COMPONENTS

Spaces that are not connected may be viewed as consisting of a (possibly infinite) number of connected pieces. This leads us to the following notion.

(2.D.1)   *Definition.*   Suppose that X is a topological space and that

A is a subspace of *X*. Then *A* is a *component* of *X* if and only if (i) *A* is connected and (ii) if *B* is a connected subspace of *X* containing *A*, then *B* = *A*. Consequently, components are maximal connected subspaces.

Of course, if *X* is connected, then the only component of *X* is *X* itself. At the other extreme, we have that if *X* is a discrete space, then each point of *X* is a component.

(2.D.2) Example. Let *X* be the subspace of the plane $\mathscr{E}^2$ consisting of vertical line segments, as shown. Then each vertical line segment is a component of *X*.

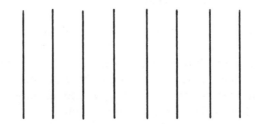

(2.D.3) Exercise. Find the components of the rational numbers with the relative topology. Do the same for the irrational numbers.

The basic properties of components are easily established.

(2.D.4) Theorem.

    (i)   Components of a space *X* are closed.
    (ii) Distinct components of a space *X* are disjoint.
    (iii) Each point of a space *X* belongs to exactly one component of *X*.
    (iv) If spaces *X* and *Y* are homeomorphic, then there is a 1–1, onto correspondence between the set of components of *X* and those of *Y*. Thus, the cardinality of the set of components is a topological invariant.

Proof. (i) This is immediate from (2.A.13). (ii) This is immediate from (2.A.10). (iii) Note that the component containing *x* is simply the union of all connected sets that contain *x*. (iv) This follows from the fact that images of connected sets under continuous maps are connected (2.A.14).

(2.D.5) Definition. A space *X* is *totally disconnected* if and only if each of its components consists of a point.

By (2.D.3) the set of rationals (or irrationals) with the relative topology is a totally disconnected space.

(2.D.6) Exercise. A connected space X is said to have an *explosion point* p (sometimes called a *dispersion point*) if and only if $X \setminus \{p\}$ is totally disconnected. Find such a space and show that a space can have at most one explosion point. [Hint: Use (2.B.10).]

(2.D.7) Definition. A subspace C of a topological space X is a *path component* of X if and only if C is path connected and is not contained in a larger path connected subspace.

(2.D.8) Exercise. Show that for open subsets of $\mathscr{E}^n$, path components and components coincide. Is the modifier "open" necessary?

E. LOCALLY CONNECTED SPACES

Consider the following example. In $\mathscr{E}^2$, for each $n \in \mathbf{Z}^+$, let $A_n = \{(x, 1/n) \mid 0 \le x \le 1\}$ and let $A_0 = \{(x,0) \mid 0 \le x \le 1\}$. The components of $X = A_0 \cup (\bigcup_{n=1}^{\infty} A_n)$ are precisely the sets $A_0, A_1, \ldots$. Although, by (2.D.4), A_0 must be closed, A_0 is clearly not open in X, and thus we see that components need not be open. (Another example of this was given in (2.D.3).) For certain types of spaces, however, components will be both open and closed. These are known as locally connected spaces, and they include the n-manifolds. The formal definition of such spaces is couched in somewhat different terms (emphasizing the local nature of the concept), but the fact that the two notions are essentially equivalent is readily established in (2.E.2).

(2.E.1) Definition. A space X is *locally connected* if and only if for each point $x \in X$ and each open set U containing x, there is a connected open set V such that $x \in V \subset U$.

Note that the foregoing definition is equivalent to saying that there is a neighborhood base at each point consisting of connected open sets.

(2.E.2) Theorem. A space X is locally connected if and only if the components of every open subset of X are open. (In particular, the components of locally connected spaces are open.)

Proof. Suppose that X is locally connected and that C is a component of an open subset U. Let c be a point in C. By (2.E.1), there is a connected

X-open subset V of U that contains c. Then V must lie in C (C is a maximal connected subset), and therefore C is open.

Conversely, let $x \in X$ and let U be any open set containing x. Then the component V of U that contains x is open and $x \in V \subset U$. Hence, X is locally connected.

(2.E.3) **Theorem.** The (finite) product of locally connected spaces is locally connected.

Proof. Suppose that $X_1, X_2, \ldots, X_n$ are locally connected and let $X = \prod_{i=1}^{n} X_i$. Suppose that $x = (x_1, x_2, \ldots, x_n) \in U$. Let $U_1 \times U_2 \times \cdots \times U_n$ be a basic open set in X that contains x and is contained in U. Since each X_i is locally connected, there are connected open sets V_i such that $x_i \in V_i \subset U_i$. Consequently, by (2.B.2), $V = V_1 \times V_2 \times \cdots \times V_n$ is a connected open set containing x and contained in U. Therefore, X is locally connected.

(2.E.4) **Corollary 1.** The space $\mathscr{E}^n$ is locally connected.

(2.E.5) **Corollary 2.** All n-manifolds are locally connected.

(2.E.6) **Exercise.** Show that local connectedness is a topological invariant, but that local connectedness is not necessarily preserved by continuous functions.

(2.E.7) **Theorem.** If X is locally connected and $f : X \to Y$ is continuous, onto, and closed, then Y is locally connected.

Proof. Suppose that U is open in Y and C is a component of U. We show that C is open. For each $x \in f^{-1}(C)$, let C_x be the component of x in $f^{-1}(U)$. By (2.E.2), C_x is open, and since $f(x) \in C$, the connected set $f(C_x)$ also lies in C. Therefore, $f^{-1}(C) = \bigcup \{C_x \mid x \in f^{-1}(C)\}$ and consequently $f^{-1}(C)$ is open. Thus, $f(X \setminus f^{-1}(C)) = Y \setminus C$ is closed (f is closed and onto), and this, of course, implies that C is open.

F. CHAINS

Paths were introduced as a means of mathematically traveling from one point to another in a topological space. A geometrically pleasing concept somewhat akin to the notion of path is that of a chain.

(2.F.1) **Definition.** A family $\{A_1, A_2, \ldots, A_n\}$ of subsets of a space

X is a *simple chain* in X in case $A_i \cap A_j \neq \varnothing$ if and only if $|i - j| \leq 1$. A simple chain $C = \{A_1, A_2, \ldots, A_n\}$ is said to *connect points a and b in X* if and only if $a \in A_1$ and $b \in A_n$.

It should be observed that the links of a simple chain need not be connected; thus the accompanying figure illustrates a perfectly acceptable simple chain.

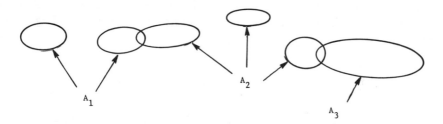

Most frequently, simple chains are forged from open links; a common procedure for their construction consists in starting with an arbitrary entanglement of open sets and then extracting a suitable simple chain from the confusion. That this may be done is a result of the following elegant and useful theorem.

(2.F.2)   Theorem.   Suppose that X is a connected space and that a and b are any two points in X. Let $\mathcal{U} = \{U_\alpha \mid \alpha \in \Lambda\}$ be a family of open sets whose union is X. Then there is a simple chain with links from $\mathcal{U}$ that connects a and b.

Proof.   Let D be the set of all points x in X such that there is a simple chain (with links in $\mathcal{U}$) that runs from a to x. The set D is certainly nonempty, since a itself is found there. We show that D is both open and closed; since X is connected, this will imply, by (2.A.4), that D is all of X.

Suppose that $x \in D$ and that $\{U_1, U_2, \ldots, U_n\}$ is a simple chain with $a \in U_1$ and $x \in U_n$. Clearly, then $U_n \subset D$, and it follows that D is open.

To see that D is also closed, we show that D contains all of its accumulation points (recall (1.E.14)). Suppose that x is an accumulation point of D and that $U \in \mathcal{U}$ contains x. Since x is an accumulation point of D, U must intersect D, and hence if z is any point in the intersection, there is a simple chain $\{U_1, \ldots, U_n\}$ of elements in $\mathcal{U}$ connecting z to x. Let r be the first integer such that $U_r \cap U \neq \varnothing$. Then $\{U_1, U_2, \ldots, U_r, U\}$ is a simple chain between a and x, and consequently $x \in D$.

(2.F.3)   Definition.   A *polygonal arc* in $\mathcal{E}^n$ is an arc whose image consists of a finite number of straight line segments.

(2.F.4) Corollary. Any two points contained in a connected open sub-set U of $\mathscr{E}^n$ may be joined by a polygonal arc in U.

   Proof. For each $x \in U$, choose ε_x small enough so that the open ball $S_{\varepsilon_x}(x)$ is contained in U. Then $\mathscr{V} = \{S_{\varepsilon_x}(x) \mid x \in U\}$ is an open cover of U, and consequently, members of $\mathscr{V}$ may be found to form a simple chain $\{V_1, V_2, \ldots, V_n\}$ running from a to b. For each i, let $z_i \in V_i \cap V_{i+1}$. Let L_1 be a line segment in V_1 that runs from a to z_1. For $i = 2, 3, \ldots, n - 1$, let L_i be a line segment in V_i from z_{i-1} to z_i. Finally, let L_n be a line segment in V_n connecting z_{n-1} to b. These segments yield a polygonal arc extending from a to b.

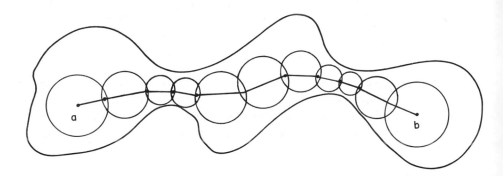

G. COMPACTNESS

In calculus, a great deal of emphasis is placed on the fact that continuous functions defined on a closed interval, say [0,1], are bounded and attain their maximum and minimum on the interval, whereas functions such as $f(x) = 1/x$ defined on (0,1] fail to be bounded. It is natural to inquire what intrinsic quality [0,1] possesses that (0,1] lacks. In topological terms, the answer turns out to be compactness—one of the most important of all topological concepts.

(2.G.1) Definition. Suppose that X is a set and that B is a subset of X. A *cover* of B is a collection of subsets of X, $\mathscr{C} = \{C_\alpha \mid \alpha \in \Lambda\}$, whose union contains B. If X is a topological space and the sets C_α are open, then the cover is said to be *open*. If Λ is a finite set, then the collection $\mathscr{C}$ is a *finite cover* of B. A *subcover* of B is a subcollection of $\mathscr{C}$ that is also a cover for B.

   The utilization of covers provides the topologist with one of his most successful tools. Covers play a prominent role in the study of such diverse areas as metrizability, dimension theory, and Čech homology. We begin by defining compactness in terms of open covers.

(2.G.2) Definition. A topological space X is *compact* if and only if every open cover of X has a finite subcover.

Any finite space is compact, no matter what its topology. More generally, any space with only a finite number of open sets is compact.

(2.G.3) Exercise. Show that every closed subspace of a compact space is compact.

(2.G.4) Example. Let $X = \mathbf{R}^1$ with the open ray topology, and let $A = [0,1]$. Then A is a compact subset of X, but A is not closed.

(2.G.5) Exercise. Suppose that $A \subset X$, A is compact, and X is T_2. Show that A is closed.

We give a nontrivial example of a compact space, by proving the previous assertion that closed and bounded intervals in $\mathscr{E}^1$ are compact. This result of course has numerous applications in elementary analysis.

(2.G.6) Theorem. Intervals that are closed and bounded in $\mathscr{E}^1$ are compact.

Proof. Let $\mathscr{U} = \{U_\alpha \mid \alpha \in \Lambda\}$ be an open cover of $[a,b]$. Set $\mathscr{V} = \{V_\beta \mid \text{for some } \alpha \in \Lambda, V_\beta \text{ is a component of } U_\alpha\}$. Clearly $\mathscr{V}$ also covers $[a,b]$, and since $[a,b]$ is locally connected, each member of $\mathscr{V}$ is open. By (2.F.2) there is a simple chain $\{V_1, \ldots, V_n\}$ of sets from $\mathscr{V}$ that connects a and b. Since $V = \bigcup_{i=1}^{n} V_i$ is a connected subset of $\mathscr{E}^1$, it follows from (2.A.8) that V is an interval. Furthermore, since $a,b \in V$, we have that $[a,b] = V$. For each i, $1 \le i \le n$, select $\alpha_i \in \Lambda$ such that $V_i \subset U_{\alpha_i}$. Then $\{U_{\alpha_1}, \ldots, U_{\alpha_n}\}$ is a finite subcover of $\mathscr{U}$.

(2.G.7) Exercise. Show that compact subsets of $\mathscr{E}^1$ are bounded and closed. Note that $\mathscr{E}^1$ itself is not compact, since a cover consisting of sets of the form $(-n,n)$ for $n \in \mathbf{Z}^+$ has no finite subcover.

The next exercise is one of the most important in the entire book. The result will be generalized at a later time with the aid of the axiom of choice, but as it stands, it represents a very substantial if not formidable task, which should more than adequately test the reader's understanding of both compactness and the product topology.

(2.G.8) Exercise. Show that if X_1 and X_2 are compact spaces, then $X_1 \times X_2$ is compact.

Compactness, like connectedness, is preserved by continuous functions.

(2.G.9) Theorem. If X is a compact topological space and $f : X \to Y$ is continuous and onto, then Y is compact.

Proof. Suppose that $\mathscr{U}$ is an open cover of Y. Then $\{f^{-1}(U) \mid U \in \mathscr{U}\}$ is an open cover of X, and the compactness of X allows us to select a finite subcover $\{f^{-1}(U_i) \mid i = 1, 2, \dots, n\}$ from $\{f^{-1}(U) \mid U \in \mathscr{U}\}$. Since f is onto, the collection $\{U_i \mid i = 1, 2, \dots, n\}$ is a finite subcover of Y.

(2.G.10) Corollary. Compactness is a topological (and a continuous) invariant.

(2.G.11) Exercise. Let f be a continuous function from a compact space into a T_2 space. Show that f is closed. If, in addition, f is 1–1, show that f is an embedding.

We now establish a theorem that gives a complete (and very applicable) characterization of compact subsets of $\mathscr{E}^n$.

(2.G.12) Theorem. Suppose that $A \subset \mathscr{E}^n$. Then A is compact if and only if A is closed and bounded.

Proof. Suppose that A is closed and bounded. Let $\mathscr{E}_i^1$ $(= \mathscr{E}^1)$ denote the i-th coordinate space of $\mathscr{E}^n$ and let $p_i : \mathscr{E}^n \to \mathscr{E}_i^1$ be the natural projection. It is easily seen that $p_i(A)$ is bounded in $\mathscr{E}_i^1$ for each i, and hence, for each $i = 1, 2, \dots, n$ there is a closed interval $[w_i, z_i] \subset \mathscr{E}_i^1$ such that $p_i(A) \subset [w_i, z_i]$. However, this implies that $A \subset [w_1, z_1] \times [w_2, z_2] \times \cdots \times [w_n, z_n]$. Since the product of closed and bounded intervals is compact by (2.G.6) and (2.G.8), it follows from (2.G.3) that the closed set A is compact.

Suppose now that A is compact. By (2.G.9), $p_i(A)$ is compact for each i, and thus by (2.G.7), $p_i(A)$ must be bounded. Since this is true in each coordinate, then A is bounded also. That A is closed follows from (2.G.5).

As an immediate corollary we have another familiar theorem from calculus.

(2.G.13) *Corollary.* Suppose that A is compact and $f : A \to \mathscr{E}^1$ is continuous. Then f attains its maximum and minimum.

(2.G.14) *Theorem.* Suppose that $A \times B$ is a compact subset of a product space $X \times Y$ and W is an open subset of $X \times Y$ which contains $A \times B$. Then there are open sets U and V in X and Y, respectively, such that $A \times B \subset U \times V \subset W$.

Proof. Let $a \in A$ and for each $b \in B$, let $U_b \times V_b$ be an open set in $X \times Y$ such that $(a,b) \in U_b \times V_b \subset W$. Then $\{V_b \mid b \in B\}$ is an open cover of B, and hence there is a finite subcover $\{V_{b_1}, V_{b_2}, \ldots, V_{b_n}\}$. Let $U^a = \bigcap_{i=1}^{n} U_{b_i}$ and $V^a = \bigcup_{i=1}^{n} V_{b_i}$. Repeat this procedure for each $a \in A$ to obtain an open cover $\{U^a \mid a \in A\}$ of A and a corresponding family of open sets $\{V^a \mid a \in A\}$ each of which contains B. Once more, we may use compactness (this time of A) to obtain a finite subcover $\{U^{a_1}, U^{a_2}, \ldots, U^{a_m}\}$ of A. Set $U = \bigcup_{i=1}^{m} U^{a_i}$ and $V = \bigcap_{i=1}^{m} U^{a_i}$ to complete the proof.

H. TWO ALTERNATE CHARACTERIZATIONS OF COMPACTNESS

Compactness may be characterized in a number of ways, one of the most important of which is given in the next theorem.

(2.H.1) *Definition.* A collection of sets $\{C_\alpha \mid \alpha \in \Lambda\}$ has the *finite intersection property* if and only if for each nonempty finite subset $K \subset \Lambda$, $\bigcap \{C_\alpha \mid \alpha \in K\} \neq \varnothing$.

(2.H.2) *Theorem.* A space X is compact if and only if for each collection $\mathscr{C} = \{C_\alpha \mid \alpha \in \Lambda\}$ of closed subsets of X with the finite intersection property, $\bigcap \{C_\alpha \mid \alpha \in \Lambda\} \neq \varnothing$.

Proof. Suppose that X is compact and $\mathscr{C} = \{C_\alpha \mid \alpha \in \Lambda\}$ is a collection of closed subsets of X with the finite intersection property. For each $\alpha \in \Lambda$, let $U_\alpha = X \setminus C_\alpha$. If $\bigcap \{C_\alpha \mid \alpha \in \Lambda\} = \varnothing$, then it is immediate from one of the De Morgan rules that $\bigcup_{\alpha \in \Lambda} U_\alpha = X$, and hence $\{U_\alpha \mid \alpha \in \Lambda\}$ is an open cover of X. Since X is compact, there is a finite subcover, $U_1, \ldots, U_n$. But an application of the other De Morgan rule yields $\bigcap_{i=1}^{n} C_i = \varnothing$, a contradiction. Thus, the intersection of the members of $\mathscr{C}$ must be nonempty.

Now suppose that X is not compact. Then there is an open cover $\{U_\alpha \mid \alpha \in \Lambda\}$ of X that has no finite subcover. For each $\alpha \in \Lambda$ let $C_\alpha = X \setminus U_\alpha$. It is easy to see that the collection $\mathscr{C} = \{C_\alpha \mid \alpha \in \Lambda\}$ has the finite intersection property, but that the intersection of all the members of $\mathscr{C}$ is empty.

The proof of the next theorem illustrates how the previous result may be used.

(2.H.3)   Theorem.   Suppose that $\{A_i \mid i \in \mathbf{Z}^+\}$ is a countable family of compact subsets of a T_2 space X, such that for each i, $A_{i+1} \subset A_i$. If there is an open set U such that $\bigcap_{i=1}^{\infty} A_i = A \subset U$, then there is an integer N such that for $i > N$, $A_i \subset U$.

Proof.   Suppose to the contrary that for each i, $C_i = A_i \setminus U$ is nonempty. Then $\{C_i \mid i \in \mathbf{Z}^+\}$ is a collection of closed sets (2.G.5) with the finite intersection property. This follows easily from the observation that if $n < m$, then $A_m \subset A_n$ and hence $A_m \setminus U \subset A_n \setminus U$. Since for each i, $C_i \subset A_1$ and A_1 is compact, we have from (2.H.2) that $\bigcap_{i=1}^{\infty} C_i \neq \varnothing$. But $\bigcap_{i=1}^{\infty} C_i = \bigcap_{i=1}^{\infty} (A_i \setminus U) = (\bigcap_{i=1}^{\infty} A_i) \setminus U = \varnothing$, a contradiction.

(2.H.4)   Exercise.   We give two generalizations of the previous theorem:

(i)   Let Λ be a partially ordered set with the property that if $\alpha, \beta \in \Lambda$, then there is a $\lambda \in \Lambda$ such that $\lambda \geq \alpha$ and $\lambda \geq \beta$. Suppose that $\{A_\alpha \mid \alpha \in \Lambda\}$ is a collection of compact subsets of a T_2 space such that $A_\alpha \subset A_\beta$ if and only if $\alpha \geq \beta$. Show that if $\bigcap \{A_\alpha \mid \alpha \in \Lambda\} \subset U$, where U is open, then there is a $\partial \in \Lambda$ such that $A_\alpha \subset U$ for each $\alpha \geq \partial$.

(ii)   Suppose that $\{A_\alpha \mid \alpha \in \Lambda\}$ is a family of closed compact sets such that $\bigcap \{A_\alpha \mid \alpha \in \Lambda\}$ is a subset of an open set U. Show that there is a finite set $\{\alpha_1, \alpha_2, \ldots, \alpha_n\} \subset \Lambda$ such that $\bigcap \{A_{\alpha_i} \mid i = 1, \ldots, n\} \subset U$.

Our second characterization of compactness, based on nests of closed subsets, is somewhat subtler than its predecessor, and some equivalent of the axiom of choice is apparently needed for its proof.

(2.H.5)   Theorem.   A space X is compact if and only if each nest of nonempty closed subsets in X has nonempty intersection.

Proof (*Catlin* [1968]). The proof of the "only if" part of the theorem is rendered trivial by (2.H.2).

Suppose then that in X each nest of closed nonempty subsets has nonempty intersection. We show that X satisfies the finite intersection property criterion for compactness.

To this end, let $\mathscr{C}$ be a family of closed subsets of X with the finite intersection property and let $\mathscr{K} = \{K \subset X \mid K$ is closed and $\mathscr{C} \cap \{K\}$ has the finite intersection property$\}$. The Kuratowski lemma (0.D.4) may be employed to extract a maximal nest $\mathscr{N}$ (with respect to inclusion) in $\mathscr{K}$. Let $D = \bigcap \{N \mid N \in \mathscr{N}\}$. By hypothesis, D is not empty; thus, to complete the proof it suffices to show that $D \subset \bigcap \{C \mid C \in \mathscr{C}\}$.

If D is not a subset of $\bigcap \{C \mid C \in \mathscr{C}\}$, then there is a $C \in \mathscr{C}$ with $C \cap D \subsetneqq D$. We obtain a contradiction by showing that $C \cap D \in \mathscr{K}$, which of course negates the maximality of $\mathscr{N}$. Note that $C \cap D \neq \varnothing$, since $C \cap D = \bigcap \{C \cap N \mid N \in \mathscr{N}\}$ and $\{C \cap N \mid N \in \mathscr{N}\}$ is a nest of closed sets ($C \cap N \neq \varnothing$, since $\mathscr{C} \bigcup \{N\}$ has the finite intersection property). To see that $C \cap D \in \mathscr{K}$, observe that whenever $C_1, C_2, \ldots, C_n \in \mathscr{C}$, then $C_1 \cap C_2 \cap \cdots \cap C_n \cap C \cap D = \bigcap \{(C_1 \cap C_2 \cap \cdots \cap C_n \cap C \cap N) \mid N \in \mathscr{N}\}$, and the latter set is nonempty, since it is an intersection of a nest of closed nonempty subsets of X. Thus, $\mathscr{C} \cup \{C \cap D\}$ has the finite intersection property, and consequently $C \cap D \in \mathscr{K}$.

I. LOCAL COMPACTNESS, COUNTABLE COMPACTNESS, AND SEQUENTIAL COMPACTNESS

Although the varieties of compactness are legion, we shall concentrate for the present on just three such mutations.

(2.I.1) *Definition.* A topological space X is *locally compact* if and only if for each $x \in X$ and for each open set U containing x, there is an open set V such that $x \in V \subset \overline{V} \subset U$ and $\overline{V}$ is compact.

(2.I.2) *Examples.*

1. For each $n \in \mathbf{Z}^+$, $\mathscr{E}^n$ is locally compact.
2. The set $\mathbf{R}^1$ with the open ray topology is not locally compact.
3. For each $i \in \mathbf{Z}^+$, let $A_i = \{(x, 1/i) \mid 0 \leq x \leq 1\}$ and let $A_0 = \{(x, 0) \mid 0 \leq x \leq 1\}$. Then $\bigcup_{i=0}^{\infty} A_i$ is locally compact.

Note that in example 3, if $A_0 = \{(0,0), (1,0)\}$, then $\bigcup_{i=0}^{\infty} A_i$ is not locally compact.

Suppose that $X = \mathbf{R}^1$ with the finite complement topology. Then X is compact but not locally compact (the closure of any open set is X). In order to obviate this problem, local compactness is sometimes defined by merely requiring that every point have a compact neighborhood. In a T_2 space, the two concepts coincide, as the reader is asked to establish in the next exercise.

(2.I.3) Exercise. Show that a T_2 space X is locally compact if and only if each point in X has a compact neighborhood.

(2.I.4) Exercise. Suppose that $X_1, X_2, \ldots, X_n$ are locally compact spaces. Show that $\prod\limits_{i=1}^{n} X_i$ is locally compact.

(2.I.5) Exercise. Show that local compactness is a topological invariant. Is it a continuous invariant?

(2.I.6) Definition. Suppose that $f : \mathbf{Z}^+ \to X$ is a sequence. A sequence $g : \mathbf{Z}^+ \to X$ is a *subsequence* of f if and only if there is a strictly increasing function $h : \mathbf{Z}^+ \to \mathbf{Z}^+$ such that $fh = g$.

(2.I.7) Definition. A topological space X is *sequentially compact* if and only if every (infinite) sequence in X has a convergent subsequence.

In the next chapter, it is shown that the notions of compactness and sequential compactness coincide in metric spaces. However, in general there is no precise relationship between these two concepts. In Chapter 4, an example is given of a compact space that is not sequentially compact. The set $[0,\Omega)$ with the order topology is sequentially compact (by (0.E.4), all sequences in $[0,\Omega)$ have a least upper bound in $[0,\Omega)$, but $[0,\Omega)$ is not compact (why?)).

Somewhat akin to the idea of sequential compactness is the notion of countable compactness.

(2.I.8) Definition. A topological space X is *countably compact* if and only if every countable open cover of X has a finite subcover.

Compactness clearly implies countable compactness; however, it follows from the next theorem and the foregoing remarks that $[0,\Omega)$ is countably compact, but not compact.

(2.I.9) Theorem. If X is sequentially compact, then X is countably compact.

Proof. Suppose that $\mathscr{U} = \{U_1, U_2, \ldots\}$ is a countable open cover of X. If no finite subcover of $\mathscr{U}$ exists, then for each n, there is a point $x_n \in X \setminus$

$\bigcup\limits_{i=1}^{n} U_i$. The sequence $\{x_n\}$ has a subsequence $\{x_{n_i}\}$ that converges to some point $x^* \in X$. Since $\mathcal{U}$ is a cover of X, there is a $U_k \in \mathcal{U}$ such that $x^* \in U_k$. However, for i sufficiently large (and greater than k), we have that $x_{n_i} \in U_k$, which is of course impossible.

We now introduce a property of topological spaces that might be viewed as a very weak form of compactness.

(2.I.10) Definition. A topological space X is *Lindelöf* if and only if every open cover of X has a countable subcover.

A principal theorem involving Lindelöf spaces is the following.

(2.I.11) Theorem (Lindelöf). If X is a second countable space, then X is Lindelöf.

Proof. Let $\mathcal{U}$ be an open cover of X and let $\mathcal{B}$ be a countable base for X. Then for each $x \in X$, there is a set $U_x \in \mathcal{U}$ and $B_x \in \mathcal{B}$ such that $x \in B_x \subset U_x$. Since $\{B_x \mid x \in X\}$ is actually a countable family and covers X (many B_x's may be duplicated), there is a corresponding countable collection of the U_x's that covers X.

Note that it follows from (1.G.10) that separable metric spaces are Lindelöf.

(2.I.12) Exercise. Show that a metric space X is Lindelöf if and only if X is second countable. [Hint: For each positive integer n, cover X with sets of the form $S_{1/n}(x)$.]

(2.I.13) Exercise. Show that the concepts of Lindelöf, second countability, and separability coincide in a metric space.

J. ONE POINT COMPACTIFICATIONS

As we proceed, the virtues of compactness will become increasingly apparent. Compactness is such a powerful property that one frequently finds it useful to embed noncompact spaces as dense subsets of compact spaces.

(2.J.1) Definition. A *compactification* of a space X is a pair (Y,h) where Y is a compact topological space and $h : X \to Y$ is an embedding whose image is a dense subset of Y.

Compactifications were originally inspired by problems in analysis where it was sometimes found desirable to add appropriate "boundary points" to a region. Actually, there are many ways of doing this; for instance, we will see in Chapter 17 (17.E.1) that every compact connected n-manifold may be viewed as a compactification of $\{x \in \mathscr{E}^n \mid |x| < 1\}$.

In this section, we are concerned with what is probably the simplest and most common means of compactifying a space, the one point (or Aleksandrov) compactification. The following situation can be considered as the motivation for the definition that eventually follows.

In $\mathscr{E}^3$ consider the sphere $S = \{(x,y,z) \mid x^2 + y^2 + (z - 1/2)^2 = (1/2)^2\}$ which rests on top of the xy plane. Denote the north pole of S by p, i.e., $p = (0,0,1)$. The following homeomorphism "wraps" the plane around the sphere by embedding $\mathscr{E}^2$ as $S \setminus \{p\}$.

$$h(x,y) = \left(\frac{x}{1 + x^2 + y^2}, \frac{y}{1 + x^2 + y^2}, \frac{x^2 + y^2}{1 + x^2 + y^2} \right)$$

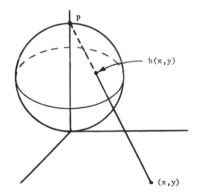

Clearly, (S,h) is a compactification of the plane.

It is instructive to consider the relation between open sets in S and the corresponding open sets in $\mathscr{E}^2$ resulting from the homeomorphism h. If U is open in $\mathscr{E}^2$, then $h(U)$ is open in S, and conversely, open sets of $S \setminus \{p\}$ have homeomorphic images in $\mathscr{E}^2$. Consider, however, an open set V of S that contains p; its complement is compact, and hence $h^{-1}(S \setminus V)$ is compact in $\mathscr{E}^2$. Thus, V may be described as being the complement of a compact set in $\mathscr{E}^2$ together with the point p.

(2.J.2) Definition. Suppose that X is a topological space, and let ∞ be a point not in X. Define a topology for $X^* = X \cup \{\infty\}$ as follows: a set $U \subset X^*$ is open if and only if either

(i) $\infty \notin U$ and U is open in X, or

(ii) $\infty \in U$ and $X^* \setminus U$ is a closed compact subset of X.

Then X^* with this topology is the *one point compactification* (*Aleksandrov compactification*) of X.

(2.J.3)   Exercise.   Show that X^* has the following properties:

(i) X^* is a topological space;

(ii) X^* is compact;

(iii) X is locally compact and T_2 if and only if X^* is T_2;

(iv) X is dense in X^* if and only if X is not compact.

A considerably more sophisticated compactification, the Stone-Čech compactification, is considered in problem A-B.10 of Chapter 7.

PROBLEMS

Sections A and B

1. (Riesz-Lennes-Hausdorff Separation Criteria). Suppose that A is a subspace of a topological space X. Show that A is disconnected if and only if there are subsets of C and D of X satisfying

    (i) $A \subset C \cup D$,

    (ii) $A \cap C \neq \emptyset$, $A \cap D \neq \emptyset$, and

    (iii) $(\bar{C} \cap D) \cap (C \cap \bar{D}) = \emptyset$.

2. Suppose that $\{U_n \mid n \in \mathbf{Z}^+\}$ is a countable collection of connected subspaces of a space X such that $U_n \cap U_{n+1} \neq \emptyset$ for each n. Show that $\bigcup \{U_n \mid n \in \mathbf{Z}^+\}$ is connected.

3. Show that an infinite set with the finite complement topology is connected.

4. Prove or disprove: if $A_1, A_2, \ldots$ is a collection of closed connected subsets of the plane such that $A_{n+1} \subset A_n$ for $n = 1, 2, \ldots$, then $\bigcap\limits_{n=1}^{\infty} A_n$ is connected.

5. Suppose that E is a proper subset of a space X, and suppose that A is a connected subset of X such that $A \cap E \neq \emptyset$ and $A \cap (X \setminus E) \neq \emptyset$. Show that $A \cap \mathrm{Fr}\, E \neq \emptyset$ and $A \cap \mathrm{Fr}(X \setminus E) \neq \emptyset$.

6. Suppose that $S \subset \mathscr{E}^1$ has the fixed point property. Show that S is a point or a closed and bounded interval.

7.* The intermediate value theorem has something of a converse. Suppose that $f : [a,b] \to \mathscr{E}^1$. Show that f is continuous if and only if

(i) whenever $x_1, x_2 \in [a,b]$ and $f(x_1) \le c \le f(x_2)$, there is an $x \in [x_1,x_2]$ such that $f(x) = c$ and
(ii) $f^{-1}(y)$ is closed for each $y \in \mathscr{E}^1$.

8.   Find an example where the intermediate value theorem holds, but the function in question is not continuous.

9.* Show that every connected open subset of $\mathscr{E}^2$ can be written as a disjoint union of open line segments.

10. Suppose that X is a T_1 space. Show that every connected subset of X containing more than one point is infinite.

11. Suppose that $A \subset \mathscr{E}^1$ is an interval and $f : A \to \mathscr{E}^1$ is continuous and 1-1. Show that f is either strictly increasing or strictly decreasing.

12. Suppose that X is a space such that each pair of points in X lie in a connected subset of X. Show that X is connected.

13. Suppose that $(X,\mathscr{U}_1)$ and $(X,\mathscr{U}_2)$ are connected. Is $(X,\mathscr{U}_1 \cap \mathscr{U}_2)$ necessarily connected?

14. Show that $\mathscr{E}^1$ with the open ray topology is connected.

15.* Suppose that $h : \mathscr{E}^1 \to \mathscr{E}^1$ is a homeomorphism and has no fixed point. Show that h^n (the composition of h with itself n times) is also fixed point free. Find a homeomorphism $h : \mathscr{E}^2 \to \mathscr{E}^2$ that is fixed point free but for which h^2 has a fixed point.

16. A metric space is said to be *well chained* if and only if for each $\varepsilon > 0$ and each pair of points x,y in X there is a finite subset $\{x_1, x_2, \ldots, x_n\}$ in X such that $x_1 = x$, $x_n = y$, and $d(x_i,x_{i+1}) < \varepsilon$ for $i = 1, 2, \ldots, n - 1$. Show that if X is connected, then X is well chained but that the converse is false.

17. Suppose that $f : \mathscr{E}^1 \to \mathscr{E}^1$ is a homeomorphism and $f^n = id$. Show that if f is order preserving, then $f = id$, and if f is order reversing, then $f^2 = id$.

18. Find a continuous function $f : \mathbf{I} \to \mathbf{I}$ with precisely two fixed points, 0 and 1.

19. Show that if a continuous function f maps a half-open interval onto itself, then f has a fixed point.

20. Show that each open subset of $\mathscr{E}^n$ can be written as a countable union of disjoint connected open sets.

Section C

1. Show that path connectivity is a topological invariant (in fact, a continuous invariant).

2. Suppose that $X_1, X_2, \ldots, X_n$ are topological spaces. Show that $\prod\limits_{i=1}^{n} X_i$ is path connected if and only if each X_i is path connected.

3. Prove or disprove: if A is a path connected subset of X, then $\bar{A}$ is path connected.

4. Prove or disprove: if A and B are path connected subsets of a space X and $A \cap \bar{B} \neq \varnothing$, then $A \cup B$ is path connected.

5. A topological space is *contractible* if and only if there is a continuous map $H : X \times I \to X$ and a point $x^* \in X$ such that $H(x,0) = x$ and $H(x,1) = x^*$ for each $x \in X$. Prove that every contractible space is path connected.

Section D

1. Define an equivalence relation on a space X by setting $x \sim y$ if and only if for any separation (U,V) of X, x and y are both in U or both in V.
   (a) Show that $\sim$ is an equivalence relation. The equivalence classes are called *quasicomponents*.
   (b) Show that quasicomponents are closed.
   (c) Show that each component of X is contained in some quasicomponent.
   (d) Describe the quasicomponents of the following subset of $\mathscr{E}^2$:
   $$X = (\bigcup_{n=1}^{\infty} A_n) \cup \{(0,0), (1,0)\}, \text{ where } A_n = \{(x,y) \mid 0 \leq x \leq 1, y = 1/n\}.$$

2. Show that a space X has an infinite number of components if and only if there is a countably infinite family of nonempty disjoint sets in X that are both open and closed.

3. Find an example in the plane where components and path components do not coincide.

4. Show that every component of a space X is a union of path components.

5. Suppose that a topological space X has the property that for every open set U and for each point $p \in U$, there is an open set V, $p \in V \subset U$ such that if $y \in V$, then y may be connected to x by a path lying in U. Show that for every open set W containing p, there is a path connected neighborhood Z with $p \in Z \subset W$.

6.* It is possible that spaces X and Y may fail to be homeomorphic even though there are continuous bijective maps $f : X \to Y$ and $g : Y \to X$. For example, let $X = (\bigcup_{n=0}^{\infty} [3n, 3n + 1)) \cup (\bigcup_{n=0}^{\infty} \{3n + 2\})$ and $Y =$

$(\bigcup_{n=0}^{\infty} (3n, \; 3n + 1)) \cup (\bigcup_{n=0}^{\infty} \{3n + 2\})$. Find continuous bijective maps $f : X \to Y$ and $g : Y \to X$, and show that X and Y are not homeomorphic.

7.* Let $X = \{x \in \mathbf{R}^1 \mid x$ is a dyadic rational$\}$, let $Y = \{y \in \mathbf{R}^1 \mid y$ is rational and $y \notin X\}$, and let $Z = \{z \in \mathbf{R}^1 \mid z$ is irrational$\}$. Then $\mathbf{R}^1 = X \cup Y \cup Z$. Define a topology for $\mathbf{R}^1$ whereby

    (i) X and Y are open,
    (ii) if $U \subset X$ or $U \subset Y$, then U is open if it is open in the usual relative topology for X or Y, and
    (iii) a neighborhood basis for a point $z \in Z$ is of the form $\{z\} \cup \{w \in X \cup Y \mid |w - z| < r\}$ $(r > 0)$.

    Show that X and Y are totally disconnected and that Z is discrete. Let $A = \mathbf{R}^1 \setminus X$ and $B = \mathbf{R}^1 \setminus Y$. Show that A and B are totally disconnected but that their union is not.

8. A function $f : X \to Y$ is *connected* if and only if the image under f of every connected set is connected. Prove or disprove: if $f : \mathscr{E}^1 \to \mathscr{E}^1$ is connected, then f is continuous.

9. Suppose that $f : X \to Y$ is connected (see problem 8) and $f^{-1}(y)$ is connected for each $y \in Y$. Show that if X and Y are T_2, then for each $y \in Y$, $f^{-1}(y)$ is closed.

10. Show that if $f : \mathscr{E}^1 \to \mathscr{E}^1$ is connected (see problem 8), and $f^{-1}(y)$ is connected for each $y \in \mathscr{E}^1$, then f is continuous.

11. Show that a path component of an n-manifold is an n-manifold.

12.* Suppose that a topological space X has an explosion point p, and that $f : X \to X$ is continuous. Show that if $f^{-1}(p)$ is finite (or empty), then f has a fixed point. [Hint: if $\{x_1, x_2, \ldots, x_n\} \subset X \setminus \{p\}$, then $X \setminus \{x_1, x_2, \ldots, x_n\}$ is connected.]

13. A function $f : X \to Y$ is *super continuous* if and only if $f^{-1}(B)$ is open for each $B \subset Y$. Show that if f is super continuous, then f is constant on each quasi-component of X.

Section E

1. Show that open subsets of locally connected spaces are locally connected. Is the same true for arbitrary subsets of locally connected spaces?

2. Suppose that $X = A \cup B$, where A and B are locally connected. Is X locally connected? If A and B are also closed, is X locally connected? If, in addition, $A \cap B$ is locally connected, is X locally connected?

3. Show that if x and y are points belonging to different components of a

locally connected topological space X, then there is a separation (U,V) of X such that $a \in U$ and $b \in V$.

4. Does the conclusion of problem 3 hold if the hypothesis "locally connected" is dropped?
5. Show that in a locally connected T_2 space, components and quasi-components coincide.
6. Prove or disprove: If a space X is locally path connected, then X is locally connected.
7.* Suppose that X is connected and locally connected and $f : X \to \mathbf{I}$ is is continuous and bijective. Show that X and $\mathbf{I}$ are homeomorphic.

Section F

1. Show that a space X is connected if and only if for each $x,y \in X$ and for each open cover $\mathscr{U}$ of X, $\mathscr{U}$ contains a simple chain from x to y.
2. Show that every two points in a connected n-manifold can be joined by a simple chain whose links are open n-cells.
3. Suppose that X is a locally connected T_1 space and p, x, and y are points in X. Show that x and y are in distinct components of $X \setminus \{p\}$ if and only if every simple chain of connected open sets from x to y has a link containing p.
4. A metric space K is *chainable* if and only if for each $\varepsilon > 0$, there is a simple chain of open sets covering K such that each link has diameter less than ε. Are the following spaces chainable?
   (a) $\mathbf{I}$
   (b) $\mathbf{I} \times \mathbf{I}$
   (c) The unit circle
   (d) The topologist's sine curve

Sections G and H

1. Show that if X is an infinite set with the finite complement topology $\mathscr{U}$, then $(X,\mathscr{U})$ is compact.
2. Suppose that $\mathscr{B}$ is a basis for a topological space $(X,\mathscr{U})$. Show that X is compact if and only if every cover of X by basic open sets has a finite subcover.
3.* Find examples to show that the intersection of two compact subsets of a topological space may fail to be compact and that the closure of a compact subset may fail to be compact.

4.* Find a space in which points are closed, but where the intersection of compact subsets is not necessarily compact.

5.* Show that in a compact space X, each compact subset is closed if and only if each continuous bijection from a compact space onto X is a homeomorphism.

6. Show that $[0,1] \subset \mathscr{E}^1$ may be covered by a family of closed intervals that has no finite subcover.

7. Suppose that C is a compact subset of a T_2 space X and that $x \in X \setminus C$. Show that there are disjoint open subsets of U and V with $x \in U$ and $C \subset V$.

8. Suppose that C and D are disjoint compact subsets of a T_2 space X. Show that there are disjoint open subsets U and V of X such that $C \subset U$ and $D \subset V$.

9.* Show that if X is T_2 and $\mathscr{C}$ is a family of compact subsets of X such that every finite intersection of members of $\mathscr{C}$ is connected, then $\bigcap \{C \mid C \in \mathscr{C}\}$ is connected.

10. Show that any first countable space is T_2 if and only if every compact subset is closed.

11.* Show that X is compact if and only if every open cover of X has an irreducible subcover (a subcover that fails to be a cover if any element is removed).

12. Suppose that $A \subset X$ and let Y be a compact space. Show that if U is a neighborhood of $A \times Y$ in $X \times Y$ then there is a neighborhood $V \subset A$ such that $V \times Y \subset U$.

13.* Show that if $f : \mathbf{I} \to X$ is continuous, onto, and open, while X is T_2 and has at least 2 points, then $\mathbf{I}$ and X are homeomorphic. [Hint: Find the smallest closed interval $[0,a]$ such that $f([0,a]) = X$.]

14. Suppose that X is T_2 and that Y is compact and T_2. Show that $f : X \to Y$ is continuous if and only if $\{(x, f(x)) \mid x \in X\}$ is closed in $X \times Y$.

15. Let $(X, \mathscr{U})$ be a compact T_2 space. Show that
(a) If $\mathscr{U} \subsetneq \mathscr{U}'$, then $(X, \mathscr{U}')$ is not compact.
(b) If $\mathscr{U}'' \subsetneq \mathscr{U}$, then $(X, \mathscr{U}'')$ is not T_2.

16. Suppose that X is a compact space and that $\mathscr{C}$ is a family of continuous functions from X into $\mathbf{I}$

(i) if $f, g \in \mathscr{C}$, then $f - g \in \mathscr{C}$, and
(ii) for each $x \in X$ there is a neighborhood U_x of x and an $f \in \mathscr{C}$ such that $f(U_x) = 0$.

Show that $f(x) = 0$ for each $f \in \mathscr{C}$ and each $x \in X$.

17. Let $X = \mathbf{Z}^+$ and let $\mathscr{U}$ be a topology for $\mathbf{Z}^+$ that consists of $\varnothing$, X, and sets of the form $\{1, 2, \ldots, n\}$. Show that $(X, \mathscr{U})$ contains no nonempty compact closed subset.

18. Let $X = X_1 \times X_2 \times \cdots \times X_n$ and suppose that a subspace A of X is

compact if and only if A is closed. Show that each X_i has the same property.

19. Suppose that X is a compact metric space. Show that X is connected if and only if X is well chained (see problem A-B.16).

20. Show that the following space X is an example of a noncompact subspace of $\mathscr{E}^2$ that has the fixed point property. For each $n \in \mathbf{Z}^+$, let $A_n = \{(x,y) \in \mathscr{E}^2 \mid x = 1/n, \ 0 \le y \le 1\}$. Let $B = \{(x,y) \in \mathscr{E}^2 \mid 0 \le x \le 1, y = 0\}$. Then $X = (\bigcup_{n=1}^{\infty} A_n) \cup B$.

21. Suppose that Y is a compact topological space and that $f : Y \to Y$ is a continuous map. Show that there is a nonempty closed set $A \subset Y$ such that $f(A) = A$.

22. Show that a compact locally connected space has only a finite number of components.

Section I

1. Suppose that X is a connected, locally connected, locally compact T_2 space. Show that if $x,y \in X$, then there is a compact connected set containing both x and y.

2.* Suppose that X is a locally compact space, Y is a T_2 space, and $f : X \to Y$ is continuous, open, and onto. Show that if C is a compact subset of Y, then there is a compact subset D of X such that $f(D) = C$.

3. Let $X = [-1,1]$ and define a topology for X by letting sets of the form $[-1,b), (a,b),$ and $(a,1]$ be open, where $a < 0 < b$ (include of course X and $\varnothing$). Show that X is compact, but that no open set (a,b) is locally compact.

4.* Suppose that X is a T_2 space and Y is a dense locally compact subspace. Show that Y is open.

5. Suppose that $A \subset X$ and define a topology for X by declaring $\varnothing$ and all sets that contain A to be open. Is this space locally compact?

6. Let C be a compact subspace of a locally compact metric space (X,d). Show that there is an $\varepsilon > 0$ such that $\{y \in X \mid d(y,C) \le \varepsilon\}$ is compact.

7. Show that every countably compact metric space is separable.

8.* Show that if every closed ball in a metric space is compact, then the space is locally compact and separable.

9.* Let $\mathscr{C} = \{\{x_n\} \mid \{x_n\}$ is a convergent sequence in $\mathscr{E}^1\}$. Define $d(\{x_n\},\{y_n\}) = \sup\{|x_n - y_n|\}$. Show that d is a metric and that $(\mathscr{C},d)$ is separable but not locally compact.

10. A space X is *pseudocompact* if and only if each continuous function $f : X \to \mathscr{E}^1$ is bounded.

   (i) Show that pseudocompactness is a continuous invariant.

(ii) Show that compact implies pseudocompact, but that the converse does not hold.

11. Let $X = \mathbf{R}^1$ and let $\mathcal{U}$ be a topology for X that consists of $\varnothing$, X, and sets of the form $(-n,n)$ where $n \in \mathbf{Z}^+$. Show that each $x \in X$ has a neighborhood base consisting of compact sets, but that not all points have a closed compact neighborhood.

12.* Show that the product of a compact space and a countably compact space is countably compact.

13.* Show that a T_1 space is countably compact if and only if every infinite subset has an accumulation point.

14. Show that the continuous image of a countably compact space is countably compact.

15. Show that a second countable T_1 space is compact if and only if it is sequentially compact.

16. Suppose that X is a countably compact T_1 space and that $U_1, U_2, \ldots$ are open subsets of X such that $\bigcap_{i=1}^{\infty} U_i = \{x\}$. Show that the sets U_i need not form a neighborhood basis at x.

17. Let $(\mathbf{Z}^+, \mathcal{U})$ have for a basis $\{\{2n-1, 2n\} \mid n = 1, 2, \ldots\}$ and let $(\mathbf{Z}^+, \mathcal{V})$ have for a basis $\{\{1,n\} \mid n = 1, 2, \ldots\}$. Show that $(\mathbf{Z}^+, \mathcal{U})$ is locally compact and $(\mathbf{Z}^+, \mathcal{V})$ is not. Define $f : (\mathbf{Z}^+, \mathcal{U}) \to (\mathbf{Z}^+, \mathcal{V})$ by $f(2n-1) = 1$ and $f(2n) = n$. Show that f is open but not continuous, and consequently local compactness is not preserved under open maps.

18. Show that a T_1 space X is countably compact if and only if every infinite open cover of X has a proper subcover.

Section J

1. Let X be the rationals with the relative topology and let X^* denote the 1-point compactification of X. Show directly that X^* is not T_2.

2. What is the 1-point compactification of $(0,1]$?

3. Show that the 1-point compactification of $\mathbf{Z}^+ \cup \{0\}$ is homeomorphic to $\{0\} \cup \{1/n \mid n \in \mathbf{Z}^+\}$.

4. Suppose that X is a locally compact T_2 space and that $f : X \to \mathscr{E}^1$ is continuous. Show that there is a continuous function $\hat{f} : X^* \to \mathscr{E}^1$ such that $\hat{f}(x) = f(x)$ for each $x \in X$ if and only if for each $\varepsilon > 0$, there is a compact subset K_ε of X such that $|f(x) - f(y)| < \varepsilon$ whenever $x,y \notin K_\varepsilon$.

5.* Suppose that X is a T_2 space and X^* is first countable. Is X necessarily locally compact?

6. Let $X = (0,1) \cup (1,2) \cup (2,3) \cup (9,10)$ be given the relative topology (with respect to $\mathscr{E}^1$). Describe the 1-point compactification of X.

Chapter 3

METRIC SPACES

In order to consolidate and extend the ideas of the previous two chapters, we now restrict our attention to metric spaces. A variety of metric spaces are considered; our study begins with those that are compact.

A. COMPACTNESS IN A METRIC SETTING

Sequences are especially useful and important in the theory of metric spaces. For example, we have already seen that in a metric context, the continuity of a function is completely determined by the behavior of convergent sequences (1.F.9). Subsequences are the key to characterizing compactness in metric spaces.

(3.A.1) *Definition.* A point x in a topological space X is a *cluster point* of a sequence $\{x_i\}$ if and only if for each neighborhood U of x and for each $i \in \mathbf{Z}^+$, there is an integer $k > i$ such that $x_k \in U$.

(3.A.2) *Exercise.* Suppose that X is a metric space and that x is a cluster point of a sequence $\{x_i\}$ in X. Show that $\{x_i\}$ contains a subsequence converging to x.

(3.A.3) *Exercise.* Find a topological space X, a sequence $\{x_i\}$ in X, and

a cluster point x of $\{x_i\}$ that is not a limit point of any subsequence of $\{x_i\}$. [Hint: cf. problem 2 of section A.]

In metric spaces the notions of compactness, sequential compactness, and countable compactness coincide. Before proving this, we introduce one additional concept that appears frequently in analysis.

(3.A.4)   Definition.   A metric space (X,d) is *totally bounded* if and only if for each $\varepsilon > 0$, there is a positive integer N_ε and a finite subset $\{x_1, x_2, \ldots, x_{N_\varepsilon}\}$ of X such that $X = \bigcup_{i=1}^{N_\varepsilon} S_\varepsilon(x_i)$.

(3.A.5)   Remark.   Total boundedness is a metric property rather than a topological property. For instance, although the open interval $(0,1)$ with the usual metric is totally bounded, it is homeomorphic to $\mathscr{E}^1$, which is not totally bounded. Nevertheless, if certain conditions are added to a metric space, then total boundedness becomes of topological interest. This is illustrated in the following theorem, which establishes a key property of sequentially compact metric spaces.

(3.A.6)   Theorem.   If (X,d) is a sequentially compact metric space, then (X,d) is totally bounded.

Proof.   Let $\varepsilon > 0$ be given. Pick an arbitrary point $x_1 \in X$. If $X = S_\varepsilon(x_1)$, we are done; if $X \neq S_\varepsilon(x_1)$, choose a point $x_2 \in X \setminus S_\varepsilon(x_1)$. Again, if $X = S_\varepsilon(x_1) \cup S_\varepsilon(x_2)$ we stop; otherwise select a point $x_3 \in X \setminus (S_\varepsilon(x_1) \cup S_\varepsilon(x_2))$. Continuing in this fashion, we will eventually encounter an integer N_ε such that $X = \bigcup_{i=1}^{N_\varepsilon} S(x_i)$, for, if not, we would have constructed a sequence that fails to have a convergent subsequence.

(3.A.7)   Corollary.   Sequentially compact metric spaces have a countable basis.

Proof.   For each $n \in \mathbf{Z}^+$, there is a family of neighborhoods $S_{1/n}(x_1), \ldots,$ $S_{1/n}(x_{k_n})$ that cover X. The collection of all these neighborhoods is clearly a countable basis for X.

(3.A.8)   Theorem.   Suppose that (X,d) is a metric space. Then (X,d) is countably compact if and only if (X,d) is sequentially compact.

Proof.   We have already seen that sequential compactness implies countable compactness (2.I.9). Hence, suppose that there is a metric space (X,d) that is countably compact but not sequentially compact. Then there is a sequence $\{x_i\} \subset X$ that has no convergent subsequence. We will construct a countable cover of X that has no finite (in fact no proper) subcover. First observe that (3.A.2) implies that the sequence $\{x_i\}$ does not have a cluster point. Let $A = \{x_i \mid i \in \mathbf{N}\}$. Clearly A is infinite (otherwise $\{x_i\}$ would have a cluster point). For each $x \in X$, there is an open neighborhood $\hat{U}_x$ of x such that $\hat{U}_x \cap A$ is finite. Since points are closed, there is an open neighborhood U_x of x such that $U_x \cap A = \varnothing$ if $x \notin A$, and $U_x \cap A = \{x\}$ if $x \in A$. Then if $U_0 = \{U_x \mid x \in X \setminus A\}$, we have that $\{\{U_a\} \mid a \in A\} \cup \{U_0\}$ is a countably infinite cover of X which has no proper subcover.

(3.A.9) **Definition.** If A is a subset of a metric space (X,d), then the *diameter of A*, denoted by *diam A*, is defined to be $\sup\{d(x,y) \mid x,y \in A\}$.

(3.A.10) **Theorem.** Suppose that $\mathcal{U}$ is an open cover of a sequentially compact metric space X. Then there is a $\lambda > 0$ such that for each $x \in X$, there is a set $U \in \mathcal{U}$ with $S_\lambda(x) \subset U$. (Equivalently, there is a $\lambda > 0$ such that whenever diam $A < \lambda$, then there is a $U \in \mathcal{U}$ that contains A.) The number λ is called a *Lebesgue number* for the cover $\mathcal{U}$.

Proof. If no such λ exists, then for each n, there is an $x_n \in X$ such that $S_{1/n}(x_n)$ fails to be in any member of $\mathcal{U}$. Since X is sequentially compact, there is a cluster point x of the sequence $\{x_n\}$ (why?). Since $\mathcal{U}$ is a cover of X, we have that $x \in U$ for some $U \in \mathcal{U}$. Let $r = d(x,X \setminus U) > 0$ and choose m large enough so that $d(x_m,x) < r/2$ and $1/m < r/4$. Then $S_{1/m}(x_m)$ is contained in U, which contradicts the way that x_m was chosen.

The following is our principal theorem.

(3.A.11) **Theorem.** Suppose that X is a metric space. Then X is compact if and only if X is sequentially compact.

Proof. If X is compact, then clearly X is countably compact. Hence by (3.A.8) we have that X is sequentially compact.

Conversely, suppose that X is sequentially compact. Let $\mathcal{U}$ be an open cover of X. By (3.A.10), there is a $\lambda > 0$ such that if $x \in X$, then $S_\lambda(x) \subset U$ for some $U \in \mathcal{U}$. Furthermore, by (3.A.6), there is a finite subset $\{x_1, \ldots, x_n\}$ of X such that $X = \bigcup_{i=1}^{n} S_\lambda(x_i)$. To complete the proof, simply choose for each $i = 1, 2, \ldots, n$, a set $U_i \in \mathcal{U}$ with $S_\lambda(x_i) \subset U_i$. Then $\{U_i \mid i = 1, 2, \ldots, n\}$ is a finite subcover.

The following corollary has many applications in analysis.

(3.A.12)   Corollary (Bolzano-Weierstrass Theorem).   Every bounded sequence of points $\{x_i\}$ in $\mathscr{E}^n$ contains a convergent subsequence.

Proof.   Let A be the set whose members are precisely the elements of the
sequence. Then $\bar{A}$ is certainly closed and bounded, and hence compact. The
corollary now follows immediately from the theorem.

(3.A.13)   Corollary.   Suppose that (X,d) is a compact metric space and that
$\mathscr{U}$ is an open cover of X. Then there is a Lebesgue number λ for $\mathscr{U}$.

Theorem (3.A.6) can be applied to prove the following basic result.

(3.1.14)   Theorem.   Suppose that A and B are disjoint subsets of a metric
space (X,d). If A is compact and B is closed, then $d(A,B) > 0$. Furthermore,
if $d(A,B) = r$, then there is an $a \in A$ such that $d(a,B) = r$.

Proof.   Suppose to the contrary that $d(A,B) = 0$. Then for each positive
integer n, there are points $a_n \in A$ and $b_n \in B$ such that $d(a_n,b_n) < 1/n$. By
the previous theorem, there is a subsequence $\{a_{n_k}\}$ of $\{a_n\}$ that converges to
some point $a \in A$. But then the corresponding subsequence $\{b_{n_k}\}$ must also
converge to a, contradicting the fact that B is closed. Therefore, $d(A,B) > 0$.
To prove the second part of the theorem, suppose that $d(A,B) = r$ and
let $d_B : A \to \mathscr{E}^1$ be defined by $d_B(a) = d(a,B)$. Then d_B is continuous (1.F.12),
and since A is compact, we have that $d_B(A)$ is compact and hence closed in $\mathscr{E}^1$.
Therefore, $r \in d_B(A)$, and consequently there is an $a \in A$ such that $d_B(a) = d(a,B) = r$.

The following example in $\mathscr{E}^2$ shows that the compactness of A is a necessary condition in (3.A.14).

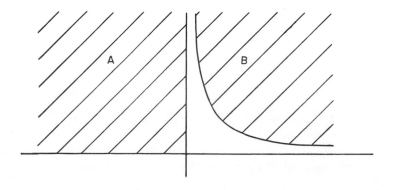

(3.A.15) Definition. A function $f : (X,d) \rightarrow (X',d')$ is *uniformly continuous* if and only if for each $\varepsilon > 0$, there is a $\delta > 0$ such that if $d(x,y) < \delta$, then $d'(f(x),f(y)) < \varepsilon$.

Note that δ depends only on ε and not on the individual points x and y. Obviously, uniform continuity implies continuity, but the converse does not hold.

(3.A.16) Exercise. Find an example of a homeomorphism between two metric spaces that is not uniformly continuous.

(3.A.17) Theorem. Suppose that (X,d) and (X',d) are metric spaces. If $f : X \rightarrow X'$ is continuous and X is compact, then f is uniformly continuous.

Proof. Let $\varepsilon > 0$ be given. For each $x \in X$, there is a $\delta_x > 0$, such that if $d(x,y) < \delta_x$, then $d(f(x),f(y)) < \varepsilon/2$. Cover X with sets of the form $S_{\delta_x}(x)$, and let λ be a Lebesgue number of this cover. Now suppose w and z are points of X such that $d(w,z) < \lambda$. Then both w and z belong to some member of the cover, say $S_{\delta_x}(x)$. If follows that $d(f(w),f(z)) < d(f(w),f(x)) + d(f(x),f(z)) < \varepsilon/2 + \varepsilon/2$.

Strangely enough, from a strictly topological point of view, all continuous functions between metric spaces are in a sense uniformly continuous. That is, a continuous function f mapping a metric space (X,d) into a metric space (Y,d') can be converted into a uniformly continuous function by switching to an appropriate equivalent metric on X. Thus, uniform continuity of a given function is a metric property rather than a topological one.

(3.A.18) Theorem. (Levine [1960]). Suppose that (X,d) and (Y,d') are metric spaces and that $f : X \rightarrow Y$ is a continuous function. Then there is a metric d^* on X, equivalent to d, that makes f uniformly continuous.

Proof. Define $d^*(a,b) = d(a,b) + d'(f(a),f(b))$. To see that d^* is a metric, we need only check that $d^*(a,c) \leq d^*(a,b) + d^*(b,c)$ for $a,b,c \in X$. But this is immediate, since $d^*(a,b) + d^*(b,c) = d(a,b) + d'(f(a),f(b)) + d(b,c) + d'(f(b),f(c)) \geq d(a,c) + d'(f(a),f(c)) = d^*(a,c)$.

It follows from (1.G.5) and (1.F.9) that d and d^* will be equivalent provided it is shown that a sequence $\{x_n\}$ converges to a point x with respect to the d metric if and only if it converges to x with respect to the d^* metric. Suppose that the sequence $\{x_n\}$ converges to x with respect to the metric d. Let $\varepsilon > 0$ be given. There is an integer N_1, such that $d(x_n,x) < \varepsilon/2$ whenever $n > N_1$. Since f is continuous, the sequence $\{f(x_n)\}$ converges to $f(x)$. Hence there is an integer N_2 such that $d'(f(x_n),f(x)) < \varepsilon/2$ whenever $n > N_2$. Thus, if $n > \max(N_1, N_2)$ we have that $x_n \in S_\varepsilon^{d*}(x)$. Conversely, if $\{x_n\}$ converges to x with respect to the metric d^*, then $\{x_n\}$ converges to x with respect to the d metric, since $d(x_n, x) \le d^*(x_n, x)$ for each n.

To see that $f : (X, d^*) \to (Y, d')$ is uniformly continuous, let $\varepsilon > 0$ be given and set $\delta = \varepsilon$. Then if $d^*(a, b) < \delta$, we have that $d(a, b) + d'(f(a), f(b)) < \delta = \varepsilon$, which of course implies that $d'(f(a), f(b)) < \varepsilon$.

B. COMPLETE METRIC SPACES

Although compactness is convenient when dealing with convergence, frequently, less stringent conditions are sufficient to achieve adequate control over sequences in metric spaces. Complete metric spaces possess enough structure to enable one to establish many important theorems that have wide applications in both topology and analysis. For instance, a key fixed point result (problem B.2), that holds in complete metric spaces can be used to prove theorems involving the existence and uniqueness of solutions to differential equations.

(3.B.1) *Definition.*   Suppose that (X, d) is a metric space. A sequence $\{x_i\}$ in X is a *Cauchy sequence* if and only if for each $\varepsilon > 0$ there is an integer N such that $d(x_m, x_n) < \varepsilon$ whenever $m, n > N$. The metric space (X, d) is *complete* if and only if for each Cauchy sequence $\{x_i\}$ in X, there is a point $x \in X$ such that $\{x_i\}$ converges to x.

(3.B.2) *Exercise.*   Give an example of a noncomplete metric space.

It is easy to show that if $\{x_i\}$ is a Cauchy sequence for which some subsequence converges to a point x, then the entire sequence must also converge to x. Thus it follows from (3.A.11) that compact metric spaces are complete. However, there is no dearth of noncompact, complete metric spaces. For instance, $\mathscr{E}^1$ is complete, as is shown in the next theorem.

(3.B.3) *Theorem.*   The space $(\mathscr{E}^1, d)$ where d is the usual metric $(d(x, y) = |x - y|)$ is complete.

Proof.   Let $\{x_n\}$ be a Cauchy sequence in $\mathscr{E}^1$. Consider $A = \{x_n \mid n \in \mathbf{Z}^+\}$. Since A is bounded (why?), $\bar{A}$ is compact (2.G.12). Now apply (3.A.11) to obtain a convergent subsequence of $\{x_n\}$. The remarks preceeding the theorem now yield that $\{x_n\}$ also converges.

We observe next that completeness is not a topological property, i.e., there are homeomorphic metric spaces one of which is complete, and the other is not. For instance, in (1.G.6), it was observed that $\mathscr{E}^1$ and the open interval $(-1,1)$ are homeomorphic; however, $(-1,1)$ is not complete. This rather distressing situation (after all, this text purports to study topological invariants) may be remedied with the introduction of the concept of topological completeness.

(3.B.4) Definition. A metric space (X,d) is *topologically complete* if and only if there is an equivalent metric d' for X such that (X,d') is complete.

Of course, a complete metric space is automatically topologically complete. A natural question to ask is whether or not all metric spaces are topologically complete. Although the answer is no [see (10.E.5)], it will be shown in Chapter 10 that any metric space may be embedded isometrically as a dense subset of complete metric space. [If (X,d) and (Y,d') are metric spaces, then a function $f : X \to Y$ is an *isometry* if and only if $d(x_1,x_2) = d'(f(x_1),f(x_2))$, for each $x_1,x_2 \in X$.]

(3.B.5) Exercise. Show that a closed subset of a complete metric space is complete.

(3.B.6) Exercise. Show that a complete subspace of a metric space is closed.

We prove later that any locally compact separable metric space is topologically complete (10.E.3). In fact, it can be shown that every locally compact metric space is topologically complete (Dugundji [1966]).

(3.B.7) Exercise. Find a locally compact metric space that is not complete.

An important property of complete metric spaces is given in the following theorem.

(3.B.8) Theorem. Suppose that $A_1 \supset A_2 \supset A_3 \supset \cdots$ is a decreasing sequence of nonempty closed subsets of a complete metric space (X,d) and that the limit of the diameters of the A_i is 0. Then $\bigcap_{i=1}^{\infty} A_i = \{x\}$ for some $x \in X$.

Proof. For each positive integer i, select a point $x_i \in A_i$. Since the diameters of the A_i's converge to 0 and the A_i's are nested, it is clear that $\{x_i\}$ is a Cauchy sequence and hence converges to a point $x \in X$. We show that

$\bigcap\limits_{i=1}^{\infty} A_i = \{x\}$. If $x \notin \bigcap\limits_{i=1}^{\infty} A_i$, let A_j be the set with smallest index that fails to contain x. By (3.A.14), $d(x,A_j) = r > 0$. However, this is impossible, since it implies that $d(x,A_i) \geq r$ for $i > j$, which prevents the sequence from converging to x. Therefore, we have that $x \in \bigcap\limits_{i=1}^{\infty} A_i$. That x is the only point in the intersection follows readily from the observation that if y is any other point in X, then $d(x,y) = s > 0$; however, for i sufficiently large, diam $A_i < s$.

(3.B.9) Corollary. Suppose that (X,d) is a complete metric space and let $D_1, D_2, \ldots$ be a sequence of open dense subsets of X. Then $\bigcap\limits_{i=1}^{\infty} D_i$ is dense in X.

Proof. Let U be an open set in X. It suffices to show that $U \cap (\bigcap\limits_{i=1}^{\infty} D_i) \neq \varnothing$. Since D_1 is open and dense, $D_1 \cap U$ must be nonempty and open. Hence, there is a positive number $\varepsilon_1 < 1$ and a point $x_1 \in X$ such that $\overline{S_{\varepsilon_1}(x_1)} \subset D_1 \cap U$. The open set $S_{\varepsilon_1}(x_1)$ has nonempty intersection with D_2, and consequently there is a point x_2 and a positive number $\varepsilon_2 < 1/2$ such that $\overline{S_{\varepsilon_2}(x_2)} \subset D_2 \cap S_{\varepsilon_1}(x_1)$. Similarly, $S_{\varepsilon_2}(x_2) \cap D_3$ is open and nonempty, and therefore there is an $\varepsilon_3 < 1/3$, etc. Repeated application of this procedure leads to a decreasing sequence of closed sets with diameters converging to 0. By the previous theorem, there is a point x common to each $\overline{S_{\varepsilon_i}(x_i)}$. Hence, x is contained in both U and $\bigcap\limits_{i=1}^{\infty} D_i$, which completes the proof.

Spaces other than complete metric spaces may also have the property described in the foregoing corollary; such spaces are known as Baire spaces.

(3.B.10) Definition. A topological space X is a *Baire space* if and only if the intersection of any countable family of open dense sets in X is dense.

(3.B.11) Exercise. Show that if X is a locally compact topological T_2 space then X is a Baire space. [Hint: Use an argument similar to that given in proof of (3.B.9).]

An important property of Baire spaces is given in the following theorem.

(3.B.12) Theorem. Suppose that X is a Baire space and $C_1, C_2, \ldots$ is a countable closed cover of X. Then at least one of the C_i's contains a nonempty open (in X) set.

Proof. This theorem follows easily from the De Morgan rules. Since $X = \bigcup_{i=1}^{\infty} C_i$, it must be the case that $\bigcap_{i=1}^{\infty} (X \setminus C_i)$ is empty. However, $X \setminus C_i$ is open for each i. Hence, since X is a Baire space, at least one of the sets, say $X \setminus C_j$, is not dense, which implies that C_j contains an open set.

(3.B.13) **Definition.** A subset A of a space X is *nowhere dense* in X if and only if $\bar{A}$ contains no open subsets of X, i.e., int $\bar{A} = \varnothing$.

(3.B.14) **Definition.** A subset A of a space X is a *set of the first category* in X if and only if A can be written as a countable union of sets nowhere dense in X. The subset A is of the *second category* if and only if A is not of the first category.

This use of category here should not be confused with the category used in the context of functors and categories (see Chapter 12).

(3.B.15) **Examples.**
1. The rationals are of the first category in $\mathscr{E}^1$.
2. The irrationals are of the second category in $\mathscr{E}^1$. If the irrationals could be written as a countable union of nowhere dense sets, then with the addition of the individual rational points as nowhere dense subsets, $\mathscr{E}^1$ itself could be represented as a countable union of nowhere dense subsets and would be of the first category (in itself). This contradicts the following classical theorem.

(3.B.16) **Theorem (Baire Category Theorem).** Any complete metric space is of the second category in itself.

Proof. Suppose that X is a complete metric space and that X is of the first category. Then X can be written as a countable union of nowhere dense sets $A_1, A_2, \cdots$. Since each A_i is nowhere dense, $X \setminus \bar{A}_i$ is both open and dense. From the completeness of X, we have by (3.B.9) that $\bigcap_{i=1}^{\infty} (X \setminus \bar{A}_i)$ is nonempty. Since $\bigcap_{i=1}^{\infty} (X \setminus \bar{A}_i) = X \setminus (\bigcup_{i=1}^{\infty} \bar{A}_i) = \varnothing$, it follow that X must be of the second category.

(3.B.17) **Definition.** A point x of a space X is *isolated* if and only if x has a neighborhood that contains no other point of X, i.e., $\{x\}$ is open in X.

(3.B.18) **Corollary.** Every countable complete metric space has an isolated point.

Note that every point of a discrete topological space is isolated. At the other end of the spectrum are the perfect spaces.

(3.B.19) Definition.   A closed subset A of a space X is *perfect* if and only if each point of A is an accumulation point of A.

The reader should note that a set A is perfect if and only if $A = A'$.

(3.B.20) Exercise.   Suppose that X and Y are topological spaces. Show that if X is perfect, then $X \times Y$ is perfect.

As an application of the Baire category theorem, we consider in detail an interesting example given by Knaster and Kuratowski [1921]. This space is of importance in dimension theory, since it is a one dimensional set that is not totally disconnected. It also is an ingenious example of a subspace of the plane possessing an explosion point.

Before constructing the example, let us first investigate one of the most unusual sets in all of mathematics, the Cantor set. To define the Cantor set, we begin by forming the following subsets of the closed interval $[0,1]$: $A_1 = [0,1/3] \cup [2/3,1]$; $A_2 = [0,1/9] \cup [2/9,1/3] \cup [2/3,7/9] \cup [8/9,1]$; $A_3 = [0,1/27] \cup [1/9,4/27] \cup [2/9,7/27] \cup [8/27,1/3] \cup [2/3,19/27] \cup [20/27,7/9] \cup [8/9,25/27] \cup [26/27,1]$; etc. In general, A_{i+1} is obtained from A_i by removing the middle third from each component of A_i. The Cantor set K is defined by $K = \bigcap\limits_{i=1}^{\infty} A_i$. Clearly, end points of the intervals in the various A_i's belong to K. It is equally obvious from the construction that for every point $x \in K$, there are points of K distinct from x that are arbitrarily close to x. Thus, K possesses no isolated points. Since K is closed (actually, K is compact) it is complete (3.B.5). Hence, we have by (3.B.18) that K is uncountable, and thus K consists of points other than the end points already noted.

We would be derelict if we failed to mention that the Cantor set was actually discovered independently by H. J. S. Smith in 1875 (Hawkins, [1970]). Nevertheless, "*Smith set*" fails to have the ring of a "*Cantor set.*"

(3.B.21) Exercise.   Show that the Cantor set is totally disconnected.

(3.B.22) Exercise.   Show that the Cantor set is homeomorphic to the set of points in $\mathscr{E}^1$ of the form $\sum\limits_{i=1}^{\infty} k_i/3^i$, where k_i is either 0 or 2.

Let us now construct the Knaster-Kuratowski example to which we previously alluded. Consider the Cantor set K in $\mathbf{I}$ as a subspace of the x axis in $\mathscr{E}^2$ and set $p = (1/2,1/2)$. For each $c \in K$, let $L(c)$ be the line segment connecting p and c. Let E be the subset of K that consists of the end points of the deleted intervals in the construction of K, and let $F = K \setminus E$. Define sets E', F', X_E, and X_F as follows: $E' = \bigcup \{L(c) \mid c \in E\}$; $F' = \bigcup \{L(c) \mid c \in F\}$; $X_E = \{(x,y) \in E' \mid y \text{ is rational}\}$; $X_F = \{(x,y) \in F' \mid y \text{ is irrational}\}$. Finally, set $X = X_E \cup X_F$.

We now show that X is connected, even though intuitively X seems too sievelike to be in "one piece."

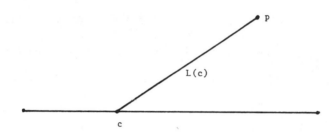

If X is not connected, then there are open sets U and V in $\mathscr{E}^2$ such that $X \subset U \cup V$ and $X \cap U \cap V = \varnothing$. We assume that $p \in U$. For each $t \in (0,1/2)$, let $K_t = \{c \in K \mid L(c) \text{ contains a boundary point } (x_c,t) \text{ of } U\}$. We claim that for each t, K_t is nowhere dense in K. Suppose not; then there are points $c_1,c_2 \in K$ such that $\varnothing \neq (c_1,c_2) \cap K \subset K_t$. Let $e \in E \cap (c_1,c_2)$. Then the boundary point (x_e,t) of U has an irrational y coordinate, i.e., t is irrational. If t were rational then (x_e,t) would be in X and hence in either U or V, both alternatives being impossible. Now suppose that $f \in F \cap (c_1,c_2)$; we conclude in a similar way that t is irrational. Hence, K_t must be nowhere dense in K.

Let $K_0 = \{c \in K \mid L(c) \text{ has no boundary points on } U\}$. Since $p \in U$, we have that for any $c \in K_0$, $L(c)$ is contained in U, and consequently, $\bar{K}_0 \neq K$ (otherwise, $V \cap X$ would be empty). Thus, there are points c_1 and c_2 in K such that $(c_1,c_2) \cap K \cap \bar{K}_0 = \varnothing$. Therefore, one may assume that $K_0 = \phi$, since otherwise the space X could be replaced with a "new" X that sits over a part of the interval (c_1,c_2) (the "new" X is a subset of X that is homeomorphic to X). With this minor point out of the way, we observe that $K = (\bigcup \{K_t \mid t \text{ is rational}\}) \cup E$, (why?), or in other words, K is a countable union of nowhere dense sets, which contradicts (3.B.16). Hence, X must be connected.

It is easy to establish that the removal of point p totally disconnects X. Each component of $X \setminus \{p\}$ must be contained in $L(c)$ for some c, since any subset of X that contains points in two $L(c)$'s can be separated by a line running between the $L(c)$'s and through the missing point p. But clearly the points of X lying in each $L(c)$ are totally disconnected, because either points with rational second coordinates or points with irrational second coordinates are missing. Thus, components consist of individual points, or, more graphically, X has been completely shattered by the loss of p.

As another vivid illustration of the category theorem in action, we prove the following rather delicate and unexpected result.

(3.B.23)   Theorem.   Suppose that X is a connected, locally connected, complete metric space. Then X cannot be written as a countable union of proper disjoint closed subsets.

Proof.   Suppose to the contrary that $X = \bigcup\limits_{i=1}^{\infty} C_i$, where the C_i's are closed and disjoint. Let $D = X \setminus (\bigcup\limits_{i=1}^{\infty} \text{int } C_i) = \bigcup\limits_{i=1}^{\infty} \text{Fr } C_i$. Then D is a closed subset of X, and, hence, by (3.B.5) and (3.B.16), D is second category (in itself).

Claim: If U is an open set in X such that $U \cap \text{Fr } C_i \neq \varnothing$, then $U \cap (D \setminus \text{Fr } C_i) \neq \varnothing$.

To establish the claim, first observe that by hypothesis it may be assumed that U is connected. Since U intersects $\text{Fr } C_i$, we have that $U \cap (X \setminus C_i)$ is nonempty, and hence, there is a positive integer $j \neq i$ such that $U \cap C_j \neq \varnothing$. If U fails to intersect $\text{Fr } C_j$, then $C_j \cap U = (\text{int } C_j) \cap U$, which implies that $U \cap C_j$ is both open and closed (in U), contradicting the connectedness of U. Therefore, $U \cap \text{Fr } C_j \neq \varnothing$, and the claim is proven.

Since K is second category, not all of the $\text{Fr } C_i$ can be nowhere dense (in K). Furthermore, by the connectedness of X, $\text{Fr } C_i \neq \varnothing$ whenever $C_i \neq \varnothing$. Thus, there is an open set V in X and an integer j such that $\varnothing \neq V \cap K \subset \overline{\text{Fr } D_j} = \text{Fr } C_j$. Therefore, we have that $V \cap \text{Fr } C_j \neq \varnothing$ and $V \cap (D \setminus \text{Fr } C_j) = \varnothing$, which contradicts the claim.

C.   CONVEX AND HAUSDORFF METRICS

A fairly strong condition that may be imposed on metric spaces is convexity. We do not deal extensively with this concept, although it should be realized that convexity serves as a backdrop for a significant amount of work in mathematics.

(3.C.1)   Definition.   A subset C of $\mathscr{E}^n$ is *convex* if and only if each pair of points a and b in C may be connected by a straight line segment that lies

entirely in C, i.e., if a and b are points of C, then the set $\{ta + (1 - t)b \mid 0 \le t \le 1\}$, is a subset of C, where $a = (a_1, a_2, \ldots, a_n)$ and $ta = (ta_1, ta_2, \ldots, ta_n)$.

The *standard n-ball*, $\mathscr{B}^n$, in $\mathscr{E}^n$ is defined to be the set of points in $\mathscr{E}^n$ whose distance from the origin is less than or equal to 1. It should be obvious that $\mathscr{B}^n$ is convex.

The following theorem will prove useful in later chapters.

(3.C.2) **Theorem.** Suppose that A is a compact convex subset of $\mathscr{E}^n$ with nonempty interior. Then A is homeomorphic to $\mathscr{B}^n$.

Proof. Let z be an interior point of A, and choose ε small enough so that the set $B_\varepsilon(z) = \{w \in \mathscr{E}^n \mid d(w,z) \le \varepsilon\}$ is contained in the interior of A. For each $x \in A \setminus \{z\}$, let $L(x)$ be the ray starting at z that passes through x. We assume for the moment that $L(x) \cap \mathrm{Fr}\ A$ consists of a single point x'. Define $H(x)$ to be the unique point on $L(x)$ with the property that $d(x,z)/d(z,x') = d(z,H(x))/\varepsilon$. A somewhat tedious argument (and hence left to the reader) can be employed to show that h is a homeomorphism between A and $B_\varepsilon(z)$. Since $B_\varepsilon(z)$ is homeomorphic to $\mathscr{B}^n$, it follows that A and $\mathscr{B}^n$ are homeomorphic. It remains to show that $L(x) \cap \mathrm{Fr}\ A$ consists of a single point for each $x \in A \setminus \{z\}$.

Suppose that x' and x'' are distinct points in $L(x) \cap \mathrm{Fr}\ A$ and that the order of the points on $L(x)$ is z, x', x''. Since $x' \in \mathrm{Fr}\ A$, there is a point w outside of A near to x' (and not on $L(x)$), such that a line segment starting at x'' and ending inside of $B_\varepsilon(z)$ passes through w. This contradicts the convexity of A; hence, no such x'' exists.

The definition of convexity may be extended to arbitrary metric spaces as follows.

(3.C.3) **Definition.** Suppose that (X,d) is a metric space and that $x,y \in X$. A point $m \in X$ is a *midpoint of x and y* if and only if $d(x,m) = d(m,y) = \frac{1}{2} d(x,y)$. The metric space X is *convex* if and only if each pair of points has at least one midpoint. The metric space X is *strongly convex* if and only if each pair has a unique midpoint. The metric space X is *without ramifications* if and only if no midpoint of x and y is also a midpoint of x' and y, where $x \ne x'$.

Consider the following subset A of the plane.

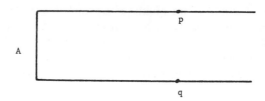

It should be apparent that the usual metric for $\mathscr{E}^2$, restricted to A, is not convex. Nevertheless, little ingenuity is required to discover a convex metric for A—simply define the distance between two points p and q to be the (usual) distance traversed when one moves along A from p to q.

Obvious query: for what spaces do convex metrics exist?

A remarkable result due to Moise [1949] and Bing [1949] ensures the existence of a convex metric for all compact, connected, locally connected metric spaces. The reader now has sufficient background to be able to follow the arguments given in those papers.

The convex metric constructed in the Bing and Moise papers need not be strongly convex. For example, the reader should be able to convince himself that $\mathscr{S}^1$ admits no strongly convex metric.

Strongly convex metric spaces without ramifications are less abundant. In fact, Rolfson [1970] has recently shown that any compact 3-dimensional strongly convex metric space without ramifications is homeomorphic to $\mathscr{B}^3 = \{(x_1,x_2,x_3) \in \mathscr{E}^3 \mid x_1^2 + x_2^2 + x_3^2 \le 1\}$. Higher dimensional analogs of this result are unknown.

We now introduce an entirely different type of metric, one that is defined on closed subsets of a given metric space. Earlier, we saw how one might define the distance between two subsets of a metric space (X,d) by setting $d(A,B) = \inf\{d(x,y) \mid x \in A \text{ and } y \in B\}$. This distance is quite useful in many contexts, but it has the disadvantage that distinct but intersecting sets always have distance 0; hence, this particular distance function does not lead to a metric for families of subsets of X. One way of creating such a metric is described as follows.

(3.C.4) Definition.   Suppose that (X,d) is a metric space with finite diameter and $\mathscr{H}$ is a collection of nonempty closed subsets of X. For $C,D \in \mathscr{H}$ let $d_C(D) = \sup\{d(x,C) \mid x \in D\}$. Then $\rho : \mathscr{H} \times \mathscr{H} \to [0,\infty)$ defined by $\rho(A,B) = \max\{d_A(B),d_B(A)\}$ is the *Hausdorff metric* for $\mathscr{H}$, and $(\mathscr{H},\rho)$ is called a *hyperspace* of X.

(3.C.5) Exercise.   Show that ρ is a metric.

Note that for points $x, y \in X$, we have that $\rho(\{x\}, \{y\}) = d(x,y)$, and hence X is isometrically embedded in $(\mathscr{H}, \rho)$. Of considerable interest is the problem of determining the actual nature of $(\mathscr{H}, \rho)$ for a given collection of closed subsets of a metric space (X, d). Let us examine what happens in a relatively simple case.

Suppose that X is the unit circle $\mathscr{S}^1$ with the usual topology, and let $\mathscr{H}$ be the collection of all closed connected subsets of X. We show that the hyperspace $(\mathscr{H}, \rho)$ is homeomorphic to the unit ball $\mathscr{B}^2 = \{(x,y) \in \mathscr{E}^2 \mid x^2 + y^2 \le 1\}$. A homeomorphism h between the two spaces is given as follows. Define $h(\mathscr{S}^1)$ to be the origin $(0,0)$. If C is a proper closed connected subset of $\mathscr{S}^1$, let c be its midpoint and let $r = (\text{length } C)/2\pi$. Then h maps C onto the point of $\mathscr{B}^2$ lying on the radius passing through c, and at a distance of $1 - r$ from the origin.

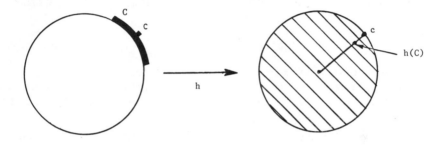

There is no problem in verifying that h is a homeomorphism.

Other examples of this sort are given in the problem set. Schori and West [1972] have recently resolved a long-standing problem involving hyperspaces by showing that if $X = I$ and $\mathscr{H}$ is the collection of all closed subsets of X, then the resulting hyperspace is the Hilbert cube. (The Hilbert cube is the space obtained by taking the product of I with itself a countable number of times and will be discussed in Chapter 7.) Amazingly, Schori went on to show in subsequent papers that the hyperspace associated with any compact, connected, locally connected metric space is also the Hilbert cube.

PROBLEMS

Section A

1. Show that a metric space is compact if and only if every infinite subset has an accumulation point (cf. chapter 2, problem I-13).

2. Suppose that X is a first countable space, $\{x_i\}$ is a sequence in X, and x is a cluster point of $\{x_i\}$. Show that x is a limit point of a subsequence of $\{x_i\}$.

3.* A subset $A \subset (X,d)$ is *metrically isolated* if and only if $d(A,B) > 0$ for each closed set B contained in $X \setminus A$. Show that if A is metrically isolated, then A is closed and has compact boundary.

4. Suppose that (X,d) and (Y,d') are metric spaces. Show that a function $f : X \to Y$ is uniformly continuous if and only if for any two sequences $\{x_n\}$ and $\{y_n\}$ in X, $\lim_{n \to \infty} d(x_n,y_n) = 0$ implies that $\lim_{n \to \infty} d'(f(x_n),f(y_n)) = 0$.

5.* Let (X,d) be a compact metric space and suppose that $f : X \to X$ is an isometry. Show that f is onto.

6. Suppose that X is a metric space with the property that whenever $f : X \to X$ is an isometry, then f is onto. Is X compact?

7.* Suppose that (X,d) is a metric space with the following two properties:

   (i) for each $x \in X$ and $r > 0$, $\{y \mid d(x,y) \leq r\}$ is compact;
   (ii) for each $x,y \in X$, there is an isometry $\psi : X \to X$ such that $\psi(x) = y$.

   Show that every isometry from X into X is onto.

8. Find an example to show that if (X,d) and (Y,d') are metric spaces and $f : X \to Y$ is continuous, there need not be an equivalent metric for Y under which f is uniformly continuous.

9. Suppose that A and B are disjoint compact subsets of a metric space (X,d). Show that there is an $a \in A$ and $b \in B$ such that $d(A,B) = d(a,b)$.

10.* Suppose that (X,d) and (Y,d') are metric spaces. Show that a function $f : X \to Y$ is uniformly continuous if and only if whenever A and B are nonempty subsets and $d(A,B) = 0$, then $d'(f(A),f(B)) = 0$.

11. Suppose that (X,d) and (Y,d') are metric spaces. If $h : X \to Y$ is a uniformly continuous homeomorphism, must h^{-1} be uniformly continuous?

12. A metrizable space $(X,\mathcal{U})$ is *topologically totally bounded* if and only if there is a metric d for X such that the topology induced by d coincides with $\mathcal{U}$ and (X,d) is totally bounded. Show that a space $(X,\mathcal{U})$ is topologically totally bounded if and only if $(X,\mathcal{U})$ is separable.

13. Determine which of the following are uniformly continuous:

   a) $f(x) = x^2$
   b) $f(x) = \sqrt{x}$
   c) $f(x) = e^x$

14. Is the sum (product) of two uniformly continuous functions uniformly continuous?

Section B

1. Show that in (3.B.8), $\bigcap_{i=1}^{\infty} A_i$ may be empty if the limit of the diameters of the A_i is not 0.

2. Let $r \in (0,1)$ and suppose that (X,d) is a complete metric space, S is a closed subset of X, and $f : S \to S$ is a map with the property that $d(f(x),f(y)) \le rd(x,y)$ for all $x,y \in S$. (Such a map is called a *contractive map*.) Show that f has a unique fixed point.

3. Suppose that d is a metric for a space X with the property that every closed set of finite diameter is compact. Show that (X,d) is a complete metric space.

4. Suppose that X is a metric space with the property that if S is any nonempty closed subset of X and $f : S \to S$ is any contractive map (see problem 2 above), then f has a fixed point. Show that X is complete. [Hint: Let $\{x_n\}$ be a nonconvergent Cauchy sequence and for each $x \in X$, let $L(x) = \inf\{d(x,x_n) \mid x_n \ne x\}$. Note that $L(x) > 0$. Choose r such that $0 < r < 1$. Define $\sigma : \mathbf{N} \to \mathbf{N}$ inductively by setting $\sigma(0) = 0$ and defining $\sigma(n)$ to be an integer $> \sigma(n - 1)$ such that $d(x_i,x_j) \le rL(x_{\sigma(n-1)})$ for all integers $i,j \ge \sigma(n)$. Let $S = \{x_{\sigma(n)} \mid n = 0, 1, 2, \ldots\}$ and let $f : S \to S$ be defined by $f(x_{\sigma(n)}) = x_{\sigma(n+1)}$.]

5. Show that any subsequence of a Cauchy sequence is Cauchy.

6.* Show that a metric space is compact if and only if it is complete in every equivalent metric.

7.* Show that a metric space is compact if and only if it is bounded in every equivalent metric.

8. Show that every subset of a first category set is first category.

9. Prove or disprove: every uncountable space is second category (in itself).

10. Show that if A is a nowhere dense subset of a topological space X, then $X \setminus A$ is dense in X.

11. Suppose that X and Y are metric spaces and that $A \subset X$ and $B \subset Y$. Show that if $A \times B$ is nowhere dense in $X \times Y$, then either A is nowhere dense in X or B is nowhere dense in Y.

12. Suppose that X is a metric space and that A is a countable subset of X. Show that A is of the first category in X if and only if A has no isolated point in X.

13. Prove or disprove: every open subset of a second category space is of the second category.

14. Show that the countable union of first category sets is of the first category.

15. Show that every countably infinite complete metric space contains an infinite number of isolated points.

16.   The Cantor set may be used as follows to describe a map $f : \mathbf{I} \to \mathscr{E}^1$ that is open but not continuous. On each component C_α of the complement of the Cantor set in $\mathbf{I}$, let $f_\alpha : C_\alpha \to \mathscr{E}^1$ be any strictly increasing continuous function. On the Cantor set itself, define f to be identically 0. Show that f is open but not continuous.

17.   Show that if U is an open subset of a space X, then Fr U is nowhere dense.

18.   Prove the following 'converse' of (3.B.8). Suppose that (X,d) is a metric space and that the intersection of each decreasing sequence of nonempty closed balls whose radii converge to zero is nonempty. Then (X,d) is complete.

19.* Another subset of the plane possessing an explosion point may be constructed as follows. As before, start with the "cone" over the Cantor set. Let $\mathscr{S}$ denote the family of all half-closed segments each connecting a point of the Cantor set with $p = (1/2,1/2)$ but excluding p, and let $\phi : [0,\Omega] \to \mathscr{S}$ be 1-1 and onto. Let $\mathscr{A}$ be the family of all compact subsets of $\mathscr{E}^2$ that intersect an uncountable number of members of $\mathscr{S}$. Let $\psi : [0,\Omega) \to \mathscr{A}$ be 1-1 and onto (why is this possible?). For each $\alpha \in [0,\Omega)$, let $\phi(\alpha) = S_\alpha$ and $\psi(\alpha) = A_\alpha$. Let α_0 be the smallest element of $[0,\Omega)$ such that $A_0 \cap S_{\alpha_0} \neq \varnothing$, and select a point c_{α_0} in this intersection. Let α_1 be the smallest element in $[0,\Omega)$ (and $\neq \alpha_0$) such that $A_1 \cap S_{\alpha_1} \neq \varnothing$. Select a point c_{α_1} in the intersection. Continuing in this fashion, show that for each $\beta \in [0,\Omega]$, exactly one point c_β is selected in S_β. Let $X = \{p\} \cup \{c_\beta \mid \beta \in [0,\Omega)\}$. Show that X is connected, but $X \setminus \{p\}$ is totally disconnected.

20.* Show that an uncountable second countable space can be written as the union of a perfect set and a countable set and that these two sets may be chosen to be disjoint.

21.   Show that a subset A of a complete metric space is countably compact if and only if A is closed and totally bounded.

22.* Show that the Knaster-Kuratowski example has the fixed point property (the fixed point need not be p.)

23.   Show that if A is a subset of a separable metric space, then A has at most a countable number of isolated points.

Section C

1.   Show that the conclusion of (3.C.2) need not hold if A has empty interior.

2.   Suppose that $p,q,r \in (X,d)$. Then q is said to be *between* p and r if and only if $d(p,r) = d(p,q) + d(q,r)$. A subset $A \subset (X,d)$ is *linear* if and only if there is an isometry (into) $\Phi : A \to \mathscr{E}^1$.

    (i) Suppose that $p,q,r \in X$. Show that $\{p,q,r\}$ is linear if and only if one of the points p, q, or r is between the other two.

    (ii) Give an example of a metric space consisting of four points that is not linear but for which every proper subset is linear.

3. An arc α contained in (X,d) is a *segment* if and only if Im α is linear. Suppose that (X,d) is a strongly convex metric space without ramifications. Let pq and pr be segments in X such that $pq \cap pr \setminus \{p\} \neq \varnothing$. Show that either $pq \subset pr$ or $pr \subset pq$. Here pq denotes the segment with endpoints p and q.

4. If (X,d) is strongly convex and $p,q,r \in X$, show that q is between p and r if and only if $q \in pr$.

5. If $U_1, U_2, \ldots, U_n$ are open subsets of a topological space X, let

$$\langle U_1, \ldots, U_n \rangle = \{A \subset X \mid \text{such that } A \subset \bigcup_{i=1}^{n} U_i \text{ and } A \cap U_i \neq \varnothing \text{ for}$$

each i}. Show that sets of this form determine a basis for a topology for $V = \{A \subset X \mid A \text{ is closed and } A \neq \varnothing\}$. This topology is called the *Vietoris topology*.

6.* Show that in compact metric spaces, the Vietoris topology and the topology generated by the Hausdorff metric coincide.

7. Let $X = \mathbf{I}$, and let $\mathscr{H}$ be the family of all subsets of $\mathbf{I}$ that consist of one or two points. Show that $\mathscr{H}$ with the Hausdorff metric is homeomorphic to a (solid) triangle.

8.* Let $X = \mathbf{I}$ and let $\mathscr{H}$ be the family of all subsets of $\mathbf{I}$ that consist of one, two, or three points. Show that $\mathscr{H}$ with the Hausdorff metric is homeomorphic to a tetrahedron.

Chapter 4

NORMALITY AND OTHER
SEPARATION PROPERTIES

In this chapter, a variety of spaces are studied—spaces that possess some, but in general not all, of the attributes of metric spaces. Normal spaces may lack a distance function, but they nevertheless have sufficient topological structure to yield some of topology's major theorems. A number of other spaces that we shall investigate are of interest precisely because they fail to have many of the properties which are inherent to metric spaces.

A. NORMAL SPACES

We have previously seen that in $\mathscr{E}^2$, the distance between closed subsets A and B may be 0, even if A and B are disjoint.

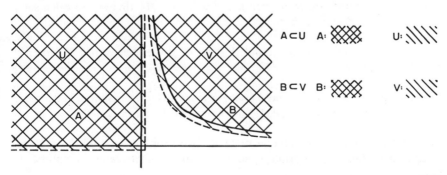

Note that in this case, however, open sets U and V can be found that serve to keep A and B apart. In fact, in metric spaces, all that is needed for such "protection" to exist is that the closure of A (respectively, B) does not intersect B (respectively, A). This is the content of the next theorem.

(4.A.1)   Theorem. If A and B are subsets of a metric space X such that $\bar{A} \cap B = \bar{B} \cap A = \varnothing$, then there are disjoint open sets U and V of X such that $A \subset U$ and $B \subset U$.

Proof. For each $a \in A$, there is a positive number r_a such that $S_{r_a}(a) \subset X \setminus \bar{B}$. Cover A with sets of the form $S_{r_a/2}(a)$ and B with similarly defined open sets of the form $S_{r_b/2}(b)$. Let U be the union of sets of the former type and V the union of sets of the latter form. It is easy to verify that U and V are the desired open sets.

This property of metric spaces turns out to be very useful, and, as is frequently the case with useful concepts, it is made the basis for a definition.

(4.A.2)   Definition. A topological space X is *completely normal* if and only if for every pair of subsets, A and B, with $\bar{A} \cap B = A \cap \bar{B} = \varnothing$, there are disjoint open sets U and V, containing A and B respectively.

A slightly weaker, but no less important notion is the following.

(4.A.3)   Definition. A topological space X is *normal* if and only if for every pair of disjoint closed subsets A and B of X there are disjoint open subsets in X that contain A and B respectively.

Clearly, any completely normal space is normal; however, the converse does not hold (an example, the Tihonov Plank, will be given later in section C of this chapter).

(4.A.4)   Exercise. Show that a space X is normal if and only if for each closed subset A and for each open set U with $A \subset U$, there is an open set V such that $A \subset V \subset \bar{V} \subset U$.

The reader might note that if X is a T_1 space that is not T_2, then X fails to be normal.

(4.A.5)   Exercise. Show that a space X is completely normal if and only if every subspace of X is normal. [Hint: Consider $X \setminus (\bar{A} \cap \bar{B})$.]

Metric spaces are of course normal. However, in view of the example cited above, there exist normal spaces that are not metrizable. A number of such examples may be derived from the following result, whose proof is an almost classic illustration of the use of compactness.

(4.A.6) *Theorem.* If X is a compact Hausdorff space, then X is normal.

Proof. Suppose that A and B are disjoint closed subsets of X. Then A and B are compact. For each pair of points a and b with $a \in A$ and $b \in B$ there are disjoint open sets $U_{(a,b)}$ and $V_{(a,b)}$ that contain a and b respectively. Fix a point $a \in A$. The family of open sets $\{V_{(a,b)} \mid b \in B\}$ forms an open cover of B from which a finite subcover $\{V_{(a,b_1)}, V_{(a,b_2)}, \ldots, V_{(a,b_n)}\}$ may be extracted. Let U_a be the intersection of the corresponding sets $U_{(a,b_1)},$ $U_{(a,b_2)}, \ldots, U_{(a,b_n)}$ and set $V_a = \bigcup_{i=1}^{n} V_{(a,b_i)}$. Repeat this procedure for each fixed $a \in A$ to obtain an open cover $\{U_a \mid a \in A\}$ of A. Let $\{U_{a_1}, \ldots, U_{a_m}\}$ be a finite subcover. For each U_{a_i} there corresponds an open set V_{a_i} that contains B. Then $\bigcup_{i=1}^{m} U_{a_i}$ and $V = \bigcap_{i=1}^{m} V_{a_i}$ are disjoint open sets that contain A and B respectively.

The following result is used in the proof of the next theorem.

(4.A.7) *Exercise.* Show that if $\{x_i\}$ is a sequence in a compact space X, then $\{x_i\}$ has a cluster point in X (compare with the remarks following (2.I.7)).

(4.A.8) *Theorem.* Suppose that $A_1 \supset A_2 \supset \cdots$ is a countable family of closed, connected, nonempty subsets of a compact Hausdorff space X. Then $A = \bigcap_{i=1}^{\infty} A_i$ is compact, connected, and nonempty.

Proof. Since A is closed it is compact. Now suppose that $A \subset U \cup V$, where U and V are disjoint open sets. By (2.H.3), there is an integer N such that if $i > N$, then $A_i \subset U \cup V$. However each A_i is connected; hence $A_i \subset U$ or $A_i \subset V$ for every $i > N$, which implies that $A \subset U$ or $A \subset V$. Thus A is connected.

We now show that $A \neq \varnothing$. For each i, let $x_i \in A_i$. Since X is compact, the sequence $\{x_i\}$ has a cluster point a in X (4.A.7). If $a \notin A$, then by (4.A.6) there are disjoint open sets U and V containing $\{a\}$ and A respectively. However it follows from (2.H.3) that there is an integer N such that if $i > N$, then $A_i \subset V$. Hence for each $i > N$, we have that $a_i \notin U$, which contradicts the fact that a is a cluster point of the sequence $\{x_i\}$.

(4.A.9) *Exercise.* Generalize the previous theorem as follows. Let Λ be an index set that is partially ordered by $\leqslant$. Suppose furthermore that for

each $\alpha, \beta \in \Lambda$, there is $\lambda \in \Lambda$ such that $\alpha \preccurlyeq \lambda$ and $\beta \preccurlyeq \lambda$ (such a set is said to be *directed*). Suppose that $\{A_\alpha \mid \alpha \in \Lambda\}$ is a family of closed, connected, non-empty subsets of a compact T_2 space X and that $A_\alpha \subset A_\beta$ if and only if $\beta \preccurlyeq \alpha$. Show that $\bigcap_{\alpha \in \Lambda} A_\alpha$ is compact and connected.

Suppose that A and B are disjoint closed subsets of a space X. Then of course if X is normal, there are disjoint open subsets U and V such that $A \subset U$ and $B \subset V$. Can the open sets U and V be chosen so that $X = U \cup V$? To do so obviously requires that X be disconnected. A sufficient condition for the existence of such a dramatic cleavage is given in our next result.

(4.A.10) Definition. Disjoint subsets A and B of a topological space X are *separated* in X if and only if there is a separation (U,V) of X such that $A \subset U$ and $B \subset V$.

(4.A.11) Theorem. Suppose that X is a compact T_2 topological space and that A and B are disjoint closed subsets of X. If no connected subset of X intersects both A and B, then A and B can be separated in X.

Proof. We first prove the theorem for the case where A and B are singletons, i.e., $A = \{x\}$ and $B = \{y\}$. Suppose that $\{x\}$ and $\{y\}$ cannot be separated. Let $\mathscr{K} = \{K_\alpha \mid \alpha \in \Lambda\}$ be the collection of all closed subsets of X in which $\{x\}$ and $\{y\}$ are not separated. Note that $X \in \mathscr{K}$. Partially order $\mathscr{K}$ by defining $K_\alpha \preceq K_\beta$ if and only if $K_\alpha \subset K_\beta$. By the Kuratowski lemma (0.D.4), there is a maximal nest $\mathscr{K}^* = \{K_\alpha \mid \alpha \in L \subset \Lambda\}$ in $\mathscr{K}$. Let $K = \bigcap_{\alpha \in L} K_\alpha$. We first show that $K \in \mathscr{K}$ (and hence by the maximality of $\mathscr{K}^*$, we will have that $K \in \mathscr{K}^*$). If $K \notin \mathscr{K}$, there are disjoint closed subsets K_1 and K_2 of K, such that $x \in K_1$, $y \in K_2$, and $K_1 \cup K_2 = K$. Since K itself is closed in the normal space X, so must be K_1 and K_2. Thus, there are disjoint open sets U and V in X with $K_1 \subset U$ and $K_2 \subset V$. By (2.H.4), some member K_β of $\mathscr{K}^*$, is contained in $U \cup V$. However, this is impossible, since it implies that $\{x\}$ and $\{y\}$ can be separated in K_β by $U \cap K_\beta$ and $V \cap K_\beta$. Consequently, we have that $K \in \mathscr{K}$.

Since x and y lie in K, it follows from the hypothesis that K cannot be connected. Therefore, K can be written as the union of disjoint closed subsets H_1 and H_2, and since K is "minimal" in $\mathscr{K}^*$, both x and y must be contained in either H_1 or in H_2, say H_1. If $\{x\}$ and $\{y\}$ could be separated in H_1 by sets H_1' and H_2'', then H_1' and $H_1'' \cup H_2$ would separate $\{x\}$ and $\{y\}$ in K, which we have just seen to be impossible. Thus, an impasse has been reached, for if $\{x\}$ and $\{y\}$ cannot be separated in H_1, then H_1 belongs to the maximal chain $\mathscr{K}^*$. It does not, and hence the special case of the theorem is settled.

Now suppose that $A = \{x\}$, B is a closed set disjoint from A, and A and B satisfy the hypothesis of the theorem. For each $b \in B$, the special case just proved is applied to obtain a separation (U_b, V_b) of X such that $x \in U_b$ and $b \in V_b$. The family $\{V_b \mid b \in B\}$ is an open cover of the compact space B and consequently yields a finite subcover $\{V_{b_1}, V_{b_2}, \ldots, V_{b_n}\}$. Then $U = \bigcap_{i=1}^{n} U_{b_i}$ and $V = \bigcup_{i=1}^{n} V_{b_i}$ form a separation of X, where $x \in U$ and $B \subset V$.

By employing the method used in the preceding paragraph, the reader should have no difficulty in establishing the general case.

A particularly elegant application of the foregoing theorem is found in the proof of the following innocent appearing, but nevertheless deep result.

(4.A.12) Theorem. Suppose that U is an open subset of a compact connected T_2 space X and that C is a component of U. Then $\bar{C}^X \cap \text{Fr } U \neq \varnothing$. (Recall that $\bar{C}^X$ is the closure of C in X.)

Proof. Suppose that $\bar{C}^X \cap \text{Fr } U = \varnothing$. Then $\bar{C}^X = C$, and hence C is closed in X. The space X is normal; therefore, there is an open set V such that $C \subset V \subset \bar{V} \subset U$. The space $\bar{V}$ is of course compact. Since no connected subset of $\bar{V}$ intersects both C and $\text{Fr } V$ (recall that C is a component of U), by the previous theorem there is a separation (F,G) of $\bar{V}$ such that $C \subset F$ and $\text{Fr } V \subset G$. Then $(F, G \cup (X \setminus V))$ is a separation of the connected space X, which is impossible.

B. URYSON'S LEMMA AND TIETZE'S THEOREM

We now prove two of the great classical theorems in general topology, Uryson's lemma and Tietze's extension theorem. Suppose that X is a space with the property that for each pair of disjoint closed subsets A and B, there is a continuous function $f_{AB} : X \to [0,1]$ such that $f_{AB}(A) = 0$ and $f_{AB}(B) = 1$. Then X is normal ($f^{-1}([0,1/3))$ and $f^{-1}((2/3,1])$ are disjoint open subsets of X that contain A and B respectively). Uryson's remarkable lemma gives us the following converse.

(4.B.1) Theorem (Uryson's Lemma). A space X is normal if and only if for each pair A and B of disjoint closed subsets of X, there is a continuous function $f : X \to [0,1]$ such that $f(a) = 0$ for each $a \in A$ and $f(b) = 1$ for each $b \in B$.

Proof. We first prove the following lemma.

(4.B.2) **Lemma.** Let D be the set of all dyadic fractions in $(0,1)$, i.e., $D = \{p/2^n \mid n,p \in \mathbf{Z}^+ \text{ and } p < 2^n\}$. Suppose that A and B are disjoint closed subsets of a normal space X. Then there is a collection of open sets $\{U_d \mid d \in D\}$ such that if $d_1 < d_2$, then $A \subset U_{d_1} \subset \overline{U_{d_1}} \subset U_{d_2} \subset \overline{U_{d_2}} \subset X \setminus B$.

Proof. By (4.A.4) there is an open set $U_{1/2}$ such that $A \subset U_{1/2} \subset \overline{U}_{1/2} \subset X \setminus B$. Another application of this exercise yields open sets $U_{1/4}$ and $U_{3/4}$ with the properties that $A \subset U_{1/4} \subset \overline{U}_{1/4} \subset U_{1/2} \subset \overline{U}_{1/2} \subset U_{3/4} \subset \overline{U}_{3/4} \subset X \setminus B$.

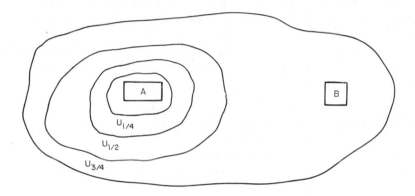

The inductive step should now be apparent. Suppose that the sets U_d (satisfying the lemma) have been defined for all members of D with denominators less than or equal to 2^k. Once again, by (4.A.4), there are open sets $U_{2j-1/2^{k+1}}$, $j = 1, 2, \ldots, 2^k$, that have the property that $A \subset U_{1/(2^{k+1})} \subset \overline{U}_{1/(2^{k+1})} \subset U_{1/2^k} \subset$ etc. The U_d's are now defined for all d's of the form $p/2^{k+1}$, and this completes the induction.

We now return to the proof of Uryson's lemma. Let $\{U_d \mid d \in D\}$ be the collection of open sets obtained in the preceding lemma. Define a function $F : X \to [0,1]$ by

$$F(x) = \begin{cases} \text{lub}\{d \mid x \notin U_d\} & \text{if } x \notin U_d \text{ for some } d \in D \\ 0 & \text{otherwise} \end{cases}$$

Note that if $x \in A$, then $x \in U_d$ for each $d \in D$ and hence $F(x) = 0$; furthermore, if $x \in B$, then $x \notin U_d$ for all $d \in D$ and therefore $F(x) = 1$. Thus, to complete the proof it suffices to establish the continuity of F.

Suppose that $x \in X$ and that $\varepsilon > 0$ is given. Assume first that $0 < F(x) < 1$. Since the dyadic rationals are dense in $(0,1)$, there are numbers $d_1, d_2 \in D$

such that $F(x) - \varepsilon < d_1 < F(x) < d_2 < F(x) + \varepsilon$. Let $V = U_{d_2} \setminus \overline{U_{d_1}}$. Then V is an open set and V contains x, since $F(x) > d_1$ implies that $x \notin \overline{U_{d_1}}$ and $F(x) < d_2$ implies that $x \in U_{d_2}$. We assert that $F(V)$ is contained in the ε-neighborhood about $F(x)$. To see this, let $y \in V$. Then since $y \in U_{d_2}$, $F(y)$ must be at most equal to d_2 and hence strictly less than $F(x) + \varepsilon$. Since $y \notin U_{d_1}$, $F(y)$ must be greater than or equal to d_1 and thus strictly greater than $F(x) - \varepsilon$. Consequently, F is continuous at x whenever $F(x)$ lies in the interval $(0,1)$. A similar proof may be supplied by the reader for the case where $F(x) = 0$ or $F(x) = 1$.

The following corollary is useful in the proof of Tietze's extension theorem.

(4.B.3)   Corollary. Suppose that A is a closed subset of a normal space X and that $f : A \to \mathscr{E}^1$ is a continuous function such that $|f(a)| \le c$ for each $a \in A$. Then there is a continuous function $h : X \to \mathscr{E}^1$ such that

(i) $|h(x)| \le (1/3)c$ for each $x \in X$, and
(ii) $|f(a) - h(a)| \le (2/3)c$ for each $a \in A$.

Proof. Consider the sets $A_+ = \{a \in A \mid f(a) \ge (1/3)c\}$ and $A_- = \{a \in A \mid f(a) \le -(1/3)c\}$. Since A_+ and A_- are closed and disjoint, there is a map $h : X \to [-(1/3)c,(1/3)c] \subset \mathscr{E}^1$ with $h(A_+) = (1/3)c$ and $h(A_-) = (-1/3)c$ (why?). A routine verification shows that h has all the requisite properties.

Uryson's lemma does not guarantee that $f^{-1}(0) = A$ and $f^{-1}(1) = B$. For this to be the case, an additional condition is imposed on the sets A and B.

(4.B.4)   Definition. A subset A of a space X is a G_δ subset of X if and only if A may be expressed as the countable intersection of open subsets of X. (G stems from *Gebiet*, the German word for "region"; δ stems from *Durchschnitt*, the German word for "intersection.")

Note that closed subsets as well as open subsets of a metric space X are G_δ sets. (If $A \subset (X,d)$ is closed, then $A = \bigcap_{n=1}^{\infty} S_{1/n}(A)$ where $S_{1/n}(A) = \{x \in X \mid d(x,A) < 1/n\}$.)

(4.B.5)   Definition. A normal T_1 space X is *perfectly normal* if and only if each closed subset of X is a G_δ.

(4.B.6) Exercise. Show that if X is perfectly normal and A and B are disjoint closed subsets of X, then there is a map $f : X \to [0,1]$ such that $f^{-1}(0) = A$ and $f^{-1}(1) = B$. [Hint: Let $A = \bigcap_{n=1}^{\infty} U_n$, where U_n is open in X and $U_{n+1} \subset U_n$. For each n, find $f_n : X \to [0,1]$ such that $f_n(A) = 0$ and $f_n(X \setminus U_n) = 1$. Define $f_A(x) = \sum_{n=1}^{\infty} \frac{f_n(x)}{2^n}$. Define a function f_B in a similar fashion and set $f = f_A/(f_A + f_B)$.]

(4.B.7) Definition. A subset A of a space X is an F_σ subset of X if and only if A may be expressed as the countable union of closed subsets of X. (F is from *Fermé*, the French word for "closed"; σ is from *Summe*, the French word for "union.")

Note that a subset A of a space X is a G_δ set if and only if $X \setminus A$ is an F_σ set.

A great number of problems in mathematics can be reduced to the question: If A is a subset of a space X and $f : A \to Y$ is a continuous map into a space Y, can f be extended to a continuous map with domain X? Uryson's lemma is one (rather trivial) solution to the extension problem. Another partial solution is given in the next theorem. For a discussion of some of the problems that are equivalent to the extension problem, see Hu [1959, pp. 1–34].

(4.B.8) Theorem (Tietze's Extension Theorem). Suppose that X is a normal space, A is a closed subset of X, and $f : A \to \mathscr{E}^1$ is a continuous function. Then there is a continuous map $F : X \to \mathscr{E}^1$ such that $F(a) = f(a)$ for each $a \in A$. Furthermore,

   (i) if $|f(a)| < c$ for each $a \in A$, then F may be chosen so that $|F(x)| < c$ for each $x \in X$, and

   (ii) if $|f(a)| \leq c$ for each $a \in A$, then F may be chosen so that $|F(x)| \leq c$ for each $x \in X$.

Proof. The proof is based on repeated applications of (4.B.3). Three cases are considered.

Case 1. Suppose that $|f(a)| \leq c$ for each $a \in A$. By (4.B.3), there is a map $g_0 : X \to \mathscr{E}^1$ such that $|f(a) - g_0(a)| \leq (2/3)c$ for each $a \in A$ and $|g_0(x)| \leq (1/3)c$, for each $x \in X$. Application of (4.B.3) to the function $f - g_0$ yields a function $g_1 : X \to \mathscr{E}^1$, where $|g_1(x)| \leq (1/3)(2/3)c$ for $x \in X$ and $|f(a) - g_0(a) - g_1(a)| \leq (2/3)(2/3)c$ for $a \in A$. Continuing inductively, we obtain a sequence of functions $g_0, g_1, g_2, \ldots$ such that for each $n \in \mathbf{N}$,

$|g_n(x)| \leq (1/3)(2/3)^n c$ on X and $|f(a) - g_0(a) - \cdots - g_n(a)| \leq (2/3)(2/3)^n c$

on A. Define $F : X \to \mathscr{E}^1$ by $F(x) = \sum_{n=0}^{\infty} g_n(x)$. The function F is seen to be

continuous as follows. For each n, let $s_n(x) = \sum_{i=0}^{n} g_i(x)$. Note that $\sum_{n=0}^{\infty} g_n(x) \leq$

$\sum_{n=0}^{\infty} (1/3)(2/3)^n c = c$, and hence the sequence of partial sums $s_n(x)$ converges
uniformly to $F(x)$, i.e., for each $\varepsilon > 0$, there is an integer N such that for
$n > N$, $|s_n(x) - F(x)| < \varepsilon$ for each $x \in X$. We give the usual advanced
calculus argument to show that F is continuous. Let $x_0 \in X$ be given. Choose
N such that for $n \geq N$, $|s_n(x) - F(x)| < \varepsilon/3$. Since $s_N(x)$ is continuous, there
is a $\delta > 0$ such that if $|x - x_0| < \delta$, then $|s_N(x) - s_N(x_0)| < \varepsilon/3$. Con-
sequently, we have that $|F(x) - F(x_0)| \leq |F(x) - s_N(x)| + |s_N(x) - s_N(x_0)|$
$+ |s_N(x_0) - F(x_0)| < \varepsilon$ whenever $|x - x_0| < \delta$, which establishes the con-
tinuity of F. That F is an extension of f and $|F(x)| \leq c$ for each $x \in X$ may be
readily verified by the reader.

Case 2. Suppose that $|f(a)| < c$ for $a \in A$. By Case 1 there is a con-
tinuous extension F of f that at least satisfies $|F(x)| \leq c$. Let $B = \{x \mid F(x) = c\}$. Note that B is closed and disjoint from A. Apply (4.B.1) to obtain a func-
tion $h : X \to \mathscr{E}^1$ such that $h(A) = 1$ and $h(B) = 0$. The desired extension
G of f is defined by pointwise multiplying the functions F and h, i.e., $G(x) = h(x) \cdot F(x)$.

Case 3. Suppose that f is not bounded. Let $h : \mathscr{E}^1 \to (-1,1)$ be an
arbitrary homeomorphism. By Case 2, the map $hf : A \to (-1,1)$ may be
extended to $G : X \to (-1,1)$; then $F = h^{-1}G$ extends f to all of X.

Euclidean 1-space $\mathscr{E}^1$ is not the only range space for which the foregoing
theorem is valid. In fact, quite a number of spaces may be used in place of $\mathscr{E}^1$,
and such spaces are called Absolute Retracts.

(4.B.9) Definition. A normal space Y is an *Absolute Retract* (AR) if and
only if for each normal space X, for each closed subspace A of X, and for
each continuous map $f : A \to Y$, there is a continuous function $F : X \to Y$
such that $F(a) = f(a)$ for each $a \in A$.

This unusual terminology will make more sense shortly, but first let us
prove a basic result concerning AR's.

(4.B.10) Theorem. If the spaces $X_1, \ldots, X_n$ are AR's, and if the product
space $\prod_{i=1}^{n} X_i$ is normal, then the product space is an AR.

Proof.   Suppose that X is a normal space and that A is a closed subset of X. Let f be a continuous map from A into $\prod\limits_{i=1}^{n} X_i$. For each i, the map $p_i f : A \to X_i$ may be extended to a map $F_i : X \to X_i$. Define $F : X \to \prod\limits_{i=1}^{n} X_i$ by $F(x) = (F_1(x), \ldots, F_n(x))$.

(4.B.11)   Corollary.   The space $\mathscr{E}^n$ is an AR.

(4.B.12)   Exercise.   Show that $\mathbf{I} \times \cdots \times \mathbf{I}$ is an AR.

(4.B.13)   Exercise.   Show that the property of being an AR is a topological invariant.

(4.B.14)   Definition.   A subset A of a space X is a *retract* of X if and only if there is a continuous function $r : X \to A$ such that $r(a) = a$ for each $a \in A$ (i.e., the identity map on A has a continuous extension r). In this case, r is called a *retraction*.

(4.B.15)   Exercise.   (a) Show that a subset A of a space X is a retract of X if and only if the following diagram commutes, where i is the inclusion map.

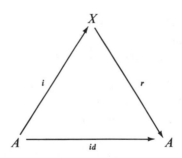

(b) Show that a subset A of X is a retract of X if and only if each continuous map $f : A \to Z$ can be extended to a continuous map $f : X \to Z$.

(4.B.16)   Examples.
1. Any point in a space X is a retract of X.
2. Any closed interval of $\mathscr{E}^1$ is a retract of $\mathscr{E}^1$.
3. If $x \in \mathscr{E}^n$, and $\varepsilon > 0$, then the closed ball $\overline{S_\varepsilon(x)}$ in $\mathscr{E}^n$ is a retract of $\mathscr{E}^n$.

(4.B.17)   Exercise.   Show that if X has the fixed point property and A is a retract of X, then A has the fixed point property.

(4.B.18) *Definition.* A normal space X is an *absolute retract* (*ar*) if and only if whenever X is homeomorphic to a closed subset B of a normal space Y, then B is a retract of Y.

(4.B.19) *Theorem.* A space X is an AR if and only if X is an *ar*.

Proof. Suppose that X is an AR. Let B be a closed subset of a normal space Y, and suppose that B and X are homeomorphic. We seek a retraction r, from Y onto B. Let $h : B \to X$ be a homeomorphism.

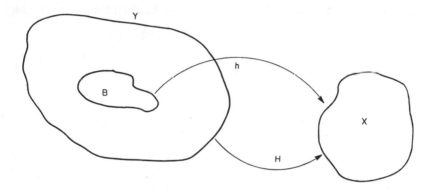

Since X is an AR, h may be extended to a continuous map $H : Y \to X$, where $H_{|B} = h$. Then $r = h^{-1}H$ is the desired retraction. The converse is considerably more difficult, and its proof will be deferred until Chapter 8.

Closely related to AR's are Absolute Neighborhood Retracts, ANR's.

(4.B.20) *Definition.* A normal space X is an *Absolute Neighborhood Retract* (*ANR*) if and only if for each normal space Y, for each closed $A \subset Y$, and for each continuous map $f : A \to X$, there is a neighborhood U of A and a continuous extension $F : U \to X$ such that $F_{|A} = f$.

Although AR's are obviously ANR's, there are numerous examples of ANR's that are not AR's. Perhaps the most prominent of these are the Euclidean n-spheres $\mathscr{S}^n = \{(x_1, \ldots, x_{n+1}) \in \mathscr{E}^{n+1} \mid x_1^2 + \cdots + x_{n+1}^2 = 1\}$.

(4.B.21) *Example.* The n-sphere $\mathscr{S}^n$ is an ANR. Suppose that Y is normal and that A is a closed subset of Y. Let $f : A \to \mathscr{S}^n$ be a continuous function. Since $\mathscr{S}^n \subset \mathscr{E}^{n+1}$, f may be regarded as a map from A into the AR, $\mathscr{E}^{n+1}$. Consequently, there is a continuous extension of f, $F : Y \to \mathscr{E}^{n+1}$. Set $U = F^{-1}(\mathscr{E}^{n+1} \setminus \{0\})$ and on U define a function F' by $F'(y) = F(y)/\|F(y)\|$, where $\|F(y)\|$ denotes the usual distance between $F(y)$ and the origin. Then F' is the

desired extension of f to a neighborhood U of A. That $\mathscr{S}^n$ is not an AR follows from the $(n + 1)$-dimensional Brouwer fixed point theorem (14.C.22); see problem 4.B.9.

The following result will be used in later sections.

(4.B.22)   Lemma (Brouwer Reduction Theorem). Suppose that X is a second countable space and that $\mathscr{C} = \{C_\alpha \mid \alpha \in \Lambda\}$ is a family of closed sub-sets of X with the property that whenever $C_1 \supset C_2 \supset C_3 \cdots$ is a countable decreasing family of members of $\mathscr{C}$, then $\bigcap_{i=1}^{\infty} C_i \in \mathscr{C}$. Then there is an irreduc-ible set C^* in $\mathscr{C}$, i.e., no proper subset of C^* belongs to $\mathscr{C}$.

Proof.   Let $U_1, U_2, \ldots$ be a countable basis for X. Let C_1 be an arbitrary member of $\mathscr{C}$ and inductively select a sequence of sets from $\mathscr{C}$ by choosing $C_n \in \mathscr{C}$ such that $C_n \subset C_{n-1} \cap (X \setminus U_n)$. (If no such set in $\mathscr{C}$ exists, then define $C_n = C_{n-1}$.)

We assert that $C^* = \bigcap_{i=1}^{\infty} C_i$ is irreducible. If this is not the case, then there is a proper subset $\hat{C}$ of C^* that belongs to $\mathscr{C}$. Clearly, there is a basis member U_n with the property that $U_n \cap \hat{C} = \varnothing$ and $U_n \cap C^* \neq \varnothing$. Since $\hat{C} \in \mathscr{C}$, and $\hat{C} \subset C^* \cap (X \setminus U_n)$, it follows that we have a situation where $C_n \subset C_{n-1} \cap (X \setminus U_n)$; however, this is impossible, since $U_n \cap C^* \neq \varnothing$ implies that $U_n \cap C_n$ is nonempty.

That $\mathscr{S}^n$ is an ANR plays a key role in the proof of the following exten-sion result, which will be used in Chapter 15.

(4.B.23)   Theorem. Suppose that K is a compact metric space, C is a closed subset of K, and f is a continuous map from C into $\mathscr{S}^n$ that cannot be extended to all of K. Then there is a closed subset C^* of K such that f cannot be extended continuously to $C^* \cup C$ but can be extended continuously to $C \cup D$, where D is any proper closed subset of C^*.

Proof.   Let $\mathscr{D} = \{D \mid D$ is closed and f can not be extended continuously to $C \cup D\}$. We show that $\mathscr{D}$ has an irreducible member. Suppose that $D_1 \supset D_2 \supset \cdots$ is a countable decreasing family of elements of $\mathscr{D}$, and let $D = \bigcap_{i=1}^{\infty} D_i$. If $D \notin \mathscr{D}$, then there is a continuous extension $F : (C \cap D) \to \mathscr{S}^n$ of f. Since $\mathscr{S}^n$ is an ANR, F can be extended continuously to an open neigh-borhood U containing $C \cup D$. However, by (2.H.3) for some integer i, we have that $D_i \subset U$, contradicting the fact that f cannot be extended con-tinuously to $C \cup D_i$. Thus, $D \in \mathscr{D}$, and hence by the Brouwer reduction theorem (4.B.22), there is an irreducible element C^* in $\mathscr{D}$; clearly, C^* is the required subset of X.

The notion of ANR is also frequently defined in a metric space context.

(4.B.24) *Definition.* A metric space X is a *metric absolute neighborhood retract*, denoted by ANR_M, if and only if for each closed subspace B of a metric space Y and for each continuous function $f : B \to X$, there is a neighborhood U of B and a continuous extension $g : U \to X$ of f.

(4.B.25) *Theorem.* Suppose that X is an ANR_M, Y is a metric space, and $h : X \to Y$ is an embedding such that $h(X)$ is closed in Y. Then there is a neighborhood U of $h(X)$ and a retraction $r : U \to h(X)$.

Proof. Consider $h^{-1} : h(X) \to X$. Since X is an ANR_M, there is a neighborhood U of $h(X)$ and an extension $g : U \to X$ of h^{-1}. Then hg is the desired retraction.

(4.B.26) *Remark.* The converse of (4.B.25) holds if X is compact (10.C.3), but its proof requires material from Chapters 8 and 10.

The following rather technical lemma is needed in the proof of (4.B.28).

(4.B.27) *Lemma.* Suppose that C_1 and C_2 are closed subspaces of a topological space $X = C_1 \cup C_2$. Let A be a subset of X and for $i = 1,2$ let U_i be an open subset of C_i that satisfies $A \cap C_i \subset U_i$. Then $U_1 \cup U_2$ is a neighborhood (in X) of A.

Proof. There are open sets V_1 and V_2 in X such that $U_1 = V_1 \cap C_1$ and $U_2 = V_2 \cap C_2$. Note that $V_1 \cap (C_1 \setminus C_2) \subset U_1$, $V_2 \cap (C_2 \setminus C_1) \subset U_2$, and $V_1 \cap V_2 = V_1 \cap V_2 \cap (C_1 \cup C_2) \subset U_1 \cup U_2$. Hence, we have that $A = (A \cap (C_1 \setminus C_2)) \cup (A \cap (C_2 \setminus C_1)) \cup (A \cap C_1 \cap C_2) \subset (V_1 \cap (C_1 \setminus C_2)) \cup (V_2 \cap (C_2 \setminus C_1)) \cup (V_1 \cap V_2) \subset U_1 \cup U_2$. Since $V_1 \cap (C_1 \setminus C_2)$, $V_2 \cap (C_2 \setminus C_1)$, and $V_1 \cap V_2$ are open subsets of X, so is their union.

(4.B.28) *Theorem* [*Borsuk, 1932*]. Suppose that C_1 and C_2 are closed subspaces of a metric space X such that $X = C_1 \cup C_2$. Suppose further that C_1, C_2, and $C_1 \cap C_2$ are ANR_M's. Then for each embedding h of X as a closed subset of a metric space Y, there is a neighborhood U of $h(X)$ and a retract $r : U \to h(X)$.

Proof. (*Borsuk [1932]*). Let Y be a metric space and $h : X \to Y$ be an embedding such that $h(X)$ is closed. By (4.B.25), there is a neighborhood U of

$h(C_1 \cap C_2)$ and a retraction $r : U \to h(C_1 \cap C_2)$. Since metric spaces are completely normal, there are open sets $\hat{U}_1$ and $\hat{U}_2$ in X such that $h(C_1) \setminus h(C_2) \subset \hat{U}_1$, $h(C_2) \setminus h(C_1) \subset \hat{U}_2$, and $\hat{U}_1 \cap \hat{U}_2 = \varnothing$. Let $U_1 = \hat{U}_1 \setminus h(C_2)$ and $U_2 = \hat{U}_2 \setminus h(C_1)$. Then we have that $h(C_1) \setminus h(C_2) \subset U_1$, $h(C_2) \setminus h(C_1) \subset U_2$, $U_1 \cap U_2 = \varnothing$, and $(U_1 \cup U_2) \cap h(C_1) \cap h(C_2) = \varnothing$. For $i = 1,2$ define sets $P_i = (U \setminus U_1 \setminus U_2) \cup h(C_i)$ and functions $r_i : P_i \to h(C_i)$ by

$$r_i(x) = \begin{cases} r(x) & \text{if} \quad x \in U \setminus U_1 \setminus U_2 \\ x & \text{if} \quad x \in h(C_i) \end{cases}$$

Clearly r_i is continuous.

Now let $E_1 = (U \cup U_1 \cup U_2) \setminus U_2$ and $E_2 = (U \cup U_1 \cup U_2) \setminus U_1$. Then for $i = 1,2$ we have that $P_i = (U \setminus U_1 \setminus U_2) \cup h(C_i) = (E_i \setminus (U_1 \cup U_2)) \cup h(C_i)$. Hence, P_i is a closed subset of E_i. By (4.B.24), there are open subsets $V_1 \subset E_1$ and $V_2 \subset E_2$ such that $P_1 \subset V_1 \subset E_1$ and $P_2 \subset V_2 \subset E_2$, and there are extensions $\hat{r}_1$ and $\hat{r}_2$ of r_1 and r_2 respectively, such that for $i = 1,2$, $\hat{r}_i : V_i \to h(C_i)$. We have that $h(C_1 \cup C_2) \subset C \cup U_1 \cup U_2 = E_1 \cup E_2$, $E_1 = (E_1 \cup E_2) \setminus U_2$, and $E_2 = (E_1 \cup E_2) \setminus U_1$, where the E_i's are closed in $E_1 \cup E_2$. Furthermore, $h(C_1 \cup C_2) \cap E_i = h(C_1 \cap C_2) \subset P_i \subset V_i$. It follows from (4.B.27) that $V = V_1 \cup V_2$ is a neighborhood of $h(C_1 \cup C_2)$ in $E_1 \cup E_2 = U \cup U_1 \cup U_2$, and since this latter set is open, V is a neighborhood of $h(C_1 \cup C_2)$ in Y. Note now that $U \setminus U_1 \setminus U_2 \subset P_1 \cap P_2 \subset V_1 \cap V_2 \subset E_1 \cap E_2 = ((U \setminus U_2) \setminus U_1) \cap ((U \setminus U_1) \cup U_2) = \cdot U \setminus U_1 \setminus U_2$, and hence $V_1 \cap V_2 = E_1 \cap E_2 = U \setminus U_1 \setminus U_2$. Since the E_i's are closed in $E_1 \cup E_2$, we have that $\overline{V}_1 \cap (V_1 \cup V_2) = V_1 \cup (\overline{V}_1 \cap V_2) \subset V_1 \cup (E_1 \cap E_2) = V_1$. Finally, since the V_i's are closed in V and by the definition of the $\hat{r}_i$'s, it follows that the function $R : V \to h(C_1 \cup C_2)$ defined by

$$R(x) = \begin{cases} \hat{r}_1(x) & \text{if} \quad x \in V_1 \\ \hat{r}_2(x) & \text{if} \quad x \in V_2 \end{cases}$$

is a retraction of V onto $h(C_1 \cup C_2)$.

(4.B.29) **Remark.** Once we have (10.C.3), it will follow that (4.B.28) may be restated so that if X is a compact metric space, then the conclusion will be that $X = C_1 \cup C_2$ is an ANR_M.

C. FURTHER RESULTS CONCERNING NORMAL SPACES

(4.C.1) **Definition.** If $\mathscr{A} = \{A_\alpha \mid \alpha \in \Lambda\}$ is a cover of a space X, then an *open shrinkage* of $\mathscr{A}$ is an open cover $\{U_\alpha \mid \alpha \in \Lambda\}$ of X such that $\overline{U}_\alpha \subset A_\alpha$ for each $\alpha \in \Lambda$.

Certain open covers of normal spaces are amenable to being shrunk. The next theorem is not only a remarkable result, but is actually quite useful in a number of distinct contexts. Its proof is dependent on the axiom of choice, which appears in its well-ordering guise.

(4.C.2) **Theorem.** Suppose that $\mathcal{U} = \{U_\alpha \mid \alpha \in \Lambda\}$ is an open cover of a normal space X such that each point in X lies in at most a finite number of the U_α (such a cover is called *point finite*). Then there is an open cover $\mathcal{V} = \{V_\alpha \mid \alpha \in \Lambda\}$ that is an open shrinkage of $\mathcal{U}$.

Proof. We assume that Λ has been well ordered with first element α_0, second element α_1, etc. We shall use transfinite induction to construct a family $\{V_\alpha \mid \alpha \in \Lambda\}$ of open sets such that

(i)   for each $\alpha \in \Lambda$, $\overline{V}_\alpha \subset U_\alpha$, and
(ii) for each $\alpha \in \Lambda$, the family $\{V_\lambda \mid \lambda < \alpha\}$ together with the family $\{U_\delta \mid \delta \geq \alpha\}$ cover X. We then show that the family $\{V_\alpha \mid \alpha \in \Lambda\}$ is an open cover of X.

For purposes of motivation observe that $X \setminus \bigcup_{\alpha_0 < \alpha} U_\alpha$ is closed and is contained in U_{α_0}. Hence by (4.A.4), there is an open set V_{α_0} such that $(X \setminus \bigcup_{\alpha_0 < \alpha} U_\alpha) \subset V_{\alpha_0} \subset \overline{V}_{\alpha_0} \subset U_{\alpha_0}$. In order to replace U_{α_1} by a similar construction, we make use of the fact that $\{V_{\alpha_0}, U_{\alpha_1}, \ldots\}$ is an open cover of X.

Now assume that for each $\beta < \alpha$, we have constructed an open set V_β such that $\overline{V}_\beta \subset U_\beta$ and that $\{V_\beta \mid \beta < \alpha\}$ and $\{U_\gamma \mid \delta \geq \alpha\}$ form a cover of X. We show that this permits the construction of an open set V_α such that

(i) $\overline{V}_\alpha \subset U_\alpha$, and
(ii) the families $\{V_\beta \mid \beta \leq \alpha\}$ and $\{U_\gamma \mid \delta > \alpha\}$ form a cover of X.

Note that $X \setminus (\cup\{V_\beta \mid \beta < \alpha\} \bigcup \cup \{U_\gamma \mid \gamma \geq \alpha\}) \subset U_\alpha$; hence, by the normality of X, there is an open set V_α such that $X \setminus (\cup\{V_\beta \mid \beta < \alpha\} \bigcup \cup \{U_\gamma \mid \gamma > \alpha\}) \subset V_\alpha \subset \overline{V}_\alpha \subset U_\alpha$. Then the combined collections $\{V_\beta \mid \beta \leq \alpha\}$ and $\{U_\gamma \mid \gamma < \alpha\}$ cover X, and this completes the inductive step. To see that the family $\{V_\beta \mid \beta \in \Lambda\}$ covers X, suppose that $x \in X$ and let $U_{\gamma_1}, U_{\gamma_2}, \ldots, U_{\gamma_n}$ be the sets in $\mathcal{U}$ that contain x. Assume that $\gamma_n > \gamma_i$ for $i = 1, 2, \ldots, n - 1$. Then $\{V_{\alpha_0}, \ldots, V_{\gamma_1}, \ldots, V_{\gamma_n}, U_{\gamma_{n+1}}, \ldots\}$ covers X, which implies that $x \in V_{\alpha_0} \cup \cdots \cup V_{\gamma_n}$.

A property of a given space is *hereditary* if and only if every subspace of the space also has the property. For example, any subspace of a metrizable space is still metrizable, subspaces of Hausdorff spaces are Hausdorff, and subspaces of first countable spaces are first countable. As we see next, normality is not hereditary.

Ordinal numbers are useful in the construction of several important examples in topology. Certain basic features of ordinals are given in the Chapter 0, and a more thorough introduction may be found in Monk [1969]. We denote the first infinite ordinal by ω and the first uncountable ordinal by Ω. Let $X = [0,\omega] = \{\alpha \mid \alpha \text{ is an ordinal and } 0 \leq \alpha \leq \omega\}$, and let $Y = [0,\Omega] = \{\alpha \mid \alpha \text{ is an ordinal and } 0 \leq \alpha \leq \Omega\}$. Each of these spaces is given the order topology. The reader may verify (see exercise below) that both X and Y are compact T_2 spaces. Consequently, $X \times Y$ is a compact Hausdorff space and hence by (4.A.6) is normal. We shall exhibit a subspace of $X \times Y$ that is not normal. (This, incidentally, implies that $X \times Y$ is not completely normal (4.A.5) and hence not metrizable (4.A.1).) For a simple example, see Problem C-5. The space $X \times Y$ with the product topology is called the *Tihonov plank*. The desired subspace is obtained by simply removing the point (ω,Ω) from the product space. Let $W = (X \times Y) \setminus \{(\omega,\Omega)\}$. Then $A = \{(\alpha,\Omega) \mid 0 \leq \alpha < \omega\}$ and $B = \{(\omega,\beta) \mid 0 \leq \beta < \Omega\}$ are disjoint closed subsets of W. We show that it is impossible to enclose A and B in disjoint open subsets. Suppose that U is any open set containing A. For each $\alpha \in X \setminus \{\omega\}$, let β_α be the least element of Y such that $(\alpha,\beta) \in U$ whenever $\beta > \beta_\alpha$. Since the collection $S = \{\beta_\alpha \mid \alpha \in X \setminus \{\omega\}\}$ is countable, the least upper bound of S, s_0, will be strictly less than Ω (0.E.4). However, $\{(\alpha,\beta) \mid \beta > s_0, \alpha < \omega\}$ is a subset of U. Now suppose that V is an open set containing B. Choose an ordinal β such that $s_0 < \beta < \Omega$.

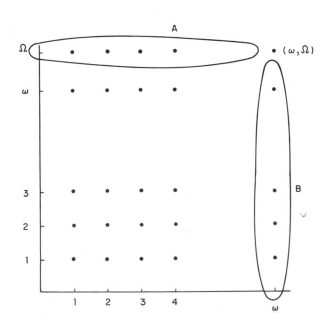

Then of course, $(\omega,\beta) \in B$, and it should now be clear that points "close to" (ω,β) must lie in both U and V; hence, W is not normal.

(4.C.3) *Exercise.* Show that $[0,\omega]$ and $[0,\Omega]$ are compact. (Hint: Consider $A = \{\alpha \in [0,\omega] \mid [0,\alpha]$ is compact$\}$ and show that if $\lambda = \sup A$, then $\lambda \in A$, and $\lambda = \omega$.)

D. THE SEPARATION AXIOMS

A number of properties other than normality have been given sufficient attention to warrant special mention. We examine a few of these.

(4.D.1) *Definition.* A topological space X is *regular* if and only if whenever $x \in X$ and F is a closed subset of X not containing x, there are disjoint open sets U and V such that $x \in U$ and $F \subset V$.

(4.D.2) *Exercise.* Show that a space X is regular if and only if for each point $x \in X$ and for each open set U containing x, there is an open set V such that $x \in V \subset \bar{V} \subset U$.

Note the similarity between this definition (4.D.1) and that for normal spaces. In fact, it should be clear that normal T_1 spaces are regular.

Requiring that points be closed is obviously not a particularly strong imposition on a topological space, but a still weaker condition is all that is needed to define an even lower form of topological life. A space X is T_0 if and only if whenever $x,y \in X$, an open set may be found that contains either x or y but not the other; there is no guarantee, however, as to whether the open set will contain specifically x or y.

A heirarchy of spaces (in increasingly restrictive structure) may be set up as follows:

T_0 space—defined above
T_1 space—defined in Chapter 1
T_2 space—defined in Chapter 1
T_3 space—a regular T_1 space
T_4 space—a normal T_1 space
T_5 space—a completely normal T_1 space
Metric spaces

The T_i's are frequently referred to as separation axioms; basically, they reflect the degree to which points or sets may be kept apart by open sets. Strangely enough, an inordinate amount of mathematical energy has been

expended in defining separation conditions that lie somewhere between the
ones listed above ($1\frac{1}{2}$, $2\frac{7}{8}$, etc.); however, apart from a few notable exceptions,
this sort of exercise would appear to be about as significant as it is interesting.

(4.D.3)   Exercise.   Show that metric $\Rightarrow T_5 \Rightarrow T_4 \Rightarrow T_3 \Rightarrow T_2 \Rightarrow T_1 \Rightarrow T_0$.

In the remainder of this chapter we shall see that none of the above
implications is reversible.

(4.D.4)   Examples.
   1.   Any set (consisting of more than one point) with the indiscrete
topology fails to be T_0.
   2.   The set $\mathbf{R}^1$ with the open ray topology is T_0 but not T_1.
   3.   The set $\mathbf{R}^1$ with the finite complement topology is T_1 but not T_2.
   4.   Let $X = \mathbf{R}^1$. For each point $x \in \mathbf{R}^1$ and each open interval U containing x, set $\hat{U} = \{y \in U \mid y \text{ is rational}\} \cup \{x\}$. The family of all such sets
$\hat{U}$ forms a basis for a topology that is T_2 but not regular.
   5.   (*Tangent disk or bubble topology*). Let $X = \{(x,y) \in \mathbf{R}^2 \mid y \geq 0\}$ and
let $L = \{(x,y) \in \mathbf{R}^2 \mid y = 0\}$. To form a topology for X, let $\mathscr{B}$ be the set of
all $U \subset X$ such that either

   (i) $U \subset X \setminus L$ and U is a member of the usual topology for $\mathscr{E}^2$, or
   (ii) U is of the form $\{x\} \cup D$ where $x \in L$ and D is an open disk tangent
to L at the point x.

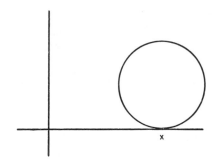

Then $\mathscr{B}$ is a basis for a topology for X that is T_3 but not T_4. The Baire
Category Theorem is used to show that X is not normal. Let $A = \{(x,0) \in
L \mid x \text{ is rational}\}$ and $B = \{(x,0) \in L \mid x \text{ is irrational}\}$. Certainly, A and B are
disjoint closed subsets of X (any subset of L is closed in X). Suppose that U
and V are open subsets of X containing A and B, respectively. We show that
U and V must intersect. For each $(x,0) \in B$, let D_x be a disk of radius r_x that
lies entirely in V and is tangent to L at $(x,0)$. For $n = 1, 2, \ldots$, set $W_n =$

$\{(x,0) \in B \mid r_x > 1/n\}$. Then $L = A \cup (\bigcup_{n=1}^{\infty} W_n)$. Since L (with the usual topology for $\mathscr{E}^1$) is a complete metric space, it follows from (3.B.16) that some W_n contains an open interval. However, since every open interval in L contains a rational point, U and V must intersect.

(4.D.5) Exercise. Show that the space X in Example 5 is regular, T_1, and separable. Show that L with the relative topology is not separable.

Under certain conditions regular spaces are normal.

(4.D.6) Theorem. A regular, Lindelöf space X is normal.

Proof. Suppose that A and B are disjoint closed subsets of X. By regularity, for each $a \in A$, there is an open set U_a containing a such that $U_a \cap B = \varnothing$ and for each $b \in B$, there is an open set V_b containing b, whose closure misses A. Clearly, we have that $A \subset \bigcup \{U_a \mid a \in A\}$ and $B \subset \bigcup \{V_b \mid b \in B\}$. Since both A and B are Lindelöf (closed subspaces of a Lindelöf space are Lindelöf), countable subcovers $\{U_1, U_2, \ldots\}$ and $\{V_1, V_2, \ldots\}$ can be found for A and B respectively. Now we separate the U's from the V's. Let $W_1 = U_1$ and $Z_1 = V_1 \setminus \overline{W_1}$. Let $W_2 = U_2 \setminus \overline{Z_1}$ and $Z_2 = V_2 \setminus (\overline{W_1 \cup W_2})$. In general, set $W_n = U_n \setminus (\overline{Z_1 \cup \cdots \cup Z_{n-1}})$ and $Z_n = V_n \setminus (\overline{W_1 \cup \cdots \cup W_n})$. It should be apparent that $W = \bigcup_{i=1}^{\infty} W_i$ and $Z = \bigcup_{i=1}^{\infty} Z_i$ are disjoint and open.

(4.D.7) Exercise. Let X be any uncountable set and let $x^* \in X$. Define a subset U of X to be open if and only if (i) $X \setminus U$ is countable, or (ii) $x^* \in X \setminus U$. Show that X with this topology is T_5, but that X is not first countable and, hence, is not metrizable.

We mention one other type of space that is of special interest in analysis. Its definition is reminiscent of Uryson's lemma.

(4.D.8) Definition. A T_1 space X is a *Tihonov space* (or $T_{3\frac{1}{2}}$) if and only if for each closed subset K of X and for each point $x \in X \setminus K$, there is a continuous function $f : X \to I$ such that $f(x) = 0$ and $f(K) = 1$.

If the requirement that X be T_1 is removed, then X is said to be *completely regular*.

(4.D.9) Exercise. Show that $T_4 \Rightarrow T_{3\frac{1}{2}} \Rightarrow T_3$.
We shall presently describe a $T_{3\frac{1}{2}}$ space that is not T_4. For a T_3 space

that fails to be $T_{3\frac{1}{2}}$, the reader may consult Steen and Seebach [1970] or attempt to devise an easier example.

(4.D.10) *Exercise.* Suppose that X and Y are topological spaces. Show that $X \times Y$ has property Q if and only if both X and Y possess Q, where Q ranges over the following properties: (1) T_0; (2) T_1; (3) T_2; (4) T_3; (5) $T_{3\frac{1}{2}}$.

(4.D.11) *Exercise.* Show that the following properties are hereditary: T_0; T_1; T_2; T_3; $T_{3\frac{1}{2}}$.

Observe that in both of the preceding exercises, we came to a halt before T_4 was reached. That "normal" space is perhaps a misnomer might well be argued from the standpoint that normality is neither hereditary nor does it behave decently under product formation. To illustrate the latter point, we consider two of the most pleasing (in their simplicity) and rewarding examples in topology.

(4.D.12) *Example (Half-Open Interval Topology).* Let $X = \mathbf{R}^1$ and let the topology for X be determined by a basis consisting of all half-open intervals of the form $[a,b)$, where $a < b$.

We show that X is T_5. That X is T_1 is clear. Suppose, then, that A and B are subsets of X such that $\bar{A} \cap B = \varnothing$ and $\bar{B} \cap A = \varnothing$. For each $x \in A$. there is a half-open interval $[x,e_x)$ contained entirely in $X \setminus \bar{B}$, and similarly for each $y \in B$ there is a half-open interval $[y,e_y) \subset X \setminus \bar{A}$. It is readily seen that the sets $U = \bigcup \{[x,e_x) \mid x \in A\}$ and $V = \bigcup \{[y,e_y) \mid y \in B\}$ are disjoint open sets containing A and B, respectively.

(4.D.13) *Exercise.* Show that $\mathbf{R}^1$ with the half-open interval topology is first countable, separable, and Lindelöf.

(4.D.14) *Exercise.* Show that $\mathbf{R}^1$ with the half-open interval topology is not second countable, and hence, by (4.D.13) and (1.G.10), is not metrizable.

Even more interesting is the next example.

(4.D.15) *Example (Sorgenfrey's Half-Open Square Topology).* With X defined to be the topological space described in the previous example, let $Y = X \times X$ (with the product topology). A typical basis element for Y is illustrated on page 121. Since X was seen to be T_5, it follows from (4.D.3), (4.D.9), and (4.D.10) that Y is $T_{3\frac{1}{2}}$.

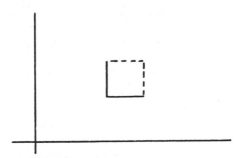

(4.D.16) Exercise. Let Y be the space defined in the previous example and let $L \subset Y$ be the diagonal line $\{(x,y) \mid y = -x\}$. Show that any subset of L is closed, and use an argument similar to that which was employed in connection with the bubble topology to demonstrate that Y is not normal. Consequently, Y is a $T_{3\frac{1}{2}}$ space that is not T_4.

PROBLEMS

Section A

1. (a) Show that if X is a normal space, Y is a topological space and if $f: X \to Y$ is continuous, closed, and onto, then Y is normal.
    (b) Is normality a continuous invariant?
    (c) Suppose that $X_1, X_2, \ldots, X_n$ are topological spaces. Show that if $\prod_{i=1}^{n} X_i$ is normal, then X_i is normal for each i.
2. Suppose that $C_1, C_2, \ldots$ are closed subsets of a normal space X such that for each $k = 1, 2, \ldots, C_k \cap (\bigcup \{C_n \mid n \neq k\}) = \varnothing$. Show that there are disjoint open sets $U_1, U_2, \ldots$ such that $C_n \subset U_n$ for each n.
3.* Does there exist a countable connected normal space that is T_1?
4. Show that a topological space X is normal if and only if for each pair of disjoint closed subsets A and B there are open sets U and V containing A and B respectively such that $\overline{U} \cap \overline{V} = \varnothing$.
5. Suppose that X is a 0-dimensional second countable set and that A and B are closed disjoint subsets of X. Show that there is a separation of X, (U,V), such that $A \subset U$ and $B \subset V$.
6. Suppose that (X,d) is a metric space. The metric d is said to be *normal* if and only if whenever A and B are closed disjoint subsets, then $d(A,B) > 0$. Show that if d is normal, then (X,d) is complete.
7.* Suppose that (X,d) is a metric space. Show that d is normal (see problem

6) if and only if every continuous function $f : (X,d) \to (Y,d')$ is uniformly continuous.

8. Show that if a space X is the union of a finite number of closed, normal subspaces, then X is normal.

Section B

1. Show that subspaces of perfectly normal spaces are perfectly normal.
2. Show that $[0,1)$ is an AR.
3. Prove Uryson's lemma using Tietze's extension theorem.
4. Suppose that X is a normal space and that A and B are disjoint closed subsets of X. Show that there is a continuous map $f : X \to [a,b]$ such that $f(A) = a$ and $f(B) = b$.
5. Let $A = \{0,1\}$. Show that a space X is path connected if and only if each continuous map $f : A \to X$ can be extended to $[0,1]$.
6. Suppose that (X,d) is a metric space. Show that (X,d) is perfectly normal.
7.* Show that a retract of a T_2 space is closed.
8. Show that if a (finite) product of ANR's is normal, then it is an ANR.
9. Assume the $(n+1)$-dimensional Brouwer fixed point theorem. Use (4.B.17) to show that $\mathscr{S}^n$ is not an AR.
10. Find a closed subset of a normal space that is not a G_δ.
11.* Show that a pseudometric space is perfectly normal.
12. Suppose that A is a retract of X and that B is a retract of Y. Show that $A \times B$ is a retract of $X \times Y$.
13. Show that a retract of an AR is an AR.
14. Show that an open normal subset of an ANR is an ANR.
15. Do the following subspaces of the plane $\mathscr{E}^2$ have the fixed point property?

Section C

1. Prove the converse of (4.C.2).
2. Show that closed subsets of a normal space are normal.
3. Show that a space X is normal if and only if for each finite open cover $\{U_1, U_2, \ldots, U_n\}$ of X there is an open cover of X, $\{V_1, V_2, \ldots, V_n\}$, with the property that $\overline{V_i} \subset U_i$ for $i = 1, 2, \ldots, n$. Avoid using the axiom of choice in your proof.

4. Suppose that X is a normal space and that $A_1, A_2, \ldots, A_n$ are closed subsets of X such that $\bigcap_{i=1}^{n} A_i = \varnothing$. Show that there are open sets $V_1, V_2, \ldots, V_n$ such that $A_i \subset V_i$ and $\bigcap_{i=1}^{n} \overline{V}_i = \varnothing$.

5. Let X_1 be an uncountable set and X_2 be any infinite set. Suppose that X_1 and X_2 have discrete topologies and let $X_1^* = X_1 \cup \{\infty_1\}$ and $X_2^* = X_2 \cup \{\infty_2\}$ be the corresponding one-point compactifications. Show that $X_1^* \times X_2^*$ is normal, but $(X_1^* \times X_2^*) \setminus \{\infty_1, \infty_2\}$ is not.

Section D

1. Show that a regular second countable space is completely normal.
2. Show that a normal space is completely regular if and only if it is regular.
3. Suppose that X is a T_2 space such that each $x \in X$ has a neighborhood V with the property that $\overline{V}$ is regular. Show that X is regular.
4.* Show that there is no countable connected $T_{3\frac{1}{2}}$ space containing more than one point.
5.* Let $X = \mathbf{Z}^+$. For each $a,b \in \mathbf{Z}^+$ which are relatively prime, let $B_{(a,b)} = \{an + b \mid n \in \mathbf{Z}^+ \cup \{0\}\}$.

   (i) Show that the collection of all such $B_{(a,b)}$ forms a basis for a T_2 topology $\mathcal{U}$ for $\mathbf{Z}^+$.
   (ii) Show that if $U = B_{(a,b)}$ and $V = B_{(c,d)}$, then $a \cdot c$ is an accumulation point of both U and V.
   (iii) Show that $(\mathbf{Z}^+, \mathcal{U})$ is connected.

6. Suppose that X is a first countable space. Show that X is T_2 if and only if every convergent sequence has a unique limit.
7. Show that every F_σ subset of a compact space is Lindelöf.
8. (i) Show that locally compact T_1 spaces are regular. [Hint: Use the one-point compactification.]
   (ii) Show that locally compact T_0 spaces are regular. [Hint: Locally compact T_0 spaces are T_2.]

9.* Suppose that X is a completely regular space, A is a compact subset of X, and B is a closed subset of X disjoint from A. Show that there is a continuous function $f : X \to [0,1]$ such that $f(A) = 0$ and $f(B) = 1$.
10. Show that F_σ subsets of a T_4 space are T_4.
11.* Suppose that X is an infinite T_2 space. Show that X has an infinite discrete subspace.
12. Suppose that X is a T_3 space and f is a continuous, open and closed function from X onto Y. Show that Y is T_2.

13. A space X is an *Uryson space* if and only if whenever $x,y \in X$ and $x \neq y$, there are open subsets U and V of X such that $x \in U$, $y \in V$, and $\overline{U} \cap \overline{V} = \emptyset$. Show that T_3 spaces are Uryson spaces and find an Uryson space that is not regular.

14.* Show that perfectly normal spaces are completely normal.

15.* Is Example (4.D.12) perfectly normal?

16. Show that every T_1 space $(X,\mathcal{U})$ can be embedded in a separable T_1 space. [Hint: Let $Y = X \cup \mathbf{Z}^+$ (disjoint union) and define a topology $\mathcal{V}$ for Y by $\mathcal{V} = \{\emptyset\} \cup \{U \cup A \mid U \in \mathcal{U}, A \subset \mathbf{Z}^+, \text{ and } \mathbf{Z}^+ \setminus A \text{ is finite}\}$.]

17. A topological space X is $T_{1\frac{3}{4}}$ if and only if for each pair of disjoint compact subspaces A and B of X, there are open sets U and V such that $A \subset U$, $B \subset V$, and $A \cap V = \emptyset$ and $B \cap U = \emptyset$. Show that $T_2 \Rightarrow T_{1\frac{3}{4}} \Rightarrow T_1$.

18. Suppose that X is a set. A *uniform structure* for X is a nonempty family $\mathcal{U}$ of subsets of $X \times X$ (relations on X) such that:

  (i) for each $U \in \mathcal{U}$, $id_X \subset U$,
  (ii) if $U \in \mathcal{U}$, then $U^{-1} \in \mathcal{U}$,
  (iii) if $U \in \mathcal{U}$ then there is a $V \in \mathcal{U}$ such that $V \circ V \subset U$,
  (iv) if $U,V \in \mathcal{U}$, then $U \cap V \in \mathcal{U}$, and
  (v) if $U \in \mathcal{U}$ and $U \subset V \subset X \times X$, then $V \in \mathcal{U}$.

  (Inverses and compositions of relations are defined in an analogous manner to that of functions).

  The *uniform topology* $T_{\mathcal{U}}$ on X determined by the uniform structure $\mathcal{U}$ is defined by: $T_{\mathcal{U}} = \{W \subset X \mid \text{for each } x \in W, \text{ there is a } U \in \mathcal{U} \text{ such that } U(x) \subset W\}$, where $U(x) = \{z \in X \mid (x,z) \in U\}$.
  (a) Show that if X is a set with a uniform structure $\mathcal{U}$, then $T_{\mathcal{U}}$ is a topology.
  (b) Find a uniform structure that yields the usual topology for $\mathbf{R}^1$, the discrete topology, and the indiscrete topology.

19. Show that if $\mathcal{U}$ is a uniform structure (see preceding problem) for a set X, then the topological space $(X,T_{\mathcal{U}})$ is completely regular. [Hint: Suppose that $x \in X$ and that C is a closed subset of X not containing x. Then there is a sequence $\{U_n \mid n \in \mathbf{Z}^+\}$ of elements of $\mathcal{U}$ such that $U_n = U_n^{-1}$, and $U_n \circ U_n \subset U_{n-1}$. Let $V_0 = id_X$. For each dyadic rational $r \in (0,1]$ expressed in the form $\sum_{k=1}^{N} 2^{-n}k$ where $0 \leq n_1 < n_2 < \cdots < n_N$, let $V_r = U_{n_N} \circ U_{n_N-1} \circ \cdots \circ U_{n_2} \circ U_{n_1}$. If $r \leq s$, then $V_r \subset V_s$ (first show that $U_n \circ V_{m2^{-n}} \subset V_{(m+1)2^{-n}}$). Define $\phi : X \to [0,1]$ by

$$\phi(y) = \begin{cases} \sup\{r \mid x \notin V_r(x)\}, & \text{if } y \neq x \\ 0, & \text{if } y = x \end{cases}$$

and show that ϕ is the desired function.]

20. Suppose that X and Y are topological spaces and that $f : X \to Y$. The function f satisfies property * if and only if for each open cover $\mathscr{U}$ of Y, $\{\operatorname{int} f^{-1}(U) \mid U \in \mathscr{U}\}$ is an open cover of X.

   (a) Show that every continuous function $g : X \to Y$ satisfies property *.

   (b) Show that if Y is T_1, then a function $g : X \to Y$ is continuous whenever g satisfies property *.

   (c) Show that property * is not equivalent to continuity.

Chapter 5

PLANE THEOREMS

A. 1-MANIFOLDS

In Chapter 1, an n-manifold M was defined to be a separable metric space with the property that each point of M is contained in an open set that is homeomorphic to $\mathscr{E}^n$. Our goal in this section is to classify connected 1-manifolds (both of them). In Chapter 16, we perform the somewhat more complicated task of classifying compact 2-manifolds. Curiously, it may be shown that a reasonable classification of 4-manifolds is impossible (Markov [1960]).

(5.A.1) Definition. A *triod* is any space homeomorphic to

$$\{(x,y) \in \mathscr{E}^2 \mid -1 \le x \le 1, y = 0\} \cup \{(x,y) \in \mathscr{E}^2 \mid x = 0, 0 \le y \le 1\}.$$

One obvious characteristic of 1-manifolds is given in the first exercise.

(5.A.2) Exercise. Show that 1-manifolds do not contain triods.

(5.A.3) Theorem. Suppose that M is a connected 1-manifold. Then either M is homeomorphic to $\mathscr{E}^1$ or M is homeomorphic to $\mathscr{S}^1 = \{(x,y) \in \mathscr{E}^2 \mid x^2 + y^2 = 1\}$.

Proof. We consider two cases.

Case 1. Suppose that M is compact. Then there is a finite open cover

127

$\mathscr{U}$ of M each of whose members is homeomorphic to the open interval $(0,1)$. We induct on the number of members in the cover. If the cover $\mathscr{U}$ contains just two sets U_1 and U_2, then it is not difficult to see from the above exercise and the compactness of M that the intersection of U_1 and U_2 consists of precisely two components C_1 and C_2, which are open (2.E.2) and homeomorphic to $(0,1)$. Let c_1 and c_2 be points in C_1 and C_2 respectively, and map the arc from c_1 to c_2 lying in U_1 onto the upper hemisphere of $\mathscr{S}^1$ and the arc from c_1 to c_2 lying in U_2 onto the lower hemisphere of $\mathscr{S}^1$, where end points are matched up in the obvious fashion. This yields the desired homeomorphism.

Suppose now that the theorem is true if the cover $\mathscr{U}$ described above consists of $n \geq 2$ sets, and let $U_1, \ldots, U_{n+1}$ be a cover of X by sets homeomorphic to $(0,1)$. It may be assumed that no U_k is contained in the union of the rest of the U_i. Pick any pair of the U_i's that intersect, say U_i and U_j. The reader should be able to see (think triod) that $U_i \cap U_j$ must be connected, and hence $U_i \cup U_j$ may be mapped homeomorphically onto $(0,1)$. Let $U_s = U_i \cup U_j$ and replace U_i and U_j by U_s in the covering set. Since this reduces the number of members of the cover, induction may be applied to finish the proof.

Case 2. Suppose that M is not compact. Let $D = \{d_1, d_2, \ldots\}$ be a countable dense subset of M. Since M is arcwise connected (2.C.12), there is a homeomorphism from $[0,1]$ onto an arc A_1 with initial point d_1 and terminal point d_2 (actually of course A_1 is the image of an arc). Let d_{n_3} be the first d_i in the sequence $d_1, d_2, \ldots$ that does not lie in A_1. Since no triods are permitted in M, an arc from d_{n_3} to A_1 must meet either d_1 or d_2 before intersecting any point in the "interior" of A_1. If the arc intersects d_1 first, map $[-1,0]$ homeomorphically onto the arc A_2 running between d_{n_3} and d_1 where 0 is matched up with d_1 and -1 with d_{n_3}. On the other hand, if d_2 is the first point struck, then $[1,2]$ is mapped homeomorphically onto the arc A_2 connecting d_{n_3} and d_2, where 1 is sent to d_2 and 2 is mapped to d_{n_3}.

Let d_{n_4} be the first member of the sequence $d_1, d_2, \ldots$ that is not contained in $A_1 \cup A_2$. An arc A_3 is sent out from d_{n_4} in search of $A_1 \cup A_2$. Again, it first meets an end point of $A_1 \cup A_2$, and depending on which one, the appropriate interval of $\mathscr{E}^1$ is mapped onto A_2 following the pattern used previously for A_2. In this fashion, a homeomorphism h may be constructed from $\mathscr{E}^1$ into M that at least covers the dense subset D. It remains to show that h is onto. Suppose that h is not onto, and let x be a point not found in the range of h. Let U be an open set containing x that is homeomorphic with $(0,1)$. Select points d and d' in D that lie on "opposite sides" of x (this is possible, since U is essentially $(0,1)$). The homeomorphism h yields an arc from d to d' that misses x. Complete this arc with another one from d to d' that lies in U to form a space S homeomorphic to $\mathscr{S}^1$ and passing through x.

However, this is impossible, since S cannot be all of M (M is not compact) and points not in S may be connected to S only by creating a triod in M. Consequently, h must be onto, and the classification of connected 1-manifolds is complete.

B. ON THE CONTRACTIBILITY OF $\mathscr{S}^1$

In this section, we establish in an elementary manner the "obvious" fact that the plane cannot be retracted onto $\mathscr{S}^1$ (5.B.6). That such a retract cannot exist is at least intuitively clear, since its existence would apparently indicate that a hole could be torn in $\mathscr{E}^2$ by a continuous function. The proof that this cannot happen is best done in an algebraic setting (see Chapter 12), but in the spirit of this chapter, we follow a lengthier but more naive path.

    The notion of homotopy is exceedingly important in topology and will be carefully examined in Chapter 12.

(5.B.1)   *Definition.*   Suppose that f and g are continuous functions from a space X into a space Y. Then f and g are *homotopic* if and only if there is a continuous function $H : X \times \mathbf{I} \to Y$ such that $H(x,0) = f(x)$ and $H(x,1) = g(x)$ for every $x \in X$. For $t \in \mathbf{I}$, let $H_t : X \to Y$ be defined by $H_t(x) = H(x,t)$.

    A homotopy enables one to "pass continuously" from one function to another.

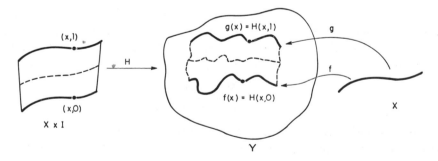

In the following theorem, $\mathscr{S}^1$ is considered to be a subset of the complex plane $\mathbf{C}$, where $\mathscr{S}^1 = \{z \in \mathbf{C} \mid |z| = 1\}$. We follow the presentation of Brown [1974].

(5.B.2)   *Theorem.*   Suppose that $f : \mathscr{S}^1 \to \mathscr{S}^1$ is homotopic to a constant map. Then there is a continuous function $\phi : \mathscr{S}^1 \to \mathscr{E}^1$ such that $f(x) = e^{i\phi(x)}$ for all $x \in \mathscr{S}^1$.

Proof. Suppose that $H : \mathscr{S}^1 \times I \to \mathscr{S}^1$ is a homotopy, where $H(x,0) = c$ and $H(x,1) = f(x)$ for each $x \in \mathscr{S}^1$. Since H is uniformly continuous, there is a $\delta > 0$ such that $|H(x,t) - H(x,t')| < 2$ whenever $|t - t'| < \delta$ and $x \in \mathscr{S}^1$. Let $0 = t_0 < t_1 < t_2 < \cdots < t_{n-1} < t_n = 1$ be a partition of I such that $|t_{i+1} - t_i| < \delta$. Note that $H_0 = c$ is of the form $e^{i\psi(x)}$, where $\psi : \mathscr{S}^1 \to \mathscr{E}^1$ is a constant map, and that $|H_{t_1}(x) - H_0(x)| < 2$ for all x. We show that H_{t_1} is of the form $e^{i\phi_1(x)}$.

Since $|H_{t_1}(x) - H_0(x)| < 2$, we have that $H_{t_1}(x) \neq -H_0(x)$ and hence $H_{t_1}(x)/H_0(x) \neq -1$. Define a function $\lambda : \mathscr{S}^1 \to \mathscr{E}^1$ by setting $\lambda(x)$ equal to the number of radians between 1 and $H_{t_1}(x)/H_0(x)$ if $H_{t_1}(x)/H_0(x)$ is on or above the x axis and to the negative of that number if $H_{t_1}(x)/H_0(x)$ is below the x axis. Then, $H_{t_1}(x)/H_0(x) = e^{i\lambda(x)}$, and consequently $H_{t_1}(x) = H_0(x)e^{i\lambda(x)} = e^{i(\psi(x)+\lambda(x))} = e^{i\phi_1(x)}$, where $\phi_1(x)$ is defined to be equal to $\psi(x) + \lambda(x)$. The same procedure may be used to show that $H_{t_2}(x) = e^{i\phi_2(x)}$, $H_{t_3} = e^{i\phi_3(x)}$, and eventually that $f(x) = H_1(x) = e^{i\phi_n(x)}$, which completes the proof.

(5.B.3) Definition. A space X is *contractible* if and only if there is a homotopy $H : X \times I \to X$ such that for each $x \in X$, $H(x,0) = x$ and $H(x,1) = c$. In other words, a space is contractible if and only if the identity map is homotopic to a constant map. The homotopy H is said to be a *contraction*.

(5.B.4) Example. Any convex set A is contractible. Let $p \in A$ and define $H : A \times I \to A$ by $H(x,t) = tp + (1 - t)x$.

(5.B.5) Theorem. The unit circle $\mathscr{S}^1$ is not contractible.

Proof. If $\mathscr{S}^1$ is contractible, then the identity map is homotopic to a constant function, and hence by the previous theorem, there is a function $\phi : \mathscr{S}^1 \to \mathscr{E}^1$ such that $x = e^{i\phi(x)}$ for each $x \in \mathscr{S}^1$. Hence, ϕ is 1–1, and in particular, $\phi(x) \neq \phi(-x)$. Define $g : \mathscr{S}^1 \to \{-1,1\}$ by

$$g(x) = \frac{\phi(x) - \phi(-x)}{|\phi(x) - \phi(-x)|}$$

Then g is obviously continuous, and furthermore, g maps $\mathscr{S}^1$ onto $\{1,-1\}$ since $g(-x) = -g(x)$. This however contradicts the connectedness of $\mathscr{S}^1$ (2.A.12).

(5.B.6) Corollary. There is no retraction from $\mathscr{E}^2$ onto $\mathscr{S}^1$.

Proof. Suppose that $r : \mathscr{E}^2 \to \mathscr{S}^1$ is a retraction. Let $p = (0,0)$ and define a homotopy $H : \mathscr{S}^1 \times I \to \mathscr{E}^2$ by $H(x,t) = tp + (1 - t)x$. Then $rH : \mathscr{S}^1 \times I \to \mathscr{S}^1$ is a contraction, which contradicts (5.B.5).

As another interesting corollary of the preceding theorem we have a slightly stronger version of the 2-dimensional Brouwer fixed point theorem.

(5.B.7) Corollary. Let $\mathscr{B}^2 = \{(x,y) \in \mathscr{E}^2 \mid x^2 + y^2 \leq 1\}$ be the unit disk and suppose $f : \mathscr{B}^2 \to \mathscr{E}^2$ is a continuous map such that $f(\mathscr{S}^1) \subset \mathscr{B}^2$. Then f has a fixed point.

Proof. Define $r : \mathscr{E}^2 \setminus \{(0,0)\} \to \mathscr{S}^1$ by $r(x) = x/|x|$. If $f(x) \neq x$ for all $x \in \mathscr{B}^2$, then $\mathscr{S}^1$ may be contracted by the homotopy

$$H(x,t) = \begin{cases} r(x - 2tf(x)) & 0 \leq t \leq \tfrac{1}{2} \\ r((2 - 2t)x - f((2 - 2t)x)) & \tfrac{1}{2} \leq t \leq 1 \end{cases}$$

which contradicts the previous theorem.

(5.B.8) Exercise. Show that the homotopy H in the preceding corollary is well defined and is in fact a contraction.

(5.B.9) Exercise. Show that contractibility is a topological invariant.

C. THE JORDAN CURVE THEOREM

A homeomorphic image of $\mathscr{S}^1$ is called a *simple closed curve*. The Jordan curve theorem, one of the most celebrated classical theorems in topology, asserts that the complement in the plane of a simple closed curve S consists of two components, each of which has S as its frontier. At first glance, it seems that this assertion is obvious; nevertheless, as we shall see, the proof is surprisingly involved. In fact, Jordan himself gave an invalid proof, and the first acceptable demonstration was presented by Veblen [1905]. If one has at hand some tools from algebraic topology, a relatively simple and elegant proof is possible. However, in order to gain experience in working with manifolds, we have chosen a longer, but more geometric approach. As usual, a number of steps have been left as exercises for the reader. To help convince the skeptical reader that there is more afoot here than might be anticipated, we have included on page 132 a relatively simple simple closed curve for his perusal. Does it have an outside or an inside?

We consider first a special type of simple closed curve, one that consists entirely of vertical and horizontal line segments. We shall refer to these curves as *rectilinear simple closed curves*. The first theorem is a simple exercise in analytic geometry, but a rigorous proof is given for the less advanced reader.

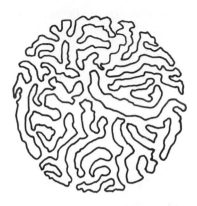

(5.C.1)   ***Theorem.***   The line L, defined by $x = 0$, separates the plane into exactly two components and is the frontier of each.

Proof.   We first show that $\mathscr{E}^2 \setminus L$ consists of at most two components. Given any three points (x_1, y_1), (x_2, y_2), and (x_3, y_3) with $x_i \neq 0$, at least two of the points must have either both first coordinates positive or both negative. Then the line segment joining them misses L, and hence they lie in the same component.

We now prove that the sets $A = \{(x, y) \in \mathscr{E}^2 \mid x < 0\}$ and $B = \{(x, y) \in \mathscr{E}^2 \mid x > 0\}$ are distinct components. Let $a = (-1, 0)$ and $b = (1, 0)$ be points in A and B, respectively. We wish to show that there does not exist a path connecting these points that misses L. Suppose that $f : \mathbf{I} \to \mathscr{E}^2$ is an arbitrary path from a to b. Then $p_1 f$ is a map from $\mathbf{I}$ into $\mathscr{E}^1$, where p_1 is the natural projection from $\mathscr{E}^2$ onto its first coordinate. Note that $p_1 f(0) = -1$ and $p_1 f(1) = 1$. By the intermediate value theorem, there is a point $c \in \mathbf{I}$ such that $p_1 f(c) = 0$, which implies that $f(c) \in L$. Since f was arbitrary, it follows that every path from a to b intersects L. Therefore, A and B are distinct path components, and thus, by (2.D.8), they are distinct components of $\mathscr{E}^2 \setminus L$.

It remains to show that L is the frontier of A and B. This, however, is easy, since any open set containing a point on L intersects both A and B; furthermore, if $p \in \mathscr{E}^2 \setminus L$, then the component containing p is an open set in $\mathscr{E}^2$ that misses L.

(5.C.2)   ***Definition.***   Suppose that A and B are homeomorphic subsets of a topological space X. Then A and B are *equivalently embedded* in X if and only if there is a homeomorphism $h : X \to X$ such that $h(A) = B$.

(5.C.3)   ***Theorem.***   Any line L' in $\mathscr{E}^2$ is equivalently embedded to the line L, defined by $x = 0$.

Proof. Suppose that the equation of L' is $x = c$. Then the space homeomorphism h defined by $h(x,y) = (x - c, y)$ is the desired map. If L has the equation $y = ax + b$, then define the homeomorphism h by $h(x,y) = (x, y - (ax + b))$.

(5.C.4) Corollary. Each line in $\mathscr{E}^2$ separates $\mathscr{E}^2$ into two components and is the frontier of each.

(5.C.5) Definition. A *rectangle* is a quadrilateral with two vertical and two horizontal sides (the "inside" of the rectangle is not included).

(5.C.6) Theorem. Suppose that R is a rectangle in $\mathscr{E}^2$. Then

   (i) R separates $\mathscr{E}^2$ into exactly two components and is the frontier of each, and
   (ii) $\mathscr{E}^2 \setminus R$ has exactly one component with compact closure (which is henceforth called the *interior* of R).

Proof. Exercise

The next goal is to prove the Jordan curve theorem for rectilinear simple closed curves. We begin by "simplifying" a given simple closed curve.

(5.C.7) Theorem. Suppose that R is a rectilinear simple closed curve in $\mathscr{E}^2$ with more than four sides. Then there is a rectangle R' and a rectilinear simple closed curve R'' such that

   (i) $R' \cap R'' = K$ is a side of R',
   (ii) $R = (\overline{R' \setminus K}) \cup (\overline{R'' \setminus K})$, and
   (iii) R'' has fewer segments than R.

Proof. Let T be the shortest (in length) horizontal line segment that has adjacent vertical segments lying below it. Then essentially one of the following four cases must occur. In each case, we indicate the side K with a dotted line.

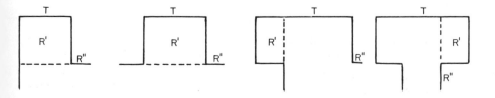

It follows from the condition imposed on T that constructions similar to the one below cannot occur. This concludes the proof of the theorem.

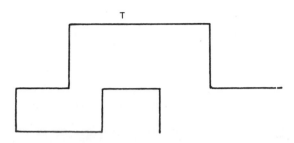

We now construct a homeomorphism of $\mathscr{E}^2$ onto itself that maps a given rectilinear simple closed curve onto one with fewer sides. We enclose the rectangle R' of the previous theorem in a pentagon as indicated below.

It should be apparent that a homeomorphism h from $\mathscr{E}^2$ onto itself may be defined that satisfies the following properties:

(i) h is the identity outside of $v_1v_2v_3v_4v_5$;

(ii)   the segments $\overline{v_5a}$, $\overline{ab}$, $\overline{bv_3}$ are moved onto the segment K and the remainder of R' and its interior is correspondingly pushed downward onto the triangle $v_3\, h(c)\, v_5$;

(iii)   the triangle $v_3v_4v_5$ is pushed by h into the area with vertices v_3, v_4, v_5, and $h(c)$.

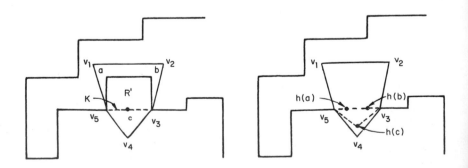

At this stage a new R', R'', and K are found for $h(R)$ (as in the previous theorem), and the process is repeated to derive a second homeomorphism whose image has still fewer line segments. Eventually, this process must stop when a rectangle is obtained. Hence, we have proven the following theorem.

(5.C.8) Theorem (Rectilinear Two-Dimensional Schönflies Theorem).
Suppose that R is a rectilinear simple closed curve lying in the plane. Then
there is a homeomorphism h from $\mathscr{E}^2$ onto itself such that $h(R) = R^*$, where
R^* is a rectangle.

As an immediate corollary we have the following special case of the
Jordan curve theorem.

(5.C.9) Corollary. Suppose that R is a rectilinear simple closed curve
in $\mathscr{E}^2$. Then $\mathscr{E}^2 \setminus R$ has exactly two components. Precisely one of the components has compact closure and R is the frontier of each component.

(5.C.10) Definition. If R is a rectilinear simple closed curve in $\mathscr{E}^2$,
then that component of $\mathscr{E}^2 \setminus R$ which has compact closure will be referred to
as the *interior* of R, denoted by $I(R)$, while the unbounded component of
$\mathscr{E}^2 \setminus R$ will be called the *exterior* of R, denoted by $E(R)$.

Note that interior (exterior) when used in this context is not the interior
(exterior) defined in Chapter 1.
In the previous chapter, we devoted considerable attention to the problem of extending continuous functions. The dilemma now facing us is that of
extending homeomorphisms to homeomorphisms, a far more formidable
task. For instance, it is clear that the subsets of the plane in the figure below,
(each of which is the union of a rectangle and a point), are homeomorphic,
but the reader should be able to prove (using perhaps a compactness and
connectedness argument) that there is no way of extending a homeomorphism
between A and B to all of $\mathscr{E}^2$.

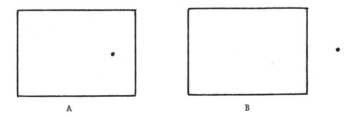

A   B

Nevertheless, homeomorphisms between rectilinear simple closed curves
may be extended to the entire plane.

(5.C.11) Theorem. Suppose that R and R' are rectilinear simple closed
curves, and that h is a homeomorphism from R onto R'. Then there is a

homeomorphism $H : \mathscr{E}^2 \to \mathscr{E}^2$ such that $H_{|R} = h$; furthermore, $H(R \cup I(R)) = R' \cup I(R')$.

Proof. If R and R' are both rectangles, the homeomorphic extension of h may be found as follows.

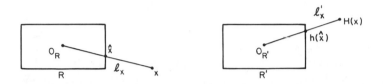

Let O_R and $O_{R'}$ denote the "centers" of R and R', respectively. Suppose that $x \in \mathscr{E}^2$. Let l_x be the ray starting at O_R and passing through x, and let $\hat{x}$ be the first point of intersection of l_x with R. Let m_x denote the length of $\overline{O_R x}$ and $m_{\hat{x}}$ the length of the segment $\overline{O_R \hat{x}}$. Let l'_x be the ray emanating from $O_{R'}$ and passing through $h(\hat{x})$. Then $H(x)$ will be the point on this ray with the property that $m_x/m_{\hat{x}} = q_x/q_{\hat{x}}$, where q_x is the length of the segment $\overline{O_{R'} H(x)}$ and $q_{\hat{x}}$ is length of $\overline{O_{R'} h(\hat{x})}$.

In the general case, let h_1 and h_2 be space homeomorphisms that take R and R' onto rectangles R_1 and R_2, respectively. Then $h_{2|R} \cdot h h_1^{-1}{}_{|R_1}$ maps R_1 homeomorphically onto R_2. By the special case for rectangles, this map may be extended to a space homeomorphism k. Then $H = h_2^{-1} k h_1$ satisfies all requirements of the theorem.

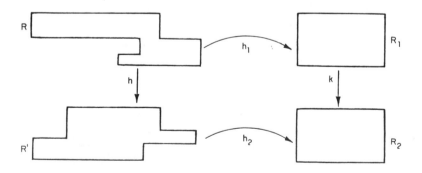

(5.C.12) Definition. A *2-cell* is any topological space homeomorphic to $\mathscr{B}^2 = \{(x,y) \in \mathscr{E}^2 \mid x^2 + y^2 \le 1\}$.

(5.C.13) Corollary. If R is a rectilinear simple closed curve, then $R \cup I(R)$ is a 2-cell.

(5.C.14) Definition. A 2-*annulus* is any space homeomorphic to $\mathscr{S}^1 \times I$.

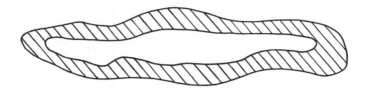

(5.C.15) Notation. If $t > 0$, then $\mathscr{R}_t$ will denote $\{(x,y) \in \mathscr{E}^2 \mid |x| \le t, |y| \le t\}$.

(5.C.16) Theorem (Rectilinear Annulus Theorem). If R and R' are rectilinear simple closed curves with $R' \subset I(R)$, then $T = (R \cup I(R)) \cap (R' \cup E(R'))$ is homeomorphic to the annulus $\mathscr{S}^1 \times I$. In fact, there is a space homeomorphism that carries T onto $\mathscr{R}_2 \setminus \text{int } \mathscr{R}_1$.

Proof. Construct a narrow rectangle connecting R and R' as indicated in the figure. Note that $AFDC$ and $BFEC$ are rectilinear simple closed curves whose intersection consists of the line segments F and C.

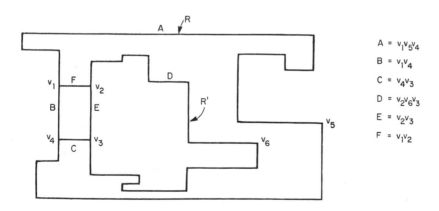

$$A = v_1 v_5 v_4$$
$$B = v_1 v_4$$
$$C = v_4 v_3$$
$$D = v_2 v_6 v_3$$
$$E = v_2 v_3$$
$$F = v_1 v_2$$

Define a homeomorphism h_1 that maps the segments A, C, D, and F onto the segments A', B', D', and F' in the figure below Then (5.C.11) may be applied to extend h_1 to a homeomorphism (which we still call h_1) that maps $AFCD \cup I(AFCD)$ onto $A'F'D'C' \cup I(A'F'D'C')$. Define a second homeomorphism h_2 that agrees with h_1 on $C \cup F$ and maps the segments B and E onto B' and E', respectively. Once more an application of (5.C.11) yields an extension of h_2 (still called h_2) that sends $BCEF \cup I(BCEF)$ homeomorphically onto $B'C'E'F' \cup I(B'C'E'F')$.

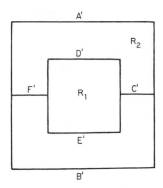

With the aid of the map gluing theorem, the homeomorphism h_1 and h_2 may be combined to form a homeomorphism h that maps $(R \cup I(R)) \cap (R' \cup E(R'))$ onto the annular region between $A'B'$ and $C'D'$, where

$$h(x) = \begin{cases} h_1(x) & \text{if} \quad x \in ACDF \cup I(ACDF) \\ h_2(x) & \text{if} \quad x \in BCEF \cup I(BCEF) \end{cases}$$

The homeomorphism h may be extended to $I(DE)$ (5.C.11). In order to extend h to $E(AB)$, let $g : \mathcal{E}^2 \to \mathcal{E}^2$ be an extension of $h_{|R}$ given by (5.C.11). Then we have that $g : E(AB) \to E(A'B')$. Now define $\hat{h} : \mathcal{E}^2 \to \mathcal{E}^2$ by

$$\hat{h}(x) = \begin{cases} h(x) & \text{if} \quad x \in R \cup I(R) \\ g(x) & \text{if} \quad x \in R \cup E(R) \end{cases}$$

(5.C.17) Exercise. Work out the details of the latter part of the above proof.

A good deal of the proof of the Jordan curve theorem depends on being able to approximate simple closed curves with rectilinear ones. This is achieved via brick partitions.

(5.C.18) Definition. A *brick partition* of $\mathcal{E}^2$ is a collection $\mathcal{T}$ of solid rectangles (rectangles together with their interiors) that satisfy the following properties:

(i) If $R \in \mathcal{T}$, then there are six line segments $A_1, A_2, \ldots, A_6$, each of which is either horizontal or vertical, such that: (a) Fr $R = A_1 \cup \cdots \cup A_6$; (b) non-end points of each A_i lie in exactly two members of $\mathcal{T}$; (c) end points of the A_i's lie in exactly three members of $\mathcal{T}$.

(ii) The union of the rectangles in $\mathcal{T}$ is all of $\mathcal{E}^2$.

(iii) If $R \in \mathcal{T}$ and Fr $R = A_1 \cup \cdots \cup A_6$ and if $R' \in \mathcal{T}$ and $R \neq R'$, then either $R \cup R' \neq \varnothing$ or for some $i = 1, 2, 3, 4, 5, 6$ $R \cap R' = A_i$.

The *mesh* of the partition is the least upper bound of diameters of the members of $\mathcal{T}$. If the least upper bound does not exist, we say that the mesh of $\mathcal{T}$ is infinite.

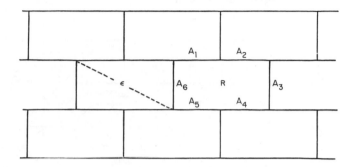

It is clear that for each positive number ε, there is a brick partition of $\mathcal{E}^2$ with mesh less than ε. The next theorem is quite useful for our purposes. Its proof is a consequence of our previous work with 1-manifolds.

(5.C.19)   Theorem.   Suppose that A is a compact subset of $\mathcal{E}^2$ and that $\mathcal{T}$ is a brick partition of $\mathcal{E}^2$. Let M be the union of the elements of $\mathcal{T}$ that intersect A. Then the frontier of M is the union of a finite number of rectilinear simple closed curves, no two of which intersect.

Proof.   Since the number of members of $\mathcal{T}$ that intersect A is finite, there can only be a finite number of components of Fr M. Note that each component is a 1-manifold, and it then follows from (5.A.3) that each component is in fact a rectilinear simple closed curve.

(5.C.20)   Theorem.   Suppose that S is a connected compact subset of $\mathcal{E}^2$ such that $\mathcal{E}^2 \setminus S$ is connected. Let U be an open set containing S. Then there is a rectilinear simple closed curve S' such that $S \subset I(S') \subset (S' \cup I(S')) \subset U$.

Proof.   Since S is compact, the distance from S to the complement of U is positive. Hence, we may choose the mesh of a brick partition of $\mathcal{E}^2$ to be small enough so that if a brick intersects S, then the brick lies entirely in U. Let M be the union of all of the partition elements that intersect S, and let R be a component of the frontier of M that contains S in its interior (why does such a component exist?). By the previous theorem, R is a rectilinear simple closed curve. If $I(R)$ is contained in U, we are done. If not, then there are a finite number of components of Fr M (other than R) that contain points of $\mathcal{E}^2 \setminus U$ in their interiors. We shall call such components holes. The holes satisfy the following conditions:

(i)   each hole is bounded by a rectilinear simple closed curve that lies in $I(R)$, and

(ii)   each hole contains points of $\mathscr{E}^2 \setminus U$.

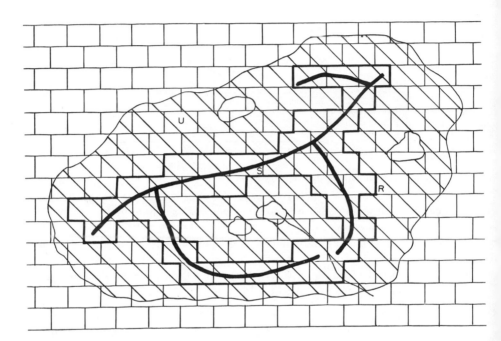

In each such hole choose a point of $\mathscr{E}^2 \setminus U$, and construct an arc from this point to ext $U \cap E(R)$ that misses S (recall that $\mathscr{E}^2 \setminus S$ is connected). The distance from S to the union of the arcs and the frontiers of the holes is positive. Hence, there is a brick partition of fine enough mesh so that each brick that hits S misses these sets (and is contained in U). We may assume that the union of the bricks that intersect S is contained in the union of the bricks of our original parirition that struck S. The boundary component of this new partition that contains S is the desired rectilinear simple closed curve S'.

(5.C.21)   **Theorem.**   There is no retraction of $\mathscr{E}^2$ onto a simple closed curve S.

Proof.   The proof mimics that of (5.B.6). Suppose that S is a simple closed curve in $\mathscr{E}^2$. By (5.B.5) and (5.B.9), S is not contractible. Suppose that

$r : \mathscr{E}^2 \to S$ is a retraction. Let p be any point in $\mathscr{E}^2$ that does not lie in S and connect each point of S with p.

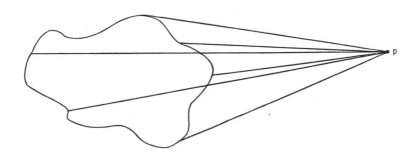

Define a homotopy $H : S \times I \to \mathscr{E}^2$ by $H(x,t) = tp + (1 - t)x$; note that H_0 is the identity and H_1 is a constant map. Then $rH : S \times I \to S$ is a contraction, which is impossible since S is not contractible.

The following theorem shows, however, that if S is a simple closed curve, then there is a neighborhood U of S that retracts to S. For every positive number t, $\mathscr{B}_t$ will denote the set $\{(x,y) \in \mathscr{E}^2 \mid x^2 + y^2 \leq t\}$. The reader should see that $\mathscr{B}_t$ is homeomorphic to $I \times I$ for each t (if not, consult (3.C.2)).

(5.C.22) **Theorem.** Suppose that S is a simple closed curve in $\mathscr{E}^2$. Then there is an open set U containing S and a map $r : U \to S$ that is the identity on S, i.e., S is a retract of U.

Proof. Let $h : S \to \mathscr{S}^1$ be a homeomorphism. Then h maps S into $\mathscr{B}_1$, and by (4.B.12) and (4.B.13), h can be extended to a continuous function H with domain $\mathscr{E}^2$. Let $V = (\text{int } \mathscr{B}^2) \setminus \mathscr{B}_{1/2}$, and let $r : V \to \mathscr{S}^1$ be a retract of V onto $\mathscr{S}^1$. Let $U = H^{-1}(V)$. Then $h^{-1}rH_{|U}$ is the desired retraction.

(5.C.23) **Theorem.** If S is a simple closed curve in $\mathscr{E}^2$, then $\mathscr{E}^2 \setminus S$ is not connected.

Proof. Suppose that S is a simple closed curve in $\mathscr{E}^2$ such that $\mathscr{E}^2 \setminus S$ is connected. We show that this leads to the existence of a retraction from $\mathscr{E}^2$ onto S, which contradicts (5.C.21).

By the previous theorem, there is an open set U containing S and a retraction r of U onto S. It follows from (5.C.20) that between S and U there is a rectilinear simple closed curve S' such that $S \subset (S' \cup I(S')) \subset U$. Furthermore, by (5.C.8), there is a space homeomorphism h which maps $S' \cup I(S')$ onto a rectangle R and its interior. Then $\phi = hr_{|(S' \cup I(S'))}h^{-1}_{|(R \cup I(R))}$

is a retraction of $R \cup I(R)$ onto the simple closed curve $h(s)$, and since ϕ can be extended easily to all of $\mathscr{E}^2$, we have contradicted (5.C.21).

Next we work toward determining the number of components of $\mathscr{E}^2 \setminus S$, and showing that S is the frontier of each component.

(5.C.24) Theorem. If A is the image of an arc in $\mathscr{E}^2$, then $\mathscr{E}^2 \setminus A$ is connected.

Proof. Suppose that $\mathscr{E}^2 \setminus A$ is not connected. Let R be a rectangle that contains A in its interior. Since $R \cup E(R)$ is connected, it must lie in some component K of $\mathscr{E}^2 \setminus A$. Note that $C = (\mathscr{E}^2 \setminus A) \setminus K$ is nonempty and its frontier lies in A. Let $x_0 \in C$. By (4.B.12), (4.B.13), and (4.B.19), there is a retraction $r : \mathscr{E}^2 \to A$. Let $p : \mathscr{E}^2 \setminus \{x_0\} \to R$ be the obvious projection of $\mathscr{E}^2$ from x_0 onto R. Define a map $h : \mathscr{E}^2 \to R$ by

$$h(x) = \begin{cases} pr(x) & \text{if} \quad x \in C \\ p(x) & \text{if} \quad x \in \mathscr{E}^2 \setminus C \end{cases}$$

Since $r_{|A}$ is the identity, we have that h is continuous; however, h is also a retraction of $\mathscr{E}^2$ onto R, which contradicts (5.C.21).

(5.C.25) Theorem. If S is a simple closed curve in $\mathscr{E}^2$ and U is a component of $\mathscr{E}^2 \setminus S$, then Fr $U = S$.

Proof. Certainly we have that Fr $U \subset S$, and that $\mathscr{E}^2 \setminus$ Fr U is not connected. If Fr $U \neq S$, then there is an arc A such that Fr $U \subset A \subset S$. Hence, it follows that $\mathscr{E}^2 \setminus A \subset \mathscr{E}^2 \setminus$ Fr U. Since $\mathscr{E}^2 \setminus A$ is connected, either $\mathscr{E}^2 \setminus A \subset U$ or $\mathscr{E}^2 \setminus A \subset \mathscr{E}^2 \setminus U$. However, $\overline{\mathscr{E}^2 \setminus A} = \mathscr{E}^2$, and consequently we have that either $\mathscr{E}^2 \subset \overline{U}$ or $\mathscr{E}^2 \subset \overline{\mathscr{E}^2 \setminus U}$. Since neither of these possibilities is viable, it must be the case that Fr $U = S$.

The following theorem will be used to ensure the existence of no more than two components in the complement of a simple closed curve. Its proof is left as an exercise.

(5.C.26) Theorem. Suppose that S is a simple closed curve in $\mathscr{E}^2$ and let x and y be points in S and $\mathscr{E}^2 \setminus S$, respectively. Then given any $\varepsilon > 0$, there is an arc A_ε such that

  (i) A_ε has end points y and z, where $z \in S$,
  (ii) $d(x,z) < \varepsilon$, and
  (iii) $A_\varepsilon \cap S = \{z\}$.

We are now ready to prove the principal result of this section.

(5.C.27) **Theorem (*Jordan Curve Theorem*).** Suppose that S is a simple closed curve in $\mathscr{E}^2$. Then $\mathscr{E}^2 \setminus S$ has exactly two components, and S is the frontier of each. Furthermore, exactly one of these components has compact closure.

Proof. It need only be established that $\mathscr{E}^2 \setminus S$ has at most two components. Suppose to the contrary that $\mathscr{E}^2 \setminus S$ has at least three components. We shall contradict (5.C.23) by constructing a simple closed curve S' in $\mathscr{E}^2$ such that $\mathscr{E}^2 \setminus S'$ is connected. Let C_1, C_2, and C_3 be components of $\mathscr{E}^2 \setminus S$. Then S is the frontier of each component. Let c_1 and c_2 be points in C_1 and C_2, respectively. In S select two points s_1 and s_2, and let $\delta = d(s_1,s_2)$. Careful repeated application of the previous theorem yields disjoint arcs $c_1 v_1$, $c_1 v_2$, $c_2 t_1$, and $c_2 t_2$ (see the following figure) such that

(i) $v_1, v_2, t_1, t_2 \in S$,
(ii) $d(s_i, v_i) < \delta/8$ for $i = 1, 2$,
(iii) $d(s_i, t_i) < \delta/8$ for $i = 1,2$,
(iv) for $i = 1,2$, $c_1 v_i$ lies in C_1 and $c_2 t_i$ lies in C_2, except for the end points v_i and t_i.

Let S' be the simple closed curve formed from the arcs $c_1 v_1$, $v_1 t_1$, $c_2 t_1$, $c_2 t_2$, $t_2 v_2$, and $c_1 v_2$. We show that the removal of S' fails to disconnect the plane.

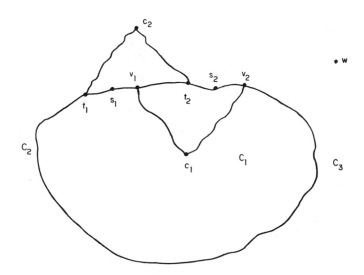

Suppose that z and z' are points in $\mathscr{E}^2 \setminus S'$. If both z and z' lie in C_1, remove a small arc A_2 from S' near c_2 (and lying in C_2). Then z and z' may be connected to a point $w \in C_3$ by arcs that miss $S' \setminus A_2$ (5.C.24). Note that these arcs intersect S before they meet S' (why?), and hence we may cross S to enter C_3 at the first point where these arcs strike S, and then continue to w. Thus z and z' are connected by an arc that misses S'. If $z \in C_1$ and $z' \in C_2$, then small arcs are removed from S' near c_1 and c_2, and a similar argument is applied. If at least one of the points z or z' lies outside C_1 and C_2, then an analogous but even easier argument may be given.

(5.C.28) *Exercise.* Supposc that K is a compact subset of a connected open subset $U \subset \mathscr{E}^2$. Show that there is a rectilinear simple closed curve R such that $K \subset I(R)$ and $R \subset \mathscr{U}$. [Hint: Use brick partitions and (5.C.8).]

D. THE SCHÖNFLIES THEOREM

Although a simple closed curve S is homeomorphic to the unit circle $\mathscr{S}^1$, does it necessarily follow that there is a homeomorphsm $h : \mathscr{E}^2 \to \mathscr{E}^2$ that carries S onto $\mathscr{S}^1$? In (5.D.9), we show that such a homeomorphism exists;

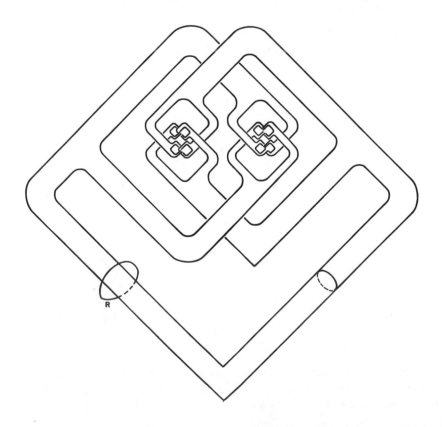

however, the proof is by no means trivial. In fact, similar theorems for higher dimensions are false without additional hypotheses. For instance, in Chapter 15, we construct an example of a 2-sphere that cannot be carried by a space homeomorphism onto the standard 2-sphere. One of the most famous examples of such a 2-sphere, the Alexander horned sphere, is shown on page 144.

Intuitively, the reason why a homeomorphism from the horned sphere onto the unit sphere cannot be extended to all of $\mathscr{E}^3$ is that it is impossible to form a "membrane" with the ring R as its boundary that misses the horned sphere. If however, a space homeomorphism $h : \mathscr{E}^3 \to \mathscr{E}^3$ mapping the horned sphere onto $\mathscr{S}^2$ were to exist, R would be "freed" and such a membrane could be found for $h(R)$ outside of $\mathscr{S}^2$. Then h^{-1} would carry this membrane back to one that is glued to R but fails to intersect the horned sphere, an impossibility.

(5.D.1) *Definition.* An n-sphere S (any space homeomorphic to $\mathscr{S}^n = \{(x_1, x_2, \ldots, x_n) \in \mathscr{E}^{n+1} \mid x_1^2 + \cdots + x_n^2 \leq 1\})$ lying in $\mathscr{E}^{n+1}$ is *tame* (in $\mathscr{E}^{n+1}$) if and only if there is a homeomorphism $h : \mathscr{E}^{n+1} \to \mathscr{E}^{n+1}$ such that $h(S) = \mathscr{S}^n$. Spheres that are not tame are *wild*.

The Alexander horned sphere is an example of a wild 2-sphere. In this terminology, the principal goal of this section is to show that all 1-spheres in the plane are tame. Bing [1963] ingeniously showed in his side approximation theorem that a tame sphere in $\mathscr{E}^3$ can be approximated either from the inside (the bounded complementary domain) or the outside (the unbounded complementary domain) by a polyhedral 2-sphere (roughly, one that is the union of a finite number of triangles).

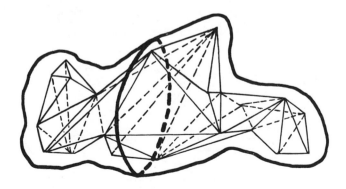

We shall eventually obtain an analogous theorem for 1-spheres (5.D.8) en route to a proof of the Schönflies theorem.

The existence of "necks" is a major obstacle in proving that an arbitrary simple closed curve may be mapped via a space homeomorphism onto $\mathscr{S}^1$. We will want to achieve very delicate approximations of simple closed curves by rectilinear curves; however, necks can prove to be troublesome, as the following figure might suggest. Note that points at the end of a neck may be quite far from the corresponding points on the approximating rectilinear curve.

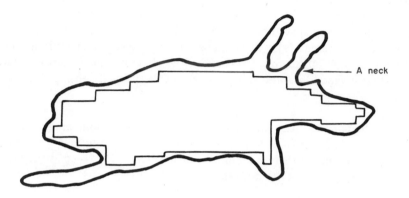

A degree of necking control is gained from the following theorem.

(5.D.2)   Theorem.   Suppose that S is a simple closed curve in $\mathscr{E}^2$ and let ε be a positive number. Then there is a positive number δ such that if $x,y \in S$ and $d(x,y) < \delta$, then at least one component of $S \setminus \{x,y\}$ has diameter less than ε. (Henceforth, we denote the smaller component of $S \setminus \{x,y\}$ by $L(x,y)$, and call δ a *necking number* for ε.)

Proof.   Let x be a point of S, and let C_x and $\hat{C}_x$ be the components containing x of $S_{\varepsilon/4}(x) \cap S$ and $S_{\varepsilon/2}(x) \cap S$, respectively.

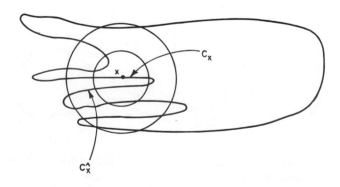

Choose $0 < \lambda_x < \varepsilon/2$ so that λ_x is less than the distance between C_x and $S \setminus \hat{C}_x$. Note that if $y \in C_x$ and z is a point of S such that $d(y,z) < \lambda_x$, then $z \in \hat{C}_x$ and at most one component of $S \setminus \{y,z\}$ has diameter greater than ε. Cover S with the sets $\{C_x \mid x \in X\}$, and select a finite subcover $\{C_{x_1}, C_{x_2}, \ldots, C_{x_n}\}$. Let δ_1 be a Lebesgue number for this cover, and let $\delta = \min\{\delta_1, \lambda_{x_1}, \ldots, \lambda_{x_n}\}$. Then it is easy to verify that δ is the required positive number.

(5.D.3) Notation. If h is a homeomorphism between subsets A and B of $\mathscr{E}^2$, then $|h|$ will denote $\sup\{d(a,h(a)) \mid a \in A\}$. We shall let $H(A,B)$ denote $\inf\{|h| \mid h \text{ is a homeomorphism from } A \text{ onto } B\}$.

(5.D.4) Theorem. Suppose that S is a simple closed curve in $\mathscr{E}^2$ and that ε is a positive number. There is a number $\delta > 0$ such that

(i) if S' is any simple closed curve with $H(S,S') < \delta$, and
(ii) if x and y are points in S' with $d(x,y) < \delta$,

then diam $L(x,y)$ is less than ε.

Proof. Corresponding to $\varepsilon/3$ there is a necking number $\hat{\delta}$, $0 < \hat{\delta} < \varepsilon$, such that if $d(x,y) < \hat{\delta}$ where $x,y \in S$, then diam $L(x,y) < \varepsilon/3$. Let $\delta = \hat{\delta}/3$ and suppose that S' is a simple closed curve such that $H(S,S') < \delta$. Let $h : S \to S'$ be any homeomorphism with $d(s,h(s)) < \delta$ for each $s \in S$. Suppose that w and z are points of S' and $d(w,z) < \delta$. We show that diam $L(w,z) < \varepsilon$. First note that $d(h^{-1}(w),h^{-1}(z)) < d(h^{-1}(w),w) + d(w,z) + d(z,h^{-1}(z)) < 3\delta$, and hence diam $L(h^{-1}(w),h^{-1}(z)) < \varepsilon/3$. Now $h(L(h^{-1}(w),h^{-1}(z))) = A$ is clearly an arc from w to z, and we need only establish that its diameter is less than ε. Suppose that $s,t \in A$. Then $d(s,t) < d(s,h^{-1}(s)) + d(h^{-1}(s),h^{-1}(t)) + d(h^{-1}(t),t) < \delta + \varepsilon/3 + \delta < \varepsilon$, and thus diam $A < \varepsilon$.

The proofs of the following important corollaries are left as nontrivial exercises, and they should give a good indication of the reader's grasp of the work done thus far.

(5.D.5) Definition. If A is a closed subset of a connected, locally connected topological space X, then any component of $X \setminus A$ is called a *complementary domain* of A.

(5.D.6) Corollary. Suppose that S is a simple closed curve in $\mathscr{E}^2$ and that $\varepsilon > 0$. There is a $\delta > 0$ such that if S' and S'' are rectilinear simple closed curves in a complementary domain of S with $S' \subset I(S'')$, $H(S,S') < \delta$, and $H(S,S'') < \delta$, then for each x,y in the annulus T determined by S' and S'', where $d(x,y) < \delta$, there is a rectilinear arc A in T such that

    (i)   diam $A < \varepsilon$, and

    (ii) A has x and y for end points.

(5.D.7) *Corollary.* Suppose that S is a simple closed curve in $\mathscr{E}^2$ and let $\varepsilon > 0$ be given. There is a $\delta > 0$ such that if S' and S'' satisfy the conditions of the previous corollary, then the annulus between S' and S'' may be written as the union of a finite number of rectilinear simple closed curves $D_1, D_2, \ldots,$ D_n together with their interiors, where

    (i) $I(D_i) \cap I(D_j) = \varnothing$ for $i \neq j$,

    (ii) diam $D_i < \varepsilon$,

    (iii) $D_i \cap D_j = \varnothing$ unless $|i - j| \leq 1$, or $|i - j| = n - 1$, and

    (iv) $D_i \cap D_{i+1}$ (and $D_1 \cap D_n$) is an arc with one end point on S', one end point on S'', and the arc minus its end points lies between S' and S''.

(5.D.8) *Theorem.* Suppose that S is a simple closed curve in $\mathscr{E}^2$, ε is a positive number, and U is a complementary domain of $\mathscr{E}^2 \setminus S$. Then there is a rectilinear simple closed curve S' contained in U such that

    (i) $H(S,S') < \varepsilon$, and

    (ii) if $U = I(S)$, then $S \subset E(S')$, and if $U = E(S)$, then $S \subset I(S')$.

    Proof. During the proof, we use the following notation. If A is the image of an arc in $\mathscr{E}^2$ with end points x and y, then we write $A = [x,y]$, and $A \setminus \{x,y\}$ will be denoted by (x,y). We first give an outline of the proof and then provide some of the necessary details.

    1. Suppose that $U = E(S)$. Let ε be a positive number. For each $x \in S$, rectilinear arcs are constructed from a point $p_x \in E(S)$ to points q_x and r_x that lie on "opposite sides" of x. The resulting simple closed curve formed by the arcs $[p_x,q_x]$, $[q_x,r_x]$, and $[r_x,p_x]$ will have diameter less than ε, and furthermore, it will intersect S only in the arc $[q_x,r_x]$.

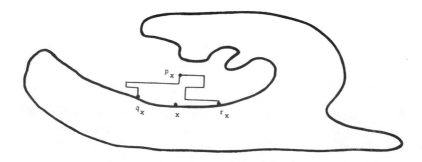

2.   Since $\{(q_x, r_x) \mid x \in S\}$ is an open cover of the compact set S, there is an irreducible finite subcover, $\mathscr{A} = \{(q_{x_i}, r_{x_i}) \mid i = 1, 2, \ldots, n\}$. ($\mathscr{A}$ is irreducible in the sense that if $(q_{x_j}, r_{x_j}) \in \mathscr{A}$, then (q_{x_j}, r_{x_j}) is not contained in the $\bigcup_{\substack{i \neq j \\ i=1}}^{n} (q_{x_i}, r_{x_i})$.)

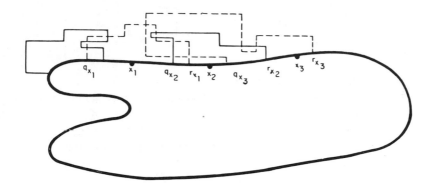

3.   For $i = 1, 2, \ldots, n$, let T_i be the simple closed curve formed by the union of $[p_{x_i}, q_{x_i}]$, $[q_{x_i}, r_{x_i}]$, and $[r_{x_i}, p_{x_i}]$. The rectilinear parts of these simple closed curves are fixed up in such a way that nonadjacent T_i's do not intersect, and that the rectilinear portions of adjacent T_i's will intersect at only one point.

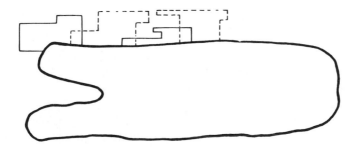

4.   We show that the rectilinear simple closed curve running around the outside of the T_i's is the desired one.

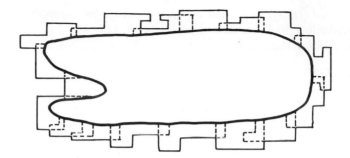

Now we fill in some of the details.

1.   Let $x \in S$ and let C_x be the component of $S \cap S_{\varepsilon/2}(x)$ that contains x. Select points a_x and b_x in C_x that lie on opposite sides of x. This may be done since there is a (uniformly continuous) homeomorphism $h : S \rightarrow \mathscr{S}^1$, and in $h(C_x)$ points $\hat{a}_x$ and $\hat{b}_x$ certainly may be chosen that lie on either side of $h(x)$. These points are mapped back by h^{-1} to a_x and b_x. (We are assuming that ε is very small compared to the diameter of S.)

Let D_x be the component of $S_{\varepsilon/2} \setminus (S \setminus (a_x,b_x))$ that contains (a_x,b_x), and select points q'_x and r'_x such that $q'_x \in (a_x,x)$ and $r'_x \in (x,b_x)$. Let $\lambda_x = \min\{d(q'_x,x), d(r'_x,x)\}$ and let δ_x be a necking number for λ_x, i.e., if $w,z \in S$ and $d(w,z) < \delta_x$, then diam $L(w,z) < \lambda_x$.

Now select a point $p_x \in D_x \cap E(S)$ and let Q_x and R_x be rectilinear arcs in $E(S) \cap D_x$ with initial point p_x and with terminal points in $S_{\delta_x}(q'_x)$ and $S_{\delta_x}(r'_x)$, respectively. From the respective terminal points, send out straight line segments toward q'_x and r'_x, and let q_x and r_x be the first points in S that these segments encounter. The reader should now be able to devise a method to obtain arcs $[p_x,q_x]$ and $[p_x,r_x]$ that have only the point p_x in common and that lie (with the exception of q_x and r_x) in $D_x \cap E(S)$.

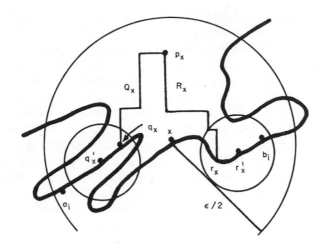

2. This step should need little additional elucidation. Relabel the arcs $[q_{x_i}, r_{x_i}]$ so that, as we go around S in one direction, points are encountered in the following order:

$$q_{x_1}, r_{x_n}, q_{x_2}, r_{x_1}, q_{x_3}, r_{x_2}, q_{x_4}, r_{x_3}, q_{x_5}, r_{x_4}, \cdots, q_{x_n}, r_{x_{n-1}}, q_{x_1}$$

For each i, let T_i denote the simple closed curve composed of the arcs $[p_{x_i}, q_{x_i}]$, $[q_{x_i}, r_{x_i}]$, and $[r_{x_i}, p_{x_i}]$. Clearly, diam $T_i < \varepsilon$, for all i.

3. With the notation thus far developed, T_i and T_j ($i \neq j$) are said to be *adjacent* if and only if $|i - j| = 1$, or $i = 1$ and $j = n$, or $j = 1$ and $i = n$; otherwise, they are *nonadjacent*. We alter the T_i's to meet the conditions described previously. All repairs are to be done in such a way that the diameters of T_i remain less than ε and that each T_i still intersects S only in the arc $[q_{x_i}, r_{x_i}]$. To illustrate the basic procedure used, we demonstrate how T_1 and T_3 may be pulled apart.

First, note that we may assume that T_1 and T_3 intersect only a finite number of times, since slight adjustments of say T_3 will alleviate the situation indicated in the next figure.

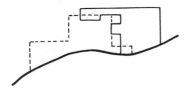
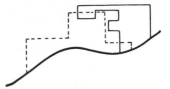

We now induct on the number of points in the intersection of T_1 and T_3. Obviously, there is nothing to do if $n = 0$ (or 1), where n represents the number of points of intersection. Assume, then, that T_1 and T_3 may be separated in the desired fashion whenever there are n or fewer points of intersection, and suppose that T_1 and T_3 intersect in $n + 1$ (actually it must be $n + 2$) points. We construct another T_3 that meets T_1 in fewer points.

Begin at $q_{x_3} \in T_3$ and proceed toward r_{x_3} (along the rectilinear portion of T_3). Let s be the first point found in common with T_1. At this point, there must be some segment of T_1 that crosses T_3. Starting at s, move along T_1 (in $I(T_3)$) until T_1 leaves $I(T_3)$ at some point s'. Now T_3 is altered as indicated by the dotted line in the next figure. We again start at q_{x_3}, but this time refuse to enter T_1 at s; instead we move toward s', always staying close to T_1, and finally we continue along T_3 until r_{x_3} is reached. The new curve has two intersections fewer with T_1, and induction takes care of the remaining ones.

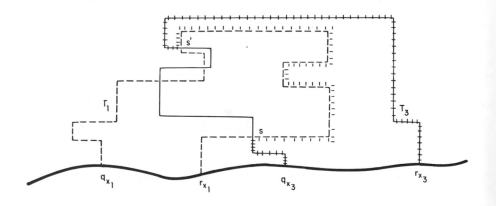

This process is repeated for all pairs of nonadjacent T_i's, and a similar procedure is applied to adjacent ones, where of course the goal is to finish with just one intersection. If sufficient care is exercised, the diameters of the T_i's will remain less than ε.

4.   We now have the following situation:

Let R be the rectilinear curve indicated in the following figure. A homeomorphism from S onto R is defined as follows.

The arc $[q_{x_1}, r_{x_1}]$ is mapped onto $[t_n, t_1]$, the arc $[r_{x_1}, q_{x_3}]$ is mapped onto $[t_1, t_2]$, the arc $[q_{x_3}, r_{x_3}]$ is mapped onto $[t_2, t_3]$, the arc $[r_{x_3}, q_{x_4}]$ is mapped onto $[t_3, t_4]$, . . . , and finally the arc $[r_{x_{n-1}}, q_{x_1}]$ is sent to $[t_{n-1}, t_n]$. By the construction, it is clear that this homeomorphism moves no point more than ε, which completes the proof.

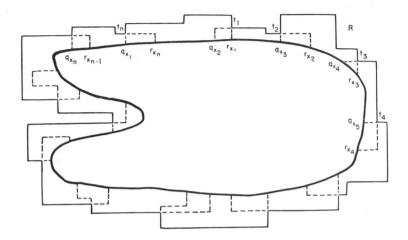

(5.D.9)   Theorem (The 2-Dimensional Schönflies Theorem).   Suppose that
S is a simple closed curve in $\mathscr{E}^2$. Then there is a homeomorphism $h : \mathscr{E}^2 \to \mathscr{E}^2$
such that $h(S) = S^1$.

Proof.   First, we define a homeomorphism from $S \cup I(S)$ onto $\mathscr{B}^2$.
To do this, we approximate S from the inside with an increasing sequence of
rectilinear simple closed curves. We then subdivide the annular region between succeeding rectilinear curves and map the pieces onto pie-shaped pieces
contained in annular regions formed by concentric circles in the interior of
$\mathscr{S}^1$. If this is done with appropriate care, a homeomorphism may be found
from S plus its interior onto the unit disk.

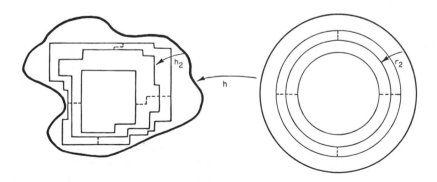

First define the following subsets of $\mathscr{S}^1$ (the points are indicated by
their radian measure): $A_1 = \{0,\pi\}$; $A_2 = \{0,\pi/2,\pi,3\pi/2\}$; $A_3 = \{0,\pi/4,\pi/2,$
$3\pi/4,\pi,5\pi/4,3\pi/2,7\pi/4\}$; etc.

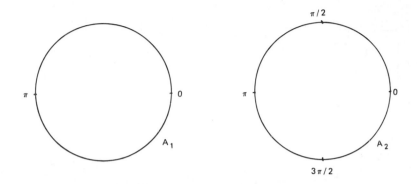

Let h be any homeomorphism from $\mathscr{S}^1$ onto S, and for each $i = 1$, 2, ... let $\lambda_i = \min\{d(h(x),h(y)) \mid x,y \in A_i\}$. Let $\{\varepsilon_i\}$ be a decreasing sequence of positive numbers such that $\varepsilon_i < \min\{\lambda_i, 1/2^i\}$ for $i = 1, 2, \cdots$. Apply (5.D.4), (5.D.6), (5.D.7), and (5.D.8) to obtain a sequence of rectilinear simple closed curves $\{S_i\}$ and a sequence of homeomorphisms $\{h_i\}$ such that

(i) $S_i \subset I(S_{i+1}) \subset (S_{i+1} \cup I(S_{i+1})) \subset I(S)$ for each $i = 1, 2, \ldots$,

(ii) $h_i : S \to S_i$ and for each $x \in S$, $d(h_i(x),x) < \varepsilon_i/2$, and

(iii) if $x_i \in S_i$ and $x_{i+1} \in S_{i+1}$ and $d(x_i,x_{i+1}) < \varepsilon_i$, then there is an arc A_i with end points x_i and x_{i+1} that (except for the end points) lies in the open region between S_i and S_{i+1} and diam $A_i < 1/i$.

Let $\{T_i\}$ be a sequence of concentric circles in $\mathscr{S}^1$, where each T_i has radius $1 - 1/(2i)$. For each i, define $r_i : \mathscr{S}^1 \to T_i$ to be the radial projection.

We begin the construction of a homeomorphism from $S \cup I(S)$ onto the unit disk by defining $\hat{h} : \bigcup_{i=1}^{\infty} S_i \to \bigcup_{i=1}^{\infty} T_i$ by $\hat{h}(s_i) = r_i h^{-1} h_i^{-1}(s_i)$ for each $s_i \in S_i$. Extend h to the interiors of the various annuli as follows. Map $r_1(0) = a_1^1$ and $r_1(\pi) = a_2^1$ into S_1 by setting $b_1^1 = \hat{h}^{-1}(a_1^1)$ and $b_2^1 = \hat{h}^{-1}(a_2^1)$. Similarly, the points of T_2, $a_1^2 = r_2(0)$, $a_2^2 = r_2(\pi)$, $a_3^2 = r_2(\pi/2)$, and $a_4^2 = r_2(3\pi/2)$, are mapped to points b_1^2, b_2^2, b_3^2, and b_4^2 of S_2 by $\hat{h}^{-1}$. Rectilinear arcs of diameter less than $1/1$ may be found that join b_1^1 to b_1^2 and b_2^1 to b_2^2 (by (ii) and (iii) above). Call these arcs B_1^1 and B_2^1. First extend $\hat{h}$ by mapping the arcs B_1^1 and B_2^1 onto the arcs A_1^1 and A_2^1 indicated in the next figure. Then use (5.C.11) to further extend $\hat{h}$ to the two 2-cells bounded by S_1, S_2, B_1^1, and B_2^1 so that they are mapped homeomorphically onto the corresponding cells in the disk bounded by T_1, T_2, A_1^1, and A_2^1. (Clearly (5.C.11) applies, even though the T_i's are not rectilinear. The skeptical reader may systematically replace the T_i's and $\mathscr{S}^1$ by squares in this proof.)

The points $a_1^3 = r_3(0)$, $a_2^3 = r_3(\pi)$, $a_3^3 = r_3(\pi/2)$, $a_4^3 = r_3(3\pi/2)$, $a_5^3 = r_3(\pi/4)$, $a_6^3 = r_3(3\pi/4)$, $a_7^3 = r_3(5\pi/4)$, and $a_8^3 = r_3(7\pi/4)$ are now mapped

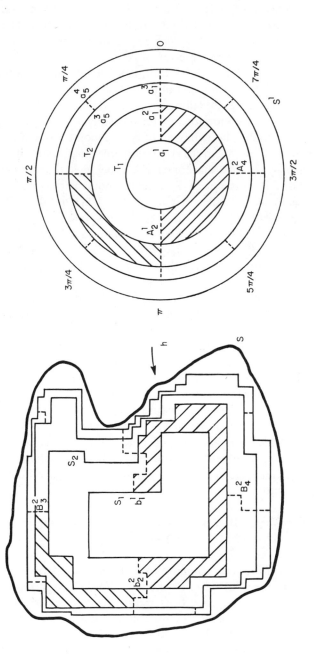

by $\hat{h}^{-1}$ onto points b_i^3, $i = 1, 2, \ldots, 8$ in S_3. Arcs of diameter less than $\frac{1}{2}$ are found and mapped onto the obvious candidates in the disk. The four pieces of the annular region between S_2 and S_3 are carried by an extension of $\hat{h}$ to the most likely wedges in the interior of S^1.

This procedure is repeated for each positive integer i, and the end result is a homeomorphism $\hat{h}$ from $I(S)$ onto $I(\mathscr{S}^1)$. Letting $\hat{h} = h^{-1}$ on S, we at least have constructed a function from $S \cup I(S)$ onto $\mathscr{S}^1 \cup I(\mathscr{S}^1)$ that is clearly 1–1, onto, and continuous except possibly at points of S. To remove any doubts concerning the behavior of $\hat{h}$ at a point $x \in S$, we show that if $\{x_i\}$ is a sequence of points in $I(S)$ that converges to x, then $\{\hat{h}(x_i)\}$ converges to $\hat{h}(x)$. It may be assumed that for each i, x_i belongs to the annulus bounded by S_i and S_{i+1}. If x_i is not in S_i, select a point $c_i \in S_i$ (c_i is in the same rectilinear region as x_i) such that $d(x_i, c_i) < 1/i$. Consider $d(\hat{h}(x), \hat{h}(x_i)) < d(\hat{h}(x), \hat{h}(c_i)) + d(\hat{h}(c_i), \hat{h}(x_i))$. Clearly, for large i, one need only worry about $d(\hat{h}(x), \hat{h}(c_i))$. Let $\varepsilon > 0$ be given. Corresponding to ε there is $\delta > 0$ such that if s_1 and $s_2 \in S$ and $d(s_1, s_2) < \delta$, then $d(h^{-1}(s_1), h^{-1}(s_2)) < \varepsilon/2$. Furthermore, there is a positive integer N such that for $i > N$, $d(h_i^{-1}(y), y) < \delta/4$ for each $y \in S_i$ and $d(r_i(x), x) < \varepsilon/2$ for each $x \in S$.

Of course, $\hat{h}(c_i)$ is equal to $r_i h^{-1} h_i^{-1}(c_i)$. We have now that for $i > N$, $d(h_i^{-1}(c_i), c_i) < \delta/4$ and $d(c_i, x) < d(c_i, x_i) + d(x_i, x) < \delta/4 + \delta/4$. Therefore, $d(h_i^{-1}(c_i), x) < d(h_i^{-1}(c_i), c_i) + d(c_i, x) < 3\delta/4$, and thus $d(h^{-1}h_i^{-1}(c_i), h^{-1}(x)) < \varepsilon/2$. Finally then we have that $d(r_i h^{-1} h_i^{-1}(c_i), h^{-1}(x)) < d(r_i h^{-1} h_i^{-1}(c_i), h^{-1}h_i^{-1}(c_i)) + d(h^{-1}h_i^{-1}(c_i), h^{-1}(x)) < \varepsilon$, and the continuity of $\hat{h}$ follows (provided that the reader understands why it was legitimate to use the sequence $\{c_i\}$ instead of the original sequence $\{x_i\}$).

A completely analogous procedure may be used to obtain a sequence $\{S_i'\}$ of rectilinear simple closed curves that converges to S from the exterior of S and whose members are mapped onto a sequence of concentric circles with radius $1 + 1/i$ and center at the origin. It is immediate from (5.C.11) that these homeomorphisms may be extended to a space homeomorphism that carries S onto $\mathscr{S}^1$.

(5.D.10)   *Corollary.*   If S is a simple closed curve lying in $\mathscr{E}^2$, then there is a homeomorphism $H : (S \cup I(S)) \rightarrow \mathscr{B}^2$.

An important property common to all Euclidean spaces $\mathscr{E}^n$ is the following: If U is an open subset of $\mathscr{E}^n$ and $h : U \rightarrow \mathscr{E}^n$ is an embedding, then $h(U)$ is open in $\mathscr{E}^n$. This property is frequently referred to as *invariance of domain*, and it is not difficult to construct examples that show that not all topological spaces satisfy this condition. Invariance of domain turns out to be one of the most powerful features of $\mathscr{E}^n$, and its proof is appropriately sophisticated. The Schönflies theorem may be used to show that invariance of domain holds in $\mathscr{E}^2$. That the invariance of domain property also holds in $\mathscr{E}^n$ for $n > 2$ is proven in Chapter 15.

(5.D.11) Exercise. Show that if U is an open subset of $\mathscr{E}^2$ and if $h : U \to \mathscr{E}^2$ is an embedding, then $h(U)$ is open in $\mathscr{E}^2$.

E. THE ANNULUS THEOREM

(5.E.1) Exercise. Suppose that R is a rectilinear simple closed curve in $\mathscr{E}^2$. Show that there is a compact set K and a homeomorphism $h : \mathscr{E}^2 \to \mathscr{E}^2$ such that $h(R) = \mathscr{S}^1$ and $h_{|\mathscr{E}^2 \setminus K} = id$. [Hint: Note that the homeomorphism constructed in (5.C.8) is equal to the identity off a compact set. Inscribe everything there in a large circle and project the rectangle obtained onto the circle.]

(5.E.2) Exercise. Show that the homeomorphism obtained in the Schönflies Theorem can be chosen so that there exists a compact set $K \subset \mathscr{E}^2$ such that $h_{|\mathscr{E}^2 \setminus K}$ is the identity.

(5.E.3) Exercise. Suppose that S and S' are simple closed curves in $\mathscr{E}^2$. Let U be an open subset of $\mathscr{E}^2$ such that U is homeomorphic to $\mathscr{E}^2$ and $S \cup S' \subset U$. Show that there exists a homeomorphism $h : \mathscr{E}^2 \to \mathscr{E}^2$ such that $h(S) = S'$ and $h_{|\mathscr{E}^2 \setminus U} = id$.

(5.E.4) Exercise. Suppose that S and S' are simple closed curves in $\mathscr{E}^2$ such that $S \subset I(S')$. Show that there is a homeomorphism $h : \mathscr{E}^2 \to \mathscr{E}^2$ such that $h(S') = \mathscr{S}^1$ and $h(S)$ is a circle with radius $\frac{1}{2}$ centered at the origin.

(5.E.5) Exercise (Annulus Theorem). Suppose that S and S' are simple closed curves in $\mathscr{E}^2$ such that $S' \subset I(S)$. Show that $\overline{I(S) \cup E(S')}$ is an annulus.

PROBLEMS

Section A

1. A *1-manifold with boundary* is a separable metric space M with the property that each point of M has a neighborhood homeomorphic to the closed unit interval. Classify all connected 1-manifolds with boundary.

2.* Find an example of a space X such that every point has a neighborhood homeomorphic to $(0,1)$ but such that X is not T_2.

3.* Let Ω be the first uncountable ordinal. Order $L = [0,\Omega) \times [0,1)$ by $(\alpha,r) < (\beta,s)$ if and only if (i) $\alpha < \beta$ or (ii) $\alpha = \beta$ and $r < s$. Then L with the corresponding order topology is called the *long line*. Essentially, we have filled in the gaps between the ordinals by the intervals $(0,1)$. Prove:
   (a) Each point of L has a neighborhood homeomorphic to $(0,1)$.
   (b) L is T_2.
   (c) L is connected.
   (d) L is not compact (in fact, not Lindelöf).
   (e) L is countably compact.
   (f) L is not separable.

Section B

1. Show that $\mathbf{I} \times \mathbf{I}$ has the fixed point property.
2. Show that there is no retraction of $\mathscr{B}^2$ onto $\mathscr{S}^1$.
3. Show that if $f : \mathscr{B}^2 \to \mathscr{E}^2$ is a continuous function such that either $f(x) = (0,0)$ or x does not lie on the ray from the origin through $f(x)$, then f has a fixed point.
4. For $n = 1, 2, 3, \ldots,$ let $A_n = \{(x,y) \in \mathscr{E}^2 \mid x = 1/n,\ 0 \le y \le 1\}$. Let $A_0 = \{(x,y) \in \mathscr{E}^2 \mid x = 0$ and $0 \le y \le 1$ or $0 \le x \le 1$ and $y = 0\}$. Let $X = \bigcup_{i=0}^{\infty} A_i$. Show that X is contractible to the point $(0,1)$, but that there is no contraction H such that $H((0,1),t) = (0,1)$ for all t.
5. Show that contractible spaces are path connected.
6. Show that any space can be embedded in a contractible space.
7. Suppose that U is a bounded open subset of $\mathscr{E}^2$ and that $h : \mathscr{E}^2 \to \mathscr{E}^2$ is a homeomorphism such that $h_{|\mathrm{Fr}\ U} = id$. Show that $h(U) = U$.

Sections C and D

1.* A polygonal simple closed curve in $\mathscr{E}^2$ is a simple closed curve in $\mathscr{E}^2$ that consists of the finite union of straight line segments (not necessarily horizontal or vertical).
   (a) Show that by a change of coordinates, things can be arranged so that each pair of vertices have distinct x and y coordinates.
   (b) Divide the bounded region into a finite number of squares and triangles by means of horizontal and vertical lines through each vertex, and prove that the bounded region is homeomorphic with a 2-cell. (This argument may be generalized to polyhedral 2-spheres in $\mathscr{E}^3$ (Moise [1952]).

2. Show that n disjoint simple closed curves in $\mathscr{E}^2$ have $n + 1$ complementary domains.

3. Let $\gamma, \gamma_1, \gamma_2, \ldots, \gamma_n$ be the images of disjoint simple closed curves in $\mathscr{E}^2$ such that for $i = 1, \ldots, n$, γ_i is contained in the bounded complementary domain D of γ. Let D_i denote the bounded complementary domain of γ_i and define $P = (\gamma \cup D) \setminus (\bigcup_{i=1}^{n} D_i)$. Show that P can be written as the union of a finite number of disks with disjoint interiors.

4. A map $f : X \to Y$ is a *local homeomorphism* if and only if for each $x \in X$, there is an open set U containing x such that $f_{|U} : U \to f(U)$ is a homeomorphism and $f(U)$ is open in Y. Show that $f : \mathscr{E}^1 \to \mathscr{S}^1$ defined by $f(x) = (\cos x, \sin x)$ is a local homeomorphism but not a homeomorphism.

5.* Let $\mathscr{B}^2 = \{(x,y) \in \mathscr{E}^2 \mid x^2 + y^2 \le 1\}$ and suppose that $f : \mathscr{B}^2 \to \mathscr{B}^2$ is a continuous function with the properties that

   (i) f maps $\mathscr{S}^1$ ($= \text{Fr } \mathscr{B}^2$) homeomorphically onto itself, and
   (ii) $f_{|\text{int } \mathscr{B}^2}$ is a local homeomorphism (see problem C.4).

   (a) Show that f is onto (recall (5.C.21)).
   (b) Suppose that $x \in \text{int } \mathscr{B}^2$, $y \in \mathscr{S}^1$, and that A is an arc between x and y. Show that $f^{-1}(A)$ consists of arcs (possibly one) whose only intersection is at $f^{-1}(y)$.
   (c) Suppose that $f(x) = f(x') = w$, where $x, x', w \in \text{int } \mathscr{B}^2$. Let w^* be the end point on $\mathscr{S}^1$ of the radius G (on any radius G in case $w = (0,0)$) that contains w. Select arbitrary points w_1 and w_2 on $\mathscr{S}^1$ that are distinct from w^* and connect these to w with straight line segments B and C, respectively. Let $\hat{x} = f^{-1}(w^*)$, $x_1 = f^{-1}(w_1)$, $x_2 = f^{-1}(w_2)$ and find arcs $A_x, A_{x'}, B_x, B_{x'}, C_x, C_{x'}$ with end points $\{x,\hat{x}\}$, $\{x',\hat{x}\}$ $\{x,x_1\}$, $\{x',x_1\}$, $\{x,x_2\}$, $\{x',x_2\}$, respectively such that $f(A_x) = f(A_{x'})$, $f(B_x) = f(B_{x'})$, and $f(C_x) = f(C_{x'})$. Finally, let D_1 and D_2 be the arcs on $\mathscr{S}^1$ with end points $\{x_1,\hat{x}\}$ and $\{x_2,\hat{x}\}$ and whose intersection consists of precisely $\hat{x}$.
      (1) Show that x' belongs to the bounded complementary domain of $D_1 \cup B_x \cup C_x \cup D_2$.
      (2) Show that x' does not belong to the bounded complementary domain of either $A_x \cup D_1 \cup B_x$ or $A_x \cup C_x \cup D_2$.
      (3) Deduce from (1) and (2) that f must be 1–1.
      (4) Show that f is a homeomorphism.

6.* Show that there are only a countable number of mutually disjoint triods in the plane (Pittman [1970]) [Hint: Let $\mathscr{D} = \{D_i \mid i \in \mathbf{Z}^+\}$ be a basis for $\mathscr{E}^1$ consisting of a countable number of open disks. If T is a triod, let $D_{T,1} \in \mathscr{D}$ such that $0 \in D_{T,1}$ and $D_{T,1} \cap \{X,Y,Z\} = \varnothing$.

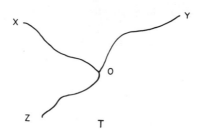

Let X_T, Y_T, Z_T be the first points of $0X \cap Bd\ D_{T,1}$, $0Y \cap Bd\ D_{T,1}$, $0Z \cap Bd\ D_{T,1}$. Show that $D_{T,1} \setminus (0X_T \cup 0Y_T \cup 0Z_T) = U_{T,2} \cup U_{T,3} \cup U_{T,4}$, where the $U_{T,i}$'s are disjoint open sets. For $i = 2$, 3, 4 select $D_{T,i} \in \mathscr{D}$ such that $D_{T,i} \subset U_{T,i}$. Thus, a quadruple of sets in $\mathscr{D}$ is associated with T.]

7.   Show that no arc separates $\mathscr{S}^2$.

8.   Show that each simple closed curve S in $\mathscr{S}^2$ has two complementary domains. Furthermore show that the closure of each complementary domain of S is homeomorphic with $\mathscr{B}^2 = \{(x,y) \mid x^2 + y^2 \leq 1\}$.

9.* A *theta curve* is any space homeomorphic to the union of the images of three arcs, each two of which intersect precisely at their end points. Show that each theta curve in $\mathscr{E}^2$ separates $\mathscr{E}^2$ into exactly three components.

Chapter 6

THE PRODUCT TOPOLOGY AND INVERSE SYSTEMS

A. THE PRODUCT TOPOLOGY REVISITED

The product topology associated with a finite number of topological spaces has been repeatedly encountered in the previous chapters. Our present task centers on extending the definition (1.D.7) given in Chapter 1. Recall from set theory that if $\mathscr{A} = \{X_\alpha \mid \alpha \in \Lambda\}$ is a family of sets, then the Cartesian product of the X_α's, $\prod_{\alpha \in \Lambda} X_\alpha$, is defined to be $\{f : \Lambda \to \bigcup X_\alpha \mid f(\alpha) \in X_\alpha \text{ for each } \alpha \in \Lambda\}$. The axiom of choice guarantees that $\prod_{\alpha \in \Lambda} X_\alpha \neq \varnothing$ if and only if $X_\alpha \neq \varnothing$ for each α. We shall denote a typical element $f \in \prod_{\alpha \in \Lambda} X_\alpha$ by $\{x_\alpha\}$, where it is understood that if $\alpha \in \Lambda$, then $f(\alpha) = x_\alpha$. The problem is to find a suitable topology for $\prod_{\alpha \in \Lambda} X_\alpha$. An initial choice for such a topology (considering the definition given when Λ is finite) might be that obtained from a basis whose elements are of the form $\prod_{\alpha \in \Lambda} U_\alpha$, where U_α is an open set in X_α for each $\alpha \in \Lambda$. However, this topology (frequently called the *box topology*) does not serve as a suitable product topology. It has more open sets than is necessary or desirable.

What, then, should be the nature of open sets in $\prod_{\alpha \in \Lambda} X_\alpha$? In the finite case, it was shown that the projection maps $p_\alpha : \prod_{\alpha \in \Lambda} X_\alpha \to X_\alpha$ are continuous. We would certainly like this to hold for the infinite product as well. However, if

161

continuity of these maps were the only desideratum for determining open sets, then assignment of the discrete topology to the product space wou.d settle the matter. However, the discrete topology has too many open sets to be of any interest. Thus, what we want is a topology for the product space that has a minimum number of open sets, but for which projection maps are still continuous. Fortunately, there is no problem in doing this. In order for p_α to be continuous, $p_\alpha^{-1}(U_\alpha)$ must be open for each open subset U_α of X_α. Let $\mathscr{S} = \{p^{-1}(U_\alpha) \mid \alpha \in \Lambda$ and U_α is open in $X_\alpha\}$. Since $\mathscr{S}$ is not closed under finite intersections, it fails to be a topology or even a basis for a topology. Nevertheless, this deficiency is easily remedied if we add to $\mathscr{S}$ all finite intersections of members of $\mathscr{S}$. The resulting family of sets meets all the requisites for a basis for a topology. Observe that basic open sets are of the form $\bigcap_{i=1}^{n} p_{\alpha_i}^{-1}(U_{\alpha_i})$ (frequently denoted by $U_{\alpha_1} \times \cdots \times U_{\alpha_n} \times \Pi \{X_\alpha \mid \alpha \in \Lambda$ and $\alpha \neq \alpha_1, \ldots, \alpha_n\}$). Clearly the definition given for the finite case (1.D.7) is simply a special case of the present one.

(6.A.1) Definition. Suppose that $\{X_\alpha \mid \alpha \in \Lambda\}$ is a family of topological spaces. Then the *product topology* for $\Pi_{\alpha \in \Lambda} X_\alpha$ is that topology which has as a basis sets of the form $\bigcap_{i=1}^{n} p_{\alpha_i}^{-1}(U_{\alpha_i})$, where U_{α_i} is open in X_{α_i}.

(6.A.2) Notation. If $\{X_\alpha \mid \alpha \in \Lambda\}$ is a collection of topological spaces, then (unless otherwise stated) $\Pi_{\alpha \in \Lambda} X_\alpha$ will denote the product of the X_α with the product topology.

In the first exercise that follows, the reader is asked to confirm that the product topology is indeed the smallest topology for which continuity of the projection functions is maintained. The second and third exercises give some additional justification for this choice of topology.

(6.A.3) Exercise. Suppose that $\{X_\alpha \mid \alpha \in \Lambda\}$ is a collection of topological spaces and that $\Pi_{\alpha \in \Lambda} X_\alpha$ has the product topology $\mathscr{U}$. Show that if $\mathscr{V}$ is another topology for $\Pi_{\alpha \in \Lambda} X_\alpha$ such that $\mathscr{V} \subsetneqq \mathscr{U}$, then not all of the projection maps are continuous.

(6.A.4) Exercise. Show that under the product topology $\mathscr{U}$ and the box topology $\mathscr{V}$, projection mappings are open maps, and that if $\mathscr{W}$ is a topology such that $\mathscr{V} \subsetneqq \mathscr{W}$, then at least one of the projection functions is not open.

(6.A.5) Exercise. Suppose that $\{X_\alpha \mid \alpha \in \Lambda\}$ and $\{Y_\alpha \mid \alpha \in \Lambda\}$ are topological spaces and that for each $\alpha \in \Lambda$, $f_\alpha : X_\alpha \to Y_\alpha$ is continuous. Define $\Pi f_\alpha : \Pi_{\alpha \in \Lambda} X_\alpha \to \Pi_{\alpha \in \Lambda} Y_\alpha$ by $\Pi f_\alpha(x) = \{f_\alpha(x_\alpha)\}$. Show that Πf_α is continuous. Show that if $X = X_\alpha$ for each α, and $\phi : X \to \Pi_{\alpha \in \Lambda} Y_\alpha$ is defined by $\phi(x) = \{f_\alpha(x)\}$, then ϕ is continuous.

The procedure used in obtaining the product topology from finite intersections of subsets of a given set X is utilized frequently enough to merit special consideration.

(6.A.6) Definition. Suppose that $(X, \mathscr{U})$ is a topological space. A *subbasis* for $(X, \mathscr{U})$ is a collection $\mathscr{S}$ of open subsets of X with the property that the collection of all finite intersections of members of $\mathscr{S}$ forms a basis for $(X, \mathscr{U})$.

Note that if $\mathscr{S}$ is a family of subsets of X whose union is X, then the collection of finite intersection of members of $\mathscr{S}$ forms a basis for a topology on X.

(6.A.7) Exercise. Show that the topology obtained from a subbasis is the smallest topology that contains the subbasic sets as open sets.

(6.A.8) Exercise. Show that sets of the form (a, ∞) and $(-\infty, b)$ form a subbasis for $\mathscr{E}^1$. Formulate and prove a similar result for the order topology.

Continuity may be expressed easily in terms of subbases.

(6.A.9) Theorem. Suppose that X and Y are topological spaces and that $\mathscr{S}$ is a subbasis for Y. Then a function $f : X \to Y$ is continuous if and only if $f^{-1}(S)$ is open for each $S \in \mathscr{S}$.

Proof. Since members of $\mathscr{S}$ are open, it follows that if f is continuous, then $f^{-1}(S)$ is open for each $S \in \mathscr{S}$. Conversely, suppose that $f^{-1}(S)$ is open for all S in $\mathscr{S}$ and let U be member of the basis for Y determined by $\mathscr{S}$. Then U is a finite intersection of members of $\mathscr{S}$, and since intersections are well behaved under inverses (0.C.7), the theorem follows.

We now determine which of the previous results relating to finite products carry over to infinite products. Perhaps the most difficult question to resolve is that of compactness. The result that the arbitrary product of compact spaces is compact is known as Tihonov's theorem, and this theorem follows readily from the following characterization of compactness.

(6.A.10) Theorem (Alexander's Lemma). Suppose that $\mathscr{S}$ is a subbasis

for the topology of a topological space X. Then X is compact if and only if every cover of X by members of $\mathscr{S}$ has a finite subcover.

Proof. Suppose that X is not compact. Then there is an open cover $\mathscr{U}$ of X from which no finite subcover can be extracted. Let **H** be the collection of all open covers of X for which no finite subcover exists. The family $\mathscr{U}$ is, of course, a member of **H**. Let $\mathscr{D}$ be a nest of covers in **H** (ordered by inclusion) and observe that $\mathscr{V} = \{V \mid V \in D \text{ and } D \in \mathscr{D}\}$ is an upper bound for $\mathscr{D}$ that lies in **H**. By Zorn's lemma (0.D.3) (some form of the axiom of choice is inevitable in this proof), there is a maximal element $\mathscr{M}$ in **H**.

The family $\mathscr{M}$ has the following peculiar property. Suppose that $M \in \mathscr{M}$ and $V_1, V_2, \ldots, V_n$ are open sets in X such that $V_1 \cap V_2 \cap \cdots \cap V_n \subset M$. Then for some i, we have that $V_i \in \mathscr{M}$. If this were not the case, then for each i, there would be sets $M_{i_1}, M_{i_2}, \ldots, M_{i_{n_i}}$ such that $V_i \cup M_{i_1} \cup \cdots \cup M_{i_{n_i}} = X$ (otherwise, $\mathscr{M}$ would be shed of its maximality). However, this is absurd, for we would then have that $X \subset (V_1 \cap \cdots \cap V_n) \cup (\bigcup_{i=1}^{n} \bigcup_{j=1}^{n} M_{ij}) \subset M \cup (\bigcup_{i=1}^{n} \bigcup_{j=1}^{n} M_{ij}))$, which implies that $\mathscr{M}$ fails to be in **H**.

Suppose now that $x \in M \in \mathscr{M}$. Since $\mathscr{S}$ is a subbasis, there are sets $S_1, S_2, \ldots, S_n \in \mathscr{S}$ such that $x \in S_1 \cap S_2 \cap \cdots \cap S_n \subset M$. By the peculiar property of $\mathscr{M}$, we have that $S_i \in \mathscr{M}$ for some i. Thus, $x \in S_i \in \mathscr{S}$ and $x \in M \in \mathscr{M}$; consequently, $\mathscr{S} \cap \mathscr{M}$ is an open cover of X. However, since $\mathscr{M}$ fails to have a finite subcover, it follows that $\mathscr{S} \cap \mathscr{M}$ has no finite subcover. On the other hand, by hypothesis, every subcollection of $\mathscr{S}$ that covers X (in particular, $\mathscr{S} \cap \mathscr{M}$) does have a finite subcover, a dilemma that can be resolved only by admitting that X was compact.

The other half of the theorem is completely trivial.

(6.A.11)   Theorem (Tihonov's Theorem).   Suppose that $\{X_\alpha \mid \alpha \in \Lambda\}$ is a family of topological spaces. Then $\prod_{\alpha \in \Lambda} X_\alpha$ is compact if and only if X_α is compact for each $\alpha \in \Lambda$.

Proof. Suppose that X_α is compact for each $\alpha \in \Lambda$. Let $\mathscr{S}$ be the subbasis of $\prod_{\alpha \in \Lambda} X_\alpha$ that consists of all sets of the form $\{p_\alpha^{-1}(U_\alpha) \mid \alpha \in \Lambda$ and U_α is open in $X_\alpha\}$. By Alexander's lemma, it suffices to show that each cover of $\prod_{\alpha \in \Lambda} X_\alpha$ by members of $\mathscr{S}$ has a finite subcover. Suppose, then, that $\mathscr{S}' \subset \mathscr{S}$ covers X, and for each α let $\mathscr{C}_\alpha = \{U \subset X_\alpha \mid p_\alpha^{-1}(U) \in \mathscr{S}'\}$. We show that for some α, $\mathscr{C}_\alpha$ covers X_α. If this is not the case, for each $\alpha \in \Lambda$ choose a point $x_\alpha \in X_\alpha \setminus (\bigcup \{U \mid U \in \mathscr{C}_\alpha\})$. Then the point $\{x_\alpha\} \in \prod_{\alpha \in \Lambda} X_\alpha$ is not covered by $\mathscr{S}'$, an impossibility. Thus, there is an $\alpha \in \Lambda$ such that $\mathscr{C}_\alpha$ is an open cover of X_α. Since X_α is compact, a finite subcover $\{U_{\alpha_1}, U_{\alpha_2}, \ldots, U_{\alpha_n}\}$ may be extracted from

$\mathscr{C}_\alpha$. The collection $\{p_\alpha^{-1}(U_{\alpha_1}), \ldots, p_\alpha^{-1}(U_{\alpha_n})\}$ constitutes a finite subcover of $\mathscr{S}'$, and the theorem follows.

The proof of the second half of this magnificent theorem is quite easy and is left to the reader.

Tihonov's theorem has been described as being the single most important theorem in General Topology. Whether or not such hyperbole is warranted might be a matter of some debate; nevertheless, its exalted position in both topology and analysis is beyond dispute. Kelley [1950] has even shown that Tihonov's theorem is equivalent to the axiom of choice.

(6.A.12) Theorem. Suppose that $\{X_\alpha \mid \alpha \in \Lambda\}$ is a collection of topological spaces. Then $A = \prod_{\alpha \in \Lambda} X_\alpha$ is connected if and only if each X_α is connected.

Proof. Define points $x = \{x_\alpha\}$ and $y = \{y_\alpha\}$ in A to be equivalent $(x \sim y)$ if and only if $\{\alpha \mid x_\alpha \neq y_\alpha\}$ is finite. The remainder of the proof is split into a number of lemmas, each of which represents a fairly trivial exercise.

Lemma 1. If $x = \{x_\alpha\} \in A$ and $D = \{y \in A \mid x \sim y\}$, then D is dense in A.

Lemma 2. Suppose that $K \subset \Lambda$ and $B = \prod_{\alpha \in \Lambda} X_\alpha$. For each $\alpha \in \Lambda \setminus K$, let a_α be an arbitrary point in X_α. Define a map $f : B \to A$ by $f(\{x_\alpha\}_{\alpha \in K}) = \{y_\alpha\}_{\alpha \in \Lambda}$, where $y_\alpha = a_\alpha$ if $\alpha \in \Lambda \setminus K$ and $y_\alpha = x_\alpha$ if $\alpha \in K$. Then f is continuous.

Lemma 3. If K is finite, then $f(B)$ is connected.

Now suppose that F is any nonempty open and closed subset of A. It must be shown that F is in fact A (2.A.4). Select a point $x = \{x_\alpha\} \in F$, and let $y = \{y_\alpha\}$ be any point in A equivalent to x in the sense of Lemma 1. We show that $y \in F$, and it will then follow from Lemma 1 that F is dense in A, and since F is closed, F will be all of A. To see this, let $K = \{\alpha_1, \alpha_2, \ldots, \alpha_n\}$ have the property that if $\alpha \in \Lambda \setminus K$, then $x_\alpha = y_\alpha$, and let $B = X_{\alpha_1} \times X_{\alpha_2} \times \cdots \times X_{\alpha_n}$. For each $\alpha \in \Lambda \setminus K$, select the point a_α to be x_α, and let $f : B \to A$ be the continuous map defined in Lemma 2. Then by Lemma 3, $f(B)$ is connected, and since $x \in f(B) \cap F$, $f(B)$ must actually be a subset of F (otherwise, $f(B) \cap F$ and $f(B) \cap (A \setminus F)$ would form a separation of $f(B)$). Thus, we have that $y \in f(B) \subset F$, which completes the proof.

More care must be exercised in dealing with families of locally compact and locally connected spaces. For example, if for each $n \in \mathbf{Z}^+$, $X_n = \{0,1\}$ (with the discrete topology), then $\prod_{n \in \mathbf{Z}^+} X_n$ is totally disconnected, and thus not locally connected (components are points and hence are not open) even though the individual spaces X_n are locally connected. The next theorem shows how this situation can be partially remedied. .

(6.A.13) Theorem. Suppose that $\{X_\alpha \mid \alpha \in \Lambda\}$ is a family of topological spaces. Then $A = \prod_{\alpha \in \Lambda} X_\alpha$ is locally connected if and only if

(i)   each X_α is locally connected, and
(ii) all but a finite number of X_α are connected.

Proof. Suppose that A is locally connected, and let $x_\alpha \in U_\alpha \subset X_\alpha$, where U_α is open in X_α. Choose any point $x \in A$ such that $p_\alpha(x) = x_\alpha$. Then $p_\alpha^{-1}(U_\alpha)$ is an open set in A containing x. Since A is locally connected, there is a connected open set V such that $x \in V \subset p_\alpha^{-1}(U_\alpha)$. Note that $p_\alpha(V)$ is an open connected set containing x_α and lying in U_α. Part (ii) follows from the fact that $p_\beta(V) = X_\beta$ for all but a finite number of β in Λ.

To prove the converse, let $K \subset \Lambda$ be a finite subset such that if $\alpha \in \Lambda \setminus K$, then X_α is connected. Suppose that $x \in V \subset \prod_{\alpha \in \Lambda} X_\alpha$, where V is open. Assume that V is of the form $V_{\alpha_1} \times V_{\alpha_2} \times \cdots \times V_{\alpha_n} \times \prod X_\gamma$, where $\gamma \in \Lambda \setminus \{\alpha_1, \alpha_2, \ldots, \alpha_n\}$. For $i = 1, 2, \ldots, n$, let U_{α_i} be a connected open set such that $x_{\alpha_i} \in U_{\alpha_i} \subset V_{\alpha_i}$. For each $\beta_j \in K$, let U_{β_j} be a connected open set in X that contains x_{β_j}. Then if $U = U_{\alpha_1} \times \cdots \times U_{\alpha_n} \times U_{\beta_1} \times \cdots \times U_{\beta_m} \times \prod X_\gamma$, where $\gamma \neq \alpha_i, \beta_j$, we have that U is open and connected. Since $x \in U \subset V$, it follows that the product space $\prod_{\alpha \in \Lambda} X_\alpha$ is locally connected.

The proof of the following theorem is left to the hyperactive reader.

(6.A.14) Theorem. Suppose that $\{X_\alpha \mid \alpha \in \Lambda\}$ is a family of topological spaces. Then $\prod_{\alpha \in \Lambda} X_\alpha$ is locally compact if and only if

(i)   each X_α is locally compact, and
(ii) all but possibly a finite number of X_α are compact.

As we shall see presently, the arbitrary product of metric spaces need not be metrizable. However, if the index set is countable, we have the following important result.

(6.A.15) Theorem. Suppose that (X_1, d_1), (X_2, d_2), $\ldots$ are metric spaces. Then $\prod_{i \in \mathbf{Z}^+} X_i$ (with the product topology) is metrizable.

Proof. Since equivalent metrics generate the same topology, one may replace each d_i with d_i', where $d_i'(x,y) = \min\{1, d_i(x,y)\}$. Let $\rho : \prod_{i \in \mathbf{Z}^+} X_i \times \prod_{i \in \mathbf{Z}^+} X_i$

$\to [0,1) \subset [0,\infty)$ be defined by $\rho(x,y) = \sum\limits_{i=1}^{\infty} d_i'(x_i,y_i)/(2^i)$, where $x = (x_1, x_2, \ldots)$ and $y = (y_1, y_2, \ldots)$. The details needed to show that ρ is a metric are omitted, and we concentrate our efforts on showing that the topology induced by the metric ρ coincides with the product topology for $\prod\limits_{i\in \mathbf{Z}^+} X_i$. Let $\mathcal{T}$ denote the topology induced by ρ, and let $\mathcal{T}'$ be the product topology.

We first show that $\mathcal{T}' \subset \mathcal{T}$. Suppose that $x = (x_1, x_2, \ldots)$ is a point in $\prod\limits_{i\in \mathbf{Z}^+} X_i$ and let $U = S_{\varepsilon_1}^{d_1'}(x_1) \times \cdots \times S_{\varepsilon_n}^{d_n'}(x_n) \times \prod\limits_{i\geq n} X_i$ be a typical basic open set containing x. Let $\varepsilon = \min\{\varepsilon_1/2, \ldots, \varepsilon_n/2^n\}$. We show that $S_\varepsilon^\rho(x) \subset U$. Suppose that $y = (y_1, y_2, \ldots) \in S_\varepsilon^\rho(x)$. Then $\rho(x,y) < \varepsilon$ and hence $\sum\limits_{i=1}^{\infty} d_i'(x_i,y_i)/2^i < \varepsilon$. Therefore, $d_i'(x_i,y_i) < \varepsilon_i$ for $i = 1, 2, \ldots, n$, and consequently, $S_\varepsilon^\rho(x) \subset U$.

Now we prove that $\mathcal{T} \subset \mathcal{T}'$. Let $U = S_\varepsilon^\rho(x)$ be a typical basic open set containing x. Choose N large enough so that $\sum\limits_{i=N}^{\infty} 1/2^i < \varepsilon/2$. Then one checks easily that $S_{\varepsilon/2^N}^{d_1'}(x_1) \times \cdots \times S_{\varepsilon/2^N}^{d_n'}(x_n) \times \prod\limits_{i\geq N+1} X_i \subset U$, which completes the proof.

(6.A.16) **Exercise.** Show that ρ is a metric.

(6.A.17) **Exercise.** Suppose that $(X_1,d_1), (X_2,d_2), \ldots$ are complete metric spaces. Show that $(\prod\limits_{i\in \mathbf{Z}^+} X_i, \rho)$ is a complete metric space.

Uncountable products are useful in building a number of counter-examples in topology. For instance, the uncountable product of $\mathbf{I}$ with itself, $\prod\limits_{i\in \mathbf{I}} \mathbf{I}_\alpha = X$, yields a compact (and hence countably compact) space that is not sequentially compact. A sequence in X that fails to have a convergent subsequence may be constructed in the following manner.

Represent each element of $\mathbf{I}$ by its binary expansion; for the dyadic rationals, choose the binary expansion that ends in a sequence of zeros. For each $n \in \mathbf{N}$, let $\alpha_n \in X$ be defined by $\alpha_n(x) = n$-th digit of the binary expansion used for $x \in \mathbf{I}$. Consider the sequence $\{\alpha_n\}$ and let $\{\alpha_{n_k}\}$ be an arbitrary subsequence of $\{\alpha_n\}$. There is at least one $x_0 \in \mathbf{I}$ such that for each k, $\alpha_{n_{2k}}(x_0) = 0$ and $\alpha_{n_{2k+1}}(x_0) = 1$. Then the subsequence $\{\alpha_{n_k}\}$ cannot converge, since its projection into the x_0-th coordinate space does not converge.

Observe also that $\prod\limits_{\alpha\in \mathbf{R}^1} \mathbf{I}_\alpha$ is not first (or second) countable, even though each factor space is. To see this, suppose that $U_1, U_2, \ldots$ is a countable basis at the point $\{x_\alpha\}$, where $x_\alpha = 0$ for each α. We may assume that for each

i, U_i has the form $U_{i_1} \times U_{i_2} \times \cdots \times U_{i_{n(i)}} \times \prod_{\alpha \neq i_j} I_\alpha$. Since $\mathbf{R}^1$ is uncountable, there is a $\beta \in \mathbf{R}^1$ that is not equal to any of the i_j for $i \in \mathbf{Z}^+$ and $1 \leq j \leq n(i)$. Let V_β be a proper open set of I_β containing 0. Then $p^{-1}(V_\beta)$ is an open set in $\prod_{\alpha \in \mathbf{R}^1} I_\alpha$ that contains $\{x_\alpha\}$ but does not contain any of the U_i.

Since $\prod_{\alpha \in \mathbf{R}_1} I_\alpha$ is not first countable, it is not metrizable. Hence, the product of metrizable spaces need not be metrizable.

The product topology may also be used to give an alternate description of the Cantor set K. Let $A = \prod_{i=1}^{\infty} X_i$, where $X_i = \{0,2\}$ with the discrete topology. Define $f : A \to K$ by $f(\{x_i\}) = \sum_{i=1}^{\infty} x_i/3^i$ (here we consider K as in (3.B.22)). Since f is 1–1 and onto, it suffices to demonstrate that f is continuous ((2.G.11) and the Tihonov theorem (6.A.11)). Let $c = \sum_{i=1}^{\infty} x_i/3^i$ be a point of K and suppose that U is an ε-neighborhood of c. Choose N large enough so that $\sum_{i>N} 2/3^i < \varepsilon$. Then the open set $\{x_1\} \times \cdots \times \{x_N\} \times \prod_{i>N} X_i$ is mapped by f into U, which establishes the continuity of f.

B. INVERSE SYSTEMS: THE PRELIMINARIES

Inverse systems have long played an important role in Algebraic Topology. For many years, however, they seem to have been regarded as a topological curiosity of rather restricted significance. Nevertheless, in the last decade a relatively large number of papers involving such systems have appeared. We shall investigate basic properties of inverse systems, and then apply our results to prove, among other things, the incredible result that every compact metric space is the continuous image of the Cantor set.

(6.B.1) Definition.   A relation $\geq$ *directs* a nonempty set D if and only if

(i) $\geq$ is reflexive and transitive, and
(ii) for each a and $b \in D$, there is a $d \in D$ such that $d \geq a$ and $d \geq b$.
The pair $(D, \geq)$ is called a *directed set*. (We shall always assume that D is infinite.)

Note that property (ii) gives a "sense of direction" to D.

(6.B.2) Definition.   Suppose that D is a directed set and that $\{X_\alpha \mid \alpha \in D\}$ is a collection of topological spaces indexed by D. Suppose that for each $\beta \geq \alpha$, there is a continuous function $f_{\alpha\beta} : X_\beta \to X_\alpha$ such that

(i) $f_{\alpha\alpha}$ is the identity, and

(ii) $f_{\alpha\beta}f_{\beta\lambda} = f_{\alpha\lambda}$ whenever $\lambda \geq \beta \geq \alpha$.

Then $(X_\alpha, f_{\alpha\beta}, D)$ is an *inverse system*.

(6.B.3) Definition. Suppose that $(X_\alpha, f_{\alpha\beta}, D)$ is an inverse system. The *inverse limit* of the system $(X_\alpha, f_{\alpha\beta}, D)$ is the subspace X_∞ of $\Pi_{\alpha \in D} X_\alpha$ consisting of those points $\{x_\alpha\}$ with the property that $x_\alpha = f_{\alpha\beta}(x_\beta)$ whenever $\beta \geq \alpha$. The elements of the inverse limit are called *threads*, and the maps $f_{\alpha\beta}$ are called *bonding maps*. The spaces X_α are called the *factor spaces* of the inverse system.

(6.B.4) Remark. If $D = \mathbf{N}$ (with the usual order), then the system $(X_n, f_{mn}, \mathbf{N})$ is usually described by $(X_n, f_n, \mathbf{N})$, where $f_n : X_n \to X_{n-1}$. In this case, the bonding maps f_{mn} are completely determined by the maps f_i, i.e., $f_{mn} = f_{m+1} \cdots f_{n-1}f_n$.

(6.B.5) Examples.

1. Let $D = \mathbf{N}$, and for each $n \in \mathbf{N}$, let $X_n = \mathscr{E}^1$. Define each bonding map f_n to be the identity function.

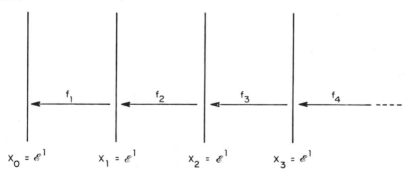

2. Let $D = \mathbf{N}$, and for each $n \in \mathbf{N}$, let $X_n = \mathbf{Z}$ with the discrete topology. Define f_n by setting $f_n(p) = p - 1$, for each $p \in \mathbf{Z}$.

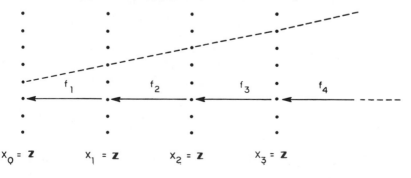

3.   Let $D = \mathbf{N}$, and for each $n \in \mathbf{N}$, let $X_n = \mathbf{N}$ with the discrete topology. Define $f_n : X_n \to X_{n-1}$ by $f_n(p) = p + 1$.

4.   Let X be any topological space and let $\{X_\alpha \subset X \mid \alpha \in D\}$ be a collection of subsets of X, where D is a set directed such that $\beta \geq \alpha$ if and only if $X_\beta \subset X_\alpha$. Define the bonding maps to be the inclusion maps.

5.   Let $D = \mathbf{N}$, and let $X_0 = \{0,1\}$, $X_1 = \{00,01,10,11\}$, $X_2 = \{000,001,010,011,100,101,110,111\}$, etc. Give each X_i the discrete topology. The bonding maps are as indicated in the following figure.

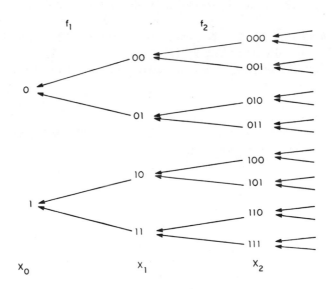

$(6.B.6)$ *Exercise.*   Determine the inverse limit in each of the preceding examples. [Hint: In the last example, try the Cantor set.]

Since X_∞ is given the relative topology inherited from $\prod\limits_{\alpha \in D} X_\alpha$, a basis for X_∞ may be defined by taking intersections of members of the basis of $\prod\limits_{\alpha \in D} X_\alpha$ with X_∞. However, a more useful description of a basis is obtained in the next theorem. If $(X_\alpha, f_{\alpha\beta}, D)$ is an inverse system and X_∞ is the inverse limit, then the restriction of the projection $p_\alpha : \prod\limits_{\alpha \in D} X_\alpha \to X_\alpha$ to X_∞ is denoted by $\hat{p}_\alpha$.

$(6.B.7)$ *Lemma.*   Suppose that $(X_\alpha, f_{\alpha\beta}, D)$ is an inverse system. Then $\hat{p}_\alpha = f_{\alpha\beta}\hat{p}_\beta$ whenever $\beta \geq \alpha$.

   Proof.   Let $\{x_\alpha\} \in X_\infty$. Then $f_{\alpha\beta}\hat{p}_\beta(\{x_\alpha\}) = f_{\alpha\beta}(x_\beta) = x_\alpha = \hat{p}_\alpha(\{x_\alpha\})$.

(6.B.8) Theorem. Suppose that $(X_\alpha, f_{\alpha\beta}, D)$ is an inverse system and that X_∞ is the inverse limit. Then $\mathbf{B} = \{\hat{p}_\alpha^{-1}(U) \mid \alpha \in D \text{ and } U \subset X_\alpha\}$ is a basis for X_∞.

Proof. It is no surprise (in fact, it is obvious) that $\mathbf{B}$ is a subbasis. We shall use properties of inverse systems to show that $\mathbf{B}$ is actually a basis. Suppose that U is an open set in X_∞ and that $\{x_\alpha\}$ is a point in U. Since X_∞ has the relative topology, there is an open set $W = U_{\alpha_1} \times \cdots \times U_{\alpha_n} \times \prod_{\alpha \neq \alpha_i} X_\alpha$ in $\prod_{\alpha \in D} X_\alpha$ such that $\{x_\alpha\} \in W \cap X_\infty \subset U$. Choose $\alpha \in D$ such that $\alpha \geq \alpha_i$ for $i = 1, 2, \ldots, n$, and let $V = \bigcap_{i=1}^{\infty} f_{\alpha_i\alpha}^{-1}(U_{\alpha_i})$. Note that V is open in X_α. Furthermore, it may be verified directly with the aid of (6.B.7) that $\hat{p}_\alpha^{-1}(V) = \bigcap_{i=1}^{n} \hat{p}_\alpha^{-1} f_{\alpha_i\alpha}^{-1}(U_{\alpha_i}) = \bigcap_{i=1}^{n} \hat{p}_\alpha^{-1}(U_{\alpha_i})$. Since $\hat{p}_\alpha(\{x_\alpha\}) = x_\alpha \in V$, we have that $\{x_\alpha\} \in \hat{p}_\alpha^{-1}(V)$. It remains to establish that $\hat{p}_\alpha^{-1}(V) \subset W \cap X_\infty$. Observe that if $\{z_\alpha\} \in \hat{p}_\alpha^{-1}(V)$, then $\hat{p}_\alpha(\{z_\alpha\}) = z_\alpha \in V$, which means that $f_{\alpha_i\alpha}(z_\alpha) = z_{\alpha_i} \in U_{\alpha_i}$ for $i = 1, 2, \ldots, n$. Therefore, $\{z_\alpha\} \in W \cap X_\infty$, and the theorem follows.

(6.B.9) Notation. Suppose that $(X_\alpha, f_{\alpha\beta}, D)$ is an inverse system. For each pair α, β in D with $\alpha < \beta$, let $A_{\alpha\beta} = \{\{x_\gamma\} \in \prod_{\alpha \in D} X_\alpha \mid x_\alpha = f_{\alpha\beta}(x_\beta)\}$.

(6.B.10) Exercise. Show that if X_α is a T_2 space for each α, then $A_{\alpha\beta}$ is closed in $\prod_{\alpha \in D} X_\alpha$ and $X_\infty = \bigcap \{A_{\alpha\beta} \mid \alpha, \beta \in D, \alpha < \beta\}$.

Inverse systems are commonly used to create rather exotic topological spaces, as we shall see, for example, in Section D of this chapter. The inverse limit of a system may (and generally does) differ radically from the factor spaces that make up the inverse system. This may even be the case when the maps themselves are quite simple. Some recent results have focused on reducing the number of distinct bonding maps while maintaining uncomplicated factor spaces. For instance, the pseudoarc (an unusual space to be defined in Chapter 9) may be viewed as the limit of the system (X_n, f_n, N), where each X_n is the interval I and all f_n's represent the same map (Henderson [1965]).

Of course, the inverse limit of a system may be rather mundane as well; in fact, the limit space may be empty even though the bonding maps are embeddings (6.B.5, example 3). However, we see next that the inverse limit cannot be empty if the factor spaces of the inverse system are compact and T_2.

(6.B.11) Theorem. Suppose that $(X_\alpha, f_{\alpha\beta}, D)$ is an inverse system, where each X_α is compact and T_2. Then X_∞ is nonempty and compact.

Proof. The finite intersection property characterization of compactness is used to find a point in X_∞. Let $\mathcal{A} = \{A_{\alpha\beta} \mid \alpha < \beta\}$. Then $\mathcal{A}$ is a collection of closed subsets of the compact space $\prod\limits_{\alpha \in D} X_\alpha$ (6.B.10). We show that $\mathcal{A}$ has the finite intersection property. Suppose that $A_{\alpha_1\beta_1}, \ldots, A_{\alpha_n\beta_n}$ is any finite subcollection of $\mathcal{A}$. Since D is directed, there is a $\lambda \in D$ such that $\lambda \geq \beta_i$ for $i = 1, 2, \ldots, n$. Pick an arbitrary point $x_\lambda \in X_\lambda$ and for $i = 1, 2, \ldots, n$ let $x_{\alpha_i} = f_{\alpha_i\lambda}(x_\lambda)$ and $x_{\beta_i} = f_{\beta_i\lambda}(x_\lambda)$. Define a point $\{y_\alpha\} \in \prod\limits_{\alpha \in D} X_\alpha$ by setting $y_{\alpha_i} = x_{\alpha_i}$ for $i = 1, 2, \ldots, n$, $y_{\beta_i} = x_{\beta_i}$ for $i = 1, 2, \ldots, n$, $y_\lambda = x_\lambda$, and letting y_α be arbitrary for the remaining coordinates. Then $\{y_\alpha\} \in \bigcap\limits_{i=1}^{\infty} A_{\alpha_i\beta_i}$, since

$$f_{\alpha_i\beta_i}(y_{\beta_i}) = f_{\alpha_i\beta_i}(x_{\beta_i}) = f_{\alpha_i\beta_i}f_{\beta_i\lambda}(x_\lambda) = f_{\alpha_i\lambda}(x_\lambda) = x_{\alpha_i} = y_{\alpha_i}.$$ Thus, $\mathcal{A}$ has the finite intersection property, and consequently it follows from (2.H.2) that $\bigcap\limits_{\alpha < \beta} A_{\alpha\beta} \neq \varnothing$. However, by (6.B.10), we have that $\bigcap\limits_{\alpha < \beta} A_{\alpha\beta} = X_\infty$. Furthermore, X_∞ is compact, since it is a closed subspace of the compact space $\prod\limits_{\alpha \in D} X_\alpha$.

(6.B.12)   Exercise.   Show that if $(X_\alpha, f_{\alpha\beta}, D)$ is an inverse system such that each $f_{\alpha\beta}$ is 1–1 and onto, then $\hat{p}_\alpha$ is 1–1 and onto.

(6.B.13)   Exercise.   Show that if (X_n, f_n, N) is an inverse system such that f_n is onto for each n, then $\hat{p}_n$ is onto for each n.

Complications arise with regard to connectedness. The following example illustrates that the inverse limit of a system of connected spaces need not be connected.

(6.B.14)   Example.   For each positive integer n, let $A_n = \{(x,y) \in \mathcal{E}^2 \mid -n < x < n, y = 0\}$, and let $X_n = \mathcal{E}^2 \setminus A_n$. Define an inverse system $(X_n, f_n, \mathbf{Z}^+)$, where each f_n is an inclusion map. It is easy to verify that the inverse limit is homeomorphic to the union of the sets $A = \{(x,y) \in \mathcal{E}^2 \mid y > 0\}$ and $B = \{(x,y) \in \mathcal{E}^2 \mid y < 0\}$.

(6.B.15)   Theorem.   Suppose that $(X_\alpha, f_{\alpha\beta}, D)$ is an inverse system of compact, connected, T_2 spaces. Furthermore, assume that $\hat{p}_\alpha$ is onto for each $\alpha \in D$. Then X_∞ is connected.

Proof.   We show that if $X_\infty = A \cup B$, where A and B are closed nonempty subsets of X_∞, then $A \cap B \neq \varnothing$. Since X_∞ is compact, so are A and B, and therefore for each α, $\hat{p}_\alpha(A)$ and $\hat{p}_\alpha(B)$ are closed subsets of X_α. However, $\hat{p}_\alpha$ is onto, and consequently, $X_\alpha = \hat{p}_\alpha(A) \cup \hat{p}_\alpha(B)$. Let $Y_\alpha = \hat{p}_\alpha(A) \cap \hat{p}_\alpha(B)$ and note that Y_α is not empty, since X_α is connected. Define $g_{\alpha\beta} = f_{\alpha\beta \mid Y_\beta}$. Then $g_{\alpha\beta}$ maps Y_β into Y_α, since $g_{\alpha\beta}(Y_\beta) = f_{\alpha\beta}(Y_\beta) = f_{\alpha\beta}(\hat{p}_\beta(A) \cap \hat{p}_\beta(B)) \subset f_{\alpha\beta}\hat{p}_\beta(A) \cap f_{\alpha\beta}\hat{p}_\beta(B) = \hat{p}_\alpha(A) \cap \hat{p}_\alpha(B) = Y_\alpha$.

Since each Y_α is compact and T_2, (6.B.11) may be applied to the inverse system $(Y_\alpha, g_{\alpha\beta}, D)$ to obtain a point $\{y_\alpha\} \in Y_\infty$. Since $\hat{p}_\alpha(\{y_\alpha\}) \in \hat{p}_\alpha(A) \cap \hat{p}_\alpha(B)$ for each α, it follows that $\{y_\alpha\} \in A \cap B$.

C. EVERY COMPACT METRIC SPACE IS A CONTINUOUS IMAGE OF THE CANTOR SET

In (6.B.5) and (6.B.6), it was seen that the Cantor set may be considered as the inverse limit of a countable number of finite discrete spaces. We shall see presently that any totally disconnected compact metric space may be so represented, and in fact, inverse systems may be employed to prove that all perfect, totally disconnected compact metric spaces are homeomorphic to the Cantor set.

(6.C.1) Theorem. If X is a totally disconnected compact T_2 space, then the family of open and closed subsets of X forms a basis for X.

Proof. The proof follows easily from (4.A.11).

(6.C.2) Definition. If X is a metric space and $\mathcal{U}$ is a cover of X, then *mesh* $\mathcal{U}$ is defined to be the least upper bound of the diameters of the sets in $\mathcal{U}$.

(6.C.3) Definition. Suppose that $\mathcal{U}$ and $\mathcal{V}$ are covers of a space X. Then $\mathcal{V}$ is a *refinement* of $\mathcal{U}$ if and only if for each $V \in \mathcal{V}$, there is a $U \in \mathcal{U}$ such that $V \subset U$. If $\mathcal{V}$ refines $\mathcal{U}$, we write $\mathcal{V} < \mathcal{U}$.

(6.C.4) Theorem. Suppose that X is a totally disconnected compact metric space. Then there is a sequence of finite covers $\mathcal{U}_0, \mathcal{U}_1, \ldots$ of X such that

(i) $\mathcal{U}_0 > \mathcal{U}_1 > \cdots$,
(ii) mesh $\mathcal{U}_n < 1/n$, (mesh $\mathcal{U}_0 < 2$),
(iii) all members of the covers are open and closed, and
(iv) for each n, if $U, V \in \mathcal{U}_n$ and $U \neq V$, then $U \cap V = \varnothing$.

Proof. We first construct $\mathcal{U}_0$. By (6.C.1), for each $x \in X$, there is an open and closed set U_x of diameter less than 2. The family of U_x's covers X, and since X is compact, there is a finite subcover $\{U_{x_1}, U_{x_2}, \ldots, U_{x_k}\}$. These sets are not necessarily disjoint, but this may be corrected by setting $U_i = U_{x_i} \setminus (\bigcup_{j=1}^{i-1} U_j)$. An inductive procedure for completing the proof should now be clear,

(6.C.5)   *Theorem.* Suppose that X is a totally disconnected compact metric space. Then there is an inverse system $(X_n, f_n, \mathbf{N})$ such that for each n, X_n is a finite discrete space and X_∞ is homeomorphic to X.

     Proof. Let $\mathcal{U}_0, \mathcal{U}_1, \ldots$ be the sequence of open and closed covers obtained in the previous theorem. For each n, let $X_n = \mathcal{U}_n$ and give X_n the discrete topology. We define the bonding maps f_n as follows. If $U_n \in \mathcal{U}_n$, let $f_n(U_n) = U_{n-1}$, where U_{n-1} is the unique member of $\mathcal{U}_{n-1}$ such that $U_n \subset U_{n-1}$.

     We now define a homeomorphism h from X onto X_∞. Suppose that $x \in X$. For each n, let U_n be the unique member of $\mathcal{U}_n$ that contains x. Observe that $U_0 \supset U_1 \supset U_2 \supset \cdots$ and, consequently, $(U_0, U_1, \ldots) \in X_\infty$. Let $h(x) = (U_0, U_1, \ldots)$. We first show that h is onto. Suppose that $(V_0, V_1, \ldots) \in X$. Then for each i, we have that $V_i \supset V_{i+1}$, and since for increasing i, diam V_i approaches 0, it follows from (3.A.11) and (3.B.8) that $\bigcap_{i=0}^{\infty} V_i$ consists of a single point z. Clearly, $h(z) = (V_0, V_1, \ldots)$. The map h is 1–1, since for distinct points x and y there are distinct sequences $U_0 \supset U_1 \supset \cdots$ and $V_0 \supset V_1 \supset \cdots$ that converge to x and y, respectively.

     It remains to establish continuity. Since X and X_∞ are compact T_2 spaces, it suffices to show that h is an open map (why?). Certainly $\mathbf{B} = \{U \mid U \in \mathcal{U}_n,\ n \in \mathbf{Z}^+\}$ is a basis for X. Note now that if $U_n \in \mathcal{U}_n$, then $h(U_n) = (X_1 \times \cdots \times X_{n-1} \times \{\mathcal{U}_n\} \times X_{n+1} \times \cdots) \cap X_\infty$, which is obviously open in X_∞.

     It is often of interest to be able to pass from one inverse system to another. This may be done whenever there are maps between the respective factor spaces which preserve the inverse structure.

(6.C.6)   *Definition.* Suppose that $\mathscr{A} = (X_\alpha, f_{\alpha\beta}, D)$ and $\mathscr{B} = (Y_\alpha, g_{\alpha\beta}, D)$ are inverse systems with the same directed set D. Suppose that for each $\alpha \in D$ there is a continuous map $h_\alpha : X_\alpha \to Y_\alpha$ such that whenever $\beta \geq \alpha$ the following diagram is commutative.

$$
\begin{array}{ccc}
X_\alpha & \xleftarrow{\;f_{\alpha\beta}\;} & X_\beta \\
{\scriptstyle h_\alpha}\downarrow & & \downarrow{\scriptstyle h_\beta} \\
Y_\alpha & \xleftarrow[\;g_{\alpha\beta}\;]{} & Y_\beta
\end{array}
$$

Then the family of maps $H = \{h_\alpha \mid \alpha \in D\}$ is called a *map* between $\mathscr{A}$ and $\mathscr{B}$.

(6.C.7)   *Definition.* Suppose that H is a map between inverse systems

$\mathscr{A}$ and $\mathscr{B}$. Define $\Pi h_\alpha : \Pi_{\alpha \in D} X_\alpha \to \Pi_{\alpha \in D} Y_\alpha$ by $\Pi h_\alpha(\{x_\alpha\}) = \{h_\alpha(x_\alpha)\}$. Then $h_\infty = \Pi h_{\alpha|X_\infty}$ is called the *limit* of H.

(6.C.8) **Theorem.** Suppose that H is a map between inverse systems $(X_\alpha, f_{\alpha\beta}, D)$ and $(Y_\alpha, g_{\alpha\beta}, D)$. Then the limit of H, h_∞, maps X_∞ into Y_∞ and is continuous.

Proof. Suppose that $\{x_\alpha\} \in X_\infty$, and let $\{y_\alpha\} = h_\infty(\{x_\alpha\})$. If $\alpha \leq \beta$, then it must be shown that $g_{\alpha\beta}(y_\beta) = y_\alpha$. However, this follows easily, since $g_{\alpha\beta}(y_\beta) = g_{\alpha\beta}h_\beta(x_\beta) = h_\alpha f_{\alpha\beta}(x_\beta) = h_\alpha(x_\alpha) = y_\alpha$, and hence $\{y_\alpha\}$ is a thread in Y_∞. Clearly, h_∞ is continuous, since it is the restriction of a continuous map.

(6.C.9) **Theorem.** Suppose that $(X_\alpha, f_{\alpha\beta}, D)$ is an inverse system.

(i)   If h_α is 1–1 for each $\alpha \in D$, then h_∞ is 1–1.
(ii) If $D = \mathbf{N}$ and each h_n is 1–1 and onto, then h_∞ is 1–1 and onto.

Proof.

(i)   This is immediate from the definition of h_∞.
(ii) To see that h_∞ is onto, let $\{y_n\} \in Y_\infty$ and for each n, set $x_n = h_n^{-1}(y_n)$. Then we have that $h_n(x_n) = y_n$ and it follows easily from the commutativity of the following diagram that $\{x_n\}$ is a thread. Hence, $h_\infty(\{x_n\}) = \{y_n\}$.

$$
\begin{array}{ccccccccc}
X_0 & \xleftarrow{\ f_1\ } & X_1 & \xleftarrow{\ f_2\ } & X_2 & \xleftarrow{\ f_3\ } & X_3 & \longleftarrow \\
\ \downarrow{\scriptstyle h_0} & & \ \downarrow{\scriptstyle h_1} & & \ \downarrow{\scriptstyle h_2} & & \ \downarrow{\scriptstyle h_3} & \\
Y_0 & \xleftarrow{\ g_1\ } & Y_1 & \xleftarrow{\ g_2\ } & Y_2 & \xleftarrow{\ g_3\ } & Y_3 & \longleftarrow
\end{array}
$$

We now proceed to show that all perfect compact totally disconnected metric spaces are homeomorphic. We begin with the following lemma, which may be established with the aid of (6.C.1) and an easy induction argument.

(6.C.10) **Lemma.** Suppose that X is a compact totally disconnected T_2 space and that n is a positive integer. Then X may be written as a disjoint union of n open and closed (nonempty) subsets.

(6.C.11) **Theorem.** If X and Y are two totally disconnected compact perfect metric spaces, then X and Y are homeomorphic. (Hence, X and Y are homeomorphic to the Cantor set.)

Proof. By (6.C.1) and the previous lemma, there are covers $\mathscr{U}_0$ and $\mathscr{U}_0'$ of X and Y, respectively, that satisfy the following conditions:

(i)   the members of each cover are open and closed;
(ii)   the elements of each cover are mutually disjoint;
(iii) the mesh of each cover is less than 2;
(iv) $\mathscr{U}_0$ and $\mathscr{U}'_0$ have the same number of elements.

Let h_0 be any 1–1 map from $\mathscr{U}_0$ onto $\mathscr{U}'_0$. Apply (6.C.1) once again to obtain refinements $\mathscr{U}_1$ and $\mathscr{U}'_1$ of $\mathscr{U}_0$ and $\mathscr{U}'_0$, respectively, whose meshes are less than 1 and which also satisfy conditions (i) and (ii) above. Lemma (6.C.10) allows us to assume that if $U_0 \in \mathscr{U}_0$, then U_0 and $h_0(U_0)$ contain the same number of elements of $\mathscr{U}_1$ and $\mathscr{U}'_1$, respectively. Hence, there is a 1–1 function h_1 from $\mathscr{U}_1$ onto $\mathscr{U}'_1$ with the property that if $U_1 \in \mathscr{U}_1$, $U_0 \in \mathscr{U}_0$, and $U_1 \subset U_0$, then $h_1(U_1) \subset h_0(U_0)$. This procedure may be repeated in the obvious inductive fashion to obtain two inverse systems and a map between them, as indicated in the following diagram. Each $\mathscr{U}_n$ is assumed to have the discrete topology and each f_i and f'_i is defined as in (6.C.5).

$$\mathscr{U}_0 \xleftarrow{\ f_1\ } \mathscr{U}_1 \xleftarrow{\ f_2\ } \mathscr{U}_2 \xleftarrow{\ f_3\ }$$

$$\Big\downarrow h_0 \qquad \Big\downarrow h_1 \qquad \Big\downarrow h_2$$

$$\mathscr{U}'_0 \xleftarrow[\ f'_1\ ]{} \mathscr{U}'_1 \xleftarrow[\ f'_2\ ]{} \mathscr{U}'_2 \xleftarrow[\ f'_3\ ]{}$$

It is clear from (6.C.8) and (6.C.9) that h_∞ is a homeomorphism. Furthermore, by (6.C.5), X is homeomorphic to $\mathscr{U}_\infty$ and Y is homeomorphic to $\mathscr{U}'_\infty$, which concludes the proof.

We now prove one of the most startling and least intuitive results in all of topology.

(6.C.12)   Theorem.   Suppose that (X,d) is a compact metric space. Then there is a continuous function f mapping the Cantor set K onto X.

Proof.   Since X is compact metric, it is easy to construct a sequence of finite covers $\mathscr{U}_0, \mathscr{U}_1, \mathscr{U}_2, \ldots$ of X with the following properties:

(i)   each $U \in \mathscr{U}_n$ is the closure of an open set;
(ii)   if $U \in \mathscr{U}_n$, then diam $U < 1/2^n$;
(iii) $\mathscr{U}_n$ refines $\mathscr{U}_{n-1}$.

Let $\mathscr{U}_0 = \{U_{01}, U_{02}, \ldots, U_{0k_0}\}$. Members of $\mathscr{U}_0$ are "disjointified" by means of the trick of letting $V_{0i} = U_{0i} \times \{i\}$ for $1 \le i \le k_0$. A subset $A \times \{i\}$ is to be open in V_{0i} if and only if A is open in U_{0i}. Let $V_0 = \bigcup_{i=1}^{k_0} V_{0i}$ and assign the disjoint union topology to V_0.

Notation soon becomes a problem, but if the reader keeps alert it will help. Let $\mathcal{U}_1 = \{U_{11}, U_{12}, \ldots, U_{1k_1}\}$. For each $U_{1j} \in \mathcal{U}_1$, choose an integer i such that $U_{1j} \subset U_{0i}$. Again we "disjointify." Let $V_{1ij} = U_{1j} \times \{i\} \times \{j\}$ (whenever $U_{1j} \subset U_{0i}$), and let V_1 be the (disjoint) union of these sets with the disjoint union topology. Define $f_1 : V_1 \to V_0$ by $f_1(u,i,j) = (u,i)$; then f_1 is obviously continuous. It should now be obvious what is afoot, but in the interest of clarity we proceed one step further.

Let $\mathcal{U}_2 = \{U_{21}, U_{22}, \ldots, U_{2k_2}\}$. For each $U_{2k} \in \mathcal{U}_2$, choose an integer j such that $U_{2k} \subset U_{1j}$. Then for each "triple" $U_{2k} \subset U_{1j} \subset U_{0i}$, define a set $V_{2ijk} = U_{2k} \times \{i\} \times \{j\} \times \{k\}$. Assign to V_{2ijk} the obvious topology, and let V_2 be the disjoint union of all these sets, again with the disjoint union topology. Define a map $f_2 : V_2 \to V_1$ by $f_2(u,i,j,k) = (u,i,j)$. Continuing in this manner, we obtain an inverse system $(V_n, f_n, \mathbf{N})$.

A second inverse system $(X_n, g_n, \mathbf{N})$ is constructed by defining $X_n = X$ for each n and setting the bonding maps g_n equal to the identity. A map between the two systems is described as follows. For each $n \in \mathbf{N}$, let $h_n : V_n \to X_n$ be defined by $h_n(u,i,j,\ldots,p) = u$. Thus we have the following commutative diagram.

Denote the respective inverse limits by V_∞ and X_∞, and the limit of the h_n by h_∞. Clearly h_∞ is onto, V_∞ is a compact metric space, and X is homeomorphic with X_∞. We now investigate properties of V_∞.

We first show that V_∞ is totally disconnected. Suppose that $x = (x_0, x_1, \ldots)$ and $y = (y_0, y_1, \ldots)$ are distinct points of V_∞. We are looking for an open and closed set in V_∞ that contains x but not y. Since x and y are distinct, there is a first integer n such that $x_n \neq y_n$. Suppose that $x_0 \neq y_0$, where $x_0 = u \times \{i\}$ and $y_0 = u' \times \{i'\}$. If $u \neq u'$, then an integer n may be chosen large enough so that any two members of $\mathcal{U}_n$ that contain x and y respectively must be disjoint. Suppose $u \in U_{np}$ and $u' \in U_{nq}$. Then the corresponding sets $V_{nij\ldots p}$ and $V_{nij\ldots q}$ are disjoint and both open and closed. Furthermore, $V_{nij\ldots p}$ is pulled back by $\hat{p}_n^{-1}$ to an open and closed set in V_∞ which contains x but not y; thus x and y have been separated in V_∞.

If $u = u'$, then $i \neq i'$. Therefore, we have $x_0 \in V_{0i}$ and $y_0 \in V_{0i'}$, furthermore V_{0i} and are $V_{0i'}$ are disjoint, open, and closed. Again, $\hat{p}_0^{-1}(V_{0i})$ is an open and closed subset of V_∞ containing x but not y.

If $x_0 = y_0$, let m be the first integer such that $x_m \neq y_m$. Suppose, for instance, that $m = 2$. Then we have $x_2 = u \times \{i\} \times \{j\} \times \{k\}$ and $y_2 = u \times \{i\} \times \{j\} \times \{k'\}$, where $k \neq k'$. Thus, we have that $x_2 \in V_{2ijk}$ and $y_2 \in V_{2ijk'}$, where V_{2ijk} and $V_{2ijk'}$ are disjoint and open and closed. Then $\hat{p}^{-1}(V_{2ijk})$ is an open and closed subset of V_∞ containing x but not y. Hence, V_∞ is totally disconnected.

Although V_∞ may not be perfect, it follows from (3.B.20) that $V_\infty \times K$ is perfect. Therefore, $V_\infty \times K$ is homeomorphic to K (6.C.11, 6.C.13), and consequently the following maps exist: a homeomorphism $H : K \to (V_\infty \times K)$, a continuous surjection $q : (V_\infty \times K) \to V_\infty$ defined by $q(v,y) = v$, a continuous surjection $h_\infty : V_\infty \to X_\infty$, and a homeomorphism $\Psi : X_\infty \to X$. Then $f = \Psi h_\infty q H$ maps K onto X.

(6.C.13) *Exercise.* Show that V_∞ is metrizable.

The Cantor set may be generalized as follows. A product space $\prod_{\alpha \in \Lambda} X_\alpha$ is a *Cantor space* if and only if for each $\alpha \in \Lambda$, $X_\alpha = \{0,1\}$ with the discrete topology. It can be shown that every compact T_2 space is the continuous image of a closed subset of some Cantor space (see Aleksandrov and Uryson [1929], for example).

D. THE DYADIC SOLENOID

To close this chapter, we present in some detail the construction of a curious space, the dyadic solenoid. We first give a geometric description of this space, and then we show how the geometric version can be realized as an inverse limit. The latter characterization is both neater and somewhat more manageable.

(6.D.1) *Definition.* A *solid torus* is any space homeomorphic to $\mathscr{S}^1 \times \mathscr{B}^2$. The standard solid torus in $\mathscr{E}^3$ is obtained by rotating the disk in the yz plane with center $(0,2,0)$ and radius 1 about the z axis.

The solenoid is constructed by successively intertwining solid tori inside of solid tori. In Euclidean 3-space, one starts with a standard solid torus T_0. In the interior of this torus, another solid torus T_1 is wrapped around twice (see the figure that follows). Then in the interior of the second torus a third one is wrapped around in precisely the same manner in which T_1 was wrapped around inside T_0. This procedure is repeated for each positive integer n, and the solenoid is defined to be the intersection of all the tori. It is assumed that the diameter of the cross sections of the tori approaches 0 with increasing i. In spite of the decreasing diameters, the intersection of the tori is nonempty (4.A.8). The successive tori T_0, T_1, ... yield an inverse system $(T_n, f_n, \mathbf{N})$, where the f_n are simply inclusion maps.

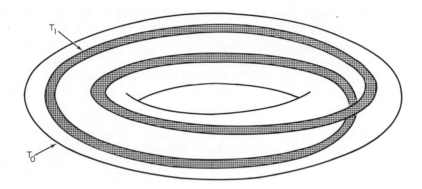

(6.D.2) *Exercise.* Show that the inverse limit T_∞ of $(T_n, f_n, \mathbf{N})$ is homeomorphic to $D = \bigcap\limits_{i=0}^{\infty} T_i$.

To describe $D = \bigcap\limits_{i=0}^{\infty} T_i$ as an inverse limit as in (6.D.2) is of little interest in itself. The following limit system proves to be more useful.

Let $(C_n, g_n, \mathbf{N})$ be the inverse system where, for each n, C_n is the unit circle represented in polar coordinates by $\{(r, \theta) \mid r = 1\}$, and $g_n : C_n \to C_{n-1}$ is defined by $g_n(1, \theta) = (1, 2\theta)$.

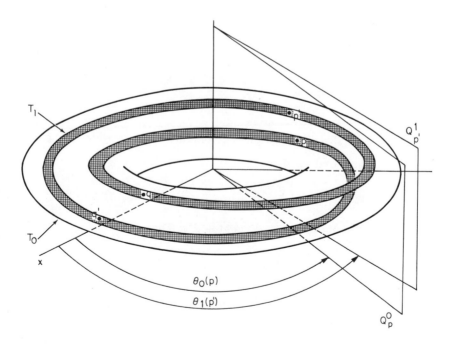

A map is constructed between the systems (T_n,f_n,N) and (C_n,g_n,N) as follows. Let $p \in T_0$ and let Q_p^0 be the plane containing the z axis and p. Define $\theta_0(p)$ to be the angle measured counterclockwise in radians between the positive x axis and Q_p^0. We assume that the range of θ_0 is $[0,2\pi)$. Now suppose that $p \in T_1$. Let Q_p^1 be the plane passing through the z axis and p. Define $\theta_1(p)$ to be the angle measured counterclockwise in radians formed between the x axis and Q_p^1 subject to the following conditions:

(i)   if $\theta_1(p) = 0$, then so does $\theta_0(p)$;
(ii) the range of θ_1 is $[0,4\pi)$, and θ_1 is continuous on $T_1 \setminus \theta_1^{-1}(0)$;
(iii) $\theta_1(p) = \theta_1(p')$ if and only if $Q_p^1 = Q_{p'}^1$ and p and p' belong to the same component of $Q_p^1 \cap T_1$ (see preceding figure).

If $p \in T_2$, then let Q_p^2 be the plane containing the z axis and p, and define $\theta_2(p)$ as the angle measured counterclockwise in radians between the x axis and Q_p^2, where the function θ_2 satisfies the conditions:

(i)   if $\theta_2(p) = 0$, then so does $\theta_1(p)$;
(ii) the range of θ_2 is $[0,8\pi)$, and θ_2 is continuous on $T_2 \setminus \theta_2^{-1}(0)$;
(iii) $\theta_2(p) = \theta_2(p')$ if and only if $Q_p^2 = Q_{p'}^2$ and p and p' belong to the same component of $Q_p^2 \cap T_2$.

Similar functions are defined for each positive integer n. Note that $\theta_n(p)$ either equals $\theta_{n-1}(p)$ or equals $\theta_{n-1}(p) + 2^n\pi$. Define a map $h_n : T_n \to C_n$ by $h_n(p) = (1,\theta_n(p)/2^n)$. Then h_n is clearly continuous, and hence we have the following diagram.

$$
\begin{array}{ccccccc}
T_0 & \xleftarrow{\ f_1\ } & T_1 & \xleftarrow{\ f_2\ } & T_2 & \longleftarrow \\
\ \Big\downarrow{\scriptstyle h_0} & & \ \Big\downarrow{\scriptstyle h_1} & & \ \Big\downarrow{\scriptstyle h_2} & \\
C_0 & \xleftarrow[\ g_1\ ]{} & C_1 & \xleftarrow[\ g_2\ ]{} & C_2 & \longleftarrow
\end{array}
$$

In order to show that the diagram is commutative, we consider two cases.

(i) Suppose that $\theta_n(p) = \theta_{n-1}(p)$. Then

$$g_n h_n(p) = g_n\left(1, \frac{\theta_n(p)}{2^n}\right) = \left(1, \frac{\theta_n(p)}{2^{n-1}}\right) = h_{n-1} f_n(p)$$

(ii) Suppose that $\theta_n(p) = \theta_{n-1}(p) + 2^n\pi$. Then

$$g_n h_n(p) = g_n\left(1, \frac{\theta_n(p)}{2^n}\right) = \left(1, \frac{\theta_n(p)}{2^{n-1}}\right)$$

$$= \left(1, \frac{\theta_{n-1}(p) + 2^n\pi}{2^{n-1}}\right) = \left(1, \frac{\theta_{n-1}(p)}{2^{n-1}} + 2\pi\right)$$

However,

$$h_{n-1}f_n(p) = \left(1, \frac{\theta_{n-1}(p)}{2^{n-1}}\right)$$

and since scurrying around the unit circle 2π radians is equivalent to having never set out, the commutativity of the diagram is established.

We now show that h is 1–1 and onto. Suppose that $t = (p, p, \ldots)$ and $t' = (p', p', \ldots)$ are distinct points in T_∞. Then clearly, $h_0(p) \neq h_0(p')$, and hence, $h_\infty(t) \neq h_\infty(t')$. Therefore, h is 1–1.

To see that h_∞ is onto, let $c = (c_0, c_1, \ldots) \in C_\infty$, where $c_n = (1, r_n)$ for each n. Note that $r_n = r_{n-1}/2$ or $r_n = (r_{n-1}/2) + \pi$. For each n, let $x_n = 2^n r_n$. Then either $x_n = x_{n-1}$ or $x_n = x_{n-1} + 2^n \pi$. Let $H_n = \theta_n^{-1}(x_n)$, i.e., H_n is the appropriate component of the cross section of T_n determined by the plane that contains the z axis and forms an angle of x_n radians with the x axis. The reader should be able to convince himself that the point $t = (p, p, \ldots)$, where $\{p\} = \bigcap\limits_{n=1}^{\infty} H_n$ is mapped by h_∞ onto c.

Recall that a topological space X is *homogeneous* if and only if, given any two points x and y in X, there is a homeomorphism $h : X \to X$ that carries x to y. Note how the use of inverse limits yields an easy proof of the following theorem:

(6.D.3) *Theorem.* The dyadic solenoid is homogeneous.

Proof. We use the above notation, where the solenoid is considered to be C_∞. Let $x = (x_0, x_1, \ldots)$ and $y = (y_0, y_1, \ldots)$ be points in C_∞, where $x_n = (1, \alpha_n)$ and $y_n = (1, \beta_n)$. Consider the mapping system

$$
\begin{array}{ccccccc}
C_0 & \xleftarrow{\;g_1\;} & C_1 & \xleftarrow{\;g_2\;} & C_2 & \longleftarrow & \\
\downarrow{\scriptstyle k_0} & & \downarrow{\scriptstyle k_1} & & \downarrow{\scriptstyle k_2} & & \\
C_0 & \xleftarrow{\;g_1\;} & C_1 & \xleftarrow{\;g_2\;} & C_2 & \longleftarrow &
\end{array}
$$

where $k_n(1, \theta) = (1, \theta + \beta_n - \alpha_n)$. It is a routine matter to check that the diagram commutes and that each k_n is a homeomorphism. Thus, by (6.C.8), (6.C.9), and (2.G.11), k_∞ is a homeomorphism, and a direct computation shows that $k_\infty(x) = y$.

More results involving inverse limit systems will be obtained in Chapter 9. At this point, the following articles should be well within the grasp of the rara avis that is interested in some of the recent developments in inverse limit theory. Brown [1960], Jolley and Rogers [1970], Nadler [1970], and Kresimar and Mardešić [1968].

PROBLEMS

Section A

1. Show that $\Pi_{\alpha \in \Lambda} X_\alpha$ has each of the following properties if and only if each X_α does:

   (i) T_0,
   (ii) T_1,
   (iii) T_2,
   (iv) regularity,
   (v)* complete regularity.

2. Suppose that $\{X_\alpha \mid \alpha \in \Lambda\}$ is a family of topological spaces, and for each α let $\mathbf{B}_\alpha$ be a basis for X_α. Describe a basis for $\Pi_{\alpha \in \Lambda} X_\alpha$ in terms of the $\mathbf{B}_\alpha$'s.

3. Show that $\Pi_{\alpha \in \Lambda} X_\alpha$ is first (second) countable if and only if each X_α is first (second) countable and all but a countable number of the X_α have the indiscrate topology.

4. Let P be a topological property and X a nonempty set. Then a topology $\mathscr{U}$ on X is *maximal with respect to P* if and only if whenever $\mathscr{V}$ is a topology for X and $\mathscr{U} \subsetneq \mathscr{V}$, then $\mathscr{V}$ does not have property P; $\mathscr{U}$ is *minimal with respect to P* if and only if whenever $\mathscr{V}$ is a topology for X and $\mathscr{V} \subsetneq \mathscr{U}$, then $\mathscr{V}$ does not have property P. Show that the product topology is maximal with respect to compactness and minimal with respect to T_2.

5. A space X is *super compact* if and only if X has a subbasis $\mathscr{S}$ with the property that any cover of X by members of $\mathscr{S}$ has a subcover consisting of just two sets.

   (a) Show that $\mathbf{I}$ is super compact; show that a triod is super compact,

   (b) Show that any super compact space is compact,

   (c) Show that the product of super compact spaces is super compact,

   (d)* Find a compact space that is not super compact.

6.* Suppose that X_1 and X_2 are nondiscrete T_1 first countable spaces. Show that if $p_2 : X_1 \times X_2 \to X_2$ is closed, then X_1 is sequentially compact (p_2 is the projection map).

7.* Show that if X is compact and Y countably compact, then $X \times Y$ is countably compact.

8. Show that sequential compactness is preserved under continuous maps.

9. Let $X = \Pi_{\alpha \in \mathbf{R}} \mathbf{I}_\alpha$, and let $Y = [0, \Omega)$ with the order topology. Show that Y is

countably compact and $X \times Y$ is countably compact, but that $X \times Y$ is neither compact nor sequentially compact.

10. Suppose that $\{X_\alpha \mid \alpha \in \Lambda\}$ is a family of topological spaces, X is a topological space, and for each $\alpha \in \Lambda$, $f_\alpha : X \to X_\alpha$ is continuous. Show that there is a unique continuous function $f : X \to \prod_{\alpha \in \Lambda} X_\alpha$ such that $p_\alpha f = f_\alpha$.

11. Let $\{X_\alpha \mid \alpha \in \Lambda\}$ be a family of spaces, and let Y be an arbitrary space. Let $\{q_\alpha : Y \to Y_\alpha \mid \alpha \in \Lambda\}$ be a family of continuous open maps from Y onto X_α. Suppose that for each topological space X and for each family of continuous maps $\{f_\alpha : X \to X_\alpha \mid \alpha \in \Lambda\}$, there is a unique continuous map $\hat{f} : X \to Y$ such that $q_\alpha \hat{f} = f_\alpha$ for each $\alpha \in \Lambda$. Show that Y is homeomorphic with $\prod_{\alpha \in \Lambda} X_\alpha$.

12. Let $\{X_\alpha \mid \alpha \in \Lambda\}$ and $\{Y_\alpha \mid \alpha \in \Lambda\}$ be two families of topological spaces and let $\{f_\alpha : X_\alpha \to Y_\alpha \mid \alpha \in \Lambda\}$ be a family of continuous functions. Show that there is a unique function $f' : \prod_{\alpha \in \Lambda} X_\alpha \to \prod_{\alpha \in \Lambda} Y_\alpha$, such that the following diagram commutes for each α.

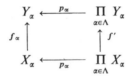

[Hint: Use problem 10 above.]

13. Suppose that $\{X_\alpha \mid \alpha \in \Lambda\}$ is a family of topological spaces and that $\Lambda = \Lambda_1 \cup \Lambda_2$ and $\Lambda_1 \cap \Lambda_2 = \varnothing$. Show that $\prod_{\alpha \in \Lambda} X_\alpha$ is homeomorphic with $\prod_{\alpha \in \Lambda_1} X_\alpha \times \prod_{\alpha \in \Lambda_2} X_\alpha$, and that $\prod_{(\beta,\gamma) \in \Lambda_1 \times \Lambda_2} (X_\beta \times X_\gamma)$ is homeomorphic with $\prod_{\alpha \in \Lambda_1} (\prod_{\alpha \in \Lambda_2} X_\alpha)$.

14. Show that the Tihonov theorem does not hold if $\prod_{\alpha \in \Lambda} X_\alpha$ is given the box topology.

15.* Suppose that $\{X_\alpha \mid \alpha \in \Lambda\}$ is a family of topological spaces. Show that $\prod_{\alpha \in \Lambda} X_\alpha$ is separable if and only if each X_α is separable and all but at most $2^{\aleph_0}$ spaces are a point ($2^{\aleph_0}$ is the number of functions from $\mathbf{Z}$ into the set $\{0,1\}$). [Hint: See Comfort [1969]].

Sections B and C

1. Show that every totally disconnected, compact metric space is homeomorphic to a subset of the Cantor set.

2. Show that there is a subspace of the irrationals in $[0,1]$ that is homeomorphic to the Cantor set.

3. Suppose that $(X_\alpha, f_{\alpha\beta}, D)$ is an inverse system such that each X_α is compact. Show that $\hat{p}_\alpha(X_\infty) = \bigcap_{\beta > \alpha} f_{\alpha\beta}(X_\beta)$.

4. Let $D = \{A \subset \mathbf{R}^1 \mid A \text{ is finite}\}$. For each $A \in D$, let $X_A = \{i : A \to \mathbf{Z} \mid i \text{ is } 1\text{--}1\}$. Order D by $A \leq B$ if and only if $A \subset B$. Observe that X_A is countable, and if $A \leq B$, then $f_{AB} : X_B \to X_A$ defined by $f_{AB}(g) = g_{|A}$ is onto. Find the inverse limit of (X_A, f_{AB}, D).

5.* Suppose that $(X_\alpha, f_{\alpha\beta}, D)$ is an inverse system of compact T_2 spaces. Show that if X_α has no more than k components for each α, then X_∞ has no more than k components.

6. Show that $\mathbf{I}$ is the inverse limit of an inverse system where all of the X_α's are $\mathscr{S}^1$.

7. Suppose that $(X_\alpha, f_{\alpha\beta}, D)$ is an inverse system and $A \subset X_\infty$ is compact. For each α, let $A_\alpha = \hat{p}_\alpha^{-1}(A)$. Let $A^* = \prod_{\alpha \in D} A_\alpha$. Define $g_{\alpha\beta} = f_{\alpha\beta | A_\alpha}$.

    (a) Show that $(A_\alpha, g_{\alpha\beta}, D)$ is an inverse system.
    (b) Let A_∞ be the inverse limit of $(A_\alpha, g_{\alpha\beta}, D)$. Show that $A_\infty = A^* \cap X_\infty = A$.

8. Let $(Y_\alpha, f_{\alpha\beta}, D)$ be an inverse system of compact T_2 spaces. Suppose that each $f_{\alpha\beta}$ is monotone (a function $f : X \to Y$ is *monotone* if and only if $f^{-1}(y)$ is connected for each $y \in Y$). Show that for each $\alpha \in D$, $\hat{p}_\alpha : X_\infty \to X_\alpha$ is monotone.

9. Suppose that $(X_\alpha, f_{\alpha\beta}, D)$ is an inverse system and that there is an integer k such that card $X_\infty < k$ for each α. Show that card $X_\infty < k$.

10.* Is the hypothesis "$\hat{p}_\alpha$ is onto" necessary in (6.B.15)?

11. Find examples that show that the hypotheses of (6.C.9) are necessary.

Section D

1. Consider the Cantor set as an inverse limit and show that it is homogeneous.

Chapter 7

FUNCTION SPACES, WEAK TOPOLOGIES AND HILBERT SPACE

Inverse systems presented us with some rather amazing examples of what can be done with subspaces of product spaces. Somewhat less exciting, but probably more ubiquitous are two further offshoots of the product topology: function spaces and the weak topology. Function spaces are essential in the study of both topology and analysis, and one version of the weak topology may be seen as an attempt to generalize certain basic properties of the product topology. Hilbert space represents an extraordinarily rich topological space, one whose properties are still the subject of substantial topological research.

A. THE POINT-OPEN TOPOLOGY

Product spaces serve as a convenient starting point for the study of function spaces. The "points" of a function space consist of the members of a family of functions. Various topologies may be given to such collections. If X and Y are topological spaces, then $\text{Hom}(X, Y)$ will denote the set of all (not necessarily continuous) functions from X into Y. One way to topologize $\text{Hom}(X, Y)$ is to observe that each of its elements may be regarded as an "X-tuple," i.e., $\text{Hom}(X, Y) = \prod_{x \in X} Y_x$, where $Y_x = Y$ for each x; thus, we may (and do) assign the product topology to $\text{Hom}(X, Y)$.

In general, a thorough understanding of a product topology lies in seeing clearly the nature of the subbasic sets. This is especially true in the present context. Note that with each point $x \in X$ and each open set $U \subset Y$, there can be associated a subbasic element of $\text{Hom}(X,Y)$, $S = \{f : X \to Y \,|\, f(x) \in U \subset Y_x\}$; furthermore, all sets in the usual subbasis for $\text{Hom}(X,Y)$ are of this form. Since such sets depend on the point x and the open subset U, the product topology when given to $\text{Hom}(X,Y)$ is called the *point-open topology*.

(7.A.1) Exercise. Show that subbasic sets of $\text{Hom}(X,Y)$ are of the form $S = \{f : X \to Y \,|\, f(x) \in U\}$, where U is an open subset of Y, and $x \in X$.

In the context of function spaces, the projection maps are of special interest. For each $x \in X$, let $e_x : \text{Hom}(X,Y) \to Y_x$ denote the natural projection and observe that if $f \in \text{Hom}(X,Y)$, then $e_x(f) = f(x)$. Thus, e_x evaluates a given function f at the point x. The projection maps e_x are called *evaluation maps*. The point-open topology is the smallest topology for which the evaluation maps are continuous and open (6.A.3). Evaluation maps will play an important role in a number of subsequent proofs.

The point-open topology is frequently given another name: the *topology of pointwise convergence*. The next theorem indicates why.

(7.A.2) Theorem. Suppose that $\text{Hom}(X,Y)$ is given the point-open topology and $\{f_n\}$ is a sequence of points in $\text{Hom}(X,Y)$. Let $f \in \text{Hom}(X,Y)$. Then the sequence $\{f_n\}$ converges to f if and only if the sequence $\{f_n(x)\}$ converges to $f(x)$ for each $x \in X$.

Proof. Suppose that $\{f_n\}$ converges to f and that $x \in X$. If U is an open set in Y containing $f(x)$, then $f \in e_x^{-1}(U)$. Since the sequence $\{f_n\}$ converges to f, there is a positive integer N such that for $n > N$, $f_n \in e_x^{-1}(U)$, which in turn implies that for $n > N$, $f_n(x) \in U$. Thus, $\{f_n(x)\}$ converges to $f(x)$.

Conversely, suppose that for each $x \in X$, $\{f_n(x)\}$ converges to $f(x)$. Let $x \in X$ be given and let $S = \{g : X \to Y \,|\, g(x) \in U \subset Y_x\}$ be a subbasic open set that contains f. Since $\{f_n(x)\}$ converges to $f(x)$, there is a positive integer N such that if $n > N$, then $f_n(x) \in U$. Consequently, if $n > N$, then $f_n \in S$, which completes the proof. (Why is it sufficient to consider only subbasic open sets in the proof?)

B. THE COMPACT-OPEN TOPOLOGY

One unsatisfactory aspect of the point-open topology is that it does not in any way reflect the topology of X. Hence, it is quite possible that a sequence $\{f_n\}$ of continuous functions in $\text{Hom}(X,Y)$ may converge to a function f in $\text{Hom}(X,Y)$ that is not continuous.

(7.B.1) Exercise. Suppose that $X = Y = [0,1]$. Find an example of a

sequence of continuous functions in $\mathrm{Hom}(X,Y)$ that converges to a function that is not continuous.

To avoid this situation, we generalize the construction that yielded the point-open topology.

(7.B.2) Definition. Suppose that X and Y are topological spaces. For each compact subset A of X and each open subset U of Y, define $S(A,U) = \{f \in \mathrm{Hom}(X,Y) \mid f(A) \subset U\}$. Then the *compact-open topology* for $\mathrm{Hom}(X,Y)$ is that topology which has a subbasis $\mathscr{S} = \{S(A,U) \mid A \text{ compact in } X \text{ and } U \text{ open in } Y\}$.

This topology turns out to be perhaps the most important and useful of the function space topologies. One should note that the point-open topology is a subset of the compact-open topology. Hence, if the evaluation maps e_x are defined as before, continuity of each e_x is immediate.

The following is an easy but mildly interesting exercise.

(7.B.3) Exercise. Suppose that $\mathrm{Hom}(X,Y)$ has the compact-open topology. Show that $\mathrm{Hom}(X,Y)$ is T_0, T_1, or T_2 if and only if Y is.

(7.B.4) Notation. Suppose that X and Y are topological spaces. Then $\mathrm{C}(X,Y)$ will denote $\{f : X \to Y \mid f \text{ is continuous}\}$.

(7.B.5) Theorem. Suppose that Y is a T_3 space and that X is a topological space. Then the compact-open topology on $\mathrm{C}(X,Y)$ is T_3.

Proof. First we show that if A is a compact subset of X, V is an open subset of Y, and $g \in W = S(A,V) \cap \mathrm{C}(X,Y)$, then there is an open set U in $\mathrm{C}(X,Y)$ such that $g \in U \subset \bar{U} \subset W$. For each $x \in A$, we have that $g(x) \in V$. Since Y is regular, there is an open set V_x in Y such that $g(x) \in V_x \subset \bar{V}_x \subset V$. Then the family $\{V_x \mid x \in A\}$ is an open cover of the compact set $g(A)$. Let $\{V_{x_1}, \ldots, V_{x_n}\}$ be a finite subcover. We will show that $U = S(A, \bigcup_{i=1}^{n} V_{x_i}) \cap \mathrm{C}(X,Y)$ is the desired open set containing g and with closure lying in W. Certainly $g \in U$. To see that $\bar{U} \subset W$, first note that since $\overline{V_{x_i}} \subset \bigcup_{i=1}^{n} V_{x_i} = \bigcup_{i=1}^{n} \overline{V_{x_i}} \subset V$, we have that $\{h \in \mathrm{C}(X,Y) \mid h(A) \subset \overline{\bigcup_{i=1}^{n} V_{x_i}}\} \subset W$. If $\hat{k} \in \mathrm{C}(X,Y)$ and $\hat{k} \notin W$, then by the previous observation there is an $a \in A$ such that $\hat{k}(a) \notin W$ and hence $\hat{k}(a) \notin \overline{\bigcup_{i=1}^{n} V_{x_i}}$. Therefore, $\hat{k} \in \{k \in \mathrm{C}(X,Y) \mid k(a) \in Y \setminus \overline{\bigcup_{i=1}^{n} V_{x_i}}\} =$

$\hat{U}$, which is an open set in $\mathbf{C}(X,Y)$ disjoint from U (if $k \in U \cap \hat{U}$, then $k(a) \in$ $(\bigcup_{i=1}^{n} V_{x_i}) \cap (Y \setminus \bigcup_{i=1}^{n} V_{x_i})$). Thus, $\hat{k} \notin \bar{U}$, and consequently $\bar{U} \subset W$. Hence, we have that $g \in U \subset \bar{U} \subset W$.

Now suppose that $W = (\bigcap_{i=1}^{n} S(A_i, V_i)) \cap \mathbf{C}(X,Y)$, where $S(A_i, V_i)$ is a member of the subbasis for $\mathrm{Hom}(X,Y)$. Suppose that $g \in W$. For each i, there is an open set U_i in $\mathbf{C}(X,Y)$ such that $g \in U_i \subset \bar{U}_i \subset S(A_i, V_i) \cap$ $\mathbf{C}(X,Y)$. Let $U = \bigcap_{i=1}^{n} U_i$. Then $g \in U \subset \bar{U} \subset W$ and hence, by (4.D.2), $\mathbf{C}(X,Y)$ is T_3.

(7.B.6) **Definition.** Suppose that X is a set and (Y,d) is a metric space. Then a function $f : X \to Y$ is *bounded* if and only if diam $f(X)$ is finite.

(7.B.7) **Notation.** Suppose that X is a topological space and (Y,d) is a metric space. Then $\mathbf{B}(X,Y)$ will denote $\{f \in \mathbf{C}(X,Y) \mid f \text{ is bounded}\}$.

Another topology, the sup topology, is commonly assigned to function spaces.

(7.B.8) **Definition.** Suppose that X is a topological space and that (Y,d) is a metric space. Then the *sup topology* for $\mathbf{B}(X,Y)$ is the topology generated by the metric $D(f,g) = \sup\{d(f(x),g(x)) \mid x \in X\}$.

(7.B.9) **Exercise.** Show that D is a metric.

The versatility of the compact-open topology is illustrated in the following theorem.

(7.B.10) **Theorem.** Suppose that (Y,d) is a metric space and X is a compact topological space. Then the sup topology on $\mathbf{C}(X,Y)$ coincides with the compact-open topology.

Proof. Let $\mathcal{T}$ be the compact-open topology for $\mathbf{C}(X,Y)$ and let $\mathcal{T}'$ be the sup topology. We first show that $\mathcal{T} \subset \mathcal{T}'$. (Compactness of X is not needed here except to guarantee that $\mathbf{C}(X,Y) = \mathbf{B}(X,Y)$.) Suppose that S is a subbasic open set of $\mathcal{T}$, i.e., $S = S(A,V) \cap \mathbf{C}(X,Y)$, where A is compact in X and V is open in Y. Let $g \in S$. We shall find an $\varepsilon > 0$ such that $S_\varepsilon(g) \subset S$. Since $g(A)$ is compact in V, there is an $\varepsilon > 0$ such that $d(g(A), Y \setminus V) = \varepsilon$. If $h \in S_\varepsilon(g)$, then $d(g(x),h(x)) < \varepsilon$ for each $x \in X$, and hence $h(x) \in V$ for each $x \in A$. Consequently, we have that $h \in S$, and therefore $S_\varepsilon(g) \subset S$.

To show that $\mathcal{T}' \subset \mathcal{T}$, suppose that $S_\varepsilon(f)$ is a basic open set of $\mathcal{T}'$. We find a finite number of subbasic open sets $S_1, \ldots, S_n$ of $\mathcal{T}$ such that $f \in$

$\bigcap_{i=1}^{n} S_i \subset S_\varepsilon(f)$. Since X is compact, so is $f(X)$, and therefore there are sets $S_{\varepsilon/6}(f(x_1)), \ldots, S_{\varepsilon/6}(f(x_n))$ that cover $f(X)$. For each i, $1 \le i \le n$, let $C_i = f^{-1}(\overline{S_{\varepsilon/6} f(x_i)})$ and $V_i = S_{\varepsilon/3}(f(x_i))$. Then C_i is compact, since it is a closed subset of the compact space X. Finally, let $S_i = S(C_i, V_i) \cap C(X,Y)$. Clearly, we have that $f \in \bigcap_{i=1}^{n} S_i$. Thus it remains to show that $\bigcap_{i=1}^{n} S_i \subset S_\varepsilon(f)$. If $g \in \bigcap_{i=1}^{n} S_i$, then $g(C_i) \subset V_i$ for each i. Let $x \in X$, and let C_j be a member of the cover $\{C_1, \ldots, C_n\}$ of X that contains x. Then both $g(x)$ and $f(x)$ are in V_j, and consequently, $d(g(x), f(x)) < 2\varepsilon/3$. Hence, we have that $\sup_{x \in X}\{d(f(x), g(x))\} < \varepsilon$ and therefore $g \in S_\varepsilon(f)$.

The sup topology (or equivalently the compact-open topology on $C(X,Y)$, when X is compact and Y is metric) is frequently referred to as the *topology of uniform convergence*. The following exercise gives the motivation for this terminology.

(7.B.11)   Exercise.   Suppose that X is a compact space and that (Y,d) is a metric space. Show that if a sequence $\{f_n\}$ converges to f in $C(X,Y)$ (with the sup topology), then $\{f_n\}$ converges to f uniformly, i.e., given $\varepsilon > 0$, there is an $N \in \mathbf{Z}^+$ such that if $n > N$, then $d(f_n(x), f(x)) < \varepsilon$ for all $x \in X$.

(7.B.12)   Definition.   Suppose that X and Y are topological spaces. Then the function $e : \mathrm{Hom}(X,Y) \times X \to Y$ defined by $e(f,x) = f(x)$ is called the *evaluation function* associated with $\mathrm{Hom}(X,Y)$.

Note that $e(f,x) = e_x(f)$, where e_x is the evaluation map discussed previously. Our principal result involving e is the following.

(7.B.13)   Theorem.   If X and Y are topological spaces and X is locally compact, then $e : C(X,Y) \times X \to Y$ is continuous where $C(X,Y)$ has the compact-open topology.

Proof.   Suppose that $g \in C(X,Y)$ and $x \in X$. Let U be an open subset of Y containing $e(g,x) = g(x)$. Since X is locally compact, there is an open set V with compact closure such that $x \in V \subset \overline{V} \subset g^{-1}(U)$. Let $A = S(\overline{V},U) \cap C(X,Y)$. Since $A \times V$ is an open subset of $C(X,Y) \times X$ containing (g,x), to complete the proof it suffices to show that $e(A \times V) \subset U$. However, this is trivial, since if $f \in A$ and $v \in V$, then $e(f,v) = f(v) \in f(\overline{V}) \subset U$.

Suppose that X, Y, and Z are topological spaces and that $f : X \times Y \to Z$ is a continuous function of y for each fixed $x \in X$. Define a function $f^* : X \to$

$C(Y,Z)$ by $(f^*(x))(y) = f(x,y)$. If $C(Y,Z)$ is given the compact-open topology, then f and f^* enjoy the following close relationship.

(7.B.14) Theorem. Suppose that $f : X \times Y \to Z$ and f^* are as defined above. Then if f is continuous, so is f^*, and if f^* is continuous and Y is locally compact, then f is continuous.

Proof. Suppose that f is continuous and that $x \in X$. Let $G = S(A,V) \cap C(Y,Z)$, where A is compact in Y and V is open in Z, and suppose that $f^*(x) \in G$. We must find an open set U in X such that $x \in U$ and $f^*(U) \subset G$. Note that this is equivalent to finding an open set U in X such that for each $u \in U, (f^*(u))(A) \subset V$ or, somewhat more aesthetically expressed, $f(U \times A) \subset V$.

Since $x \in f^{*-1}(G)$, we have that $\{x\} \times A \subset f^{-1}(V)$, and therefore, by (2.G.14), there is an open neighborhood U of x with the property that $U \times A \subset f^{-1}(V)$. Hence, $f^*(U) \subset V$.

Suppose now that f^* is continuous. Let $h : X \times Y \to C(Y,Z) \times Y$ be defined by $h(x,y) = (f^*(x),y)$. Then h is continuous, and the continuity of f follows from the fact that $f = eh$, where e is the evaluation function.

(7.B.15) Definition. Suppose that $\mathcal{F} = \{f_\alpha : X \to X_\alpha \mid \alpha \in \Lambda\}$ is a collection of functions from a space X into arbitrary spaces X_α. Then $\mathcal{F}$ *separates points from closed sets* in X if and only if whenever B is a closed subset of X and $x \in X \setminus B$, there is an $\alpha \in \Lambda$ such that $f_\alpha(x) \notin \overline{f_\alpha(B)}$.

The following result will prove to be central in proofs of a number of metrization theorems as well as in our brief examination of Stone-Čech compactifications (see problem A-B.9 at the end of this chapter).

(7.B.16) Theorem. Suppose that X is a T_1 space and $\mathcal{F} = \{f_\alpha : X \to X_\alpha \mid \alpha \in \Lambda\}$ is a family of continuous functions that separates points from closed subsets. Then $ê : X \to \prod_{\alpha \in \Lambda} X_\alpha$ defined by $ê(x) = \{f_\alpha(x)\}$ is an embedding.

Proof. By (6.A.5), $ê$ is continuous. If $x \neq y$, there is an $\alpha \in \Lambda$ such that $f_\alpha(x) \neq f_\alpha(y)$ (points are closed and $\mathcal{F}$ separates points from closed sets). Hence, $ê$ is 1–1. To complete the proof, we show that $ê$ is an open mapping. It clearly suffices to show that if U is a member of a basis for X, then $ê(U)$ is open in $ê(X)$. First, we construct a basis suitable for our purposes. Let $\mathbf{B} = \{f_\alpha^{-1}(V) \mid \alpha \in \Lambda, V \text{ open in } X_\alpha\}$. That $\mathbf{B}$ is a basis may be seen as follows. Suppose that $x \in W$, where W is an arbitrary open set in X. Obviously, x does not belong to the closed set $X \setminus W$, and hence there is an $\alpha \in \Lambda$ such that $f_\alpha(x) \notin \overline{f_\alpha(X \setminus W)}$. Let $V = X_\alpha \setminus \overline{f_\alpha(X \setminus W)}$. Then it is easily established that $x \in f_\alpha^{-1}(V) \subset W$; therefore, $\mathbf{B}$ is a basis.

Now consider $\hat{e}(f_\alpha^{-1}(V))$. First observe that $f_\alpha = p_\alpha \hat{e}$, where p_α is the natural projection from the product space into the α-th factor. Note that $\hat{e}(f_\alpha^{-1}(V)) = p_\alpha^{-1}(V) \cap \hat{e}(X)$ and that $p_\alpha^{-1}(V) \cap \hat{e}(X)$ is an open set in $\hat{e}(X)$. Thus, $\hat{e}$ is an open map, and consequently $\hat{e}$ is an embedding.

As an interesting corollary, we have that any Tihonov space can be embedded in a product of unit intervals.

(7.B.17) Corollary. Suppose that X is a Tihonov space. Let $\mathscr{F} = \{f : X \to \mathbf{I} \mid f \text{ is continuous.}\}$ Then $\hat{e} : X \to \underset{f \in F}{\Pi} \mathbf{I}_f$ is an embedding (where $\mathbf{I}_f = \mathbf{I}$ for each $f \in \mathscr{F}$, and the f-th coordinate of $\hat{e}(x)$ is $f(x)$).

(7.B.18) Exercise. Suppose that X and Y are Tihonov spaces and that $h : X \to Y$ is continuous. Let $\mathscr{F} = \{f : X \to \mathbf{I} \mid f \text{ is continuous}\}$ and let $\mathscr{G} = \{g : Y \to \mathbf{I} \mid g \text{ is continuous}\}$. Show that there is a map $\psi : \underset{f \in \mathscr{F}}{\Pi} \mathbf{I}_f \to \underset{g \in \mathscr{G}}{\Pi} \mathbf{I}_g$ such that the following diagram is commutative.

$$
\begin{array}{ccc}
X & \xrightarrow{\ h\ } & Y \\
\hat{e} \downarrow & & \downarrow \hat{e}' \\
\underset{f \in \mathscr{F}}{\Pi} \mathbf{I}_f & \xrightarrow{\ \psi\ } & \underset{g \in \mathscr{G}}{\Pi} \mathbf{I}_g
\end{array}
$$

[Hint: If $\{z_f\} \in \underset{F \in \mathscr{F}}{\Pi} \mathbf{I}_f$, let the g-th coordinate of $\psi(\{z_f\})$ be the (gh)-th coordinate of $\{z_f\}$.] Show also that $\psi(\overline{\hat{e}(X)}) \subset \overline{\hat{e}'(Y)}$.

C.   THE WEAK TOPOLOGY: I

Suppose that $\{X_\alpha \mid \alpha \in \Lambda\}$ is a family of topological spaces. A good deal of care was taken when we defined the product topology for $\underset{\alpha \in \Lambda}{\Pi} X_\alpha$ to ensure that the projection maps $p_\alpha : \underset{\alpha \in \Lambda}{\Pi} X_\alpha \to X_\alpha$ would be continuous, but yet that $\underset{\alpha \in \Lambda}{\Pi} X_\alpha$ would have as few open sets as possible. This procedure is generalized as follows.

Suppose that $\{X_\alpha \mid \alpha \in \Lambda\}$ is a family of spaces and that X is an arbitrary set. Suppose, furthermore, that for each $\alpha \in \Lambda$, there is a function $f_\alpha : X \to X_\alpha$. Our goal is to define a topology for X that guarantees the continuity of each f_α. Of course, if X is given the discrete topology, then each f_α will be continuous. However, as was the case with the product topology, we would like to keep the topology as small as possible. We do this by applying the procedure used in defining the product topology. Let $\mathscr{S}$ be the family of all sets of the form $f_\alpha^{-1}(U_\alpha)$, where U_α is open in X_α and $\alpha \in \Lambda$. Then $\mathscr{S}$ is a subbasis for a topology for X, called the *weak topology* (or *initial topology*) *induced by the family* $\{f_\alpha \mid \alpha \in \Lambda\}$. It should be noted that in this terminology, the product topology for $Y = \prod_{\alpha \in \Lambda} X_\alpha$ is just the weak topology for Y induced by the natural projection maps $\{p_\alpha \mid \alpha \in \Lambda\}$.

One of the most useful properties of the weak topology (and hence of the product topology) is given in the next theorem.

(7.C.1)   Theorem.   Suppose that X is given the weak topology induced by the family $\{f_\alpha : X \rightarrow X_\alpha \mid \alpha \in \Lambda\}$. Let $g : Y \rightarrow X$, where Y is an arbitrary topological space. Then g is continuous if and only if $f_\alpha g$ is continuous for each $\alpha \in \Lambda$.

Proof.   Obviously, if g is continuous, then so is $f_\alpha g$. Suppose, then, that $f_\alpha g$ is continuous for each $\alpha \in \Lambda$, and let U be an open set in X. It suffices to show that $g^{-1}(U)$ is open in Y. By (6.A.9), it may be assumed that $U = f_\alpha^{-1}(U_\alpha)$ for some $\alpha \in \Lambda$, where U_α is open in X_α. Since $g^{-1}(U)$ equals $g^{-1}(f_\alpha^{-1}(U_\alpha))$, which is open by hypothesis, it follows that g is continuous.

We mention that (7.B.16) may easily be generalized (with practically no change in the proof) to handle the case where X has been given the weak topology induced by some family of functions.

D.   THE WEAK TOPOLOGY: II

In the literature, one frequently encounters another topology called the weak topology, a topology that is induced by subsets of a given set rather than by a family of functions. Suppose that X is a union of a collection of sets $\{A_\alpha \mid \alpha \in \Lambda\}$, where each A_α enjoys its own topology. We have already seen that if the A_α are disjoint, then X may be given the free union topology. Suppose, however, that the A_α are not necessarily disjoint, but that

(i)   the topologies of A_α and A_β coincide on $A_\alpha \cap A_\beta$ for each $\alpha, \beta \in \Lambda$, i.e., the relative topology that $A_\alpha \cap A_\beta$ inherits from A_α is the same as that which it inherits from A_β, and

(ii)   either (a) $A_\alpha \cap A_\beta$ is open in both A_α and A_β for all $\alpha, \beta \in \Lambda$ or (b) $A_\alpha \cap A_\beta$ is closed in both A_α and A_β for all $\alpha, \beta \in \Lambda$.

Then $\{U \subset X \mid U \cap A_\alpha$ is open in A_α for each $\alpha \in \Lambda\}$ is a topology that is called the *weak topology for X associated with* (or *induced by*) *the spaces*

$\{A_\alpha \mid \alpha \in \Lambda\}$. It should be apparent that the free union topology is simply a special case of this topology.

The weak topology is used in the construction of what are known as cell complexes (to be studied in Chapter 8). Suppose that during an idle moment one started to build a space by adding to an interval, a triangle, to the triangle, a tetrahedron, to the tetrahedron, some four-dimensional analog, etc. Since each piece of the resulting space has its respective Euclidean topology, the weak topology may be assigned to the union of the pieces.

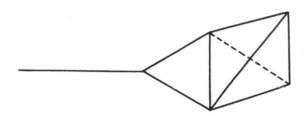

One virtue of this particular version of the weak topology is given by the following theorem.

(7.D.1) **Theorem.** Suppose that X has the weak topology induced by a family of subsets $\{A_\alpha \mid \alpha \in \Lambda\}$ and that Y is an arbitrary space. Then a function $f : X \to Y$ is continuous if and only if $f|_{A_\alpha} : A_\alpha \to Y$ is continuous for each $\alpha \in \Lambda$.

Proof. Since restrictions of continuous maps are continuous, it suffices to show that if all the restrictions are continuous, then so is f. Suppose that U is an open subset of Y. Since $f^{-1}(U) \cap A_\alpha$ equals $f_{|A_\alpha}^{-1}(U)$, which is open in A_α for each α, it follows from the definition of the weak topology that $f^{-1}(U)$ is open in X. Therefore, f is continuous.

The following theorem is needed in the next chapter.

(7.D.2) **Theorem.** Suppose that $X_1 \subset X_2 \subset X_3 \subset \cdots$ is a sequence of topological spaces and that $X = \bigcup_{n=1}^{\infty} X_n$ is given the weak topology induced by the X_n. If each X_n is normal and closed in X, then X is normal.

Proof. Suppose that A and B are closed disjoint subsets of X. To show that X is normal, we construct a continuous function $f : X \to [0,1]$ such that $f(A) = 0$ and $f(B) = 1$ (4.B.1). Choose N large enough so that for each $n > N$, we have $X_n \cap A \neq \varnothing$ and $X_n \cap B \neq \varnothing$. Since X_N is normal, Uryson's lemma yields a continuous function $f_N : X_N \to [0,1]$ such that $f_N(X_N \cap A) = 0$ and

$f_N(X_N \cap B) = 1$. Define $\hat{f}_N : X_N \cup (X_{N+1} \cap A) \cup (X_{N+1} \cap B) \to [0,1]$ by

$$\hat{f}_N(x) = \begin{cases} 0 & \text{if} & x \in X_{N+1} \cap A \\ 1 & \text{if} & x \in X_{N+1} \cap B \\ f_N(x) & \text{if} & x \in X_N \end{cases}$$

By the map gluing theorem, $\hat{f}_N$ is continuous. Tietze's theorem may be applied to extend $\hat{f}_N$ to a continuous function $f_{N+1} : X_{N+1} \to [0,1]$, where of course $f_{N+1}(X_{N+1} \cap A) = 0$ and $f_{N+1}(X_{N+1} \cap B) = 1$.

Inductively, for $i \geq N$ we obtain a function $f_i : X_i \to [0,1]$ such that $f_{i|X_j} = f_j$ whenever $N \leq j \leq i$ and furthermore $f_i(X_i \cap A) = 0$ and $f_i(X_i \cap B) = 1$. Define $f : X \to [0,1]$ by $f(x) = f_n(x)$ whenever $x \in X_n$. Then clearly, f is well defined, $f(A) = 0$, and $f(B) = 1$. It remains to show that f is continuous. Let U be an open subset of $[0,1]$ and observe that $f^{-1}(U) = \bigcup_{n=N}^{\infty} f_n^{-1}(U)$ and that each $f_n^{-1}(U)$ is open in X_n. However, $f_n^{-1}(U) = X_n \cap f^{-1}(U)$ and hence by the definition of the weak topology induced by the family $\{X_n \mid n \in \mathbf{Z}^+\}$, $f^{-1}(U)$ is open in X. Therefore, f is continuous and by (4.B.1) X is normal.

E.   HILBERT SPACE

One of the more interesting spaces in topology, both because of its versatility and because of its unpredictability, is the classical Hilbert space. Hilbert space serves as an important bridge in spanning the hiatus between analysis and topology; we, however, will focus our attention on some of its topological properties. Although it has recently been shown that Hilbert space is homeomorphic to the countable product of $\mathscr{E}^1$ with itself (Anderson [1966]), we begin with the classical description of Hilbert space.

Let $\mathscr{H}$ denote the set of all infinite sequences $\{x_1, x_2, \ldots\}$ in $\mathbf{R}^1$ that are square summable (i.e., $\sum_{i=1}^{\infty} x_i^2 < \infty$). We shall frequently denote such sequences by $\langle x_n \rangle$ or $\langle x_1, x_2, \ldots \rangle$. A metric ρ is defined on $\mathscr{H}$ in a manner entirely analogous to that used to define the usual Euclidean metric for $\mathscr{E}^n$. If $x = \langle x_n \rangle$ and $y = \langle y_n \rangle$ are elements of $\mathscr{H}$, then define $\rho(x,y) = \sqrt{\sum_{n=1}^{\infty} (x_n - y_n)^2}$. That ρ is a metric is an easy consequence of (0.F.4). (Square summable sequences are used to ensure that $\rho(x,y)$ is a real number.) The metric space $(\mathscr{H}, \rho)$ is called *Hilbert space*.

To gain an idea of the "size" of $(\mathscr{H}, \rho)$, note that for each n, $\mathscr{E}^n$ can be embedded isometrically in $\mathscr{H}$ via the map $\psi_n : \mathscr{E}^n \to \mathscr{H}$ defined by $\psi_n((x_1, x_2, \ldots, x_n)) = \langle x_1, \ldots, x_n, 0, 0, \ldots \rangle$. Further indication of the vastness of $\mathscr{H}$ is exemplified by the fact that any separable metric space may be embedded in a compact subset of $\mathscr{H}$ that has empty interior (7.E.8).

Before turning to the more unusual features of $(\mathscr{H},\rho)$, we first show that it is both separable and complete.

(7.E.1)   Theorem.   Hilbert Space $(\mathscr{H},\rho)$ is a separable metric space.

Proof.   We exploit the separability of $\mathscr{E}^n$ to find a countable dense subset of $\mathscr{H}$. For each $n \in \mathbf{Z}^+$, let $D_n = \{\langle r_1, r_2, \ldots, r_n, 0, 0, \ldots\rangle \in \mathscr{H} \mid r_i$ is rational$\}$, and let $D = \bigcup_{n=1}^{\infty} D_n$. It is clear that D is countable, and thus it remains to show that D is dense. Let $\varepsilon > 0$ be given and suppose that $x = \langle x_n\rangle \in \mathscr{H}$. We show that $D \cap S_\varepsilon(x) \neq \varnothing$. Choose N large enough so that $\sum_{i=N+1}^{\infty} x_i^2 < \varepsilon^2/4$. For $i = 1, 2, \ldots, N$, let r_i be a rational number with the property that $|x_i - r_i| < \varepsilon/\sqrt{4N}$. Then if $r = \langle r_1, r_2, \ldots, r_n, 0, 0, \ldots\rangle$, we have that $\rho(x,r) \leq \sqrt{\sum_{i=1}^{n} (x_i - r_i)^2} + \sqrt{\sum_{i=N+1}^{\infty} (x_i^2)} < \varepsilon/2 + \varepsilon/2$, and hence $r \in D \cap S_\varepsilon(x)$.

(7.E.2)   Theorem.   Hilbert space $(\mathscr{H},\rho)$ is a complete metric space.

Proof.   Suppose that $x^1 = \langle x_1^1, x_2^1, \ldots\rangle$, $x^2 = \langle x_1^2, x_2^2, \ldots\rangle, \ldots$ is a Cauchy sequence in $(\mathscr{H},\rho)$. For each k, every sequence of the form $\{x_k^i\}_{i=1}^{\infty}$ is a Cauchy sequence in $\mathscr{E}^1$, since for each i and j, we have that $|x_k^i - x_k^j| \leq \rho(x^i,x^j)$. Consequently, for each k, the sequence $\{x_k^i\}_{i=1}^{\infty}$ converges to a point $y_k \in \mathscr{E}^1$ (3.B.3). Let $y = \langle y_1, y_2, \ldots\rangle$. We simultaneously show that $y \in \mathscr{H}$ and that y is the limit of the sequence $x^1, x^2, \ldots.$

First note that $\lim_{m \to \infty} \sum_{i=1}^{p} (x_i^n - x_i^m)^2 = \sum_{i=1}^{p} (x_i^n - \lim_{m \to \infty} x_i^m)^2 = \sum_{i=1}^{p} (x_i^n - y_i)^2$. Since $x^1, x^2, \ldots$ is a Cauchy sequence, there is a positive integer N such that if $m,n > N$, then $\sum_{i=1}^{p} (x_i^n - x_i^m)^2 < (\varepsilon/2)^2$ for all integers p. Consequently, for $n > N$ we have that $\lim_{m \to \infty} \sum_{i=1}^{p} (x_i^n - y_i)^2 \leq (\varepsilon/2)^2$.

We apply the Cauchy-Schwarz inequality (0.F.1) to conclude that $\sqrt{\sum_{i=1}^{p} (0 - y_i)^2} \leq \sqrt{\sum_{i=1}^{p} (0 - x_i^n)^2} + \sqrt{\sum_{i=1}^{p} (x_i^n - y_i)^2}$ and hence $\sqrt{\sum_{i=1}^{p} y_i^2}$ is bounded despite increasing p. Thus, y is in $\mathscr{H}$. Furthermore, since for $n > N$, we have that $\sqrt{\sum_{i=1}^{\infty} (x_i^n - y_i)^2} < \varepsilon/2$, it follows that $\rho(x^n,y)$ approaches 0 as n increases. Therefore, the sequence $\{x^n\}$ converges to y.

(7.E.3) Theorem. Hilbert Space $(\mathcal{H},\rho)$ is strongly convex without ramifications.

Proof. We first show that $\mathcal{H}$ is strongly convex. Suppose that $x = \langle x_n \rangle$ and $y = \langle y_n \rangle$ are elements of $\mathcal{H}$. The obvious candidate for a midpoint between x and y is $m = \langle \frac{1}{2}(x_n + y_n) \rangle$: confirmation that m is in this position is trivial, but the uniqueness of m must be verified. Suppose that b also lies midway between x and y. Then $\rho(x,b) = \rho(y,b) = \frac{1}{2}\rho(x,y)$, and hence

$$\sqrt{\sum_{n=1}^{\infty} (x_n - b_n)^2} = \sqrt{\sum_{n=1}^{\infty} (y_n - b_n)^2} = \frac{1}{2}\sqrt{\sum_{n=1}^{\infty} (x_n - y_n)^2} \quad \text{or equivalently}$$

$4 \sum_{n=1}^{\infty} (x_n - b_n)^2 = 4 \sum_{n=1}^{\infty} (y_n - b_n)^2 = \sum_{n=1}^{\infty} (x_n - y_n)^2$. The following well-known identity now comes immediately to mind:

$$(u + v - 2w)^2 = 2(u - w)^2 + 2(v - w)^2 - (u - v)^2$$

Consequently, we have that $\sum_{n=1}^{\infty} (x_n + y_n - 2b_n)^2 = 2 \sum_{n=1}^{\infty} (x_n - b_n)^2 +$

$2 \sum_{n=1}^{\infty} (y_n - b_n)^2 - \sum_{n=1}^{\infty} (x_n - y_n)^2 = 0$, which leads us to conclude that $b_n = (1/2)(x_n + y_n)$; hence, m is the unique midpoint.

We now show that $\mathcal{H}$ is without ramifications. Suppose that m is a midpoint of x and y and also of x and z. Then $\rho(x,m) = \rho(z,m) = \rho(y,m)$. From the preceding paragraph, we have that $\frac{1}{2}(x_n + y_n) = m_n = \frac{1}{2}(x_n + z_n)$ for all n. Therefore, it follows that $y_n = z_n$ for all n, and hence $y = z$.

From the definition of Hilbert Space and from the foregoing theorems, one might be led to believe that $(\mathcal{H},\rho)$ has the same properties as Euclidean space.

The next two theorems should dispel any such notions.

(7.E.4) Theorem. Hilbert Space $(\mathcal{H},\rho)$ is not locally compact. In fact, for each $x \in \mathcal{H}$ and each $\varepsilon > 0$, $\overline{S_\varepsilon(x)}$ fails to be compact.

Proof. Suppose that $x = \langle x_1, x_2, ... \rangle \in \mathcal{H}$ and let $\varepsilon > 0$ be given. For $n = 1, 2, \ldots$ let $y_n = \langle x_1, x_2, \ldots, x_{n-1}, x_n + \varepsilon, x_{n+1}, \ldots \rangle$. Note that for each n, $y_n \in \overline{S_\varepsilon(x)}$, but for $i \neq j$, $\rho(y_i,y_j) = \sqrt{2}\varepsilon$. Thus, the sequence $\{y_n\}$ has no convergent subsequence, and therefore $\overline{S_\varepsilon(x)}$ is not compact.

Invariance of domain fails badly in $(\mathcal{H},\rho)$, i.e., open sets in $\mathcal{H}$ may be mapped homeomorphically onto nonopen subsets of $\mathcal{H}$. In fact, $\mathcal{H}$ itself can be embedded (even isometrically) into itself as a nowhere dense subset.

(7.E.5) Theorem. Let $A = \{\langle x_1, x_2, ... \rangle \in \mathcal{H} \mid x_1 = 0\}$ and define

$\psi : \mathcal{H} \to A$ by $\psi(\langle x_1, x_2, \ldots \rangle) = \langle 0, x_1, x_2, \ldots \rangle$. Then ψ is an isometry and int $\bar{A} = \varnothing$.

Proof. Clearly ψ is an isometry. To see that A is nowhere dense in $\mathcal{H}$, let $x = \langle x_1, x_2, \ldots \rangle \in \mathcal{H}$ and let $\varepsilon > 0$ be given; we show that $S_\varepsilon(x)$ is not contained in $\bar{A}$ and hence $\bar{A}$ has empty interior. Choose $b \neq 0$ so that $|x_1 - b| < \varepsilon/2$ and let $y = \langle b, x_2, x_3, \ldots \rangle$. The proof will be complete once we establish that if $\delta = \min\{|b|, \varepsilon/2\}$, then $S_\delta(y) \subset S_\varepsilon(x)$ and $S_\delta(y) \cap A = \varnothing$.

Suppose that $z \in S_\delta(y)$. Then we have that $\rho(x,z) \leq \rho(x,y) + \rho(y,z) < \varepsilon/2 + \varepsilon/2$, and consequently, $S_\delta(y) \subset S_\varepsilon(x)$. If $w = \langle w_1, w_2, \ldots \rangle \in S_\delta(y) \cap A$, then $w_1 = 0$, and hence $\rho(w,y) \geq |b| \geq \delta > \rho(w,y)$, an unusual occurrence at best.

Another strange property of Hilbert Space is that compact subsets of $\mathcal{H}$ fail to have interior points. (This follows easily from (7.E.4).) Furthermore, if K is a compact subset of $\mathcal{H}$, then $\mathcal{H}$ is homeomorphic to $\mathcal{H} \setminus K$. In fact, if $K_1, K_2, \ldots$ is a sequence of compact subsets of $\mathcal{H}$, then $\mathcal{H} \setminus \bigcup_{i=1}^{\infty} K_i$ is still homeomorphic to $\mathcal{H}$. We shall not prove these latter oddities, but the interested reader might begin a serious study of Hilbert space with the previously mentioned article of Anderson. In this article the relationship between $\mathcal{H}$ and $\prod_{i=1}^{\infty} \mathscr{E}_i^1$ ($\mathscr{E}_i^1 = \mathscr{E}^1$, for each i) is investigated and, it is shown that $\mathcal{H}$ and $\prod_{i=1}^{\infty} \mathscr{E}_i^1$ are homeomorphic. Some hint that this might be a nontrivial result is seen from the following considerations. Let $\mathcal{U}_{\mathcal{H}}$ denote the usual topology for $(\mathcal{H}, \rho)$. Let $\hat{\mathcal{H}} = \{(x_1, x_2, \ldots) \in \prod_{i=1}^{\infty} \mathscr{E}_i^1 \mid \prod_{i=1}^{\infty} x_1^2 < \infty\}$, and let $\mathcal{U}_{\hat{\mathcal{H}}}$ be the relative topology for $\hat{\mathcal{H}}$ considered as a subset of the product space $\prod_{i=1}^{\infty} \mathscr{E}_i^1$.

Then $\mathcal{U}_{\mathcal{H}} \neq \mathcal{U}_{\hat{\mathcal{H}}}$. To see this, let $e_1 = (1, 0, 0, \ldots)$, $e_2 = (0, 1, 0, 0, \ldots)$, etc. clearly, there are no limit points of the sequence $\{e_1, e_2, \ldots\}$ in $(\mathcal{H}, \mathcal{U}_{\mathcal{H}})$. On the other hand, it follows readily from the nature of basic open sets in the product topology that the sequence $\{e_1, e_2, \ldots\}$ does converge to $(0, 0, \ldots)$ in $(\hat{\mathcal{H}}, \mathcal{U}_{\hat{\mathcal{H}}})$.

(7.E.6) **Exercise.** Show that $i : \hat{\mathcal{H}} \to \mathcal{H}$ is continuous but that i^{-1} is not.

One compact subspace of $(\mathcal{H}, \rho)$, the *Hilbert cube*, $\mathcal{H}_c$, is of special importance. We define $\mathcal{H}_c$ to be $\{\langle x_n \rangle \in \mathcal{H} \mid 0 \leq x_n \leq 1/n \text{ for } n \in \mathbf{Z}^+\}$. Although $\mathcal{H}_c$ is compact (7.E.7) and hence lacks interior points in $\mathcal{H}$, it nevertheless has sufficient capacity to swallow any separable metric space (7.E.8).

Unlike $\mathcal{H}$, $\mathcal{H}_c$ may easily be realized as a product space.

(7.E.7)   Theorem.   The Hilbert cube $\mathcal{H}_C$ is homeomorphic to $\prod\limits_{i=1}^{\infty} \mathbf{I}_i$, where for each i, $\mathbf{I}_i = \mathbf{I}$.

Proof.   Let $\psi : \prod\limits_{i=1}^{\infty} \mathbf{I}_i \to \mathcal{H}_C$ be defined by $\psi((x_1, x_2, \ldots)) = \langle x_1, x_2/2, x_3/3, \ldots \rangle$. Clearly, ψ is 1–1 and onto. That ψ is continuous may be seen as follows. Suppose that $x = (x_1, x_2, \ldots)$, $\varepsilon > 0$, and $V = S_\varepsilon(\psi(x))$. Choose an integer N large enough so that $\sum\limits_{n > N} 1/n^2 < \varepsilon^2/4$, and let $U = U_1 \times U_2 \times \cdots \times U_n \times \mathcal{E}^1 \times \mathcal{E}^1 \times \cdots$ be a neighborhood of x, where $U_i = \{z_i \in \mathbf{I} \mid |x_i - z_i| < \varepsilon/\sqrt{4N}\}$ for each $i = 1, 2, \ldots, N$. Observe that if $z = (z_1, z_2, \ldots) \in U$, then $\sum\limits_{i=1}^{N} ((x_i/i) - (z_i/i))^2 < \sum\limits_{i=1}^{N} (x_i - z_i)^2 < \varepsilon^2/4$. Furthermore, $\sum\limits_{i=N+1}^{\infty} ((x_i/i) - (z_i/i))^2 < \varepsilon^2/4$. Thus, we have that $\psi(U) \subset V$, and hence ψ is continuous. That ψ is a homeomorphism follows from (2.G.11).

Next we show that any separable metric space can be embedded in $\mathcal{H}_C$.

(7.E.8)   Theorem.   If (X,d) is a separable metric space, then there is an embedding $\hat{e} : X \to \mathcal{H}_C$.

Proof.   Since (X,d) is separable, it has a countable basis $\mathbf{B}$. Let $\mathscr{A} = \{(U,V) \mid U,V \in \mathbf{B} \text{ and } \overline{U} \subset V\}$. Since X is normal, Uryson's lemma may be applied to each pair $(U,V) \in \mathscr{A}$ to obtain a map $f_{UV} : X \to \mathbf{I}$ with the property that $f_{UV}(\overline{U}) = 0$ and $f_{UV}(X \setminus V) = 1$. Note that $\mathscr{F} = \{f_{UV} \mid (U,V) \in \mathscr{A}\}$ is a countable family of functions that separates points and closed sets (7.B.15). For each $f \in \mathscr{F}$, let $\mathbf{I}_f$ be the unit interval $\mathbf{I}$ and define $\hat{e} : X \to \prod\limits_{f \in \mathscr{F}} \mathbf{I}_f$ by setting $\hat{e}(x) = \{f(x)\}_{f \in \mathscr{F}}$. Since $\prod\limits_{f \in \mathscr{F}} \mathbf{I}_f$ is homeomorphic to $\mathcal{H}_C$ and $\hat{e}$ is an embedding (7.B.16), the proof is complete.

As an encore to the preceding beautiful result, we show that $\mathcal{H}_C$ has the fixed point property. We will utilize the Brouwer fixed point theorem for $\mathbf{I}^n$ (to be proven in Chapter 14).

(7.E.9)   Exercise.   Suppose that (X,d) is a compact metric space and that $f : X \to X$ is continuous. Show that if for each $\varepsilon > 0$, there is a point $x_\varepsilon \in X$ such that $d(f(x_\varepsilon), x_\varepsilon) < \varepsilon$, then f leaves some point of X fixed.

(7.E.10)   Theorem.   The Hilbert cube, $\mathcal{H}_C$, has the fixed point property.

Proof.   Consider $\mathcal{H}_C$ as the product space $[0,1] \times [0,1/2] \times [0,1/2^2] \times$

$\cdots$ (why is this possible?). Let f be a continuous map from H_C into H_C, and let $\varepsilon > 0$ be given. Choose N large enough so that $\sum\limits_{n > N} 1/n^2 < \varepsilon$, and let $A = [0,1] \times [0,1/2] \times \cdots \times [0,1/2^N]$. Define a map $\hat{f} : A \to A$ by $\hat{f}(x_1, x_2, \ldots, x_N) = (y_1, y_2, \ldots, y_N)$, where $f(x_1, x_2, \ldots, x_N, 0, 0, \ldots) = (y_1, y_2, \ldots, y_N, y_{N+1}, y_{N+2}, \ldots)$. Then, by the Brouwer fixed point theorem (14.C.22), $\hat{f}$ has a fixed point, $a = (a_1, a_2, \ldots, a_N)$. Let $a^* = (a_1, a_2, \ldots, a_N, 0, 0, \ldots)$. Then $\rho(f(a^*),a^*)) < \varepsilon$, and consequently, by (7.E.9), $\mathscr{H}_C$ has the fixed point property.

A more general result is obtained in problem E.1.

PROBLEMS

Sections A and B

1. Suppose that X and Y are topological spaces and that card X is infinite. Show that $\mathrm{Hom}(X,Y)$ (with the point-open topology) is locally connected if and only if Y is locally connected and connected.

2. Suppose that X and Y are topological spaces. Show that if $C(X,Y)$ (with the compact-open topology) is T_3, then so is Y.

3. Show that if X is a locally compact T_2 space, and X and Y are second countable, then $C(X,Y)$ (with the compact-open topology) is second countable.

4. Show that the sup topology on $C(X,Y)$ depends not only on the topologies of X and Y but also on the metric chosen for Y (construct examples).

5. Suppose that X is a topological space, Y is a bounded metric space, and $\mathscr{K}$ is the collection of all compact subsets of X. For each $f \in C(X,Y)$, $K \in \mathscr{K}$, and $\varepsilon > 0$, let $S(f,K,\varepsilon) = \{g \in C(X,Y) \mid d(f(x),g(x)) < \varepsilon \text{ for all } x \in K\}$. Then $\{S(f,K,\varepsilon) \mid f \in C(X,Y), K \in \mathscr{K}, \varepsilon > 0\}$ is a subbasis for a topology called the *topology of uniform convergence on compacta*. Show that the topology of uniform convergence on compacta coincides with the compact-open topology.

6. Suppose that X and Y are topological spaces. Let $j : Y \to C(X,Y)$ be defined by $j(y) = c_y$ where $c_y : X \to Y$ is the constant map from X onto y. Then j is called the *natural injection* of Y into $C(X,Y)$. ($C(X,Y)$ has the compact-open topology.) Prove:
   (a) j is an embedding of Y into $C(X,Y)$;
   (b) if Y is T_2, then $j(Y)$ is closed;
   (c) $j(Y)$ is a retract of $C(X,Y)$.

7.* Show that if X is a completely regular, T_2 space and (Y,d) is a bounded metric space containing a nondegenerate path, then the compact-open topology on $C(X,Y)$ coincides with the sup topology on $C(X,Y)$ if and only if X is compact.

8. Show that if X is a compact space and Y is a metric space, then $C(X,Y)$ with the compact-open topology is an AR if and only if Y is.

9.* Suppose that X is a Tihonov space. Show that $\mathscr{A} = \{f : X \to \mathscr{E}^1 \mid f$ bounded and continuous$\}$ separates points from closed sets, and therefore, by (7.B.16), $\hat{e} : X \to \prod_{f \in \mathscr{A}} I_f$ is an embedding. The *Stone-Čech compactification* of X is the closure βX of $\hat{e}(X)$ in $\prod_{f \in \mathscr{A}} I_f$. Show that if $f : X \to Y$ is continuous and Y is a compact, T_2 space, then there is a unique extension of f to βX. [Hint: Use (7.B.18)].

10. A space X is *real compact* if and only if X can be embedded as a closed subset of a product of copies of $\mathscr{E}^1$.
   (a) Show that compact, T_2 spaces are real compact.
   (b) Show that an arbitrary product of real compact spaces is real compact.

11.* Suppose that X is a topological space and Y is a metric space. Let $B(X,Y)$ be defined as in (7.B.8). Show that if Y is complete, then so is $B(X,Y)$.

Section C

1. Let $S = \{x \in \mathscr{E}^1 \mid x \geq 0\}$ and let $f : \mathbf{R}^1 \to S$ be defined by $f(x) = x^2$. Determine the weak topology for $\mathbf{R}^1$ induced by f.

2. Suppose that Y is a topological space and that $f : X \to Y$. Let $\mathscr{W}$ be the weak topology for X induced by f. Show that $A \subset X$ is closed (with respect to the weak topology) if and only if $A = f^{-1}(B)$, where B is closed in Y.

3.* Generalize in some way the result of problem 2, where the weak topology for X is induced by a family of functions $\{f_\alpha : X \to Y_\alpha \mid \alpha \in \Lambda\}$.

4.* Suppose that $(X,\mathscr{U})$ is a $T_{3\frac{1}{2}}$ space. Show that $\mathscr{U}$ coincides with the weak topology generated by $\{f : X \to \mathscr{E}^1 \mid f$ is continuous$\}$.

5. Suppose that S is an arbitrary set and that $f : S \to T$ is an onto function, where T is a connected topological space. Show that if S has the weak topology induced by f, then S is connected.

6. Repeat problem 5, replacing "connected" with "compact" and "countably compact."

7. Suppose that A is a subset of a space X. Show that relative topology for A coincides with the weak topology for A generated by the inclusion function $i : A \to X$.

8. Suppose that $(X,\mathscr{U})$ is a topological space. Let $\{Y_\alpha \mid \alpha \in \Lambda\}$ be a family of spaces and $\{f_\alpha : X \to Y_\alpha \mid \alpha \in \Lambda\}$ a corresponding family of functions

with the property that for each space Z and for each function $f : Z \to X$, f is continuous if and only if $f_\alpha f$ is continuous for each $\alpha \in \Lambda$. Show that $\mathcal{U}$ is the weak topology for X generated by $\{f_\alpha : X \to Y_\alpha \mid \alpha \in \Lambda\}$.

Section D

1. Suppose that X is a topological space and that $\{A_\alpha \mid \alpha \in \Lambda\}$ is an open cover of X. Show that the weak topology induced by the cover $\{A_\alpha\}$ coincides with the given topology on X.

2. Suppose that X is a topological space and that $\{A_\alpha \mid \alpha \in \Lambda\}$ is a closed cover of X. Show that if $\{A_\alpha \mid \alpha \in \Lambda\}$ is locally finite (each $x \in X$ is contained in a neighborhood which intersects only a finite number of the A_α), then the weak topology induced by $\{A_\alpha \mid \alpha \in \Lambda\}$ coincides with the given topology.

3. Suppose that (A_1, d_1) and (A_2, d_2) are metric spaces bounded by 1. Suppose that $d_{1|A_1 \cap A_2} = d_{2|A_1 \cap A_2}$ and $A_1 \cap A_2 \neq \varnothing$ is closed in A_1 and A_2. Define a function d by

$$d(x,y) = \begin{cases} d_1(x,y) & \text{if} \quad x,y \in A_1 \\ d_2(x,y) & \text{if} \quad x,y \in A_2 \\ \inf\{d_1(x,z) + d_2(z,y) \mid z \in A_1 \cap A_2\} \\ \qquad \text{if} \quad x \in A_1 \setminus A_2 \quad \text{and} \quad y \in A_2 \setminus A_1. \end{cases}$$

   (a) Prove that d is a metric.
   (b) Show that the metric topology coincides with the weak topology induced by $\{A_1, A_2\}$.

4. Suppose that $X = \bigcup_{\alpha \in \Lambda} A_\alpha$ where the A_α's satisfy the necessary conditions for X to be given the weak topology induced by the family $\{A_\alpha \mid \alpha \in \Lambda\}$. Prove or disprove:
   (a) If each A_α is connected, then so is X.
   (b) If each A_α is compact, then so is X.
   (c) If each A_α is second countable, then so is X.
   (d) If each A_α is first countable, then so is X.
   (e) If each A_α is separable, then so is X.
   (f) If each A_α is T_0, T_1, etc., then so is X.

5. Show that if every space X_n in a given expanding sequence of spaces $X_1 \subset X_2 \subset \cdots$ is normal and closed in X_{n+1}, then $X = \bigcup_{n=1}^{\infty} X_n$ is normal (X has the weak topology induced by X_n).

6.* Suppose that X is a topological space and $A_1 \subset A_2 \subset \cdots$ is a sequence of subspaces of X such that $X = \bigcup_{n=1}^{\infty} A_n$. Show that the weak topology on

X induced by the A_n does not necessarily coincide with the original topology for X.

Section E

1.* Suppose that $\{A_\alpha \mid \alpha \in \Lambda\}$ is a collection of compact T_2 spaces with the property that for each finite subcollection $\{A_{\alpha_1}, \ldots, A_{\alpha_n}\}$, $\prod_{i=1}^{n} A_{\alpha_i}$ has the fixed point property. Show that $\prod_{\alpha \in \Lambda} A_\alpha$ has the fixed point property.

2. Show that int $\mathcal{H}_C = \varnothing$.

3. Let $X = \{u,v,w,y\}$, and define a metric for X by setting $d(u,y) = d(v,y) = d(w,y) = 1$ and $d(u,v) = d(u,w) = d(v,w) = 2$. Show that (X,d) cannot be isometrically embedded in $\mathcal{H}$.

4. Let $X = \prod_{i=1}^{\infty} \mathscr{E}_i^1$, where $\mathscr{E}_i^1 = \mathscr{E}^1$ for each i. Define a metric

$$d(x,y) = \sum_{i=1}^{\infty} \frac{|x_i - y_i|}{(1 + |x_i - y_i|)2^i}$$

Show that the topology induced by d coincides with the product topology.

5. Show that $\prod_{i=1}^{\infty} \mathscr{E}_i^1$ with the metric given in problem E.4 contains all separable metric spaces X via the function $f(x) = \langle d(x,x_i) \rangle$, where $x \in X$ and $\{x_i\}$ is a countable dense subset of X.

6. Find an example of an infinite set of points A in $\mathcal{H}$ with the property that $\rho(x,y) = 2$ for each $x,y \in A$.

7. Show that in $(\mathcal{H},\rho)$ there is a point x and a closed subset C such that $\rho(x,C) = \varepsilon > 0$, but that $\rho(x,c) > \varepsilon$ for all $c \in C$.

8. Prove or disprove: If X, Y, and Z are topological spaces such that $X \times Y$ is homeomorphic to $X \times Z$, then Y is homeomorphic to Z.

9. Let $X = \prod_{i=1}^{\infty} \mathscr{E}_i^1$, where $\mathscr{E}_i^1 = \mathscr{E}^1$ for each i. Define a metric

$$d(x,y) = \sum_{i=1}^{\infty} \frac{1}{i!} \frac{|x_i - y_i|}{1 + |x_i - y_i|}$$

Show that (X,d) is a complete, separable metric space that is not totally bounded.

10.* Show that (X,d) in problem 9 is homeomorphic to a subspace of Hilbert space, and Hilbert space is homeomorphic to a subspace of (X,d).

Chapter 8

QUOTIENT SPACES

A. THE QUOTIENT TOPOLOGY

A *partition* of a set X is a collection G of mutually disjoint nonempty subsets of X whose union is X. Partitions arise quite naturally in mathematics and are quite important in topology. They lead to the creation of many complex and interesting spaces.

Suppose that G is a partition of a topological space X. Define X/G to be the set whose "points" are the members of the given partition G. Then X/G is called a *quotient* or *decomposition set*. The function $P : X \to X/G$ that maps a point x to the unique set of G that contains x is called the *quotient map*. We shall frequently denote $P(x)$ by $[x]$. Note that $P(x) = P(y)$ if and only if x and y belong to the same member of the partition. We topologize X/G in such a manner that (i) P is continuous, and (ii) X/G has a maximal number of open sets; contrast this with the rationale behind the definition of the product topology.

(8.A.1) **Definition.** Suppose that X is a topological space, G is a partition of X, and that $P : X \to X/G$ is the quotient map. Then the topology $\mathcal{U} = \{ U \subset X/G \mid P^{-1}(U) \text{ is open in } X \}$ is called the *quotient topology* for X/G.

(8.A.2) **Exercise.** Show that

   (i)   the quotient topology is a topology;

   (ii) if X/G has the quotient topology, then P is continuous;

(iii)   if $\mathcal{U}$ is the quotient topology and $\mathcal{V}$ is any other topology for which P is continuous, then $\mathcal{V} \subset \mathcal{U}$;

(iv)   if X/G has the quotient topology, $A \subset X/G$, and $P^{-1}(A)$ is closed in X, then A is closed in X/G.

Suppose that $f : X \rightarrow Y$ is a continuous onto map. Since f is a function, it is clear that $G_f = \{f^{-1}(y) \mid y \in Y\}$ is a partition of X. If $x \in X$, note that $P(x) = [x] = f^{-1}(f(x))$. In this context, the function $\Phi_f : X/G_f \rightarrow Y$ defined by $\Phi_f([x]) = f(x)$ is of special interest (why is Φ_f well defined?).

(8.A.3)   Theorem.   Suppose that $f : X \rightarrow Y$ is a continuous onto map and that X/G_f has the quotient topology. Then the function Φ_f is continuous.

Proof.   Let U be an open subset of Y. Then $\Phi_f^{-1}(U) = Pf^{-1}(U)$. Since $P^{-1}(\Phi_f^{-1}(U)) = P^{-1}(Pf^{-1}(U)) = f^{-1}(U)$ and $f^{-1}(U)$ is open in X, it follows from the definition of the quotient topology that $\Phi_f^{-1}(U)$ is open in X/G_f. Consequently, Φ_f is continuous.

(8.A.4)   Remark.   There is a striking parallel between quotient spaces and quotient groups. If G_1 and G_2 are groups and $h : G_1 \rightarrow G_2$ is a homomorphism from G_1 onto G_2, then G_1 may be partitioned into sets of the form $\{h^{-1}(y) \mid y \in G_2\}$. Let G_1/G_2 denote the resulting quotient set. There is a natural group structure that may be assigned to G_1/G_2 whereby the function $\Phi_h : G_1/G_2 \rightarrow G_2$ becomes an isomorphism (first isomorphism theorem). In the topological context, it now becomes reasonable to ask whether or not Φ_f is the topological counterpart of an isomorphism, i.e., a homeomorphism. It is clear that Φ_f is 1–1 and onto. Although in many situations Φ_f is a homeomorphism, the following example illustrates what can go awry.

(8.A.5)   Example.   Let $X = [0, 2\pi)$, and suppose that $f : X \rightarrow \mathcal{S}^1$ is defined by $f(x) = (\cos x, \sin x)$. Then f is a continuous 1–1 map from X onto $\mathcal{S}^1$, and since G_f is essentially X, it follows that X/G_f is homeomorphic to X. However, X is not homeomorphic to $\mathcal{S}^1$, since $\mathcal{S}^1$ is compact and X is not. Therefore, Φ_f is not a homeomorphism.

The following theorem gives two conditions under which Φ_f is a homeomorphism.

(8.A.6)   Theorem.   Suppose that $f : X \to Y$ is a continuous function from X onto Y and that f is either open or closed. Then $\Phi_f : X/G_f \to Y$ is a homeomorphism.

Proof.   Suppose that f is closed, and let A be a closed subset of X/G_f. It is sufficient to show that Φ_f is closed in Y. This is immediate, however, since $\Phi_f(A) = fP^{-1}(A)$. The case where f is open is no more difficult.

(8.A.7)   Corollary.   If X is compact, Y is T_2, and $f : X \to Y$ is continuous and onto, then Φ_f is a homeomorphism.

B.   IDENTIFICATIONS

Partitions of a space X are most commonly formed by "identifying" certain points in X with others. An equivalence relation $\sim$ on X leads quite naturally to such a partition, where the members of the partition are the equivalence classes under $\sim$. In this case, we shall denote the resulting decomposition or quotient space by $X/\sim$. For instance, if $X = [0,2\pi]$, we can define an equivalence relation $\sim$ on X by declaring $0 \sim 2\pi$, and $x \sim y$ if and only if $x = y$ for $x,y \in (0,2\pi)$. The equivalence classes consist precisely of single points in the open interval $(0,2\pi)$ together with the set $\{0,2\pi\}$. What does the corresponding quotient space look like? Essentially, the end points of the interval have been "glued" together, and intuitively it would seem that the resulting space should be a circle; that this is indeed the case may be seen from the following commutative diagram, where $f(x) = (\cos x, \sin x)$.

Since $[0,2\pi]$ is compact, we have by (8.A.7) that Φ_f is a homeomorphism, and consequently, the quotient space is topologically a circle.

Many interesting topological spaces may be constructed in a similar fashion. For example, although the torus is usually defined to be the product space $\mathscr{S}^1 \times \mathscr{S}^1$, an alternate description is obtained from quotient spaces. Start with the square X in the xy plane ($\mathscr{E}^2$) whose vertices are the points $(0,0)$, $(0,2)$, $(2,0)$ and $(2,2)$. Identifying each point in X of the form $(0,y)$ with the corresponding point $(2,y)$ on the opposite side of the square intuitively has the effect of rolling up the square into a tube. Gluing the ends of the tube together by identifying points of the form $(x,0)$ with $(x,2)$ yields a quotient space homeomorphic to the torus $\mathscr{S}^1 \times \mathscr{S}^1$.

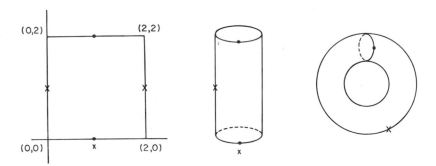

This may be seen from the following diagram, where $f(x,y) = ((\cos \pi x, \sin \pi x), (\cos \pi y, \sin \pi y))$ and X/G_f is the quotient space just described.

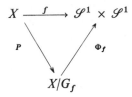

More exotic spaces may also be constructed. For instance, the Klein bottle is obtained by forming a tube and then identifying points of the form $(x,0)$ with $(2 - x, 2)$. This has the effect intuitively of twisting the tube before gluing. It can be shown that the resulting space is not embeddable in $\mathscr{E}^3$, although it can be embedded in $\mathscr{E}^4$.

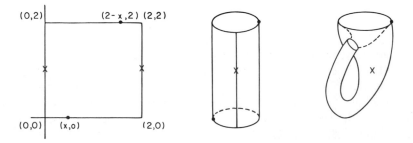

In the problem section, the reader is asked to construct a number of other curious spaces using the techniques just described. Notice that each of the spaces that we have created by identifying certain boundary points of the square is a 2-manifold (why?). In Chapter 16, it is shown that all connected compact 2-manifolds may be obtained in this fashion, although of course identifications of boundary points of a square may be made so that the associated quotient space is not a 2-manifold.

Quotient spaces are also used to construct the "cone" over a given space.

(8.B.1) *Definition.* Suppose that X is a topological space. Let G be the equivalence relation on $X \times [0,1]$ determined by: $(x,t) \sim (y,s)$ if and only if either $(x,t) = (y,s)$ or $t = s = 1$. The *cone over* X is the quotient space $(X \times [0,1])/G$.

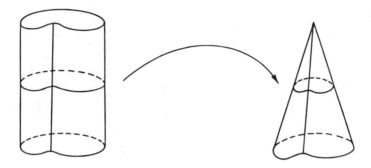

(8.B.2) *Definition.* Suppose that X is a topological space. The *suspension of* X is the quotient space obtained by identifying in $X \times [-1,1]$ all points of the form $(x,1)$ with each other, and points of the form $(x,-1)$ with each other.

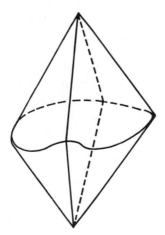

Cones and suspensions also may be derived as special cases of an important technique that involves sewing spaces together with the aid of a given continuous function and an appropriate equivalence relation.

Recall from Chapter 1 that if X and Y are disjoint spaces, then the free union of X and Y is the topological space $Z = X \cup Y$, where a set $W \subset Z$ is open if and only if $W \cap X$ and $W \cap Y$ are open in X and Y, respectively. Suppose that X and Y are disjoint spaces and that A is a closed subset of X. Let $f : A \to Y$ be a continuous function, and in $X \cup Y$ define an equivalence relation by identifying each point $x \in f^{-1}(y)$ with y. Denote the resulting quotient space by $X \cup_f Y$. Thus A has been "glued" to its image $f(A)$, and in the process, X and Y have been sewn together. The space $X \cup_f Y$ is called the *adjunction of X and Y by f.* (Of course, if X and Y are not initially disjoint, then the usual procedure may be utilized to "disjointify" them.)

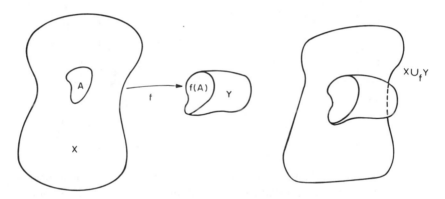

The cone over X may now be obtained as follows. Let p be any point not in X. Set $Z = X \times I$, $A = X \times \{1\}$, and let $f : A \to \{p\}$. Then $Z \cup_f \{p\}$ is clearly homeomorphic to the cone over X described in (8.B.1). Suspensions may be formed in a similar manner.

(8.B.3)   Exercise. Suppose that X and Y are disjoint spaces, A is a closed subset of X, and $f : A \to Y$ is continuous. Let $P : X \cup Y \to X \cup_f Y$ be the quotient map.Show that $P|_{X \setminus A}$ is an embedding, $P|_Y$ is an embedding, $P(X \setminus A)$ is open in $X \cup_f Y$, and $P(Y)$ is closed in $X \cup_f Y$.

The proof of the following theorem is surprisingly intricate.

(8.B.4)   Theorem. Suppose that X and Y are normal, A is a closed subset of X, and $f : A \to Y$ is continuous. Then $X \cup_f Y$ is normal.

Proof. Suppose that A_1 and A_2 are disjoint closed subsets of $Z = X \cup_f Y$. Then $A_1 \cap P(Y)$ and $A_2 \cap P(Y)$ are disjoint closed sets in $P(Y)$, and by (8.B.3), $P(Y)$ is homeomorphic to Y. Since Y is normal, there are open sets U_1 and U_2 in $P(Y)$ with disjoint closures such that $A_1 \cap P(Y) \subset U_1$ and $A_2 \cap P(Y) \subset U_2$. Furthermore, $B_1 = (P^{-1}(A_1 \cup \overline{U}_1)) \cap X$ and $B_2 = (P^{-1}(A_2 \cup \overline{U}_2)) \cap X$ are disjoint closed subsets of the normal space X. Let V_1 and V_2 be disjoint open sets in X containing B_1 and B_2 respectively and set $W_1 = P(V_1 \setminus A) \cup U_1$ and $W_2 = P(V_2 \setminus A) \cup U_2$.

Clearly, W_1 and W_2 are disjoint and $A_1 \subset W_1$ and $A_2 \subset W_2$. Hence, it remains to show that W_1 and W_2 are open in $X \cup_f Y$. The set W_1 will be open if and only if $P^{-1}(W_1) \cap X$ and $P^{-1}(W_1) \cap Y$ are open in X and Y, respectively. Since $P^{-1}(W_1) = P^{-1}(P(V_1 \setminus A) \cup U_1) = (V_1 \setminus A) \cup P^{-1}(U_1)$, we have that $P^{-1}(W_1) \cap X = (V_1 \setminus A) \cup f^{-1}(U_1)$. Of course, $V_1 \setminus A$ is open in X and $f^{-1}(U_1)$ may be written as $Z \cap A$, where Z is open in X and $Z \subset V_1$. Consequently, we have that $P^{-1}(W_1) \cap X = (V_1 \setminus A) \cup (Z \cap A) = (V_1 \setminus A) \cup (Z \cap V_1 \cap A) = (V_1 \setminus A) \cup (Z \cap V_1)$, and hence $P^{-1}(W_1) \cap X$ is open in X. Since $P^{-1}(W_1) \cap Y = U_1$, it follows that W_1 is open in $X \cup_f Y$. In a similar manner, it may be shown that W_2 is open, which concludes the proof.

The reader's mounting concern that (4.B.19) would be forever bereft of proof can now be allayed. There, it remained to show that absolute retracts are Absolute Retracts. To this end, suppose that Y is an absolute retract, X is a normal space, A is a closed subset of X, and $f : A \to Y$ is a continuous map. We find a continuous extension of f to X. Form the space $X \cup_f Y$ and let $P : X \cup Y \to X \cup_f Y$ be the quotient map. By (8.B.4) and (8.B.3), $X \cup_f Y$ is normal, and Y is embedded as a closed subset of $X \cup_f Y$. Since Y is an absolute retract, there is a retraction $r : X \cup_f Y \to P(Y)$ that leaves points in $P(Y)$ fixed. Now define $F : X \to Y$ by $F = (P|_Y)^{-1} r P|_X$. Then F is the desired extension.

(8.B.5)   *Exercise.*   Suppose that X and Y are T_1 spaces and that $X \cup_f Y$ is an adjunction space. Show that $X \cup_f Y$ is T_1.

C.   IDENTIFICATION MAPS

(8.C.1)   *Definition.*   Suppose that X and Y are topological spaces and that $f : X \to Y$ is continuous and onto. Then f is an *identification map* if and only if the topology for Y coincides with $\mathscr{U} = \{U \subset Y \mid f^{-1}(U) \text{ is open in } X\}$.

Note that the quotient map $P : X \to X/G$ is an identification map and that open and closed mappings are also identification maps.

A key property of identification maps is given in the next theorem.

(8.C.2)   *Theorem.*   Suppose that X, Y, and Z are topological spaces,

$f : X \rightarrow Y$ is an identification map, and that $g : Y \rightarrow Z$. Then g is continuous if and only if gf is continuous.

Proof. Certainly if g is continuous, then so is gf. Suppose, then, that gf is continuous. Let U be an open set in Z. Since f is an identification map, to show that $g^{-1}(U)$ is open it suffices to demonstrate that $f^{-1}g^{-1}(U)$ is open in X. However, this is immediate, since $f^{-1}g^{-1}(U) = (gf)^{-1}(U)$ and gf is continuous.

(8.C.3) *Exercise.* Obtain a characterization of identification maps by proving a converse of (8.C.2).

(8.C.4) *Exercise.* Suppose that X, Y, and Z are topological spaces. Suppose further that $f : X \rightarrow Y$ is an identification map and that $h : X \rightarrow Z$ is continuous. Show that if hf^{-1} is a function, then hf^{-1} is continuous.

We now generalize (8.A.6) slightly.

(8.C.5) *Theorem.* Suppose that $f : X \rightarrow Y$ is an identification map. Then $\Phi_f : X/G_f \rightarrow Y$ is a homeomorphism (with the notation defined as in Section A).

Proof. We show that Φ_f is open. Let U be an open subset of X/G_f. Since f is an identification map, it suffices to show that $f^{-1}(\Phi_f(U))$ is open in X. However, $f^{-1}(\Phi_f(U)) = P^{-1}(U)$ and since P is continuous, the result follows.

We may also extend (2.E.7).

(8.C.6) *Theorem.* Suppose that X is a locally connected space and that $f : X \rightarrow Y$ is an identification map. Then Y is locally connected.

Proof. By (2.E.2), it suffices to show that components of open sets of Y are themselves open. Let C be a component of an open set $U \subset Y$. Since f is an identification map, we need only show that $f^{-1}(C)$ is open. Suppose that $x \in f^{-1}(C)$ and let D_x be the component of $f^{-1}(U)$ that contains x. Then $f(D_x)$ is connected, lies in U, and intersects C. Therefore, we have that $f(D_x) \subset C$, and hence $D_x \subset f^{-1}(C)$. Since $f^{-1}(U)$ is open and X is locally connected, it follows that D_x must be open (2.E.2). Thus, $f^{-1}(C)$ is a union of open sets, and consequently Y is locally connected.

D. THE STRONG TOPOLOGY AND k-SPACES

The weak (or initial) topology induced by a family of functions was introduced as a generalization of the product topology. In a similar manner, a topology (called the strong or final topology) may be defined that generalizes the notion of the quotient topology. Suppose that $\mathscr{A} = \{X_\alpha \mid \alpha \in \Lambda\}$ is a family of topological spaces, Y is a set, and that for each $\alpha \in \Lambda$, there is a function $f_\alpha : X_\alpha \to Y$. As the reader can surely guess by now, we will define a topology for Y in such a way that each f_α is continuous and Y has a maximum number of open sets. This is accomplished by declaring a subset U of Y to be open if and only if $f^{-1}(U)$ is open in X_α for each α. Since inverses are well behaved, the resulting collection of subsets of Y is easily seen to be a topology for Y that meets all of our requisites. This topology is called the *strong* (or *final*) *topology associated with the family of functions* $\{f_\alpha : X_\alpha \to Y \mid \alpha \in \Lambda\}$.

(8.D.1) Exercise. Show that if the word "open" is replaced by "closed" in the definition of the strong topology, then the resulting topology coincides with the strong topology.

(8.D.2) Exercise. Suppose that Y has given the strong topology associated with a family of functions $\{f_\alpha : X_\alpha \to Y \mid \alpha \in \Lambda\}$. Let Z be an arbitrary space. Show that a function $f : Y \to Z$ is continuous if and only if $ff_\alpha : X_\alpha \to Z$ is continuous for each α.

The strong topology is used to establish a class of spaces known as k-spaces.

(8.D.3) Definition. Suppose that $(X,\mathscr{U})$ is a topological space and that $\mathscr{C} = \{C \subset X \mid C \text{ is compact}\}$. For each $C \in \mathscr{C}$, let $i_C : C \to X$ be the inclusion map. Then X is a k-*space* if and only if the strong topology $\mathscr{K}$ for X induced by $\{i_C \mid C \in \mathscr{C}\}$ coincides with $\mathscr{U}$. The topology $\mathscr{K}$ is called the k-*topology for* X *associated with* $\mathscr{U}$.

Proofs of the following remarks are trivial.

(8.D.4) Remarks.
   1. If $(X,\mathscr{U})$ is a topological space, then $\mathscr{U}$ is always a subset of the k-topology for X associated with $\mathscr{U}$.
   2. If X is a k-space and Y is an arbitrary space, then a function $f : X \to Y$ is continuous if and only if $f|_C : C \to Y$ is continuous for each compact subset C of X (8.D.2).

(8.D.5) Theorem. Suppose that $(X,\mathscr{U})$ is a topological space and that $\mathscr{C} = \{C \mid C \text{ is a compact subset of } (X,\mathscr{U})\}$. Then $(X,\mathscr{U})$ is a k-space if and only if a subset A of $(X,\mathscr{U})$ is closed whenever $A \cap C$ is closed in C for each $C \in \mathscr{C}$.

Proof. Suppose that $(X,\mathscr{U})$ is a k-space and let A be a subset of X such that $A \cap C$ is closed in C for each $C \in \mathscr{C}$. Since $A \cap C = i_C^{-1}(A)$, it follows that $i_C^{-1}(A)$ is closed in C for all $C \in \mathscr{C}$. Therefore, by (8.D.1), A is closed in $(X, \mathscr{U})$ since $(X, \mathscr{U})$ is a k-space.

Conversely, suppose that a subset A of $(X, \mathscr{U})$ is closed whenever $A \cap C$ is closed in C for each $C \in \mathscr{C}$. Let $\mathscr{K}$ be the k-topology for X associated with $\mathscr{U}$. By Remark (8.D.4), we have that $\mathscr{U} \subset \mathscr{K}$. Suppose then that A is a closed subset of $(X, \mathscr{K})$. We show that A is a closed subset of $(X, \mathscr{U})$. Since A is closed in $(X, \mathscr{K})$, it follows from (8.D.1) that $i_C^{-1}(A)$ is closed in C for each $C \in \mathscr{C}$, where C has the relative topology inherited from $(X, \mathscr{U})$. Since $i_C^{-1}(A) = A \cap C$, it is now immediate by the hypothesis that A is closed in $(X, \mathscr{U})$. Hence, we have that $\mathscr{K} \subset \mathscr{U}$.

(8.D.6) *Exercise.* Show that "open" may be substituted for "closed" in the hypothesis of the preceding theorem.

The k-spaces encompass a wide range of spaces.

(8.D.7) *Theorem.* (i) Locally compact spaces are k-spaces. (ii) First countable spaces are k-spaces. (In particular, metric spaces are k-spaces.)

Proof. Suppose that X is locally compact. We apply (8.D.6) to show that X is a k-space. Suopose that $U \subset X$ and that $U \cap C$ is open in C for each compact subset C of X. Let $x \in U$. Since X is locally compact, there is an open subset V of X that contains x and has compact closure. Since $\bar{V}$ is compact, we have that $U \cap \bar{V}$ is open in $\bar{V}$ and hence $U \cap \bar{V} \cap V = U \cap V$ is open in $\bar{V} \cap V = V$. Therefore, $U \cap V$ is open in X (V is open in X) and contains x. Consequently, U may be written as a union of open sets and hence is open in X. Thus, X is a k-space.

Suppose that X is first countable and that $F \subset X$ has the property that $F \cap C$ is closed for each compact subset C of X. We show that F is closed in X and then apply (8.D.5). Suppose that $x \in \bar{F}$ and that $\{x_n\}$ is a sequence in F that converges to x (why does such a sequence exist?). Then $C = \{x_n \mid n \in \mathbf{Z}^+\} \cup \{x\}$ is compact and thus has closed intersection with F. However, this implies that $x \in F \cap C$ (why?). Hence, $\bar{F} \subset F$ and the result follows.

All of the usual questions concerning products, subspaces, etc. may be asked about k-spaces. A few of these are taken up in the problem section, but most are left to the reader.

The reader should be cautioned that terminology involving weak and strong topologies is not always consistent. Especially lamentable is the fact that the weak topology induced by a family of spaces is disturbingly close to the concept of the strong topology. In fact, if $\{A_\alpha \mid \alpha \in \Lambda\}$ is a family of disjoint topological spaces, and if $X = \bigcup \{A_\alpha \mid \alpha \in \Lambda\}$, then the weak topology for X induced by the A_α coincides with the strong topology for X generated by the inclusion maps $i_A : A \rightarrow X$ (which in turn is nothing more than the free union topology). Thus, in reading the literature in topology (as in theology) one should be quite careful to check definitions before trying to apply the results.

E. CW-COMPLEXES

(8.E.1) Notation.

$$\mathscr{B}^n = \{(x_1, x_2, \ldots, x_n) \in \mathscr{E}^n \mid x_1^2 + \cdots + x_n^2 \le 1\}$$

(the *standard n-cell*);

$$I(\mathscr{B}^n) = \{(x_1, x_2, \ldots, x_n) \in \mathscr{E}^n \mid x_1^2 + \cdots + x_n^2 < 1\};$$

$$\mathscr{S}^n = \{(x_1, x_2, \ldots, x_{n+1}) \in \mathscr{E}^{n+1} \mid x_1^2 + \cdots + x_{n+1}^2 = 1\}$$

(the *standard n-sphere*);

$$\mathscr{B}^0 = I(\mathscr{B}^0) = \{0\} \in \mathscr{E}^1$$

$$\mathscr{S}^{-1} = \varnothing.$$

The *CW*-complexes offer an excellent illustration of how distinct topologies (in this case, the quotient topology and the weak topology induced by a family of spaces) may be successfully intertwined to produce spaces considerably more interesting than their progenitors. The notion of *CW*-complex was introduced by Whitehead [1949]. Essentially, *CW*-complexes are formed by gluing together standard *n*-cells. For instance, to a collection of 0-cells (points) one might carefully glue a number of 1-cells, and then to the resulting space a number of 2-cells might be adjoined, etc.

(8.E.2) Example.

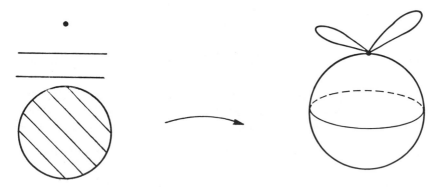

The attachings indicated in the figure are effected through a series of mappings. We initiate formalities with the definition of a cell complex, a concept slightly more general than that of a CW-complex.

(8.E.3) Definition. Suppose that X is a set. Then a *cell structure* on X is a pair $(X,\mathbf{H})$, where $\mathbf{H} = \{h_\alpha : \mathscr{B}^{n_\alpha} \to X \mid \alpha \in \Lambda\}$ is a family of functions each of which maps the appropriate standard n_α-cell into X, subject to the following conditions:

(i) $h_\alpha|_{I(\mathscr{B}^{n_\alpha})}$ is one to one;
(ii) $\{h_\alpha(I(\mathscr{B}^{n_\alpha})) \mid \alpha \in \Lambda\}$ forms a partition of X;
(iii) if for each $n \in \mathbf{N}$, $X^n = \bigcup_{n_\alpha \leqslant n} h_\alpha(\mathscr{B}^{n_\alpha})$, then $h_\alpha(\mathscr{S}^{n_\alpha - 1}) \subset X^{n_\alpha - 1}$.

We shall eventually topologize X with the aid of the cell structure.

(8.E.4) Notation. The set X^n defined above is called the *n-skeleton* of X and the maps $\{h_\alpha \mid \alpha \in \Lambda\}$ are called the *characteristic maps*. We shall denote $h_\alpha(\mathscr{B}^{n_\alpha})$ by C^{n_α}, and e^{n_α} will denote $h_\alpha(I(\mathscr{B}^{n_\alpha}))$.

Unfortunately, the following nomenclature is standard. For each $\alpha \in \Lambda$, the set C^{n_α} is called a *closed n-cell* and e^{n_α} is referred to as an *open n-cell*. Recall, however, that we have previously defined a 2-cell to be any space homeomorphic to $\mathscr{B}^2$. Similarly, in Chapter 17, n-cells are defined to be any space homeomorphic to $\mathscr{B}^n$. In the context of cell complexes, a closed n-cell need not be an n-cell at all!

(8.E.5) Exercise. Show that the set X in (8.E.2) has a cell structure; create some additional sets with cell structures.

Suppose that $(X,\mathbf{H})$ is a cell structure, where $\mathbf{H} = \{h_\alpha : \mathscr{B}^{n_\alpha} \to X \mid \alpha \in \Lambda\}$. For $n \in \mathbf{N}$, let X^n be the n-skeleton of X and set $X^{-1} = \varnothing$. For each $n \in \mathbf{N}$, give $\Lambda_n = \{\alpha \in \Lambda \mid n_\alpha = n\}$ the discrete topology, and define functions

$$p_0 : (\Lambda_0 \times \mathscr{B}^0) \cup X^{-1} \to X^0$$
$$p_1 : (\Lambda_1 \times \mathscr{B}^1) \cup X^0 \to X^1$$
$$\vdots$$
$$p_n : (\Lambda_n \times \mathscr{B}^n) \cup X^{n-1} \to X^n$$
$$\vdots$$

by $p_i(\alpha,b) = h_\alpha(b)$ if $(\alpha,b) \in \Lambda_i \times \mathscr{B}^i$, and $p_i(x) = x$ if $x \in X^{i-1}$. Let X^0 be given the quotient topology associated with p_0 (note that X^0 is a discrete space). Inductively, define a topology for each X^n by assigning to X^n the quotient topology generated by p_n. Finally let $X = \bigcup_{n=1}^{\infty} X^n$ be given the weak topology $\mathcal{W}$ induced by the spaces X^n. The topological space $(X,\mathcal{W})$ is called a *CW-complex*. If each p_n is restricted to $\Lambda_n \times \mathcal{B}^{n_\alpha}$, then X^n may be viewed as the adjunction space of X^{n-1} and $\Lambda_n \times \mathcal{B}^{n_\alpha}$ associated with the map $p_n|_{\Lambda_n \times \operatorname{Fr} \mathcal{B}^{n_\alpha}}$. Thus, if $(X,\mathbf{H})$ is a *CW*-complex, then X^{n-1} is a closed subset of X^n. Furthermore, for each n, X^n is closed in X (why?).

(8.E.6) *Examples.*

1. See (8.E.2). The topology for X considered as a *CW*-complex coincides with the relative topology from $\mathscr{E}^3$.

2. A cell structure whose corresponding *CW*-complex fails to have the "right topology." Let $\Lambda_0 = \{x \mid x \in \mathbf{I}\}$, and let $\Lambda_i = \varnothing$ for $i > 0\}$. Define $\mathcal{B}^{n_r} = \mathcal{B}^0$ for each $r \in \mathbf{I}$ and define $h_r : \mathcal{B}^{n_r} \to \mathbf{I}$ by $h_r(0) = r$. Then $(\mathbf{I},\mathbf{H})$ is a cell structure, where $\mathbf{H} = \{h_r \mid r \in \mathbf{I}\}$. However, the resulting *CW*-complex is not homeomorphic to $\mathbf{I}$ with the usual topology.

(8.E.7) *Exercise.* Show that $\mathbf{I}$ with the usual topology can be regarded as a *CW*-complex.

(8.E.8) *Theorem.* If $(X,\mathbf{H})$ is a *CW*-complex, then X is normal and T_2.

Proof. The normality of each X^n is an immediate consequence of (8.B.4) and the inductive procedure for assigning topologies to the X^n. The normality of X now follows from (7.D.2).

Since X^0 is discrete, it is clearly T_1. By (8.B.5) and an inductive argument, it follows that X^n is T_1 for each n. Suppose that $x \in X$; then $x \in X^n$ for some n, and hence $\{x\}$ is closed in X^n. Since X^n is closed in X, we have that $\{x\}$ is closed in X and therefore X is T_1. Since X is T_1 and normal, it is also T_2 and normal (why?).

The weak topology used for *CW*-complexes could have been defined in terms of the closed n-cells.

(8.E.9) *Theorem.* Suppose that $(X,\mathbf{H})$ is a *CW*-complex. Then a set $U \subset X$ is open if and only if $U \cap C^{n_\alpha}$ is open in C^{n_α} for each closed n-cell C^{n_α} in X.

Proof. The necessity is obvious. Suppose then that $U \cap C^{n_\alpha}$ is open in C^{n_α} for each $\alpha \in \Lambda$. Since X has the weak topology, it suffices to show that $U \cap X^n$ is open in X^n for each n. We proceed inductively. The case where $n = 0$ presents no problem, since X^0 has the discrete topology. Suppose that $U \cap X^k$ is open in X^k for $k < n$, and consider $U \cap X^n$. The set $U \cap X^n$ will be open in X^n if and only if $p_n^{-1}(U \cap X^n)$ is open in $X^{n-1} \cup (\Lambda_n \times \mathcal{B}^n)$, and this in turn will be true if we can show that $p_n^{-1}(U \cap X^n) \cap X^{n-1}$ is open in

X^{n-1} and that $p_n^{-1}(U \cap X^n) \cap (\Lambda_n \times \mathscr{B}^n)$ is open in $(\Lambda_n \times \mathscr{B}^n)$. Since $p_n^{-1}(U \cap X^n) \cap X^{n-1} = U \cap X^{n-1}$, it follows from the inductive hypothesis that $p^{-1}(U \cap X^n) \cap X^{n-1}$ is open in X^{n-1}.

To complete the proof, it must be established that $p_n^{-1}(U \cap X^n) \cap (\Lambda_n \times \mathscr{B}^n)$ is open in $(\Lambda_n \times \mathscr{B}^n)$. This we do by showing that $p_n^{-1}(U \cap X^n) \cap (\alpha \times \mathscr{B}^n)$ is open in $(\alpha \times \mathscr{B}^n)$ for each $\alpha \in \Lambda_n$. Note that $p_n^{-1}(U \cap X^n) \cap (\alpha \times \mathscr{B}^n) = (p_n^{-1}(U \cap C^{n\alpha})) \cap (\alpha \times \mathscr{B}^n)$. By hypothesis, $U \cap C^{n\alpha}$ is open in $C^{n\alpha}$, and consequently $p_n^{-1}(U \cap X^n) \cap (\alpha \times \mathscr{B}^n)$ is an open subset of $(\alpha \times \mathscr{B}^n)$, which concludes the proof.

(8.E.10) Definition. Suppose that $(X,\mathbf{H})$ is a cell structure and that A is a subset of X. Let $\Lambda_A = \{\alpha \in \Lambda \mid h_\alpha(\mathscr{B}^{n\alpha}) \subset A\}$. Then A is a *subcomplex* of X if and only if $\mathbf{H}_A = \{h_\alpha : \mathscr{B}^{n\alpha} \to A \mid \alpha \in \Lambda_A\}$ is a cell structure for A.

(8.E.11) Exercise. Suppose that $(X,\mathbf{H})$ is a cell structure and that $A \subset X$. Show that A is a subcomplex of X if and only if whenever an open cell $e^{n\alpha}$ intersects A, then the corresponding closed cell $C^{n\alpha}$ is contained in A.

(8.E.12) Definition. Suppose that $(X,\mathbf{H})$ is a cell structure. Then a subcomplex A of X is *finite* if and only if it consists of a finite number of open cells (and hence of a finite number of closed cells).

(8.E.13) Lemma. Suppose that $(X,\mathbf{H})$ is a CW-complex. Then for each n,

(i)   each closed cell $C^{n\alpha}$ in X^n is contained in the union of a finite number of sets $e^{m\beta}$ where $m \le n$, and

(ii)   each compact subset of X^n is contained in a finite subcomplex of X^n.

Proof. We prove simultaneously by induction that for each n, $C^{n\alpha}$ is the union of a finite number of cells and that every compact subset K of X^n is contained in a finite subcomplex of X^n.

Since $C^{0\alpha} = e^{0\alpha}$, it is clear that $C^{0\alpha}$ is the union of a finite (namely one) number of cells, and since X^0 is discrete, it is equally clear that every compact subset of X^0 is contained in (in fact, is equal to) a finite subcomplex of X^0.

Suppose that both hypotheses are true for all $k \le n - 1$. Let $C^{n\alpha}$ be a closed cell in X^n. Then $C^{n\alpha} = p_n(\alpha \times I(\mathscr{B}^n)) \cup p_n(\alpha \times \mathscr{S}^{n-1}) = e^{n\alpha} \cup p_n(\alpha \times \mathscr{S}^{n-1})$. Since $p_n(\alpha \times \mathscr{S}^{n-1})$ is a compact subset of X^{n-1}, it is contained in a finite subcomplex of X^{n-1}. Furthermore, each of the finite number of closed cells in the subcomplex is the union of a finite number of open cells. Now let K be a compact subset of X^n. Then $K \cap X^{n-1}$ is contained in a finite subcomplex. For each $C^{n\alpha}$ in X^n such that $e^{n\alpha} \cap K \ne \varnothing$, pick a point $q_{n\alpha} \in e^{n\alpha} \cap K$. Since $e^{n\alpha} \cap e^{n\beta} = \varnothing$ whenever $\alpha \ne \beta$, it follows that $q_{n\alpha} \ne q_{n\beta}$.

Assign $A = \{q_{n_\alpha} \mid C^{n_\alpha} \cap K \neq \varnothing\}$ the relative topology. Then A is discrete. We show that A is closed, hence compact, and hence finite, which will imply that K is contained in a finite union of closed n-cells, which of course implies that K is contained in a finite subcomplex. It is easy to see that $p_n^{-1}(A)$ is closed, and since p_n is an identification map, it follows that A is closed in X^n. However, X^n is closed in X, and therefore A is closed in X.

(8.E.14)   Theorem.   A CW-complex $(X,\mathbf{H})$ is compact if and only if X is a finite complex (i.e., consists of a finite number of cells).

Proof.   Suppose that X is finite. Then $X = \bigcup_{i=1}^{k} e^{n_{\alpha i}} = \bigcup_{i=1}^{k} C^{n_{\alpha i}}$. Since each $C^{n_{\alpha i}}$ is compact, so is X.

Conversely, suppose that X is compact and that $\{e^{n_\alpha} \mid \alpha \in \Lambda\}$ is the family of open cells whose union is X. Select a point q_{n_α} in each e^{n_α} and let $A = \{q_{n_\alpha} \mid \alpha \in \Lambda\}$. We show that A is closed. From (8.E.13) and (8.E.3.i and ii), it follows that $A \cap C^{n_\alpha}$ is finite, and hence is closed. Thus, $p_n^{-1}(A \cap C^{n_\alpha})$ is closed for each α, and therefore $p_n^{-1}(A)$ is closed (since Λ_n has the discrete topology). Consequently, A is closed in X, and hence A is compact and finite. (Actually one needs here a simple induction argument since $p_n^{-1}(A \cap C^{n_\alpha}) = (p_n^{-1}(A \cap C^{n_\alpha}) \cap \{\Lambda_{n_\alpha} \times \mathscr{B}^n\}) \cap (p_n^{-1}(A \cap C^{n_\alpha}) \cap X^{n-1}).$)

(8.E.15)   Theorem.   If X is a CW-complex and A is a subcomplex, then A is closed in X.

Proof.   We show inductively that $A \cap X^n$ is closed for each n. We have that $p_0^{-1}(A \cap X^0)$ is closed, since X^0 is discrete. Suppose then that $A \cap X^{n-1}$ is closed. We show that $A \cap X^n$ is closed. Consider $p_n^{-1}(A \cap X^n) = p_n^{-1}(A \cap X^{n-1}) \cup p_n^{-1}(\bigcup \{C^{n_\alpha} \mid \alpha \in \Lambda_{A_n}\})$, where $\{C^{n_\alpha} \mid \alpha \in \Lambda_{A_n}\}$ is the collection of closed n-cells in A. Since for each α, $p_n^{-1}(C^{n_\alpha})$ is the union of $(\alpha \times \mathscr{B}^n)$ and a subset of $p_n^{-1}(A \times X^{n-1})$, it follows that $p_n^{-1}(A \cap X^n) = (\bigcup \{(\alpha \times \mathscr{B}^n) \mid \alpha \in \Lambda_{A_n}\}) \cup p^{-1}(A \cap X^{n-1})$.

Therefore, $A \cap X^n$ is closed in X^n, and since X has the weak topology associated with the spaces X^n, it follows that A is closed in X.

(8.E.16)   Definition.   A CW-complex is *locally finite* if and only if each $x \in X$ has a neighborhood that lies in a finite subcomplex of X (finite in the sense that Λ_A is finite: see (8.E.10)).

(8.E.17)   Examples.
1.   Any CW-complex that is locally compact is locally finite (see the next theorem).
2.   As an example of a CW-complex that fails to be locally finite, for

each $n \in \mathbf{Z}^+$ let $\mathbf{I}_n = \mathbf{I}$, and let X be the space obtained by identifying all the left-hand end points of the $\mathbf{I}_n$ together. Then X with the obvious cell structure is not locally finite.

Locally finite CW-complexes are easily characterized.

(8.E.18)   Theorem.   A CW-complex X is locally finite if and only if it is locally compact.

Proof.   Suppose that X is locally compact and $x \in X$. Then X has a compact neighborhood which by (8.E.13) can intersect only a finite number of the open cells comprising X. The union of the corresponding closed cells yields the desired subcomplex.

Since finite subcomplexes are clearly compact, local finiteness obviously implies local compactness.

It is not surprising, in view of the construction of the CW-complexes, that any CW-complex may itself be considered as a quotient space. Suppose that $(X,\mathbf{H})$ is a CW-complex. For $n \in \mathbf{N}$, let $\hat{X}_n = (\Lambda_n \times \mathscr{B}^n) \cup X^{n-1}$, where Λ_n and X^{n-1} are as defined previously. Let $\hat{X}_{-1} = \varnothing$ and let $\hat{X}$ be the free union of the spaces $\hat{X}^n$. Define a map $p : \hat{X} \to X$ by setting $p(x) = p_n(x)$ if $x \in \hat{X}^n$, where p_n is the map described after (8.E.5). It is easy to see that p is continuous. We show next that p is an identification.

(8.E.19)   Theorem.   The map $p : \hat{X} \to X$ is an identification.

Proof.   Suppose that $U \subset X$ and that $p^{-1}(U)$ is open in $\hat{X}$. We show that U is open by verifying that $U \cap X^n$ is open in X^n for each n. Since the maps p_n are identifications, it follows that $U \cap X^n$ will be open if and only if $p_n^{-1}(U \cap X^n)$ is open in $X^{n-1} \times (\Lambda_n \times \mathscr{B}^n)$. Since $p_n^{-1}(U \cap X^n) = p^{-1}(U) \cap ((\Lambda_n \times \mathscr{B}^n) \cap X^{n-1})$, the result follows.

It is now immediate from (8.C.5) that X is homeomorphic to $\hat{X}/G_p$. The next result is used in a later chapter to show that locally finite CW-complexes are metrizable.

(8.E.20)   Theorem.   If X is a locally finite CW-complex, then

   (i) p is a closed map, and
   (ii) $p^{-1}(x)$ is compact for each $x \in X$.

Proof.   We prove the first part of the theorem and leave the second half to the intrepid reader. Suppose that K is a closed subset of X. We show that $\overline{p(K)} \subset p(K)$. Suppose that $x \in \overline{p(K)}$. Since X is locally finite, there is a neighborhood U of x that is the union of only a finite number of closed cells. Let $T = p(K) \cap U$. Note that $x \in \overline{T}$. For each $n \geq 0$ and $\alpha \in \Lambda_n$, let $K_\alpha = K \cap (\alpha \times \mathscr{B}^n)$, $G_\alpha = p(K_\alpha)$, and $V_\alpha = p^{-1}(U) \cap (\alpha \times \mathscr{B}^n)$. It follows from (8.E.16) that with the exception of a finite number $\alpha \in \Lambda$, all the V_α are empty. Therefore, only a finite number of the G_α can intersect U. If S denotes the union of these G_α's, then clearly S is compact and furthermore $T = p(K) \cap U = S \cap U$. Thus, T is compact and hence closed. Since $x \in \overline{T} = T \subset p(K)$, the first part of the theorem is established.

F. UPPER SEMICONTINUOUS DECOMPOSITIONS: AN INTRODUCTION

In the past several years, there has been considerable interest in quotient spaces resulting from what are called upper semicontinuous decompositions. Decompositions of this sort will be considered in greater depth in Chapter 18.

(8.F.1)   *Definition.*   Let $G = \{g_\alpha \mid \alpha \in \Lambda\}$ be a partition of a topological space X. Then G is an is *upper semicontinuous decomposition* of X if and only if for each $g_\alpha \in G$ and for each open set U containing g_α there is a saturated open set V such that $g_\alpha \subset V \subset U$. (A subset V of X is *saturated* if and only if $V = P^{-1}P(V)$, where P is the quotient map of X to X/G.)

Motivation for this terminology might be based on a consideration of decompositions arising from the graphs of upper semicontinuous functions.

(8.F.2)   *Definition.*   A function $f : X \to \mathscr{E}^1$ is *upper semicontinuous* if and only if for each $a \in \mathscr{E}^1$, $\{x \mid f(x) < a\}$ is open.

A typical graph of such a function where $X = \mathscr{E}^1$ is seen in the following figure.

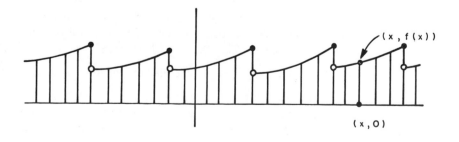

$(x, f(x))$

$(x, 0)$

The collection of segments connecting points $(x,0)$ with their image $(x,f(x))$, together with individual points of $\mathscr{E}^2$ that do not lie in any of these segments yields a decomposition that is upper semicontinuous.

Geometrically, the preceding example of an upper semicontinuous decomposition illustrates quite well what is happening: little sets in the decomposition may approach big ones, but the converse is not allowed. That such decompositions are not at all uncommon may be seen from the following exercise.

(8.F.3)   Exercise.   Suppose that G is a decomposition of a space X, and P is the quotient map. Show that G is upper semicontinuous if and only if P is closed.

If the members of a decomposition G are compact and connected, then G is said to be a *monotone decomposition.* Monotone decompositions are of interest, since it may be shown that the associated decomposition spaces often retain many of the properties of the original spaces. For example, if G is a monotone upper semicontinuous decomposition of a space X and if X has any of the following properties—metrizability, normality, connectedness, compactness, local connectedness, local compactness, separability—then so will X/G. Thus for monotone decompositions, the decomposition space is in many ways similar to the original space. On the other hand, it is shown in Chapter 18 that if K is any compact metric space, then there is a monotone decomposition G of $\mathscr{E}^3$ such that K may be embedded in $\mathscr{E}^3/G$. Hence, such spaces as the Hilbert Cube may be embedded in a decomposition space of $\mathscr{E}^3$. The celebrated dog bone space, a decomposition of $\mathscr{E}^3$ unearthed by Bing [1957a] leads us to another curious result involving upper semicontinuous decompositions.

The nondegenerate elements of this decomposition (members of the decomposition that are not points) are constructed as follows. Start with a solid double torus T_0 lying in $\mathscr{E}^3$ (see next figure). Inside T_0 intertwine four solid double tori as indicated in the figure. Let T_{11}, T_{12}, T_{13} and T_{14} denote these tori, and let $T_1 = \bigcup_{i=1}^{4} T_{1i}$. In each of the T_{1i}, intertwine four more tori in exactly the same fashion as the previous four tori were embedded in T_0. Let T_2 be the union of these sixteen tori. This procedure is repeated inductively to obtain a sequence of sets $T_0, T_1, \ldots$, where each T_i is the union of 4^i tori. It may be shown that $\bigcap_{i=0}^{\infty} T_i$ is a collection of arcs (actually, a "Cantor set" of arcs in the sense that the intersection of the nondegenerate elements and a vertical plane yields a subspace homeomorphic to the Cantor set). The *dog bone decomposition* G of $\mathscr{E}^3$ consists of these arcs, together with single points in the complement of $\bigcap_{i=0}^{\infty} T_i$. Bing showed that $\mathscr{E}^3/G$ is not homeomorphic to $\mathscr{E}^3$, which seemed remarkable enough at the time, but a few years later he showed that $\mathscr{E}^3/G \times \mathscr{E}^1$ is homeomorphic to $\mathscr{E}^4$. How does that strike your intuition?

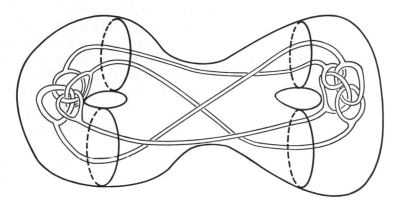

PROBLEMS

Sections A and B

1. Suppose that (X,d) is a pseudometric space. Define an equivalence relation on X by setting $x \sim y$ if and only if $d(x,y) = 0$. Show that $X/\sim$ is metrizable.

2. Suppose that X and Y are topological spaces, $f : X \to Y$ is continuous and onto, and that $\mathscr{V}$ is a topology for X/G_f such that $\mathscr{V} \subsetneq \mathscr{U}$, where $\mathscr{U}$ is the quotient topology. Show that ϕ_f is not continuous with respect to $\mathscr{V}$.

3. In $\mathscr{E}^2$, declare two points equivalent if and only if they lie on the same circle with center at the origin. What is $\mathscr{E}^2/\sim$?

4.* Describe how a double torus (not solid) can be derived as a quotient space.

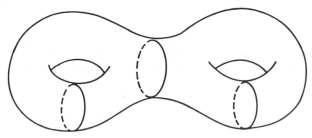

5. Describe how a Moebius strip can be viewed as a quotient space.

6. Suppose that $\sim$ is an equivalence relation on a space X, $\approx$ is an equivalence relation on a space Y, and $f : X \to Y$ is a continuous map with the property that if $x_1 \sim x_2$, then $f(x_1) \approx f(x_2)$. Show that $\hat{f} : X/\sim \to Y/\approx$ defined by $\hat{f}([x]) = [f(x)]$ is well defined and continuous.

7. Is the hypothesis that Y is T_2 necessary in (8.A.7)?

8.* The *projective plane* P_2 is the quotient space obtained from $\mathscr{B}^2$ by identifying antipodal points of $\mathscr{S}^1$. Show that P_2 is homeomorphic to the quotient space obtained from $\mathscr{E}^3 \setminus \{\mathbf{O}\}$ by identifying points x and y if and only if they lie on the same line passing through the origin.

9. Suppose $X = \bigcup_{\alpha \in \Lambda} X_\alpha$ is given the weak topology (induced by the sets $\{X_\alpha \mid \alpha \in \Lambda\}$. Let Y be the free union of $\{X_\alpha \mid \alpha \in \Lambda\}$, and let $h : Y \to X$ be the obvious map. Show that h is continuous and Y/G_h is homeomorphic with X.

10. Let $\mathscr{B}^2$ be the unit disk in $\mathscr{E}^2$ and $\mathscr{S}^1$ its boundary. Suppose that $f : \mathscr{S}^1 \to \mathscr{S}^1 \subset \mathscr{B}^2$ is continuous and onto. Is $\mathscr{B}^2 \cup_f \mathscr{B}^2$ a 2-manifold? If f is a homeomorphism, is $\mathscr{B}^2 \cup_f \mathscr{B}^2$ a 2-manifold?

11. In $\mathscr{S}^2$, every simple closed curve separates $\mathscr{S}^2$. Is the same true in the torus, the Klein bottle, the projective plane? (See problem 8).

12.* Show that the space obtained by sewing a disk homeomorphically along its boundary to the boundary of a Moebius band is homeomorphic to the projective plane. (See problem 8).

13. Show that none of the T_i properties hold up under quotient spaces.

14.* Show that the following spaces are not homeomorphic: A is the cone over $\{(n,0) \mid n \in \mathbf{Z}^+\}$; B is the union of $\{(n,0) \mid n \in \mathbf{Z}^+\}$ and all points in the plane lying on line segments connecting points of $\{(n,0) \mid n \in \mathbf{Z}^+\}$ with $(0,1)$. [Hint: Use first countability.]

15. Let $X = (0,1) \times [0,1]$ and identify all points of the form $(x,1)$ with each other. Show that the resulting quotient space is not homeomorphic to Y, where Y is union of the interior of the triangle with vertices $(0,0)$, $(1,0)$, and $(0,1)$, the open line segment $\{(x,0) \mid 0 < x < 1\}$, and the point $(0,1)$.

Section C

1. Show that a retract is an identification map.

2. Suppose that $f : X \to Y$ is 1–1. Show that f is an identification map if and only if f is a homeomorphism.

3. For each $\alpha \in \Lambda$, let $f_\alpha : X_\alpha \to Y_\alpha$ be continuous, open, and onto. Show that $\Pi f_\alpha : \Pi_{\alpha \in \Lambda} X_\alpha \to \Pi_{\alpha \in \Lambda} Y_\alpha$ is an identification.

4. Find an identification map that is neither open nor closed.

Section D

1. Find an example of a space that is not a k-space.
2. Suppose that X is a k-space, Y is a topological space, and $f : X \rightarrow Y$ is an identification. Show that Y is a k-space.
3. Suppose that G is a partition of a locally compact space X. Show that if X/G is T_2, then X/G is a k-space.
4. Suppose that X is a T_2 k-space. Let $\{C_\alpha \mid \alpha \in \Lambda\}$ be the collection of compact subsets of X. Let Y be the disjoint union of the C_α with the disjoint union topology. Show that Y is locally compact and X is homeomorphic to a quotient space of Y.
5. Show that the product of uncountably many copies of $\mathscr{E}^1$ is not a k-space. (Consider $A = \{\{x_\alpha\} \in \prod\limits_{\alpha \in \Lambda} \mathscr{E}^1 \mid$ for some integer $n \geq 0$, $x = n$ for all but at most n coordinates and $x_\alpha = 0$ otherwise$\}$.)
6. Show that if $f : X \rightarrow Y$ is continuous and open, then Y has the strong topology.
7. Suppose that X is a set and $(Y,\mathscr{U})$ is a topological space. Let $f : X \rightarrow Y$ be onto and give X the weak topology, $\mathscr{W}$. Let $\mathscr{S}$ be the strong topology for Y (with respect to f and $\mathscr{W}$). Compare $\mathscr{U}$ and $\mathscr{S}$.
8. Are closed subsets of k-spaces necessarily k-spaces?
9.* Suppose that X is a space with the property that every compact subset of X is closed. Let X^* be the 1-point compactification of X. Show that every compact subset of X^* is closed if and only if X^* is a k-space.
10.* Suppose that $(X,\mathscr{U})$ is a topological space and let $\mathscr{K}$ be the k-topology associated with $(X,\mathscr{U})$. Is $(X,\mathscr{K})$ a k-space?

Section E

1. Let $X = \{(x,y,z) \in \mathscr{E}^3 \mid x^2 + y^2 + z^2 = 1\}$. Find
    (a) two distinct cell structures for X such that the resulting CW-complex has the relative topology, and
    (b) two distinct cell structures for X such that the resulting CW-complexes yield different topologies for X, neither of which is the relative topology.
2. Show that $\mathscr{E}^n$ with the usual topology is a CW-complex.
3. Show that $\mathscr{S}^1$ is a CW-complex.
4. Is the following space a CW-complex (with the relative topology in $\mathscr{E}^2$)?

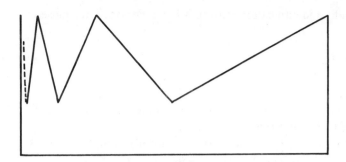

5. Is there a locally finite CW-complex X with the property that for each n, $X^n \setminus X^{n-1} \neq \varnothing$.
6. Represent the Moebius band as a CW-complex.
7. Represent the torus and the Klein bottle as CW-complexes.
8. Prove or disprove: if $(X,\mathbf{H})$ is a CW-complex and $(A,\mathbf{H}_A)$ and $(B,\mathbf{H}_B)$ are subcomplexes, then $(A \cap B,\ \mathbf{H}_A \cap \mathbf{H}_B)$ is a subcomplex and $(A \cup B,\ \mathbf{H}_A \cup \mathbf{H}_B)$ is a subcomplex.
9.* Show that a CW-complex is connected if and only if X^1 is connected.

Section F

1. Show that if X is normal and if G is an upper semicontinuous decomposition of X into compact sets, then X/G is normal.
2. Suppose that X is a metric space and G is a decomposition of X into compact subsets. Show that the following conditions are equivalent:

    (i) G is upper semicontinuous;
    (ii) if whenever $x_i, y_i \in g_i \in G$ $(i \in \mathbf{Z}^+)$ and $\{x_i\}$ converges to a point $x \in g \in G$, then there is a subsequence $\{y_{n_i}\}$ of $\{y_i\}$ that converges to a point $y \in g$.

3. Show that if X is separable and G is an upper semicontinuous decomposition of X, then X/G is separable.
4.* Show that if X is second countable and G is an upper semicontinuous decomposition of X into compact subsets of X, then X/G is second countable.
5. Suppose that X is a space and G is a decomposition of X with only finitely many nondegenerate elements, all of which are closed. Show that G is upper semicontinuous.
6. Show that a decomposition G of a space X is upper semicontinuous if and only if for each open U in X, $\{g \in G \mid g \subset U\}$ is open.

7.* Let G be a decomposition of $\mathcal{E}^2$, whose only nondegenerate element is the x axis. Show that X/G is neither first countable nor locally compact.

8.* Suppose that G is a monotone upper semicontinuous decomposition of a space X and that A is a compact connected subset of X/G. Show that $P^{-1}(A)$ is compact and connected.

9. Show that if G is an upper semicontinuous decomposition of a regular space X into compact sets, then X/G is regular.

Chapter 9

CONTINUA

This chapter is devoted to an important part of point-set topology: the study of continua. A *continuum* is a connected compact topological space. Although the literature involving continua is extensive, we limit our presentation to establishing a few of the better known theorems and to examining some of the more bizarre examples of continua that arise.

A. THE ARC

Our study of continua begins with an investigation of arcs. Recall that an arc is defined to be a homeomorphism from the closed interval I into a space X. However, here, as is often the case in mathematics, the definition is not always easy to apply. The primary goal of this section is to obtain a characterization of the arc that is sometimes more convenient than the definition.

An arc is obviously a metric continuum, but then so are a host of other spaces that are not arcs (as usual, whenever convenient the arc will be confused with its image). Intuitively, it would seem that an arc is as "skinny" as a continuum can be. However, circles, trees, etc., are just as slender. The question, then, is how do these spaces differ from an arc? Notice that if any point is removed from a circle, the resulting space is connected. Furthermore, various points can be extracted from a tree without disconnecting it. On the other hand, if any point is plucked from an arc (with the exception of the two end points), the resulting space is not connected. Trivial as this ob-

A tree

servation may be, it is the keystone of the arc. We shall work toward proving that any metric continuum with exactly two "non-cut points" is an arc.

(9.A.1) *Definition.* Suppose that X is a connected topological space and $x \in X$. Then x is a *cut point* of X if and only if $X \setminus \{x\}$ is disconnected. Points that are not cut points are called *non-cut points*. A cut point x is said to *separate points a and b in X* if and only if a and b lie in different members of some separation of $X \setminus \{x\}$.

A basic property of **I** is that it is linearly ordered; furthermore, the usual topology for **I** coincides with the topology induced by the usual order relation. Hence, it should not be too surprising that efforts to obtain a characterization of the arc will revolve around the introduction of a linear order on certain metric continua—one for which the original topology and the order topology are identical.

(9.A.2) *Definition.* Suppose that X is a connected topological space and that $a,b \in X$. Let $S(a,b)$ be the set $\{a,b\} \cup \{p \in X \mid p \text{ separates } a \text{ and } b \text{ in } X\}$. The *separation order* on $S(a,b)$ is defined as follows. Suppose that $s_1, s_2 \in S(a,b)$. Then $s_1 \le s_2$ if and only if

  (i) $s_1 = a$ or $s_2 = b$, or
  (ii) s_1 separates a and s_2 in X, or
  (iii) $s_1 = s_2$.

One should note that the separation order when applied to **I** yields the usual order.

(9.A.3) *Theorem.* The separation order $\le$ is a linear order on $S(a,b)$.

Proof. The proof is in two steps.

I. The separation order $\leq$ is a partial order.

We prove only the transitivity of $\leq$. Antisymmetry is established by Exercise (9.A.4). Suppose that $s_1 \leq s_2$ and $s_2 \leq s_3$. Only the trivial cases are omitted if we assume that $s_1, s_2 \in S(a,b) \setminus \{a,b\}$ and $s_1 \neq s_2 \neq s_3$. Since $s_1 < s_2$, there are disjoint open sets U and V such that $X \setminus \{s_1\} = U \cup V$, where $a \in U$ and $s_2 \in V$. Similarly, there is a separation (U',V') of $X \setminus \{s_2\}$, where $a \in U'$ and $s_3 \in V'$. We show that $s_3 \in V$, which of course implies that s_1 separates a from s_3. Suppose that $s_3 \in U$. Then by (2.B.10), we have that $U \cup \{s_1\}$ is connected, and hence $U \cup \{s_1\}$ must lie in V', since $s_3 \in U \cap V'$ (2.A.6). However, this is impossible, since it implies that a is in V'.

II. The separation order $\leq$ is a linear order.

Suppose that $s_1, s_2 \in S(a,b)$. It must be shown that either $s_1 \leq s_2$ or $s_2 \leq s_1$. Again it will be assumed that $s_1, s_2 \in S(a,b) \setminus \{a,b\}$, and that $s_1 \neq s_2$. Let (U,V) be a separation of $X \setminus \{s_1\}$, where $a \in U$ and $b \in V$. If $s_2 \in V$, then $s_1 \leq s_2$. If $s_2 \in U$, then $X \setminus \{s_2\} = U' \cup V'$, where $a \in U'$ and $b \in V'$, and U' and V' are disjoint open sets. Observe that the connected set $V \cup \{s_1\}$ must be contained in V', since $b \in V \cap V'$. Thus, $s_2 \leq s_1$, and the proof is complete.

(9.A.4) **Exercise.** Show that if $s_1 \leq s_2$ and $s_2 \leq s_1$, then $s_1 = s_2$.

Since the separation order is a linear order, there are two natural topologies for $S(a,b)$: the order topology and the relative topology. In general, these topologies need not be the same, but if K is a T_1 continuum with exactly two non-cut points, then the two topologies agree, as we see in the next theorem.

(9.A.5) **Theorem.** Suppose that K is a T_1 continuum with exactly two non-cut points a and b. Then $S(a,b) = K$, and the order topology induced by the separation order is precisely the original topology on K.

The proof of the theorem is based on the following lemma.

(9.A.6) **Lemma.** Suppose that K is a T_1 continuum, p is a cut point of K, and (U,V) is a separation of $K \setminus \{p\}$. Then each of U and V contains at least one non-cut point of K.

Proof. Suppose that U consists entirely of cut points. For each $x \in U$, let (U_x, V_x) be a separation of $K \setminus \{x\}$, where $p \in V_x$. Partially order $\mathcal{U} = \{U_x \mid x \in U\}$ by inclusion and let $\mathcal{W} = \{U_{x_\alpha} \mid \alpha \in \Lambda\}$ be a maximal nest in $\mathcal{U}$ (0.D.4). We show that $\bigcap \{U_{x_\alpha} \mid \alpha \in \Lambda\} = \emptyset$. To see this, suppose that $z \in \bigcap \{U_{x_\alpha} \mid \alpha \in \Lambda\}$, and note that

(i)   since $z \in U_{x_\alpha}$, we have that $V_{x_\alpha} \cup \{x_\alpha\} \subset U_z \cup V_z$;

(ii) $p \in V_{x_\alpha} \cap V_z$, and therefore the connected set $V_{x_\alpha} \cup \{x_\alpha\}$ is contained in V_z, and hence, $x_\alpha \in V_z$.

Consequently, $U_z \cup \{z\}$ is contained in $U_{x_\alpha} \cup V_{x_\alpha}$, and since $z \in U_{x_\alpha}$, it follows that $U_z \subsetneq U_{x_\alpha}$. This contradicts the maximality of the nest $\mathscr{W}$. Thus, $\bigcap \{U_{x_\alpha} \mid \alpha \in \Lambda\} = \varnothing$, and furthermore, $\bigcap \{U_{x_\alpha} \cup \{x_\alpha\} \mid \alpha \in \Lambda\} = \varnothing$ (why?). Therefore, it follows that $\{V_{x_\alpha} \mid \alpha \in \Lambda\}$ is an open cover of the compact set K that has no finite subcover, and hence we conclude that U must have a non-cut point.

Proof of 9.A.5.   That $K = S(a,b)$ is obvious, since if $p \in K \setminus \{a,b\}$, then p is a cut point and thus there is a separation (U,V) of $K \setminus \{p\}$. The lemma (together with the hypothesis of the theorem) guarantees that a and b are separated by p. Therefore, we have that $p \in S(a,b)$.

Let $\mathscr{T}$ denote the given topology for K, and let $\mathscr{T}'$ denote the order topology. To show that $\mathscr{T}' \subset \mathscr{T}$, it is sufficient to prove that sets of the form $W = \{x \mid x < p\}$ and $Y = \{x \mid x > p\}$ are in $\mathscr{T}$. Suppose that $p \in K \setminus \{a,b\}$. Then there is a separation (U_p, V_p) of $K \setminus \{p\}$ such that $a \in U_p$ and $b \in V_p$. We show that $W = U_p$, and a similar argument may be used to show that $Y = V_p$.

Suppose that $x \in W$. If $x \in V_p$, then (by the definition of the separation order) $x > p$. This being absurd, we have that $W \subset U_p$.

Now suppose that $z \in U_p$, and let (U_z, V_z) be a separation of $K \setminus \{z\}$ such that $a \in U_z$ and $b \in V_z$. We consider two cases. If $p \in V_z$, then z separates a from p and hence $z < p$. Therefore, we have that $z \in W$. If $p \in U_z$, we arrive at a contradiction as follows. Since $V_z \cup \{z\} \subset U_p \cup V_p$ and since $b \in V_z \cap V_p$, it follows that $V_z \cup \{z\}$ is actually contained in V_p. Consequently, we have that $z \in V_p$, which is contrary to our original assumption.

We now show that $\mathscr{T} \subset \mathscr{T}'$. Let U be an element of the given topology $\mathscr{T}$ for K. If $x \in U$, then it is necessary to find an $(\leq)$-interval (c,d) such that $x \in (c,d) \subset U$. If no such interval exists, then the collection $\{[p,q] \cap (K \setminus U) \mid x \in (p,q)\}$ has the finite intersection property, and hence the intersection of all the members of this family is nonempty and is contained in $K \setminus U$. (Note that $[p,q]$ is closed, since $\mathscr{T}' \subset \mathscr{T}$.) However, it is easy to see that $\bigcap \{[p,q]\} \mid x \in (p,q)\} = \{x\} \subset U$. This should strike the reader as being a bit unusual. Therefore, the desired open interval exists, and consequently $\mathscr{T} \subset \mathscr{T}'$. (The pedant may wish to check the cases where $x = a$ or $x = b$.)

Now that the candidate for arc status has been successfully ordered, the task remains of defining a map from a suitable dense subset of K onto the dyadic rationals in $\mathbf{I}$ and then of extending the map in the obvious manner.

(9.A.7) **Lemma.** Suppose that A is a countable linearly ordered set such that

    (i) A has no smallest or largest element, and

    (ii) if $a,b \in A$ with $a < b$, then there is a $c \in A$ such that $a < c < b$.

Then there is an order-preserving bijection $h : A \to D$, where $D = \{k/2^n \mid 0 < k < 2^n, k,n \in \mathbf{Z}^+\}$.

    Proof. Let $A = \{a_1, a_2, \ldots\}$ and define $h(a_1) = 1/2$. Let n_1 be the first integer such that $a_{n_1} < a_1$ and n_2 be the first integer such that $a_1 < a_{n_1}$. Let $h(a_{n_1}) = 1/2^2$ and $h(a_{n_1}) = 3/2^2$. Et cetera.

(9.A.8) **Theorem.** Suppose that K is a metric continuum with exactly two non-cut points a and b. Then K is homeomorphic to **I**.

    Proof. Let A be a countable dense subset of $K \setminus \{a,b\}$. Since K is connected, it follows that A satisfies the two conditions of the hypothesis of the preceding lemma. Consequently, there is a function h mapping A onto D that preserves order. Let $h(a) = 0$ and $h(b) = 1$. Suppose that $p \in K \setminus (A \cup \{a,b\})$. It must be decided where p is to be sent. Since p is a cut point, there is a separation (U_p, V_p) of $K \setminus \{p\}$, such that $a \in U_p$ and $b \in V_p$. Note that $0 < \sup\{h(x) \mid x \in A \cap U_p\} \leq \inf\{h(x) \mid x \in A \cap V_p\} < 1$, and an elementary argument yields that $\sup\{h(x) \mid x \in A \cap U_p\} = \inf\{h(x) \mid x \in A \cap V_p\}$. Let $h(x)$ be this common value. It is not difficult to check that h is the desired homeomorphism.

    Characterizations are not only aesthetically pleasing, but are often useful in proving theorems. For example, we use (9.A.8) in the proof of (9.B.1).

    In the problem section, a characterization of 1-spheres is obtained. The 2-spheres, however, are somewhat more difficult to handle. Perhaps the most elegant characterization of the 2-sphere along these lines is due to Bing [1946], who showed that a locally connected metric continuum X is homeomorphic to $\mathscr{S}^2$ if and only if X is separated by each simple closed curve contained in X and is not separated by any pair of points in X. This is known as the Kline sphere characterization.

    Can 3-spheres be so neatly categorized? One of the great unsolved problems in mathematics deals specifically with this problem. The Poincaré conjecture asserts (in terminology that will become comprehensible once Chapter 12 is completed) that every compact, simply connected 3-manifold is homeomorphic to $\mathscr{S}^3$. Since the early 1900s, this conjecture has plagued many of the world's outstanding mathematicians, and as yet no one has been able to prove or disprove it. This is an excellent illustration of one of mathematics more intriguing aspects: the simplicity of the statement of a proposition often belies the difficulties encountered in its resolution.

B.   PEANO CONTINUA

We now turn our attention to a major class of continua, the Peano continua. A *Peano continuum* is a locally connected metric continuum. Arcs are Peano continua, but hardly representative of them. The result toward which we are heading is the venerable (1913) Hahn-Mazurkiewicz theorem, which states that every Peano continuum is a continuous image of the arc I. Thus, there is a continuous map from I onto the n-dimensional cube, or even more amusing, there is a continuous function mapping I onto the Hilbert cube. If the reader suddenly feels uncertain with regard to his intuition concerning continuous functions, he is in good company. Peano's amazing discovery in 1890 that I could be mapped continuously onto the unit square created havoc in the mathematical world (but scarcely anywhere else, needless to say). Among the major casualties brought about by Peano's result was the then existing concept of dimension.

As a first step toward proving the Hahn-Mazurkiewicz theorem, it will be shown that Peano continua are arc connected, an interesting result in itself. Since the proof is based on the notion of simple chains, we recall that if a and b are points in a topological space X, then a *simple chain* of sets from a to b is a finite collection of subsets of X, $\{U_1, U_2, \ldots, U_n\}$, where $a \in U_1$, $b \in U_n$ and $U_i \cap U_j \neq \varnothing$ if and only if $|i - j| \leq 1$.

(9.B.1)   Theorem.   Suppose that P is a Peano continuum. Then P is arc connected.

Proof.   Let a and b be points in P. An arc must be constructed in P with end points a and b. We form a decreasing sequence of simple chains from a to b, and show that the intersection of these chains is an arc with end points a and b.

Apply (2.F.2) to obtain a simple chain $\mathscr{C}_1 = \{U_{11}, \ldots, U_{1k_1}\}$ of open connected sets from a to b whose links have diameter less than 1. Let $\mathscr{C}_1^* = \bigcup \{U_{1i} \mid 1 \leq i \leq k_1\}$. Then each $x \in \mathscr{C}_1^*$ is contained in at most two links of $\mathscr{C}_1$, and hence for each $x \in \mathscr{C}_1^*$ there is an open connected set V_x with diameter less than 1/2 whose closure lies in each of these links. We extract a simple chain from the family of V_x's. Since (2.F.2) is applicable only to connected sets, we first construct simple chains in each of the U_{1i}'s and then patch these up to obtain a simple chain from a to b. To this end, select points $x_i \in U_{1i} \cap U_{1(i+1)}$ for $i = 1, 2, \ldots, k_1 - 1$, and form simple chains from a to x_1, x_1 to x_2, $\ldots$, and x_{k_1-1} to b that lie in U_{11}, $U_{12}, \ldots$, and $U_{1(k_1-1)}$, respectively. Now we patch. Retain all of the links of the first such chain up up to and including the first link that intersects a link of the second chain. Throw out the remaining members of the first chain, and continue with the second one until a link of the third chain is encountered. Remove the rest of the second chain, and keep going with the third, etc. This will eventually lead to a simple chain that runs from a to b. The whole procedure is now inductively repeated to obtain a sequence of simple chains $\mathscr{C}_1$, $\mathscr{C}_2, \ldots$, each of which is composed of connected open links, such that

(i)   if U_{ni} is a link of $\mathscr{C}_n$, then $\operatorname{diam} U_{ni} < 1/2^n$, and
(ii) if U_{ni} is a link of $\mathscr{C}_n$, then $\overline{U}_{ni} \subset U_{n-1,j}$ for some link $U_{n-1,j}$ in $\mathscr{C}_{n-1}$.

For each $n > 1$, let $\mathscr{C}_n^*$ denote the union of the closures of the links in $\mathscr{C}_n$. Then $A = \bigcap_{n=1}^{\infty} \mathscr{C}_n^*$ is a metric continuum containing a and b. To see that A is an arc with end points a and b, we need only check that if $x \in A \setminus \{a,b\}$, then x is a cut point. Suppose that $x \in A \setminus \{a,b\}$. Then x belongs to each $\mathscr{C}_n^*$ and is contained in at most two links of $\mathscr{C}_n$. Let D_n be the union of all links in $\mathscr{C}_n$ that precede these one or two links, and let E_n be the union of those following them. Then the sets $D = \bigcup_{n=1}^{\infty} (D_n \cap A)$ and $E = \bigcup_{n=1}^{\infty} (E_n \cap A)$ form a separation of $A \setminus \{x\}$. Clearly, $a \in D$ and $b \in E$.

Before proving the Hahn-Mazurkiewicz theorem, we present the reader with the following exercise.

(9.B.2) Exercise. Suppose that P is a Peano continuum and let $\varepsilon > 0$ be given. Show that there is a $\delta > 0$ such that if $x,y \in P$ and $d(x,y) < \delta$, then there is an arc from x to y with diameter less than ε.

(9.B.3) Theorem (*Hahn* [*1914*]; *Mazurkiewicz* [*1920*]). If P is a Peano continuum, then P is a continuous image of **I**. Conversely, if P is a T_2 space and $f : \mathbf{I} \to P$ is continuous and onto, then P is a Peano continuum.

Proof. Since, logically speaking, one miracle is equivalent to another, it should not be surprising to find that the first half of the theorem depends on a previous result: Every compact metric space is a continuous image of the Cantor set. We shall extend a map from the Cantor set onto P to all of **I**.

Suppose that $f : K \to P$ is the map given by (6.C.12). Note that $\mathbf{I} \setminus K$ consists of a countable collection of disjoint open intervals, $\mathscr{I} = \{I_1, I_2, \ldots\}$. Denote I_n by (a_n, b_n), and assume that the collection $\mathscr{I}$ has been linearly ordered in a manner that respects decreasing size.

If $f(a_n) = f(b_n)$, then define $f(x) = f(a_n)$ for each $x \in (a_n, b_n)$. If $f(a_n) \neq f(b_n)$, the extension is slightly more complicated. From (9.B.2), corresponding to $\varepsilon = 1/2^n$, there is a δ_n such that if $d(x,y) < \delta_n$, then arcs may be found from x to y with diameter less than $1/2^n$. Since $f : K \to P$ is uniformly continuous, corresponding to δ_n, there is a γ_n such that if $|x - y| < \gamma_n$, then $d(f(x), f(y)) < \delta_n$. Assume that $\gamma_1 > \gamma_2 > \gamma_3 \cdots$.

Corresponding to γ_1 there are at most a finite number of intervals I_m with diameter greater than γ_1. For each such interval I_j, let A_j be the image of any arc α_j in P with domain $[a_j, b_j]$ running from $f(a_j)$ to $f(b_j)$ (9.B.1), and extend the map f to I_j by setting $f(x) = \alpha_j(x)$ for $x \in (a_j, b_j)$. For any interval I_m with $\gamma_2 \leq \operatorname{diam} I_m < \gamma_1$, there is an arc A_m of diameter less than $1/2$ that connects $f(a_m)$ and $f(b_m)$. Map I_m onto A_m in the obvious fashion. If I_m is such that $\gamma_3 \leq \operatorname{diam} I_m < \gamma_2$, there is an arc A_m with diameter less than $1/2^2$ that joins $f(a_m)$ to $f(b_m)$. Again extend f to map I_m onto A_m. An easy inductive argument yields an extension F of f that maps $\mathbf{I}$ onto P. We leave to the reader the task of showing that F is continuous.

The converse follows easily from (2.G.11) and (2.E.7) except for one detail: the proof of the metrizability of P, which is deferred until the next chapter (10.C.8).

The foregoing theorems may be combined to yield an easy proof of the following result.

(9.B.4)   Theorem.   Suppose that X is a T_2 space. Then X is arc connected if and only if X is path connected.

Proof.   Certainly arc connectedness implies path connectedness. Suppose then that $a, b \in X$ and let $f : \mathbf{I} \to X$ be a path with endpoints a and b. Since $f(\mathbf{I})$ is a Peano continuum (9.B.3), there is an arc in $f(\mathbf{I})$ from a to b (9.B.1), and consequently X is arc connected.

The Hahn-Mazurkiewicz theorem is typical of many existence type theorems in that it is of no particular help in actually defining a map from $\mathbf{I}$ onto a Peano continuum P. It is interesting to see how Peano constructed a map from $\mathbf{I}$ onto $P = \mathbf{I} \times \mathbf{I}$.

Divide $\mathbf{I} \times \mathbf{I}$ into nine squares and partition $\mathbf{I}$ into nine subintervals. A linear map f_1 from $\mathbf{I}$ onto the subset of $\mathbf{I} \times \mathbf{I}$ indicated in the figure is easily defined.

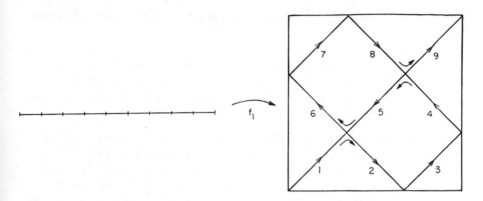

Now divide each of the nine squares into nine more squares and partition **I** into 81 pieces. Define a map f_2 from **I** into **I** $\times$ **I** by following the bouncing arrows.

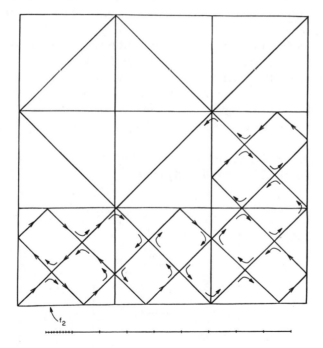

A pattern has begun to emerge which the reader should be able to distinguish.

(9.B.5) Exercise. Show that the sequence of functions $f_1, f_2, \ldots$ above converges uniformly to a (continuous) function $f : \mathbf{I} \to P$.

A rather neat characterization of Peano continua is found in the next theorem.

(9.B.6)   Theorem (*Fraser* [*1972*]). Suppose that (X,d) is a compact metric space. For each $x \in X$ and each $\varepsilon > 0$, let $B_\varepsilon(x) = \{y \in X \mid d(x,y) \le \varepsilon\}$. If for each $x \in X$ and each $\varepsilon > 0$, $\overline{S_\varepsilon^d(x)} = B_\varepsilon(x)$, then X is a Peano continuum (cf. problem B.13).

Proof.   It must be shown that X is connected and locally connected; this will be accomplished if we can establish that $B_\varepsilon(x)$ is connected for all $x \in X$ and all $\varepsilon > 0$. Suppose that $B_\varepsilon(x)$ is not connected for some $x \in X$ and some $\varepsilon > 0$. Then there are disjoint nonempty closed sets A_1 and A_2 such that $B_\varepsilon(x) = A_1 \cup A_2$. Suppose that $x \in A_1$. By (3.A.14), there is a $y \in A_2$ and an $\hat{\varepsilon} \le \varepsilon$ such that $d(x,y) = d(x,A_2) = \hat{\varepsilon} > 0$. Then $y \in B_\varepsilon(x)$, but $y \notin \overline{S_{\hat{\varepsilon}}(x)}$ which contradicts the hypothesis. Thus, $B_\varepsilon(x)$ is connected.

We mention that Fraser has also established a partial converse: if X is a Peano continuum, then X has a metric d under which $B_\varepsilon(x) = \overline{S_\varepsilon(x)}$ for all $x \in X$ and $\varepsilon > 0$.

Before leaving the topic of Peano continua, we construct a special member of this class: the Sierpinski or universal plane curve, C. This continuum is universal in the sense that it can be shown (Sierpinski, 1916) that if M is any one-dimensional Peano continuum embeddable in $\mathscr{E}^2$, then M can be embedded in C, even though C is one dimensional. The construction of the Sierpinski curve is based on a Cantor like procedure, which we describe geometrically.

Start with the unit square $\mathbf{I} \times \mathbf{I}$ and from it extract the interior of the rectangle R indicated in the following figure.

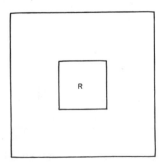

The second step consists of the removal of the interiors of eight more squares, and in the third step 64 such interiors are eliminated, etc. That which remains after the removal of all these squares is called the *universal plane curve*.

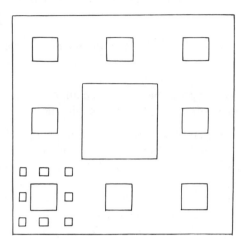

If instead of removing squares from the unit square, one removes cubes (in an analogous fashion) from the unit cube, then the resulting space is called the *universal curve*. The universal curve is a 1-dimensional Peano continuum with the property that any 1-dimensional continuum can be embedded in it. Interestingly, it can be shown that simple closed curves and the universal curve are the only 1-dimensional Peano continua that are homogeneous Anderson [1958].

C. CHAINABLE CONTINUA

We have already seen the wide applicability of arguments using simple chains of sets. Our study now centers on spaces that may actually be defined in terms of simple chains.

(9.C.1) Definition. A simple chain $\mathscr{C}$ of open sets is called an *ε-chain* if and only if each link of $\mathscr{C}$ has diameter less than ε.

(9.C.2) Definition. A metric space X is *chainable* if and only if for each $\varepsilon > 0$, there is an ε-chain that covers X. If $a,b \in X$, then X is *chainable from a to b* if and only if for each $\varepsilon > 0$, there is an ε-chain $\{C_1, \ldots, C_n\}$ covering X such that $a \in C_1$ and $b \in C_n$.

(9.C.3) Examples.
1. The unit interval **I** is chainable from 0 to 1;
2. {0,1} with the relative topology is not chainable;
3. (0,1) is chainable, but not from a to b for any $a,b \in (0,1)$;
4. the topologist's sine curve is chainable.

In spite of examples 2 and 3, chainability does have some hereditary traits.

(9.C.4) Theorem. Suppose that X is a chainable metric continuum, and that $K \subset X$ is a subcontinuum. Then K is chainable.

Proof. Suppose that $\varepsilon > 0$. Since X is chainable, there is an ε-chain $\{C_1, C_2, \ldots, C_n\}$ that covers X. Let i be the first integer such that $C_i \cap K \neq \varnothing$ and let j be the largest integer such that $C_j \cap K \neq \varnothing$. We shall show that $\{C_i \cap K, C_{i+1} \cap K, \ldots, C_j \cap K\}$ is an ε-chain in K that covers K. This is clearly the case unless there are links C_p, C_{p+1} with $i \leq p < j$ such that $C_p \cap K$ and $C_{p+1} \cap K$ fail to intersect. However, if $(C_p \cap K) \cap (C_{p+1} \cap K) = \varnothing$, then $\bigcup_{i \leqslant m < p} (C_m \cap K)$ and $\bigcup_{p+1 \leqslant m < j} (C_m \cap K)$ form a separation of K, which is impossible.

Intuitively, it seems that chainable continua are not very thick. In fact, it can be demonstrated that any chainable continuum is homeomorphic to a plane continuum, or in other words, all chainable continuum are embeddable in the plane (Bing [1951a]). Furthermore, Bing showed (in the same paper) that if K is a chainable continuum, then there is an uncountable collection of disjoint continua lying in the plane, each of which is homeomorphic to K.

A rather surprising result is that chainable continua have the fixed point property.

(9.C.5) Definition. Suppose that K is a chainable continuum. Then a sequence of simple chains $\mathscr{C}_1, \mathscr{C}_2, \ldots$ each of which covers K, is called a *defining sequence of chains* for K if and only if

(i) $\mathscr{C}_n$ is a $(1/2^n)$-chain with the property that disjoint links have disjoint closure, and
(ii) $\mathscr{C}_{n+1}$ is a proper refinement of $\mathscr{C}_n$, i.e., the closure of each link in $\mathscr{C}_{n+1}$ is contained in some link of $\mathscr{C}_n$.

(9.C.6) Exercise. Show that each chainable continuum has a defining sequence of chains.

(9.C.7) **Theorem** *(Hamilton [1951])*. If K is a chainable continuum, then K has the fixed point property.

Proof. Suppose that f is a continuous function from K into K. Let $\mathscr{C}_1, \mathscr{C}_2, \ldots$ be a defining sequence of chains for K, and for each $k \in \mathbf{Z}^+$ let $\mathscr{C}_k = \{C_1^k, \ldots, C_n^k\}$. By (7.E.9), it suffices to find, for each $\varepsilon > 0$, a point $p \in K$ such that $d(p, f(p)) < \varepsilon$. Let $\varepsilon > 0$ be given. Choose an integer k large enough so that $1/2^k < \varepsilon$. Now define:

$$A = \{x \in K \mid \text{if } f(x) \in \overline{C_i^k} \text{ and } x \in \overline{C_j^k}, \text{ then } j < i\}$$

$$B = \{x \in K \mid \text{there is an integer } i \text{ such that } x \in \overline{C_i^k} \text{ and } f(x) \in \overline{C_i^k}\}$$

$$C = \{x \in K \mid \text{if } f(x) \in \overline{C_i^k} \text{ and } x \in \overline{C_j^k}, \text{ then } j > i\}$$

Note that A and C are closed (their complements are open: why?) If $B = \varnothing$, then it is easy to see that A and C are disjoint closed sets whose union is K, which is impossible since K is connected. Hence, $B \neq \varnothing$ and any point in B will serve as the point p.

We mention here that Dyer [1956] has shown that an arbitrary product of chainable continua has the fixed point property. An important consequence of this result is a rather novel approach to proving the Brouwer fixed point theorem. Most readers should have little difficulty with Dyer's article, which illustrates nicely how chainable continua may be manipulated.

Distant relatives of the chainable continua are the irreducible continua.

(9.C.8) **Definition.** Suppose that X is a topological space and that H_1 and H_2 are closed subsets of X. Then a continuum $K \subset X$ is *irreducible from H_1 to H_2* if and only if

(i) $K \cap H_1 \neq \varnothing$ and $K \cap H_2 \neq \varnothing$, and
(ii) for each proper subcontinuum L of K, either $L \cap H_1 = \varnothing$ or $L \cap H_2 = \varnothing$.

One should note that an arc with end points a and b is irreducible from a to b. More generally, we have the following result.

(9.C.9) **Exercise.** Suppose that K is a metric continuum and $a, b \in K$. Show that if K is chainable from a to b, then K is irreducible from a to b.

That the converse of the preceding exercise is false may be seen from our semiuniversal example, the topologist's sine curve X.

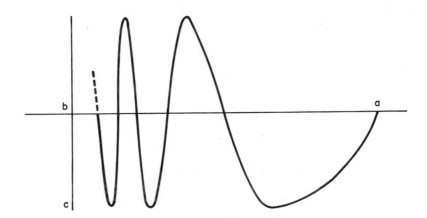

The space X is irreducible from a to b but not chainable between these points (although one might observe that X is chainable between a and c).

That irreducible continua are common enough is evident from the next theorem.

(9.C.10) Theorem. Suppose that X is a Hausdorff space and that H_1 and H_2 are disjoint closed subsets of X. Then if M^* is any continuum intersecting both H_1 and H_2, there is a subcontinuum K of M^* that is irreducible from H_1 to H_2.

Proof. Let $\mathscr{M} = \{M \mid M$ is a continuum intersecting H_1 and H_2 and $M \subset M^*\}$. Order $\mathscr{M}$ by inclusion $(M_1 \leq M_2$ if and only if $M_1 \subset M_2)$ and let $\mathscr{N}$ be a maximal nest in $\mathscr{M}$ (0.D.4). Then $K = \bigcap \{M \mid M \in \mathscr{N}\}$ is a continuum, and, since for each $i \in \mathbf{Z}^+$, we have that $K \cap H_i = \bigcap \{M \cap H_i \mid M \in \mathscr{N}\}$ is nonempty (2.H.2), it follows that K is the desired irreducible continuum.

The concept of unicoherence is taken up next. Further consideration is given to this topic in Chapter 15.

(9.C.11) Definition. A connected space X is *unicoherent* if and only if whenever $X = H_1 \cup H_2$ where H_1 and H_2 are closed and connected, then $H_1 \cap H_2$ is connected.

It is easy to see that arcs are unicoherent, and that circles are not. It is not so obvious what the status of $\mathscr{E}^n$, disks, spheres, etc., should be. The next result throws little light on this problem, although it does yield a number of examples of unicoherent spaces.

(9.C.12) **Theorem.** Every chainable continuum is unicoherent.

Proof. Suppose that K is a chainable continuum that is not unicoherent. Then K may be written as the union of closed connected sets C and D, where $C \cap D = A \cup B$, and A and B are nonempty, disjoint, and closed. Let $r = d(A,B)$. By (9.C.6), there is a simple chain $\mathcal{C}_1$ covering K of mesh less than $r/4$ with the property that if C_i and C_j are nonadjacent links in $\mathcal{C}_1$, then $\bar{C}_i \cap \bar{C}_j = \varnothing$. It follows from the proof of (9.C.4) that the links of $\mathcal{C}_1$ that intersect C form a subchain $\mathcal{C}'_1$ that covers C. Let C^1_j be the first link of $\mathcal{C}'_1$ that intersects either A or B. If A was struck first, let $C^1_{m_1}$ be the last link of $\mathcal{C}'_1$ that intersects B. It follows from the connectedness of C that there is a link $C^1_{k_1} \in \mathcal{C}'_1$ that misses $A \cup B$ and $j_1 < k_1 < m_1$.

Let r_1 be a Lebesgue number for $\mathcal{C}_1$. Then there is a simple chain $\mathcal{C}_2$ of mesh less than $\min\{r_1, 1/2^2\}$ that refines $\mathcal{C}_1$, covers C, and has links $C^2_{j_2}$, $C^2_{k_2}$, $C^2_{m_2}$ $(j_2 < k_2 < m_2)$ where

(i) $C^2_{j_2} \cap A \neq \varnothing$ and $C^2_{m_2} \cap B \neq \varnothing$, and
(ii) $C^2_{k_2} \subset C^1_{k_1}$.

The reader should be able to convince himself that $\mathcal{C}_2$ exists since C is connected, and no harm is done if the numbering of links is reversed.

This procedure is repeated inductively to obtain a decreasing sequence of links $C^1_{k_1}$, $C^2_{k_2}$, ... that do not intersect $A \cup B$. Let $p = \bigcap_{i=1}^{\infty} C^i_{k_i}$ and set $s = d(p,D)$. If n is any positive integer such that $1/n < s$, then the link $C^n_{k_n}$ misses D; however, this is impossible, since there are links in $\mathcal{C}_n$ preceding $C^n_{k_n}$ that intersect $A \cap D$, and ones following $C^n_{k_n}$ that meet $B \cap D$.

An important property of locally connected unicoherent spaces is the following.

(9.C.13) **Theorem.** Suppose that X is a locally connected unicoherent space. Then if a closed subset A separates two points p and q in X, so must some component of A.

Proof. Three lemmas are needed. The proofs are due to Stone [1949].

(9.C.14) **Lemma.** Suppose that D_1 and D_2 are disjoint connected subsets of a locally connected unicoherent space X such that Fr $D_1 \subset$ Fr D_2. Then Fr D_1 is connected.

Proof. Let $\{C_\alpha \mid \alpha \in \Lambda\}$ be the set of components of $X \setminus (D_1 \cup D_2)$. Then for each $\alpha \in \Lambda$, we have that Fr $C_\alpha \subset$ Fr $(D_1 \cup D_2) \subset$ Fr $D_1 \cup$ Fr D_2 = Fr D_2. Hence, $\bar{C}_\alpha \cap \bar{D}_2 \neq \varnothing$ for each α, and so by (2.A.11) it follows that

$A = \bar{D}_2 \cup \bigcup_{\alpha \in \Lambda} \bar{C}_\alpha$ is connected. Note that $X \setminus D_1 \subset A \subset \overline{X \setminus D_1}$; therefore, $\bar{A} = \overline{X \setminus D_1}$ and consequently, $\overline{X \setminus D_1}$ is connected. Since $X = \overline{X \setminus D_1} \cup \bar{D}_1$ is unicoherent, we have that $\overline{X \setminus D_1} \cap \bar{D}_1 = \mathrm{Fr}\, D_1$ is connected.

(9.C.15) Lemma. If D_1 and D_2 are connected open subsets of a locally connected unicoherent space X such that $\mathrm{Fr}\, D_1 \cap \mathrm{Fr}\, D_2 = \varnothing$, then $D_1 \cap D_2$ is connected.

Proof. Suppose that $D_1 \cap D_2$ is not connected, and let C be a component of $D_1 \cap D_2$. Let x be a point in $(D_1 \cap D_2) \setminus C$. Since C is closed in $D_1 \cap D_2$, it follows that $x \notin \bar{C}$. Let C^* be the component of $X \setminus C$ that contains x. Note that C^* is open (X is locally connected) and that $C \cap C^* = \varnothing$, and hence $C \cap \bar{C}^* = \varnothing$.

Let $\hat{C}$ be the component of $X \setminus \bar{C}^*$ containing C. Then $\mathrm{Fr}\, C^* = \mathrm{Fr}\, \hat{C}$ and hence, by (9.C.14), $\mathrm{Fr}\, C^*$ is connected. On the other hand, we have that $\mathrm{Fr}\, C^* \subset \mathrm{Fr}\, C \subset \mathrm{Fr}(D_1 \cup D_2) \subset \mathrm{Fr}\, D_1 \cup \mathrm{Fr}\, D_2$. Since $\mathrm{Fr}\, D_1 \cup \mathrm{Fr}\, D_2$ is the union of two disjoint closed sets, either $\mathrm{Fr}\, C^* \subset \mathrm{Fr}\, D_1$ or $\mathrm{Fr}\, C^* \subset \mathrm{Fr}\, D_2$, say $\mathrm{Fr}\, C^* \subset \mathrm{Fr}\, D_1$. However, $c \in D_1 \cap C^*$ and $C \subset D_1 \cap (X \setminus C^*)$. Therefore, $D_1 \cup \mathrm{Fr}\, C^* \neq \varnothing$ and it follows that $D_1 \cap \mathrm{Fr}\, D_1 \neq \varnothing$, which contradicts the openness of D_1.

(9.C.16) Lemma. If D_1 and D_2 are disjoint closed subsets of a locally connected unicoherent space X neither of which separates points p and q in X, then $D_1 \cup D_2$ does not separate p and q.

Proof. The proof is left as an exercise for the reader.

Proof of (9.C.13). Suppose that A is a closed subset of X that separates points p and q. Let D be the component of $X \setminus A$ containing p. Clearly, $q \notin D$, and furthermore, since $\mathrm{Fr}\, D \subset \mathrm{Fr}\, A \subset A$, it follows that $q \notin \bar{D}$. Let C be the component of $X \setminus \bar{D}$ that contains q. Then C and D are disjoint open sets and $\mathrm{Fr}\, C \subset \mathrm{Fr}\, D$.

We show that $\mathrm{Fr}\, C$ is connected. Suppose that $\mathrm{Fr}\, C$ is not connected. Then $\mathrm{Fr}\, C = F_1 \cup F_2$, where F_1 and F_2 are nonempty disjoint closed subsets of X. Observe that $C \cup F_2 \cup D$ is connected (2.A.13), contains p and q, and does not intersect F_1. Similarly, $C \cup F_1 \cup D$ is connected, contains p and q, and does not intersect F_2. Thus, neither F_1 nor F_2 separates p and q, and so by (9.C.16), $F_1 \cup F_2 = \mathrm{Fr}\, C$ does not separate p and q, which is obviously absurd. Hence, $\mathrm{Fr}\, C$ is connected.

To conclude the proof, let K be the component of A that contains $\mathrm{Fr}\, C$. Then K separates p and q in X.

In the problem section, the reader is asked to prove the converse of this theorem.

D. DECOMPOSABLE AND INDECOMPOSABLE CONTINUA

This topic will eventually lead to some rather picturesque examples of pathological topological spaces. A continuum X is *decomposable* if and only if X can be written as the union of two proper subcontinua; X is *indecomposable* if and only if X is not decomposable. The more familiar continua are certainly decomposable, and the reader would probably be hard pressed to find an example of a continuum that is indecomposable (other than a point).

(9.D.1) Exercise. Suppose that X is a T_2 continuum. Show that X is decomposable if and only if X contains a proper subcontinuum with non-empty interior (in X). Thus all nontrivial Peano continua are decomposable.

The first step in constructing an indecomposable continuum is to introduce the idea of composant.

(9.D.2) Definition. Suppose that X is a continuum and that $x \in X$. Then the *composant* of x, K_x, is the union of all proper subcontinua of X that contain x.

(9.D.3) Examples.
    1. Suppose that X is an arc with end points a and b. Then $K_a = X \setminus \{b\}$, $K_b = X \setminus \{a\}$, and if c is an interior point of X, $K_c = X$.
    2. If $X = \mathscr{S}^1$ or $\mathscr{S}^2$, then $K_x = X$ for each $x \in X$.

(9.D.4) Exercise. Find the composants of a triod and of the topologist's sine curve.

(9.D.5) Exercise. Show that if X is a decomposable T_2 continuum, then there is an $x \in X$ such that $K_x = X$.

Composants are characterized as follows.

(9.D.6) Theorem. Suppose that X is a continuum and that $x \in X$. Then $K_x = X \setminus \{y \mid X$ is irreducible from x to $y\}$.

    Proof. Suppose that $y \in K_x$. Then there is a proper subcontinuum C containing x and y, and hence X is not irreducible from x to y.
    The remainder of the proof is somewhat easier, and is omitted.

(9.D.7) Theorem. Suppose that X is a T_2 continuum and that $x \in X$. Then $\overline{K}_x = X$, and hence composants are dense subsets.

Proof. Suppose that U is an arbitrary open set in X. We shall show that $U \cap K_x \neq \emptyset$. Since X is regular, there is a nonempty open set V such that $\overline{V} \subset U$. If $x \in \overline{V}$, then $x \in U \cap K_x$ and we are through. If $x \notin \overline{V}$, let C be the component of $X \setminus \overline{V}$ containing x. Then $\overline{C}$ is a proper subcontinuum of X that contains x, and hence $\overline{C} \subset K_x$. However, by (4.A.12), we have that $\overline{C} \cap \overline{V} \neq \emptyset$, and therefore K_x intersects U.

In (9.D.3), it was observed that an arc has only three distinct composants. This represents a special case of the following theorem.

(9.D.8) ***Theorem.*** Suppose that X is a decomposable T_1 continuum and that $a,b \in X$. If X is irreducible from a to b, then X has precisely three distinct composants.

Proof. Suppose that $X = A \cup B$, where A and B are proper subcontinua of X with $a \in X \setminus B$ and $b \in X \setminus A$. It is trivial to verify that X has at least three composants, namely K_a, K_b, and K_z, where z is any point in $A \cap B$. (Note that $K_z = X$, since both $A \subset K_z$ and $B \subset K_z$.)

Suppose then that $y \in A$ and that $y \neq a, b$, or z. We show that either $K_y = K_z$ or $K_y = K_a$. If $b \in K_y$, then there is a proper subcontinuum D of X such that both y and b lie in D. Hence, $A \cap D \neq \emptyset$, and consequently $A \cup D$ is a continuum containing a and b. Since X is irreducible from a to b, it follows that $A \cup D = X$. Furthermore, we have that $A \subset K_y$ and $B \subset K_y$, and therefore $K_y = X = K_z$.

Now suppose that $b \notin K_y$. To see that $K_y = K_a$, let $x \in K_a$ and let E be a proper subcontinuum of X containing x and a. Note that $b \notin E$. Since $a \in A \cap E$, we have that $A \cup E$ is a proper subcontinuum of X in which are found x and y. Therefore, $x \in K_y$ and hence $K_a \subset K_y$. Now suppose that $w \in K_y$. Then there is a proper subcontinuum F of X such that $w, y \in F$ and $b \notin F$. Since $y \in A \cap F$, it follows that $A \cup F$ is contained in K_a, and thus $w \in K_a$.

As a corollary to the next theorem, we have the unsettling result that indecomposable continua possess an uncountable number of distinct composants.

(9.D.9) ***Theorem.*** Suppose that X is a T_2 continuum with a countable basis. Then each composant K_x of X is the union of a countable number of proper subcontinua of X.

Proof. Suppose that $x \in X$, and that $U_1, U_2, \ldots$ is a countable basis for $X \setminus \{x\}$. For each positive integer i, let C_i be the component of $X \setminus U_i$ containing x (if such exists), and observe that $\overline{C_i} \subset K_x$, since $\overline{C_i}$ is a proper subcontinuum. We claim that $K_x = \bigcup\limits_{i=1}^{\infty} \bar{C}_i$. To see this, suppose that y is an arbitrary point in K_x and that D is a proper subcontinuum containing both x and y. Let U_j be a member of the basis with the property that $\overline{U}_j \subset X \setminus D$. Then since D is connected, D must lie in C_j, which completes the proof.

(9.D.10) **Corollary.** If X is an indecomposable T_2 continuum with a countable basis, then X has uncountably many distinct composants.

Proof. It will be shown in Chapter 10 (10.C.1) that a T_2 continuum with a countable basis is metrizable. Since such a continuum is compact by definition, it may be considered as a complete metric space, and hence the Baire category theorem applies. Suppose that X could be written as the union of a countable number of composants $K_{x_1}, K_{x_2}, \cdots$. Since each composant is the union of a countable number of proper subcontinua, so must be X. However, by (3.B.12), at least one of these proper subcontinua has nonempty interior, and hence, by (9.D.1), X is decomposable. Thus, X has uncountably many composants.

The next exercise says that the situation is even worse.

(9.D.11) **Exercise.** Show that if X is an indecomposable T_2 continuum, then distinct composants are disjoint.

A major clue to the construction of an indecomposable continuum is revealed in the following theorem.

(9.D.12) **Theorem.** Suppose that X is a T_2 continuum with a countable basis. Then X is indecomposable if and only if there are points a, b, and c in X such that X is irreducible between each pair of these points.

Proof. Suppose that X is indecomposable and let K_a, K_b, and K_c be distinct (and therefore disjoint) composants of X. If D is a proper subcontinuum of X containing a and b, then $D \subset K_a \cap K_b$, which contradicts the disjointness of K_a and K_b. Thus, X is irreducible from a to b. The same argument applies to b and c, and a and c.

Suppose now that points a, b, and c exist such that X is irreducible between each pair of them. Consider the corresponding composants K_a, K_b, and K_c. If X is decomposable, then it follows from the proof of (9.D.8) that K_a, K_b, and K_c are distinct. It also follows from this proof that each of these composants must in fact be X. Thus, X is indecomposable.

Hence, the problem of finding an indecomposable continuum has now been reduced to starting with three points and constructing a continuum that is irreducible between each pair of them. How might one create an irreducible continuum of this nature? The answer is easy to obtain, since it follows from (9.C.9) that if a continuum is chainable between points a and b, then it is also irreducible from a to b. The obvious way to build a chainable continuum is to employ simple chains, and this is what we do next.

The entire construction will be carried out in the plane. Select any three noncollinear points a, b, and c in $\mathscr{E}^2$, and construct a simple chain $\mathscr{C}_1$ consisting of open disks that starts at a, passes by b, and ends at c. Inside $\mathscr{C}_1$, construct a simple chain $\mathscr{C}_2$ of open disks that starts at b, passes through c, and ends at a. Then inside $\mathscr{C}_2$ construct a third chain $\mathscr{C}_3$ that begins at c, runs through a, and terminates at b (see the figure that follows). The whole process starts over again with a simple chain $\mathscr{C}_4$ that lies inside $\mathscr{C}_3$ and follows the pattern a-b-c. In general, for each positive integer n, construct a simple chain, $\mathscr{C}_{3n+1}$ that runs the a-b-c route, $\mathscr{C}_{3n+2}$ that passes from b to c to a, and $\mathscr{C}_{3n+3}$ that proceeds from c to a to b. Clearly, for each $n \in \mathbf{Z}^+$, we may assume that mesh $\mathscr{C}_n < 1/n$.

For every $n \in \mathbf{Z}^+$, let $D_n = \overline{\bigcup \{C \mid C \in \mathscr{C}_n\}}$. Then $X = \bigcap_{n=1}^{\infty} D_n$ is an indecomposable continuum.

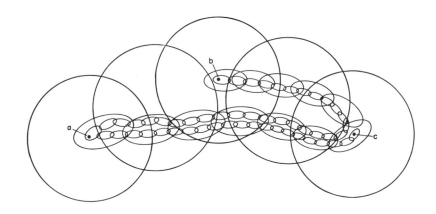

(9.D.13) Exercise. Show that the space just constructed is an indecomposable continuum.

Unknown to the reader, we have previously exhibited another indecomposable continuum, the dyadic solenoid. We shall prove in detail, using inverse systems, that the solenoid is indecomposable.

Consider the dyadic solenoid, S, as an inverse limit of the inverse system

$(C_i, f_i, \mathbf{N})$, where C_i is the unit circle in the complex plane ($\mathscr{E}^2$) and a typical bonding map $f_i : C_i \to C_{i-1}$ is defined by $f_i(z) = z^2$. Let $\hat{p}_i : S \to C_i$ be the customary projection map restricted to the inverse limit S. The following ingenious proof of the indecomposability of S is due to Nadler [1973].

(9.D.14) **Theorem.** The dyadic solenoid S is an indecomposable continuum.

Proof. Let A and B be subcontinua of S such that $A \cap B = S$. We show that either A or B fails to be a proper subset of S, or equivalently, that $A \subset B$ or $B \subset A$. The heart of the matter lies in the following assertion: For each positive integer i, either $\hat{p}_i(A) \subset \hat{p}_i(B)$ or $\hat{p}_i(B) \subset \hat{p}_i(A)$. Suppose that the assertion is false for some integer k, and let $a \in \hat{p}_k(A) \setminus \hat{p}_k(B)$ and $b \in \hat{p}_k(B) \setminus \hat{p}_k(A)$. Since each bonding map f_i is onto, it follows from (6.B.13) that $\hat{p}_i$ is also onto for all i. Thus, we have that $\sqrt{a} \in \hat{p}_{k+1}(A) \cup \hat{p}_{k+1}(B)$. If $\sqrt{a} \in \hat{p}_{k+1}(B)$, then $a = f_{k+1}(\sqrt{a}) \in f_{k+1}\hat{p}_{k+1}(B) = \hat{p}_k(B)$, which is impossible. Therefore, $\sqrt{a} \in \hat{p}_{k+1}(A) \setminus \hat{p}_{k+1}(B)$. Using similar arguments, we may conclude that $-\sqrt{a} \in \hat{p}_{k+1}(A) \setminus \hat{p}_{k+1}(B)$ and $\sqrt{b}, -\sqrt{b} \in \hat{p}_{k+1}(B) \setminus \hat{p}_{k+1}(A)$.

It is immediate from the connectedness of A and the fact that $\sqrt{a}$ and $-\sqrt{a}$ are antipodal that $\hat{p}_{k+1}(A)$ includes at least one of the semicircles of C_{k+1} with end points $\sqrt{a}$ and $-\sqrt{a}$. But either $\sqrt{b}$ or $-\sqrt{b}$ must also fall in this semicircle, contradicting the fact that both $\sqrt{b}$ and $-\sqrt{b}$ lie in the complement of $\hat{p}_{k+1}(A)$. Consequently, the assertion is established.

There are now two possibilities: $\hat{p}_i(A) \subset \hat{p}_i(B)$ for infinitely many i, or $\hat{p}_i(B) \subset \hat{p}_i(A)$ for infinitely many i. It follows from the first exercise below that in the former case, $\hat{p}_i(A) \subset \hat{p}_i(B)$ for all positive integers i, and, of course, in the latter case, $\hat{p}_i(B) \subset \hat{p}_i(A)$ for each i. The proof is completed by observing that A is the inverse limit of $(\hat{p}_i(A), f_{i+1|\hat{p}_{i+1}(A)}, \mathbf{N})$ and B is the inverse limit of $(\hat{p}_i(B), f_{i+1|\hat{p}_{i+1}(B)}, \mathbf{N})$, and therefore $A \subset B$ or $B \subset A$ (9.D.16). ∎

(9.D.15) **Exercise.** Show that if $\hat{p}_i(A) \subset \hat{p}_i(B)$ for infinitely many i, then $\hat{p}_i(A) \subset \hat{p}_i(B)$ for all i.

(9.D.16) **Exercise.** Finish the proof of the previous theorem by showing that $A \subset B$ or $B \subset A$.

E. THE PSEUDO-ARC

In this section, an important addition is made to our small band of indecomposable continua. The pseudo-arc (actually there are many, but they are all homeomorphic) has probably been the most scrutinized of all the indecomposable continua, and its properties are for the most part as spectacular as they are difficult to prove. The idea behind the construction of a pseudo-arc is somewhat similar to the trick employed in creating our first example of an indecomposable continuum: simple chains are used to define the space. In the previous example, simple chains were carefully woven between three points in the plane; in the case of the pseudo-arc, chains are stretched out between two points. The stretching, however, is of a very contorted nature (similar to this explanation) and it will soon become clear that "pseudo-arc" is an appropriate label for this space.

Any space that results from the following construction is called a *pseudo-arc*.

Let a and b be any two points in the plane, and let $\mathscr{C}_1$, $\mathscr{C}_2$, ... be a sequence of simple chains of connected open sets satisfying the following conditions:

(i) the point a belongs to the first link of each chain, and b to the last;

(ii) the closure of each link of $\mathscr{C}_{n+1}$ is contained in some link of $\mathscr{C}_n$;

(iii) mesh $\mathscr{C}_n < 1/n$ for each positive integer n;

(iv) if C_k^{n+1} and C_m^{n+1} are links of $\mathscr{C}_{n+1}$ ($k < m$), and $C_k^{n+1} \subset C_p^n$ and $C_m^{n+1} \subset C_p^n$, where $|p - q| > 2$, then there are links C_s^{n+1} and C_t^{n+1}, where $k < s < t < m$, such that C_s^{n+1} is contained in a link adjacent to C_q^n and C_t^{n+1} is contained in a link adjacent to C_n (note carefully that $s < t$).

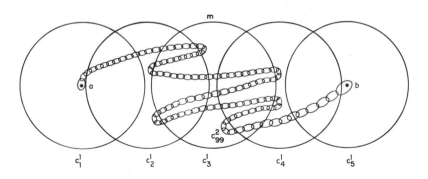

If $A_n = \bigcup \{\bar{C} \mid C$ is a link of $\mathscr{C}_n\}$, then $A = \bigcap\limits_{n=1}^{\infty} A_n$ is the *pseudo-arc* associated with the chains $\mathscr{C}_1, \mathscr{C}_2, \cdots$.

We first show that A is indecomposable. In fact, we prove something even more astonishing: Every subcontinuum of A is indecomposable.

(9.E.1) Theorem. The pseudo-arc A is hereditarily indecomposable.

Proof. Suppose that there is a subcontinuum B of A that is decomposable. Then B may be written as the union of two proper subcontinua H_1 and H_2. Use is made of the zigzagging property of the chains to show that H_1 is not connected.

It is obvious that there are points x and y in B and an integer n such that $d(x,H_1) > 2/n$ and $d(y,H_2) > 2/n$. This of course implies that $x \in H_2$ and $y \in H_1$. Consider the parts of the chains $\mathscr{C}_n$ and $\mathscr{C}_{n+1}$ that run between x and y. Denote these by $\{C_k^n, C_{k+1}^n, ..., C_v^n\}$ and $\{C_p^{n+1}, C_{p+1}^{n+1}, ..., C_w^{n+1}\}$. Assume that $x \in C_k^n$. Since the diameter of each link of $\mathscr{C}_n$ is less than $1/n$, it follows that $C_{k+1}^n \cap H_1 = \varnothing$, and hence $C_{k+1}^n \cap H_2 \neq \varnothing$. Similarly, $C_{v-1}^n \cap H_2 = \varnothing$ and $C_{v-1}^n \cap H_1 \neq \varnothing$. The twisting of the chains now leads to a contradiction.

As a result of $\mathscr{C}_{n+1}$'s kinky relationship with $\mathscr{C}_n$, there are links C_c^{n+1}, C_d^{n+1}, and C_e^{n+1} ($c < d < e$) such that $C_d^{n+1} \subset C_{k+1}^n$ and $C_c^{n+1} \subset C_e^{n+1} \subset C_{v-1}^n$. Thus, C_d^{n+1} misses H_1, but the other two links do not, which implies that H_1 is not connected and establishes our contradiction.

Moise [1948] has proven an even stronger result: Every subcontinuum of A is homeomorphic to A. Naturally, arcs also possess this property. However, a very striking and dramatic difference between arcs and pseudo-arcs arises from a result of Bing [1948]: The pseudo-arc is homogeneous. Moreover, Bing [1951b] has shown that any homogeneous hereditarily indecomposable continuum is a pseudo-arc, as is any chainable hereditarily indecomposable continuum.

This latter result is used to show that there are many pseudo-arcs. In fact, if X is either $\mathscr{E}^n$ ($n > 1$) or Hilbert space, then most of the continua lying in X are pseudo-arcs.

(9.E.2) **Theorem.** Suppose that $X = \mathscr{E}^n$ ($n > 1$) or $X = $ Hilbert space. (Assume X has a bounded metric.) Let $\mathscr{C}(X)$ be the family of all continua in X and $\mathscr{A}(X)$ the subfamily of pseudo-arcs. If $\mathscr{C}(X)$ is given the Hausdorff metric, then $\mathscr{A}(X)$ is a dense G_δ subset of $\mathscr{C}(X)$.

Proof. We show first that $\mathscr{A}(X)$ is dense in $\mathscr{C}(X)$. Suppose that $C \in \mathscr{C}(X)$ and that $\varepsilon > 0$. It suffices to find a pseudo-arc A such that $\rho(A,C) < \varepsilon$. There is a broken line $\overline{ab}$ in X such that $\rho(\overline{ab},C) < \varepsilon/2$ (why?). Let $\mathscr{D}$ be an $(\varepsilon/2)$-chain from a to b that covers $\overline{ab}$ and whose links are open balls in X. Inside $\bigcup \{D \mid D \in \mathscr{D}\}$, construct a pseudo-arc A containing the points a and b. Then $\rho(A,C) < \varepsilon$.

A broken line

Next we show that $\mathscr{A}(X)$ is a G_δ subset of $\mathscr{C}(X)$. For each $i \in \mathbf{Z}^+$, let $\mathscr{D}_i$ be the collection of all members $\mathscr{C}(X)$ that cannot be covered by a $(1/i)$-chain. Then $\mathscr{D}_i$ is a closed subset of $\mathscr{C}(X)$, since if any sequence $C_1, C_2, \ldots$ in $\mathscr{D}_i$ converges to a continuum C, C must lie in $\mathscr{D}_i$, for otherwise, a $(1/i)$-chain covering C would also contain some member of the sequence C_1, $C_2, \cdots$. It is clear that $\mathscr{C}(X) \setminus (\bigcup_{i=1}^{\infty} \mathscr{D}_i)$ consists of all the chainable continua in $\mathscr{C}(X)$.

Let $\mathscr{G}_i$ be the subfamily of $\mathscr{C}(X)$ consisting of sets K with the property that K contains a subcontinuum K' that is the union of two continua K_1 and K_2, where

    (i)   there is an $x \in K_1$ such that $d(x,K_2) \geq 1/i$, and
    (ii) there is a $y \in K_2$ such that $d(y,K_1) \geq 1/i$.

Note that $\mathscr{G}_i$ is closed in $\mathscr{C}(X)$, and that if $H \in \mathscr{C}(X) \setminus (\bigcup_{i=1}^{\infty} \mathscr{G}_i)$, then H is hereditarily indecomposable.

Finally, observe that if $\mathscr{P}$ represents the collection of points in X, then $\mathscr{P}$ is a closed subset of $\mathscr{C}(X)$.

Let $\mathscr{W} = \mathscr{C}(X) \setminus (\mathscr{P} \cup \bigcup_{i=1}^{\infty} \mathscr{D}_i \cup \bigcup_{i=1}^{\infty} \mathscr{G}_i)$. Then $\mathscr{W}$ is a dense G_δ subset of $\mathscr{C}(X)$.

We conclude this chapter with a few additional curiosa involving continua. Although it was a nontrivial matter to construct just one hereditarily indecomposable continuum, the one produced is hardly unique. Bing [1951b] has shown that there are as many topologically different hereditarily indecomposable continua as there are real numbers.

Two results that are reminiscent of the fact that compact metric spaces are continuous images of the Cantor set have been established by Mazurkiewicz [1920] and Bellamy [1971]. Mazurkiewicz proved that every chainable continuum is a continuous image of the pseudo-arc, and Bellamy recently demonstrated that every metric continuum is a continuous image of some indecomposable metric continuum, but that there is no continuum of which every indecomposable continuum is a continuous image

Schori [1965] used inverse systems to construct a universal chainable continuum, universal in the sense that if X is any chainable continuum, there is a subcontinuum of U homeomorphic to X. Anderson and Choquet [1959] have exploited inverse systems to concoct a plane continuum with the property that no two of its subcontinua are homeomorphic. Needless to say, the literature is replete with such aberrant examples and peculiar theorems. The curious reader might wish to consult the following articles for further enlightenment: Bing [1948, 1951a,b], Fugate [1965], Burgess [1959, 1961], and Jones [1951].

PROBLEMS

Section A

1. Describe the possibilities for $S(a,b)$ when X is a circle, a tree, or a circle with a tail.
2. Find an example of a space X containing points a and b for which the relative topology on $S(a,b)$ does not coincide with the order topology.
3.* Suppose that K is a metric continuum such that for each two distinct points $x,y \in K$, $K \setminus \{x,y\}$ is not connected.
    (a) Show that no point separates K.
    (b) Show that if (U,V) is a separation of $K \setminus \{x,y\}$, then $U \cup \{x,y\}$ and $V \cup \{x,y\}$ are connected and at least one of these sets is an arc.
    (c) Show that both $U \cup \{x,y\}$ and $V \cup \{x,y\}$ are arcs.
    (d) Show that K is homeomorphic to $\mathscr{S}^1$.
4. Prove or disprove the following assertion. If K is a T_2 continuum containing a point x such that $K \setminus \{x\}$ contains two components but there is no point y such that $K \setminus \{y\}$ has more than two components, then K is an arc.

Section B

1. A metric space X is *uniformly locally connected* if and only if for each $\varepsilon > 0$, there is a $\delta > 0$ such that if $d(x,y) < \delta$, then x and y lie in a connected set of diameter less than ε. Show that a compact locally connected metric space is uniformly locally connected.
2. Show that uniformly locally connected spaces are locally connected.
3. A metric continuum K has property S if and only if for each $\varepsilon > 0$, K is the union of a finite collection of connected sets, each of diameter less than ε. Show that a metric continuum K is a Peano continuum if and only if K has property S.
4. A space X is *connected im Kleinen* at a point $p \in K$ if and only if for each open set U containing p, there is an open set V such that $p \in V \subset U$ and any pair of points in V lies in a connected subset of U. Show that if X is connected im Kleinen at each point $p \in X$, then X is locally connected.
5. Show that the following space is connected im Kleinen at x, but that there is no basis of connected open sets at 0.

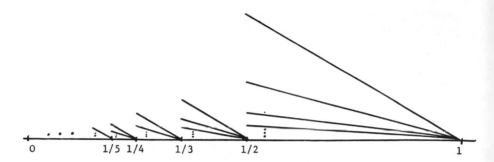

6. Suppose that K is a nondegenerate Peano continuum and L is an arbitrary Peano continuum. Show that there is a continuous onto map $f : K \to L$.

7. Show that connected open subsets of Peano continua are arc connected.

8.* Let S be the unit square. Show that if $f : S \to \mathbf{I}$, then there are disjoint arcs A and B in S each with at most one boundary point of S such that $f(A \cup B) = \mathbf{I}$.

9.* Let A be set of all vertical line segments lying between two disks. Show that there is an arc in $\mathcal{E}^3$ whose image intersects each line segment, but that there is no arc in $\mathcal{E}^3$ whose image intersects each line segment exactly once.

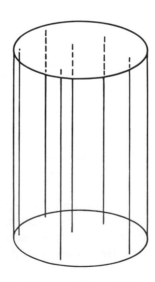

10. Let $A_1, A_2, \dots$ be a sequence of subsets of a space X. Then

(i) lim sup $A_n = \{x \in X \mid$ each open set containing x intersects infinitely many $A_n\}$, and

(ii) lim inf $A_n = \{x \in X \mid$ each open set containing x intersects all but finitely many $A_n\}$.

Find lim sup and lim inf of the following sequence of subsets of the plane.

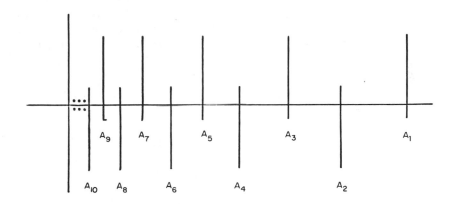

11. Suppose that $A_1, A_2, \ldots$ is a sequence of subsets of a space X. Show that lim inf A_n and lim sup A_n are closed subsets of X.

12. Suppose that $A_1, A_2, \ldots$ is a sequence of subsets of a space X. If lim inf $A_n =$ lim sup A_n, we write lim A_n for their common value. Show that if X is compact T_2 and A_n is connected for every n, then lim A_n exists and is connected.

13. Show that the topologist's sine curve is not a counterexample for Theorem (9.B.6).

Section C

1. Show that a chainable continuum cannot contain a 1-sphere.

2. Show that the arc is the only nondegenerate metric locally connected chainable continuum. [Hint: Use problem B.7 and the characterization of the arc given in Section A.]

3. Suppose that K is a chainable continuum and that $\mathscr{C}$ is a simple chain of open sets covering K. Show that there is a $\delta > 0$ such that any δ-chain $\mathscr{D}$ properly refines $\mathscr{C}$ (i.e., if $D \in \mathscr{D}$, there is $C \in \mathscr{C}$ such that $\bar{D} \subset C$).

4. Show that the topologist's sine curve has the fixed point property.

5. A continuum K is a *triod* if and only if there is a proper subcontinuum

$H \subset K$ such that $K \setminus H$ is the union of three nonempty pairwise separated sets. Show that a chainable continuum contains no triods.

6. Define a space to be *open-unicoherent* if the word "closed" in the definition of unicoherence is replaced by "open." Show that if a locally connected, connected space is unicoherent, then it is open-unicoherent.

7.* Suppose that X is locally connected and connected with the property that if a closed subset A separates points p and q in S, then so does some component of A. Use the following steps to show that X is unicoherent.

(a) Suppose that A and B are closed and connected, $A \cup B = X$, and $A \cap B = H \cup K$, where H and K are disjoint, closed, and nonempty. Show that there is a component D of $X \setminus A$ such that Fr D is not connected.

(b) Show that $X \setminus D$ is connected.

(c) Show that there is a component C of $X \setminus \bar{D}$ such that Fr C is not connected.

(d) Show that Fr $C \subset$ Fr D.

(e) Let $p \in C$, $q \in D$, and let F be a component of Fr C that separates p and q. Let $x \in (\text{Fr } C) \setminus F$ and show that $C \cup \{x\} \cup D$ is connected, contains p and q, and misses F to conclude the proof.

8. Suppose that X is a unicoherent T_2 continuum and Y is a T_2 space. Show that if $f : X \to Y$ is a continuous onto map such that $f^{-1}(C)$ is connected for each connected subspace C of Y, then Y is unicoherent.

Sections D and E

1. Suppose that K is a metric continuum. Show that there is an uncountable set $A \subset X$ such that K is irreducible between any two points in A.

2. Suppose that K is T_2 continuum and M is a subcontinuum. Show that if $L \subset M$ is a subcontinuum of M such that $M \setminus L$ is not connected, then M is decomposable.

3. Suppose that K is a T_2 continuum such that every subcontinuum M of K has the property that no subcontinuum of M separates M. Show that K is hereditarily indecomposable.

4.* Suppose that X is a hereditarily decomposable, hereditarily unicoherent continuum that contains no triods. Let $\mathscr{F}$ be any collection of subcontinua of X. Show that $\bigcap \{F \mid F \in \mathscr{F}\}$ is a subcontinuum.

5. Suppose that K is an indecomposable continuum and let $p \in K$. Show that $\{x \in K \mid K$ is irreducible between p and $x\}$ is dense in K.

6. Show that the union of a countable number of proper subcontinua of an indecomposable continuum K can not separate K.

7.* Show that each composant of a metric continuum is an F_σ set.

8.* Let K be the Cantor set. Let S be the union of all semicircles in the upper half-plane with both end points in K and such that the end points are symmetric with respect to the line $x = 1/2$. For $i = 1, 2, \ldots$, let S_i be the union of all semicircles in the lower half-plane whose ends are on K and are symmetric with respect to the line $x = 5/(3^i \cdot 2)$. Let $M = \bigcup_{i=0}^{\infty} S_i$. Show that M is chainable and M is indecomposable.

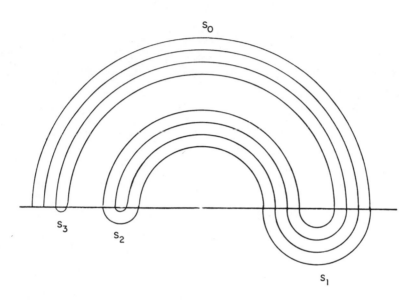

9. Show that the plane contains uncountably many pairwise disjoint non-degenerate continua, none of which contains an arc [Hint: Use the pseudo-arc.]
10. Construct the first two stages of a pseudo-arc starting with six links in the initial chain.

Chapter 10

PARACOMPACTNESS AND METRIZABILITY

Most results involving paracompactness and metrizability are derived from judicious refinements of open covers. Paracompactness is defined in terms of such coverings, and metrization theory is almost wholly dependent on the successful exploitation of appropriate covers.

The notion of paracompactness is a relatively recent one as topological ideas go, and was first introduced by Dieudonné [1944]. In the hierarchy of topological spaces, paracompactness lies between normality and metrizability. It retains many of the virtues of compactness, but is somewhat more general. Besides its usefulness in the theory of metrizability, paracompactness served as a catalyst in the liberation of dimension theory from the confines of separable metric spaces.

We give an admittedly somewhat cavalier treatment of metrization: only a few of the more celebrated results are stated, and even fewer proofs are undertaken. Presumably, however, enough material is presented to either whet or extinguish the reader's appetite. The reader is invited to consult Nagami [1970] for a rather terse but nevertheless more complete and elegant presentation.

In Sections D and E, Moore spaces and the completion of metric spaces are dealt with briefly.

A. PARACOMPACTNESS

(10.A.1) **Definition.** A family $\mathscr{C}$ of subsets of a space X is *locally finite* if

and only if each $x \in X$ lies in a neighborhood that intersects at most a finite number of members of $\mathscr{C}$.

As an easy but important application of this definition, we have that the union of a locally finite family of closed sets is closed.

(10.A.2) Exercise.

(i) Suppose that $\{A_\alpha \mid \alpha \in \Lambda\}$ is a locally finite collection of closed subsets of a space X. Show that $\bigcup_{\alpha \in \Lambda} A_\alpha$ is closed.

(ii) Suppose that $\{A_\alpha \mid \alpha \in \Lambda\}$ is a locally finite collection of subsets of a space X. Show that $\{\bar{A}_\alpha \mid \alpha \in \Lambda\}$ is locally finite.

The simplest type of cover refinement is defined next.

(10.A.3) Definition. If $\mathscr{U}$ is a cover of a space X, then a cover $\mathscr{V}$ of X *refines* $\mathscr{U}$ if and only if for each $V \in \mathscr{V}$ there is a $U \in \mathscr{U}$ such that $V \subset U$. The cover $\mathscr{V}$ is an *open* (*closed*) *refinement* of $\mathscr{U}$ if and only if $\mathscr{V}$ is an open (closed) cover.

(10.A.4) Example. Suppose that $\mathscr{U}$ is an open cover of a compact metric space X, and that δ is a Lebesgue number for $\mathscr{U}$. Then any cover $\mathscr{V}$ of X with mesh less than δ refines $\mathscr{U}$.

(10.A.5) Definition. A T_2 space X is *paracompact* if and only if each open cover of X has a locally finite open refinement.

If an open locally finite refinement of an open cover exists, then it can be chosen in a particularly elegant fashion.

(10.A.6) Theorem. Suppose that $\mathscr{U} = \{U_\alpha \mid \alpha \in \Lambda\}$ is an open cover of a topological space X and that $\mathscr{W} = \{W_\beta \mid \beta \in \Gamma\}$ is an open locally finite refinement in $\mathscr{U}$. Then there is a locally finite open refinement $\mathscr{V} = \{V_\alpha \mid \alpha \in \Lambda\}$ of $\mathscr{U}$ such that $V_\alpha \subset U_\alpha$ for each α. (Such a refinement is called a *precise locally finite refinement of $\mathscr{U}$*.)

Proof. For each $\beta \in \Gamma$, let $f(\beta)$ be a member of Λ with the property that $W_\beta \subset U_{f(\beta)}$. Then for each $\alpha \in \Lambda$, let $V_\alpha = \bigcup \{W_\beta \mid f(\beta) = \alpha\}$. There is of course no guarantee that V_α is nonempty, but this is immaterial, since in any case $\mathscr{V} = \{V_\alpha \mid \alpha \in \Lambda\}$ forms an open cover of X that refines $\mathscr{U}$ and in fact, we have that $V_\alpha \subset U_\alpha$ for each $\alpha \in \Lambda$. Since $\mathscr{W}$ is locally finite, so is $\mathscr{V}$.

Certainly, compact T_2 spaces are paracompact. The compactness condition may be relaxed somewhat if the space is T_3.

(10.A.7)   Theorem.   If X is a Lindelöf, T_3 space, then X is paracompact.

Proof.   Suppose that $\mathscr{U} = \{U_\alpha \mid \alpha \in \Lambda\}$ is an open cover of X. For each $x \in X$, select an $\alpha_x \in \Lambda$ such that $x \in U_{\alpha_x}$. Since X is regular there are open sets V_x and W_x such that $x \in V_x \subset \overline{V}_x \subset W_x \subset \overline{W}_x \subset U_{\alpha_x}$. Then $\mathscr{V} = \{V_x \mid x \in X\}$ is an open cover of X, and since X is Lindelöf, we may extract a countable subcover $\{V_{x_1}, V_{x_2}, \ldots\}$ from $\mathscr{V}$. Let $T_1 = W_{x_1}$, and for each integer $n > 1$, set $T_n = W_{x_n} \cap (X \setminus \overline{V}_{x_1}) \cap \cdots \cap (X \setminus \overline{V}_{x_{n-1}})$. We show that $\mathscr{T} = \{T_n \mid n \in \mathbf{Z}^+\}$ is a locally finite open refinement of $\mathscr{U}$ that covers X.

That $\mathscr{T}$ is a cover of X may be seen as follows. Suppose that $x \in X$. If $x \notin T_1 = W_{x_1}$, let $n \in \mathbf{Z}^+$ be the first integer such that $x \in W_{x_n}$. Then since $x \notin W_{x_1} \cup \cdots \cup W_{x_{n-1}}$ and since $\overline{V}_{x_1} \cup \cdots \cup \overline{V}_{x_{n-1}} \subset W_{x_1} \cup \cdots \cup W_{x_n}$, we have that $x \in T_n$. Hence, $\mathscr{T}$ is an open cover of X.

Clearly $\mathscr{T}$ refines $\mathscr{U}$; thus it remains to establish that $\mathscr{T}$ is locally finite. Suppose that $x \in X$. Then $x \in V_{x_n}$ for some n, and since $T_{n+j} \cap V_{x_n} = \varnothing$ for each $j \in \mathbf{Z}^+$, it follows that $\mathscr{T}$ is locally finite.

A far more significant result is that metric spaces are paracompact. This result was established by Stone [1948]. Stone's demonstration of this theorem is exceptionally intricate and has since been greatly simplified by Rudin [1969], whose proof we give here.

(10.A.8)   Theorem.   If (X,d) is a metric space, then X is paracompact.

Proof.   Suppose that $\mathscr{U} = \{U_\alpha \mid \alpha \in \Lambda\}$ is an open cover of (X,d) and that Λ is well ordered (0.D.6). For each $n \in \mathbf{Z}^+$, define V_{α_n} (by induction on n) to be the union of all sets of the form $S^d_{1/2^n}(x)$ such that

  (i) α is the first member of Λ with $x \in U_\alpha$,
  (ii) $x \notin V_{\beta j}$ if $j < n$, and
  (iii) $S^d_{3/2^n}(x) \subset U_\alpha$.

We show that the family $\mathscr{V} = \{V_{\alpha n} \mid n \in \mathbf{Z}^+\}$ is a locally finite open refinement of $\mathscr{U}$ that covers X.

Clearly, $\mathscr{V}$ refines $\mathscr{U}$. To see that $\mathscr{V}$ covers X, observe that if $x \in X$, there is a first member α of Λ such that $x \in U_\alpha$ and an n so large that (iii) holds. Then by (ii), $x \in V_{\beta j}$ for some $j \leq n$.

To complete the proof, we show that $\mathscr{V}$ is locally finite. Suppose that $x \in V_{\alpha n}$ for some n and choose j large enough so that $S_{1/2^j}(x) \subset V_{\alpha n}$. Clearly, $\mathscr{V}$ will be locally finite if we can establish

(a) if $i \geq n + j$, then $S^d_{1/2^{n+j}}(x)$ intersects no $V_{\beta i}$, and

(b) if $i < n + j$, then $S^d_{1/2^{n+j}}(x)$ intersects $V_{\beta i}$ for at most one β.

To see that (a) holds, note that since $i > n$, by (ii) each ball of radius $1/2^i$ used in the definition of $V_{\beta i}$ has its center y outside of $V_{\alpha n}$. Since $S^d_{1/2^j}(x) \subset V_{\alpha n_j}$ we have that $d(x,y) \geq 1/2^j$. However, since $i \geq j + 1$ and $n + j \geq j + 1$, it follows that $S^d_{1/2^{n+j}}(x) \cap S^d_{1/2^i}(y) = \varnothing$.

We establish part (b) as follows. Suppose that $p \in V_{\beta i}$ and $q \in V_{\gamma i}$. It suffices to show that $d(p,q) > 1/2^{n+j-1}$. There are points y and z such that $p \in S^d_{1/2^i}(y) \subset V_{\beta i}$ and $q \in S^d_{1/2^i}(z) \subset V_{\gamma i}$. By (iii) we have that $S^d_{1/2^i}(y) \subset U_\beta$, and by (ii) we have that $z \notin U_\beta$. Hence, it follows that $d(y,z) \geq 3/2^i$ and consequently $d(p,q) > 1/2^i \geq 1/2^{n+j-1}$, which concludes the proof.

Although metrizability ensures paracompactness, the next lower species in the pecking order of topological spaces, T_5, is not quite strong enough (see Problem A.10). The converse of Stone's Theorem is false (any nonmetrizable compact T_2 space will do—for instance, an uncountable product of unit intervals). However, paracompactness is sufficiently powerful to guarantee normality.

(10.A.9) Theorem. If X is a paracompact space, then X is normal.

Proof. We show that X is regular, and the reader may extend the techniques slightly to obtain the normality of X. Suppose that $x \in X$ and that F is a closed subset of X not containing x. For each $p \in F$, let U_p be an open set containing p whose closure misses x (X is T_2). Then $\{U_p \mid p \in F\}$ together with $X \setminus F$ forms an open cover $\mathcal{U}$ of X. Since X is paracompact, there is a locally finite open refinement $\mathcal{V} = \{V_\alpha \mid \alpha \in \Lambda\}$ of $\mathcal{U}$. Note that $x \notin \overline{V}_\alpha$ for each $\alpha \in \Lambda$ for which $V_\alpha \cap F \neq \varnothing$. Let $W = \bigcup \{V_\alpha \in \mathcal{V} \mid V_\alpha \cup F \neq \varnothing\}$. Then $F \subset W$, W is open, and $x \notin W$. Furthermore, since $\mathcal{V}$ is locally finite, it follows from (10.A.2) that $x \notin \overline{W}$.

(10.A.10) Definition. If $\mathcal{U}$ is a cover of X and $A \subset X$, then the *star of A with respect to $\mathcal{U}$*, denoted by $\mathrm{St}(A,\mathcal{U})$, is defined to be the union of all sets in $\mathcal{U}$ that intersect A.

(10.A.11) Definition. Suppose that $\mathcal{U}$ is a cover of a space X. A cover $\mathcal{V}$ of X is called a *star refinement* of $\mathcal{U}$ if and only if for each $V \in \mathcal{V}$ there is a $U \in \mathcal{U}$ such that $\mathrm{St}(V,\mathcal{V}) \subset U$, i.e., $\{\mathrm{St}(V,\mathcal{V}) \mid V \in \mathcal{V}\}$ refines $\mathcal{U}$.

A slightly different but equally useful concept is that of barycentric refinement.

(10.A.12) Definition. Suppose that $\mathcal{U}$ is a cover of a space X. Then a

cover $\mathscr{V}$ of X is a *barycentric refinement* of $\mathscr{U}$ if and only if $\{\mathrm{St}(x,\mathscr{V}) \mid x \in X\}$ refines $\mathscr{U}$.

(10.A.13)   Exercise.   Show that a barycentric refinement of a barycentric refinement is a star refinement.

Our principal result is the following.

(10.A.14)   Theorem.   If X is paracompact, then each open cover $\mathscr{U}$ of X has an open star refinement.

Proof.   By the previous exercise, it suffices to show that $\mathscr{U}$ possesses a barycentric refinement. Since X is paracompact, there is an open locally finite refinement $\mathscr{V}^* = \{V_\alpha^* \mid \alpha \in \Lambda\}$ of $\mathscr{U}$. By (10.A.9), X is normal, and hence by (4.C.2), there is an open cover $\mathscr{V} = \{V_\alpha \mid \alpha \in \Lambda\}$ such that $\overline{V}_\alpha \subset V_\alpha^*$ for each $\alpha \in \Lambda$. Clearly, $\mathscr{V}$ is also locally finite. For each $x \in X$, define $W_x = \bigcap \{V_\alpha^* \mid x \in \overline{V}_\alpha\}$. It is not difficult to see that this is a "finite" intersection, and hence W_x is open.

For each $x \in X$, let $F_x = \bigcup \{\overline{V}_\alpha \mid x \notin \overline{V}_\alpha\}$, and note once again that by (10.A.2) F_x is closed. Finally, set Z_x equal to $W_x \setminus F_x$, and then show that $\mathscr{Z} = \{Z_x \mid x \in X\}$ is indeed a barycentric refinement of $\mathscr{U}$.

We note in passing that the converse of this theorem also holds, but do not belabor (or prove) the point.

The following inequality might well be more than a frivolous conjecture: card (known mathematicians) $\leq$ card (known varieties of compactness).

(10.A.15)   Definition.   A locally compact T_2 space X is *σ-compact* if and only if X can be expressed as a countable union of compact subsets.

Certainly, $\mathscr{E}^n$ is σ-compact for all n, and equally obvious is the fact that an uncountable discrete metric space is not σ-compact. Two basic properties of σ-compact spaces are given by the next theorems.

(10.A.16)   Theorem.   If a space X is σ-compact, then there is a sequence $U_1, U_2, \ldots$ of open subsets of X each with compact closure such that $X = \bigcup\limits_{n=1}^{\infty} U_n$ and, for each n, $\overline{U}_n \subset U_{n+1}$.

Proof.   Since X is σ-compact, we have that $X = \bigcup\limits_{n=1}^{\infty} C_n$, where each C_n is compact. For each $x \in C_1$, let V_x be an open neighborhood of x with compact closure (X is locally compact). Then $\mathscr{V} = \{V_x \mid x \in C_1\}$ covers C_1 and a finite subcover may be extracted from $\mathscr{V}$. Let U_1 be the union of the members of this finite subcover. Note that $\overline{U}_1$ is compact. The same procedure is repeated to obtain an open set U_2 with compact closure that contains the compact set $\overline{U}_1 \cup C_2$. The proof may now be concluded with an easy inductive argument.

(10.A.17) Theorem. If X is a σ-compact space, then X is Lindelöf.

Proof. Suppose that X is σ-compact and that $\mathscr{U}$ is an open cover of X. Since X is σ-compact, we have that $X = \bigcup\limits_{n=1}^{\infty} C_n$, where each C_n is compact. For each n, $\mathscr{U}$ is a cover of C_n and hence there is finite number of sets $U_{n_1}, U_{n_2}, \ldots ,$ $U_{n_{k(n)}}$ in $\mathscr{U}$ that cover C_n. Then $\{U_{n_i} \mid n \in \mathbf{Z}^+, 1 \le i \le k(n)\}$ is a countable subcover of $\mathscr{U}$.

The notions of paracompactness and σ-compactness are closely related. Since T_3, Lindelöf spaces are paracompact, it follows from (10.A.17) and the fact that locally compact T_2 spaces are T_3, that σ-compact spaces are paracompact. There is even a partial converse that will prove useful in Chapter 17.

(10.A.18) Theorem. If X is a locally compact paracompact space, then X may be expressed as a free union of σ-compact spaces.

Proof. For each $x \in X$, let U_x be an open neighborhood of x with compact closure. Let $\mathscr{U} = \{U_x \mid x \in X\}$. By (10.A.6), there is a precise open locally finite refinement $\mathscr{V} = \{V_x \mid x \in X\}$ of $\mathscr{U}$. Partition X into disjoint subsets by means of the following equivalence relation, $\sim$. Two points $a,b \in X$ are equivalent if and only if a and b can be connected by a simple chain with links in $\mathscr{V}$.

Let $\mathscr{C} = \{C_\alpha \mid \alpha \in \Lambda\}$ denote the set of equivalence classes derived from $\sim$. Note that each C_α is open (it is a union of V's from $\mathscr{V}$) and hence is locally compact. To see that C_α is σ-compact, let W_1 be any member of $\mathscr{V}$ lying in C_α and set $W_2 = \bigcup \{V \in \mathscr{V} \mid V \cap W_1 \ne \varnothing\}$. Then we have that $W_2 \subset C_\alpha$; furthermore, W_2 is a finite union ($\overline{W}_1$ is compact and $\mathscr{V}$ is locally finite), and consequently, $\overline{W}_2$ is compact. Inductively define $W_n = \bigcup \{V \in \mathscr{V} \mid V \cap W_{n-1} \ne \varnothing\}$. Clearly, $\overline{W}_n$ is compact and $C_\alpha = \bigcup\limits_{n=1}^{\infty} \overline{W}_n$.

B. PARTITIONS OF UNITY

As a diversion before taking up the subject of metrizability, we consider briefly the idea of a partition of unity.

(10.B.1) Definition. A *partition of unity* of a space X is a family $\{f_\alpha \mid \alpha \in \Lambda\}$ of continuous functions each of which maps X into $[0,1]$ such that

(i) for each $x \in X$, $\{\alpha \mid f_\alpha(x) \neq 0\}$ is finite, and
(ii) $\sum_{\alpha \in \Lambda} f_\alpha(x) = 1$ (whence the terminology).

A partition of unity $\{f_\alpha \mid \alpha \in \Lambda\}$ is *subordinate to an open cover* $\mathscr{U} = \{U_\beta \mid \beta \in \Gamma\}$ of X if and only if for each $\alpha \in \Lambda$, there is a $\beta \in \Gamma$ such that $f_{\alpha \mid X \setminus U_\beta} = 0$.

(10.B.2) Theorem. If X is paracompact, then each open cover of X has a partition of unity subordinate to it.

Proof. Let $\mathscr{U} = \{U_\beta \mid \beta \in \Gamma\}$ be an open cover of X. By (10.A.9), X is normal. Therefore, we may apply (4.C.2) twice to obtain open covers $\mathscr{V} = \{V_\beta \mid \beta \in \Gamma\}$ and $\mathscr{W} = \{W_\beta \mid \beta \in \Gamma\}$ such that for each $\beta \in \Gamma$, $\overline{W_\beta} \subset V_\beta \subset \overline{V_\beta} \subset U_\beta$. Tietze's extension theorem is used to generate functions $g_\beta : X \to [0,1]$, where $g_{\beta \mid W_\beta} = 1$ and $g_{\beta \mid X \setminus V_\beta} = 0$. Since X is paracompact, we may assume (with a little care) that $\mathscr{V}$ and $\mathscr{W}$ are locally finite. The g_β's are normalized by defining, for each β, a function $f_\beta : X \to [0,1]$, where $f_\beta(x) = g_\beta(x)/(\sum_{\beta \in \Gamma} g_\beta(x))$. (Why does $\sum_{\beta \in \Gamma} g_\beta$ make sense?) Then the family $\{f_\beta \mid \beta \in \Gamma\}$ is the desired partition of unity.

(10.B.3) Definition. A function $f : X \to \mathscr{E}^1$ is *lower semicontinuous* if and only if for each $\alpha \in \mathscr{E}^1$, $\{x \mid f(x) > a\}$ is open.

The following result is a particularly useful consequence of (10.B.2).

(10.B.4) Theorem. Suppose that X is paracompact and that $f : X \to \mathscr{E}^1$ and $g : X \to \mathscr{E}^1$ are upper and lower semicontinuous real valued functions, respectively, such that $f(x) < g(x)$ for each $x \in X$. Then there is a continuous function $h : X \to \mathscr{E}^1$ such that $f(x) < h(x) < g(x)$ for each $x \in X$.

Proof. We first construct an open cover of X and then find a partition of unity subordinate to it. For every rational r, let $U_r = \{x \mid f(x) < r\} \cap \{x \mid g(x) > r\}$. Since f is upper semicontinuous and g is lower semicontinuous, it follows that U_r is open. Furthermore, it follows from the hypothesis that $\mathscr{U} = \{U_r \mid r \text{ is rational}\}$ covers X. Hence, by the previous theorem, there is a partition of unity $\{f_r \mid r \text{ is rational}\}$ subordinate to $\mathscr{U}$. Then the function h defined by $h(x) = \sum_r r f_r(x)$ is the desired mapping.

C. A SAMPLING OF METRIZABILITY THEORY

Recall that a topological space $(X, \mathcal{U})$ is metrizable if there is a metric for X such that the induced metric topology coincides with $\mathcal{U}$. Although metric spaces lead naturally to topological spaces, not all topological spaces are metrizable: for instance, nonnormal spaces are not metrizable (metric spaces are normal, and normality is a topological invariant). This leads one to the problem of finding conditions sufficient to ensure the metrizability of a space. The purpose of this section is to provide some fairly representative solutions to these problems.

Perhaps the major classical metrization result and still one of the most useful is the following theorem due to Uryson.

(10.C.1)   Theorem (Uryson's Metrization Theorem). If a space X is second countable and T_3, then X is separable and metrizable.

Proof.   The trick used is simple (and is precisely that employed in (7.E.8)). First we embed X in the Hilbert cube, and then we conclude the proof by using the fact that subspaces of a separable metrizable space retain both the metrizability and the separability.

Let **B** be a countable basis for X and define $\mathscr{A}$ to be $\{(U,V) \mid U,V \in \mathbf{B}$ and $\overline{U} \subset V\}$. Since X is normal ((2.I.11) and (4.D.6)), Tietze's extension theorem can be exploited to obtain for each $(U,V) \in \mathscr{A}$ a continuous function $f_{UV} :$ $X \to [0,1]$ that maps $\overline{U}$ to 0 and $(X \setminus V)$ to 1. The remainder of the proof mimics that of (7.E.8) in every detail, and is omitted here.

(10.C.2)   Corollary. Suppose that X is a locally compact separable metric space. Then the one point compactification X^* of X is metrizable.

Proof.   Since X is a separable metric, it is Lindelöf (2.I.13), and therefore, since X is also locally compact, X may be written as a countable union of open subsets with compact closure; thus, X is σ-compact. By (10.A.16), $X = \bigcup_{n=1}^{\infty} U_n$, where $\overline{U}_n \subset U_{n+1}$ and each U_n is open and has compact closure. Observe that $\{X \setminus \overline{U}_n \mid n \in \mathbf{Z}^+\}$ forms a countable basis at ∞ ($X^* = X \cup \{\infty\}$), since any compact subset of X is contained in a finite number of the U_n. Consequently, X^* is second countable and T_3; therefore, by the Uryson metrization theorem, X^* is metrizable.

(10.C.3)   Exercise. Suppose that X is a compact metric space such that whenever $h : X \to Y$ is an embedding of X into a metric space Y, then there

is a neighborhood U of $h(X)$ in Y and a retraction $r : U \to h(X)$. Show that X is an ANR_M.

Uryson's metrization theorem is adequate for working with separable metric spaces, but for many purposes more general metrization results are needed.

(10.C.4) **Definition.** Suppose that X is a topological space. A sequence of open covers $\mathscr{U}_1, \mathscr{U}_2, \ldots$ of X is a *development* for X if and only if for each $x \in X$, $\{\operatorname{St}(x, \mathscr{U}_n) \mid n = 1, 2, \ldots\}$ is a neighborhood basis at x. A T_1 space X is *developable* if and only if X has a development.

Historically, the first documented metrization theorem appears to be the following.

(10.C.5) **Theorem** (*Aleksandrov and Uryson* [*1923*]). A T_0 space X is metrizable if and only if X has a development $\mathscr{U}_1, \mathscr{U}_2, \ldots$ such that if $U, V \in \mathscr{U}_n$ and $U \cap V \neq \varnothing$, then $U \cup V \subset W$ for some $W \in \mathscr{U}_{n-1}$.

Rather than prove this theorem we show how others have taken advantage of it to establish their own metrization results. For instance, Mrs. Frink [1937] utilized the Aleksandrov-Uryson theorem to derive the following surprisingly useful theorem.

(10.C.6) **Theorem.** A T_1 space X is metrizable if and only if for each $x \in X$, there is a decreasing sequence of open sets $U_1^x \supset U_2^x \supset \cdots$ such that

(i) $\{U_i^x \mid i \in \mathbf{Z}^+\}$ forms a neighborhood basis at x, and

(ii) for each $x \in X$ and each integer i, there is an integer $n_{(x,i)} > i$ such that for each $y \in X$, if $U_{n(x,i)}^x \cap U_{n(x,i)}^y \neq \varnothing$, then $U_{n(x,i)}^y \subset U_i^x$.

Proof. For each $x \in X$ select a sequence of positive integers as follows. Let $a_1^x = 1$, $a_2^x = n(x, a_1^x)$, $a_3^x = n(x, a_2^x)$, etc. Note that $a_1^x < a_2^x < \cdots$. For each i, let $V_i^x = U_{a_i^x}^x$, and set $\mathscr{V}_i = \{V_i^x \mid x \in X\}$. Clearly, $\{V_i^x \mid i = 1, 2, \ldots\}$ forms a neighborhood basis at x.

We show that $\mathscr{V}_1, \mathscr{V}_2, \ldots$ is a development for X and satisfies the hypothesis of the Aleksandrov-Uryson Theorem. Suppose that $x \in X$ and $i \in \mathbf{Z}^+$. To establish that the $\mathscr{V}_i$'s form a development, an integer $k \in \mathbf{Z}^+$ must be found such that $\operatorname{St}(x, \mathscr{V}_k) \subset V_i(x)$. Let $k = a_{i+1}^x$ and suppose that $x \in V_k^y$. Since $a_k^x \geq k = a_{i+1}^x$, we have that $V_k^y = U_{a_k^y}^y \subset U_{a_{i+1}^x}^y$. Therefore, $x \in U_{a_{i+1}^x}^x \cap U_{a_{i+1}^x}^y$ and consequently by hypothesis, it follows that $V_k^y = U_{a_{i+1}^x}^y \subset U_{a_i^x}^x = V_i^x$ ($a_{i+1}^x = n(x, a_i^x)$). Thus, $\mathscr{V}_1, \mathscr{V}_2, \ldots$ is a development for X.

Now suppose that $V_{i+1}^x \cap V_{i+1}^y \neq \varnothing$, or equivalently, that $U_{n(x,a_i^x)}^x \cap$

$U^y_{n(y,a^x_i)} \neq \varnothing$. Then if $n(x,a^x_i) \leq n(y,a^y_i)$, we have that $U^x_{n(x,a^x_i)} \cap U^y_{n(x,a^x_i)} \neq \varnothing$, which by hypothesis implies that $V^y_{i+1} = U^y_{a^y_{i+1}} = U^y_{n(x,a^x_i)} \subset U^x_{a^x_i} = V^x_i$. An analogous argument results in an analogous conclusion if $n(y,a^y_i) \leq n(x,a^x_i)$. Hence, by the Aleksandrov-Uryson theorem, X is metrizable. The converse is straightforward and left as a problem (see Problem C.4).

The following question is of considerable interest. Suppose that X is metrizable and $f : X \to Y$ is continuous and onto. Is Y metrizable? Stone used Frink's result to obtain a particularly elegant solution to this problem.

(10.C.7)   Theorem.   Suppose that (X,d) is a metric space and $f : X \to Y$ is continuous, closed, and onto. If Fr $f^{-1}(y)$ is compact for each $y \in Y$, then Y is metrizable.

Proof (Stone [1956]).   For each $y \in Y$ and $n \in \mathbf{Z}^+$, let $N^y_n = S^d_{1/n}(\text{Fr} f^{-1}(y))$, $U^y_n = N^y_n \cup \text{int} f^{-1}(y)$, $V^y_n = \bigcup \{f^{-1}(z) \mid f^{-1}(z) \subset U^y_n\}$, and $W^y_n = f(V^y_n)$. Note that $W^y_n = Y \setminus f(X \setminus U^y_n)$, and since f is closed, both W^y_n and V^y_n are open.

We first show that for each $y \in Y$, $\{W^y_n \mid n \in \mathbf{Z}^+\}$ satisfies the conditions of (10.C.6). Obviously, the W^y_n's form a decreasing sequence of open sets, and verification that they comprise a neighborhood basis at the point y can be safely entrusted to the reader. Establishment of condition (ii) of Frink's theorem is considerably trickier. Suppose that $y \in Y$ and $i \in \mathbf{Z}^+$. If int $f^{-1}(y) \neq \varnothing$, select an arbitrary point $x_y \in \text{int} f^{-1}(y)$ (if int $f^{-1}(y) = \varnothing$, don't worry). Now choose n to be sufficiently large so that

(i) $n > 2i$,
(ii)   if Fr $f^{-1}(y) \neq \varnothing$, then $d(\text{Fr} f^{-1}(y), X \setminus V^y_{2i}) > 2/n$, and
(iii)   if int $f^{-1}(y) \neq \varnothing$, then $S_{1/n}(x_y) \subset f^{-1}(y)$.

We assert that $n = n(y,i)$. Suppose that $W^y_n \cap W^z_n \neq \varnothing$. Two steps are needed to show that $W^z_n \subset W^y_i$.

Step 1.   We show that $f^{-1}(z) \subset V^y_{2i}$. Let $p \in W^z_n \cap W^y_n$, and select $w \in f^{-1}(p) \subset V^z_n \cap V^y_n \subset U^z_n \cap U^y_n$. If $w \in \text{int} f^{-1}(z)$, then $f^{-1}(z) \cap V^y_n \neq \varnothing$, and hence, $f^{-1}(z) \subset V^y_n \subset V^y_{2i}$.

If $w \notin \text{int} f^{-1}(z)$, then $w \in N^z_n$. Note that in this case, $w \in X \setminus \text{int} f^{-1}(y)$, for if not, we have that (1) $f^{-1}(y) \subset V^z_n$, (2) $x_y \in V^z_n \subset (\text{int} f^{-1}(z) \cup N^z_n)$, and since $x_y \notin f^{-1}(z)$, it follows that (3) $x_y \in N^z_n$. This, however, contradicts condition (iii) above. Consequently, $w \in N^z_n \cap N^y_n$, and there are points $u \in$ Fr $f^{-1}(y)$ and $v \in$ Fr $f^{-1}(z)$ such that $d(w,u) < 1/n$ and $d(w,v) < 1/n$; hence, $d(u,v) < 2/n$. Therefore, by (ii) above, we have that $v \in V^y_{2i}$, and since $f^{-1}(z) \cap V^y_{2i} \neq \varnothing$, it follows that $f^{-1}(z) \subset V^y_{2i}$.

Step 2.   We show that $U^z_n \subset U^y_i$. By step 1, we know that $f^{-1}(z) \subset U^y_{2i} = N^y_{2i} \cup \text{int} f^{-1}(y)$, and hence, $f^{-1}(z) \subset N^y_{2i}$. Let $u \in U^z_n$. Choose $v \in$

$f^{-1}(z)$ such that $d(u,v) < 1/n < 1/2i$. Since $v \in N^y_{2i}$, there is an $s \in \mathrm{Fr}\, f^{-1}(y)$ such that $d(s,v) < 1/2i$. Hence, $d(s,u) < 1/i$ and $u \in N^y_i \subset U^y_i$. Therefore, $U^z_n \subset U^y_i$.

To complete the proof of the theorem, we merely observe that it is immediate from step 2 that $V^z_n \subset V^y_i$, and consequently we have that $W^z_n \subset W^y_i$.

An important corollary is the following (which was used in the proof of (9.B.3)).

(10.C.8) Corollary. Suppose that X is a compact metric space and Y is T_2. If there is a continuous, onto function $f : X \to Y$, then Y is metrizable.

(10.C.9) Corollary. Locally finite *CW*-complexes are metrizable (see (8.E.20)).

(10.C.10) Definition. A space X is *fully normal* if and only if each open cover of X has an open barycentric refinement.

(10.C.11) Remark. It follows from the proof of (10.A.14) that any paracompact space is fully normal. Stone [1959] has shown that T_2 fully normal spaces are paracompact.

We use the Aleksandrov-Uryson theorem to prove that fully normal developable spaces are metrizable.

(10.C.12) Theorem. A space X is metrizable if and only if X is fully normal and developable.

Proof. The following notation is used: if $\mathcal{U}$ and $\mathcal{V}$ are covers of X, then $\mathcal{U} \cap \mathcal{V} = \{U \cap V \mid U \in \mathcal{U} \text{ and } V \in \mathcal{V}\}$. Suppose that $\mathcal{D}_1, \mathcal{D}_2, \ldots$ is a development for X. Let $\mathcal{H}_1$ be a barycentric refinement of $\mathcal{D}_1$ and set $\mathcal{U}_2 = \mathcal{H}_1 \cap \mathcal{D}_2$. Note that $\mathcal{U}_2$ refines $\mathcal{D}_2$ and barycentrically refines $\mathcal{D}_1$.

Let $\mathcal{H}_2$ be a barycentric refinement of $\mathcal{U}_2$ and set $\mathcal{U}_3 = \mathcal{H}_2 \cap \mathcal{D}_3$. Continuing in this fashion, we obtain a sequence of open covers $\mathcal{U}_2, \mathcal{U}_3, \ldots$, where each $\mathcal{U}_i$ refines $\mathcal{D}_1, \mathcal{D}_2, \ldots, \mathcal{D}_i$ and for each i, $\mathcal{U}_{i+1}$ is a barycentric refinement of $\mathcal{U}_i$. It is not difficult to see that $\{\mathcal{U}_i \mid i = 2, 3, \ldots\}$ satisfies the requirements of the Aleksandrov-Uryson theorem. Again, the easy converse is left to the reader (problem C.4).

Bing has given an especially convenient metrization theorem in terms of a condition that is somewhat stronger than normality.

(10.C.13) *Definition.* A family of subsets $\{H_\alpha \mid \alpha \in \Lambda\}$ of a space X is *discrete* if and only if $\{\overline{H}_\alpha \mid \alpha \in \Lambda\}$ is locally finite and the $\overline{H}_\alpha$'s are mutually disjoint.

(10.C.14) *Definition.* A topological space X is *collectionwise normal* if and only if, for each discrete family of subsets $\{H_\alpha \mid \alpha \in \Lambda\}$, there are mutually disjoint open subsets $\{G_\alpha \mid \alpha \in \Lambda\}$ such that $H_\alpha \subset G_\alpha$ for each $\alpha \in \Lambda$.

A proof of the next theorem may be based on showing that developable, collectionwise normal spaces are paracompact.

(10.C.15) *Theorem* (*Bing* [*1947*]). A space X is metrizable if and only if it is collectionwise normal and developable.

(10.C.16) *Definition.* A cover $\mathscr{U}$ of a space is *σ-locally finite* if and only if $\mathscr{U}$ can be expressed as the union of a countable collection of families, each of which is locally finite.

Uryson's metrization theorem is an easy corollary of the following result, which is one of the most powerful of all the metrization theorems. A proof of this result may be found in Kelley [1955].

(10.C.17) *Theorem.* (*Nagata* [*1950*], *and Smirnov* [*1953*]). A space X is metrizable if and only if X is T_3 and has a σ-locally finite basis.

D. MOORE SPACES

There has been some recent interest in investigating an old concept: Moore spaces. Moore spaces are generalizations of metric spaces.

(10.D.1) *Definition.* A T_3 space with a development is a *Moore space*.

Clearly, metric spaces are Moore spaces; however, there are examples of Moore spaces that are not metrizable. One such example is given below. That Moore spaces are "almost" metric spaces can be inferred from the following theorems, which hold in both metric spaces and in Moore spaces.

(i)   Suppose that M is a Moore Space and $A \subset M$. Then A is compact if and only if each infinite sequence in A has an accumulation point in M.

(ii)   If a Moore space is Lindelöf, then it is second countable.

(iii)   If M is a Moore space and every uncountable subset of M has an accumulation point, then M is second countable.

Thus far, we have the following relationships:

Paracompact + developable $\xrightarrow[(10.C.11)]{}$ developable + fully normal

$\xrightarrow[(10.C.12)]{}$ metrizable

Hence, in particular, a paracompact Moore space is metrizable. Can the requisite of paracompactness be weakened? This leads to the celebrated

Moore Space Conjecture.

Every normal Moore space is metrizable.

Jones [1937] has shown that every separable normal Moore space is metrizable; however, the general conjecture has deep set theory ramifications and may even be in some sense "unsolvable" (Bing [1965]; Heath [1964]). We now give an example of a Moore space that is not metrizable.

The space to be constructed is connected and locally Euclidean (and if it were metrizable, it would be a 2-manifold). In $\mathscr{E}^3$, start with a Cantor set

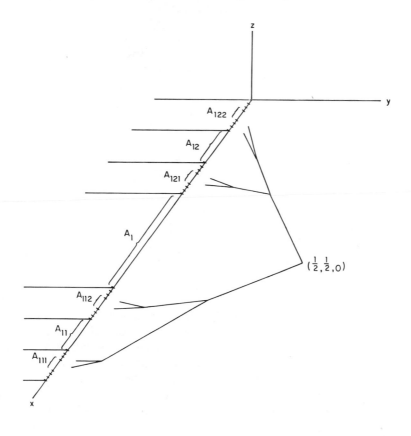

on the x axis and denote the "middle thirds" by A_1; A_{11}, A_{12}; A_{111}, A_{112}; A_{121}, A_{122}; $\cdots$.

From the point $(\frac{1}{2}, \frac{1}{2}, 0)$, a network of roads is sent out, all of which are directed toward the Cantor set in such a way that every point of the Cantor set is a limit of precisely one route, and each route eventually "arrives" at precisely one point of the Cantor set.

From each point of the Cantor set, extend a ray in the negative y direction parallel to the y axis, as indicated in the preceding figure. Let M be the union of these rays and the road system, and let $N = M \times \mathbf{I}$.

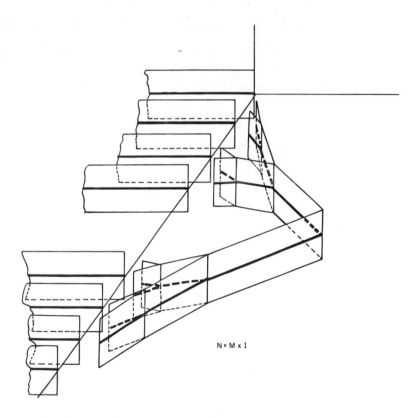

$N = M \times 1$

To obtain a space that is locally $\mathscr{E}^2$, something must be done about the branch points. Consider a cylinder about a typical branch point p.

This "neighborhood" of p is replaced by a hexagon. The sides ℓ_1, ℓ_2, ℓ_3 are glued to ℓ_1', ℓ_2' and ℓ_3' respectively, and thus the "fat Y's" around a branch point are replaced by a twisted two-dimensional hexagon. This procedure is repeated for each branch point. Let X be the space derived from N in this manner. What is the topology for X? For points outside $K \times \mathbf{I}$, the usual

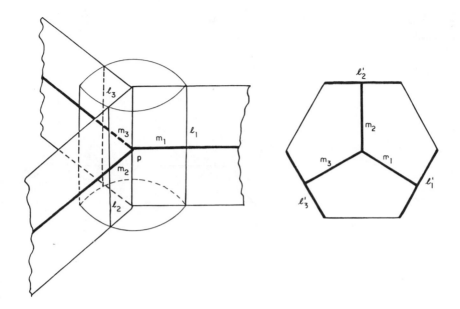

relative plane topology is used, i.e., disks form a basis at each point. Suppose $w \in K \times \mathbf{I}$, and say $w = (x,0)$. A typical neighborhood for w will have the following form. Let $\varepsilon > 0$ be given. Starting at a point a distance ε from the Cantor set, we select the route in X that leads to w. Near a point (which is not a branch point), we have the following situation (a).

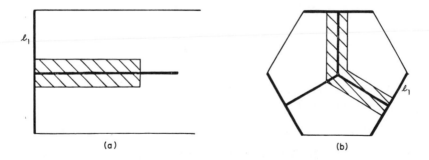

(a)   (b)

As we enter a hexagon, we simply proceed as shown in (b), where the darkened lines indicate which branch was taken en route to w and the shaded area denotes the neighborhood in question. This gives us one half of the neighborhood of w.

The "other half" of the neighborhood is easily described as simply consisting of half of an open disk that lies in the product of $[-1,1]$ and the ray emanating from w.

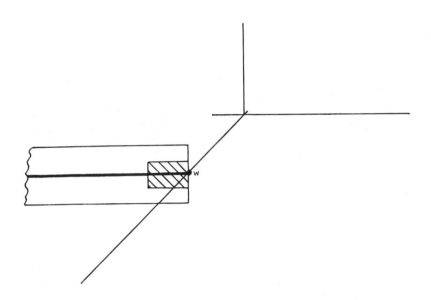

It is easy to see that the two halves form a whole disk (without boundary).

(10.D.2) Exercise. Show that X with the topology described above is connected (even arc connected), regular, T_2, and that each point has a neighborhood homeomorphic to $\mathscr{E}^2$.

(10.D.3) Exercise. Show that X is developable.

(10.D.4) Exercise. Show that X is not metric. [Hint: If X is metric, then by (10.A.18), X is σ-compact. However, this implies that X is separable metric, which it certainly is not.]

E. COMPLETION OF METRIC SPACES

We deal briefly with the idea of completing a metric space. A quick review of Chapter 3 might be in order here. In that chapter, it was noted that although completeness fails to be a topological invariant, topological completeness by definition is preserved under homeomorphisms. There do exist, however, metric spaces that are not topologically complete. Nevertheless, all metric spaces are "almost" complete in the sense that they can be embedded as dense subsets of complete metric spaces. This result is a generalization (both in content and in proof) of one of the standard constructions of the real numbers from the rationals. Recall that a map $f : (X,d) \to (Y,d')$ is an *isometry* if and only if $d(x_1,x_2) = d'(f(x_1),f(x_2))$ for each $x_1,x_2 \in X$.

(10.E.1)   Theorem.   Every metric space can be isometrically embedded as a dense subset of a complete metric space.

Proof.   The details of the proof, although not difficult tend to be tedious. We suppose that the reader has had previous experience with similar details in the Cauchy sequence approach to the construction of the real line, and hence we merely outline the major steps in the proof.

Suppose that (X,d) is a metric space.

Step 1.   Define $\mathscr{C}$ to be the class of all Cauchy sequences in X.

Step 2.   Define $\hat{d} : \mathscr{C} \times \mathscr{C} \to [0,\infty]$ by $\hat{d}(\{s_n\},\{t_n\}) = \lim_{n \to \infty} d(s_n,t_n)$.

Step 3.   Note that $\hat{d}$ has all of the properties of a metric except that $\hat{d}(\{s_n\},\{t_n\}) = 0$ does not necessarily imply that $\{s_n\} = \{t_n\}$ ($\hat{d}$ is a pseudo-metric).

Step 4.   Rectify the problem in step 3 by defining equivalence classes where $\{s_n\} \sim \{t_n\}$ if and only if $\hat{d}(\{s_n\},\{t_n\}) = 0$. Let $\hat{\mathscr{C}}$ be the set of these equivalence classes, and redefine $\hat{d}$ in the obvious manner so that $\hat{d}$ is now a metric function.

Step 5.   Define $\theta : X \to \hat{\mathscr{C}}$, where $\theta(x)$ is the equivalence class representing the constant sequence with all elements equal to x.

Step 6.   Note that $\theta(X)$ is dense in $\hat{\mathscr{C}}$, since if $[\{s_n\}] \in \hat{\mathscr{C}}$, then for large k, we have that $\theta(s_k)$ is "close to" $\{s_n\}$.

Step 7.   Show that $\hat{\mathscr{C}}$ is complete. Note that since $\theta(X)$ is dense in $\hat{\mathscr{C}}$, it suffices to show that Cauchy sequences in $\theta(X)$ converge in $\hat{\mathscr{C}}$.

Although a subspace of a complete metric space is not necessarily complete, is it topologically complete? The reader should realize that this need not be the case. The following theorem, however, gives a surprisingly general condition for which the answer is yes.

(10.E.2)   Theorem.   Suppose that (X,d) is a complete metric space and that U is an open subset of X. Then U is topologically complete.

Proof.   We prove the theorem by establishing that U is homeomorphic to a closed subset of a complete metric space, and hence is topologically complete. Define a function $f : U \to \mathcal{E}^1$ by $f(u) = 1/d(u, X \setminus U)$, and let $W = \{(u, f(u)) \in X \times \mathcal{E}^1 \mid u \in U\}$. Then it is not difficult to show that W and U are homeomorphic and W is closed in $X \times \mathcal{E}^1$.

(10.E.3) Corollary. Locally compact separable metric spaces are topologically complete.

Proof. Suppose that X is a locally compact separable metric space. Then by (10.C.2), the one point compactification of X, X^*, is a compact metric space and hence complete. However, X is an open subset of X^*, and the result follows.

(10.E.4)   Theorem.   If A is a G_δ subset of a complete metric space X, then A is topologically complete.

Proof. The trick is exactly the same as that used in the previous theorem: the set A is embedded as a closed subset of a complete metric space. Since A is a G_δ subset of X, A may be written as an intersection of open subsets of X, say $A = \bigcap_{n=1}^{\infty} U_n$. By the previous theorem, each U_n is topologically complete, and hence by (6.A.17) the product $\prod_{n=1}^{\infty} U_n$ is a complete metric space. Embed A in $\prod_{n=1}^{\infty} U_n$ by $\psi : A \to \prod_{n=1}^{\infty} U_n$, where $\psi(a) = (a, a, \ldots)$. Again, it is not unreasonable to ask the reader to check that A and $\psi(A)$ are homeomorphic and that $\psi(A)$ is closed in $\prod_{n=1}^{\infty} U_n$.

(10.E.5) Exercise. Show that topologically complete spaces are Baire spaces and hence the rationals are not topologically complete.

(10.E.6) Exercise. Show that the irrationals are topologically complete.

If one can complete metric spaces, can one "complete" Moore Spaces? The question is meaningless unless a reasonable definition can be given for "complete Moore space." Theorem (3.B.8) is used to motivate the following definitions.

(10.E.7) Definition. Suppose that $C = \{\mathcal{C}_1, \mathcal{C}_2, \ldots\}$ is a nested development for a Moore space M (i.e., $\mathcal{C}_1 \supset \mathcal{C}_2 \cdots$). Then M is *complete relative to* C if and only if for any decreasing sequence of closed sets $F_1 \supset F_2 \supset \cdots$

with the property that for each n, $F_n \subset G_n$ for some $G_n \in \mathscr{C}_n$, the intersection $\bigcap_{n=1}^{\infty} F_n \neq \varnothing$.

(10.E.8) *Definition.* A T_1 space M is a *complete Moore space* if and only if M has a nested development **C** such that M is complete relative to **C**.

In closing, we mention (without proof) a few theorems concerning complete metric spaces that also hold for complete Moore spaces.

(i)   The countable intersection of dense open sets in a complete Moore space is dense in M.

(ii)   If M is a locally connected complete Baire space, then connected open subsets are arcwise connected.

(iii)   If M is a complete Moore space for which there do not exist uncountably many disjoint open sets, then M is separable.

There are examples of Moore spaces that cannot be embedded in complete Moore spaces, and this has prompted a great deal of investigation of additional properties that a Moore space must possess in order to have a completion. A very readable account of this problem may be found in Armentrout [1967].

PROBLEMS

Section A

1.   Show that a closed subset of a paracompact space is paracompact.
2.   Show that a regular second countable T_1 space is paracompact.
3.   Show that if each open subset of a paracompact space X is paracompact, then every subset of X is paracompact.
4.*   Show that $[0,\Omega]$ is paracompact, but contains a nonparacompact subset. [Hint: Try $[0,\Omega)$.]
5.   Show that the product of a compact T_2 space and a paracompact space is paracompact.
6.   Show that if $X \times Y$ is paracompact, then X and Y are paracompact.
7.   Show that if X is a T_2 space such that each open cover of X has a locally finite closed refinement, then X is paracompact.
8.   A space X is *metacompact* if and only if each open cover of X has a point-finite open refinement (i.e., if $\mathscr{U}$ is an open cover of X, there is an open refinement $\mathscr{V}$ of $\mathscr{U}$ such that each x in X belongs to at most a finite

number of members of $\mathscr{V}$.) Show that a space is compact if and only if it is both countably compact and metacompact. [Hint: Use Zorn's lemma to show that every point-finite cover has an irreducible sub-cover.]

9.* Show that a paracompact space X is compact if and only if X is count-ably compact.

10. Show that the space $[0,\Omega)$ is T_5 but not paracompact.

Section B

1. Suppose that for every locally finite cover $\mathscr{U}$ of a T_2 space X, there is a partition of unity subordinate to $\mathscr{U}$. Show that X is normal.

2. Suppose that X is paracompact and $\mathscr{U} = \{U_\alpha \mid \alpha \in \Lambda\}$ is a locally finite open cover of X. Show that if $\psi : \Lambda \to (0,\infty)$ is a function, then there is a map $f : X \to (0,\infty)$ such that for each $x \in X$, $f(x) \le \sup\{\psi(\alpha) \mid x \in U_\alpha\}$.

3. Show that if $\{U_1, U_2, \ldots, U_n\}$ is a finite open cover of a normal space X, then there are continuous functions $f_1, f_2, \ldots, f_n$ such that:
   (a) $f_i : X \to [0,1]$ for each $i = 1, 2, \ldots, n$
   (b) $\{x \mid f_i(x) \ne \emptyset\} \subset U_i$ for each $i = 1, 2, \ldots, n$
   (c) $\sum_{i=1}^{n} f_i(x) = 1$.
   [Hint: Use (4.C.2) and (4.B.1).]

4. Use Problem 3 to show that every compact n-manifold M can be embedded in $\mathscr{E}^m$ for some $m \in \mathbf{Z}^+$. [Hint: Find an open cover $\{U_1, U_2, \ldots, U_t\}$ of M such that for each i, there is a homeomorphism $g_i : U_i \to \mathscr{E}^n$. Find continuous functions $f_1, f_2, \ldots, f_t$ as in Problem 3, and for $i = 1, 2, \ldots, t$ define

$$\phi_i(x) = \begin{cases} f_i(x) \cdot g_i(x) \text{ (scalar product) if } x \in U_i \\ \mathbf{0}, \qquad\qquad\qquad\quad \text{if } x \in M \setminus \{y \mid f_i(y) \ne 0\}. \end{cases}$$

Finally define $H : M \to \mathscr{E}^t \times \mathscr{E}^{nt}$ by $H(x) = (f_1(x), \ldots, f_t(x), \phi_1(x), \ldots, \phi_t(x))$ and show that H is the desired embedding.]

Section C

1. Let Λ be an index set, and for each $\alpha \in \Lambda$, let $I_\alpha = \mathbf{I}$. Let X be the space obtained from $\{I_\alpha \mid \alpha \in \Lambda\}$ by identifying all left-handed end points to a common point x^* (this is called the *porcupine space*). The *starfish metric* d for X is defined as follows. If $x \in I_\alpha$, $y \in I_\beta$, and $\alpha \ne \beta$, then $d(x,y) = |x - x^*| + |y - y^*|$; if $\alpha = \beta$, then $d(x,y) = |x - y|$. Does the

topology induced by the starfish metric coincide with the porcupine topology?

2. Show that Uryson's metrization theorem is a corollary of Bing's metrization theorem.

3.* Show that if a space has a development, then it has a nested development (i.e., a development $\mathscr{C}_1$, $\mathscr{C}_2$, ... with the property that $\mathscr{C}_i \supset \mathscr{C}_{i+1}$ for each i).

4. Prove the converses of Theorems (10.C.6) and (10.C.12).

5. Show that every metric space is developable. Find a T_1 space that is not developable.

6. Suppose that X is an uncountable set with the discrete metric. Is X^* metrizable?

7.* Use the Nagata-Smirnov metrization theorem to show that if X is a normal space and has a locally finite open cover $\mathscr{U}$ with the property that each $U \in \mathscr{U}$ is metrizable, then X is metrizable.

8.* Show that a locally metrizable, T_2 space X is metrizable if and only if X is paracompact. (A space X is *locally metrizable* if and only if each point x has an open neighborhood that is metrizable.)

9. Suppose that X is a metric space, Y is T_2, and $f : X \to Y$ is closed, continuous, and onto. Show that Y is metrizable if and only if the set of accumulation points of X is compact.

10. Suppose that X is a metric space, Y is T_2, and $f : X \to Y$ is continuous and onto. Show that Y is metrizable if and only if X is compact.

11. Find a σ-locally finite basis for $\mathscr{E}^1$.

12.* Prove or disprove: if X is a topological space and A and B are metrizable subspaces of X such that $X = A \cup B$, then X is metrizable.

13. Show that metric spaces are collectionwise normal.

Section D

1. Show that the "bubble" space described in Chapter 4 is a Moore space that is not metrizable.

2. Let K be the Cantor set and order the intervals of $\mathbf{I} \setminus K$ by length (and from left to right for those of the same length). Denote the intervals by A_1; A_{11}, A_{12}; A_{111}, A_{112}; A_{121}, A_{122}; $\cdots$. Let a_1 be the point in the plane one unit below the midpoint of A_1. Draw segments emanating from a_1 that go half way toward the centers of A_{11} and A_{12}. Let a_{11} and a_{12} be end points of these segments. Continue as indicated in the figure. Let M be the union of these segments and K. If $x \in K$, then starting at a_1 there is only one arc in M that arrives at x. Denote this arc by A_x. Topologize M by defining those sets to be open which consist of half-

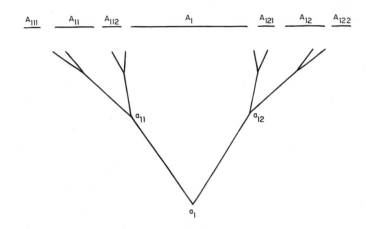

open arcs $(w,x]$ lying in A_x together with the obvious half-open arcs
leaving each branch point (but not including the next branch point).

(a) Show that M with this topology is a Moore space.
(b) Show that M is not metrizable (note that M is separable but not
    second countable).
3.* Show that Moore spaces have the following property: There is a count-
    able family **C** of open covers of X such that if H and K are disjoint
    closed subsets of X, one of which is finite, then there is a cover $\mathscr{C} \in \mathbf{C}$
    such that no element of $\mathscr{C}$ intersects both H and K.
4. Prove (i) in the text following (10.D.1).
5. Prove (ii) in the text following (10.D.1).
6.* Prove (iii) in the text following (10.D.1) (Jones [1937]).

Section E

1. Define a *completion* of a metric space X to be any complete metric space in which X is embedded isometrically as a dense subset. Show that all completions of a given metric space are isometric.

2. Show that the completion of a metric space X is separable if and only if X is separable.

3. Suppose that (X,d) is a compact metric space, A is a dense subset of X, and $f : A \to Y$ maps A continuously into a complete metric space Y. Show that f may be extended continuously to X if and only if f is uniformly continuous.

4. Give an example of a topologically complete metric space that is not locally compact.

5. Suppose that (X,d) is a metric space and that A is a subset of X such that $(A,d_{|A})$ is complete. Show that A is closed.

6. Suppose that Y is a completion of (X,d). Show that Y is second countable if and only if (X,d) is.

7. Suppose that X is a dense subset of a complete metric space (Y,d'). Let $d = d'_{|X \times X}$. Show that there is an isometry $\psi : (\hat{C},\hat{d}) \to (Y,d')$, where $\psi(\theta(x)) = x$ for each $x \in X$. (Notation is as in 10.E.1.)

8. Suppose that X and Y are complete metric spaces and that D and E are dense subsets of X and Y, respectively. Show that an isometry $f : D \to E$ can be extended uniquely to an isometry $F : X \to Y$.

9. Show that the completion of a metric space (X,d) is a compactification of X if and only if (X,d) is totally bounded. [Hint: Show that if a metric space X has a totally bounded dense subset, then X is totally bounded.]

Chapter 11

NETS AND FILTERS

A. ON THE FAILINGS OF SEQUENCES

Sequences in a nonmetric setting may behave in an erratic fashion. To illustrate their eccentricities, we list four propositions involving sequences that are valid for metric spaces but which are partially false in a more general context. Examples are then given to demonstrate what can go awry.

(11.A.1) *Proposition.* Suppose that X is a metric space and that $x \in X$ is a cluster point of a sequence $\{x_n\}$ in X. Then there is a subsequence of $\{x_n\}$ that converges to x.

(11.A.2) *Proposition.* Suppose that X and Y are metric spaces. Then a function $f : X \to Y$ is continuous at a point x if and only if the sequence $\{f(x_n)\}$ converges to $f(x)$ whenever the sequence $\{x_n\}$ converges to x.

(11.A.3) *Proposition.* A subset U of a metric space X is open if and only if whenever a sequence $\{x_n\}$ converges to $x \in U$, there is an integer N such that for $n > N$, $x_n \in U$.

(11.A.4) *Proposition.* A subset C of a metric space X is compact if and only if each sequence in C has a cluster point in C.

(11.A.5) *Example.* Let $X = (\mathbf{Z}^+ \times \mathbf{Z}^+) \cup \{(0,0)\}$. For each $i \in \mathbf{Z}^+$, let $C_i = \{(i,n) \mid n \in \mathbf{Z}^+\}$. Define open sets as follows:

(i) $\{(m,n)\}$ is open for each $(m,n) \in \mathbf{Z}^+ \times \mathbf{Z}^+$;

(ii)   a set U containing $(0,0)$ is open if and only if U has the property that there is an integer N_U such that $(X \setminus U) \cap C_i$ is finite (or empty) for each $i > N_U$.

Let $f : \mathbf{Z}^+ \to X \setminus \{(0,0)\}$ be any onto map, and set $x_n = f(n)$ for each $n \in \mathbf{Z}^+$. It is clear that $(0,0)$ is a cluster point of the sequence $\{x_n\}$, but it will be shown that no subsequence of $\{x_n\}$ converges to $(0,0)$.

Suppose to the contrary that $\{x_{n_i}\}$ is a subsequence of $\{x_n\}$ converging to $(0,0)$. Note that if infinitely many of the x_{n_i} lie in C_k for some $k \in \mathbf{Z}^+$, then the sequence $\{x_{n_i}\}$ fails eventually to be in the open neighborhood $\bigcup_{i \neq k} C_i \cup$ $\{(0,0)\}$ of $(0,0)$. Thus, each C_k contains at most a finite number of members of the sequence $\{x_{n_i}\}$. Let $U = \{(0,0)\} \cup (X \setminus \bigcup \{\{x_{n_i}\} \mid i \in \mathbf{Z}^+\})$, and note that U is an open neighborhood of $(0,0)$ that does not contain a single member of the "converging" subsequence $\{x_{n_i}\}$.

(11.A.6)   Exercise.   Show that X is T_2 and Lindelöf.

(11.A.7)   Example.   Suppose that $X = \mathbf{R}^1$ has the following topology: a set $U \subset X$ is open if and only if $X \setminus U$ is countable or $U = \varnothing$. Note that the only convergent sequences are the ones that are eventually constant. Let $Y = \mathscr{E}^1$ and let $f : X \to Y$ be the identity function. If $x \in X$ and a sequence $\{x_n\}$ converges to x, then the sequence $\{f(x_n)\}$ will converge to $f(x)$, but f is not continuous at any point x.

(11.A.8)   Example.   Let $X = \mathbf{R}^1$ have the topology described in (11.A.7). Let U be any nonopen subset of X and note that whenever a sequence $\{x_n\}$ converges to a point $x \in U$, there is an integer N such that for $n > N$, $x_n \in U$. Hence, (11.A.3) need not hold in a nonmetric setting.

(11.A.9)   Example.   Let $X = [0,\Omega)$ with the order topology. Then every infinite sequence in X has a cluster point, but X is not compact.

We shall not expose all of the failings of sequences, but the foregoing examples should be enough to convince the reader that sequences are somewhat less than adequate. Nets and filters were invented to obtain theorems that should hold for sequences, but do not.

B.   NETS

The notion of sequence may be generalized as follows.

(11.B.1) *Definition.* Suppose that D is a directed set (6.B.1) and that X is an arbitrary set. A *net in X* is a pair (N,D), where N is a function mapping D into X.

When D is given, we often refer to the net (N,D) by simply writing $N : D \to X$, or just N. If $\alpha \in D$, then $N(\alpha)$ will be denoted by x_α; when there is little chance for confusion, the function N is denoted by its range $\{x_\alpha\}$ or $\{x_\alpha\}_{\alpha \in D}$.

The following example shows that sequences are special cases of nets.

(11.B.2) *Example.* Let $D = \mathbf{Z}^+$ with the usual ordering. Then any function $N : D \to X$ is both a sequence and a net.

(11.B.3) *Example.* Let $X = [0,\infty)$ and let $D = \mathscr{P}(X)$ be ordered by inclusion. Define $N : D \to X$ by $N(A) = \inf\{x \mid x \in A\}$. Then N is a net but it is clearly not a sequence.

As was the case with sequences, the most important property associated with nets is convergence.

(11.B.4) *Definition.* If $N : D \to X$ is a net, then D is *eventually in a set* $A \subset X$ if and only if there is an $\alpha \in D$ such that for each $\beta \geq \alpha$, $N(\beta) \in A$. A net N *converges* to a point $x \in X$ if and only if N is eventually in each open set containing x.

Note that this definition, when specialized to sequences, agrees with (1.F.8). Does the net in (11.B.3) converge?

Before we attempt to discover the interplay between nets, continuous functions, and topological spaces, we introduce the concept of subnet.

(11.B.5) *Definition.* Suppose that $N : D \to X$ is a net. A *subnet* of N is a net (N',E), where there is a function $P : E \to D$ such that

(i) $NP = N'$, and
(ii) for each $d \in D$, there is an $e \in E$ such that $P(z) \geq d$ for all $z \geq e$.

Note that while P need not be strictly increasing, the tag end of P eventually surpasses any given element of D.

(11.B.6) *Exercise.* Show that subsequences are subnets.

(11.B.7) *Definition.* A net $N : D \to X$ is *frequently* in a set $A \subset X$ if and only if for each $\alpha \in D$, there is a $\beta \in D$, $\beta > \alpha$, such that $N(\beta) \in A$. A point $x \in X$ is a *cluster point of N* if and only if N is frequently in every neighborhood of x.

The reader should note the similarity between cluster points of nets and cluster points of sequences.

We now proceed to show how nets may be used to generalize the propositions stated for metric spaces at the beginning of the chapter.

(11.B.8)   *Theorem.*   Suppose that $N : D \to X$ is a net in X. Then a point $x \in X$ is a cluster point of a net $\{x_\alpha\}$ if and only if there is a subnet of $\{x_\alpha\}$ that converges to x.

Proof.   Suppose that x is a cluster point of $\{x_\alpha\}$. To find a convergent subnet, we first construct an appropriate directed set E. Define E to be $\{(\alpha, U) \mid \alpha \in D,\ U \text{ is an open neighborhood of } x \text{ containing } x_\alpha\}$. The fact that x is a cluster point ensures that E is nonempty. We order E in the following way: $(\alpha_1, U) \le (\alpha_2, V)$ if and only if $\alpha_1 \le \alpha_2$ and $V \subset U$. Let $P : E \to D$ be defined by $P(\alpha, Y) = \alpha$. It is clear that (NP, E) is a subnet of $\{x_\alpha\}$. To see that this subnet converges to y, suppose that W is any open neighborhood of x and choose $\alpha^* \in D$ such that $x_{\alpha^*} \in W$. Then $(\alpha^*, W) \in E$. Note that if $(\alpha, U) \ge (\alpha^*, W)$, then $U \subset W$, and since $x_\alpha \in U$, it follows that $NP(\alpha, U) = N(\alpha) = x_\alpha \in U$.

Conversely, suppose that $N : D \to X$ is a net containing a subnet $NP : E \to X$ that converges to a point $x \in X$. Suppose that U is an arbitrary neighborhood of x and that $\alpha \in D$. We shall find an $\alpha' \in D$, $\alpha' > \alpha$, such that $N(\alpha') \in U$. There is an $e_1 \in E$ such that $P(e_1) > \alpha$, and an $e_2 \in E$ such that $e_2 \ge e_1$ and $NP(e_2) \in U$. If $\alpha' = P(e_2)$, we have that $N(\alpha') \in U$, and hence x is a cluster point of (N, D).

Resolution of the problems encountered in Examples (11.A.7) and (11.A.8) above is quite routine and left to the reader. We now consider the relationship between nets and compactness.

(11.B.9)   *Theorem.*   A space X is compact if and only if each net in X has a convergent subnet.

Proof.   By (11.B.8), it suffices to show that X is compact if and only if each net in X has a cluster point. Suppose that every net in X has a cluster point. We use the finite intersection property characterization of compactness to prove that X is compact. Let $\mathscr{A}$ be a collection of closed subsets of X with the finite intersection property. Expand $\mathscr{A}$ to a family $\mathscr{B}$ that includes all finite intersections of members of $\mathscr{A}$. Direct $\mathscr{B}$ by containment ($B \ge B'$ if and only if $B \subset B'$). For each $B \in \mathscr{B}$, select a point $x_B \in B$ (the axiom of choice permits this) and define $N : \mathscr{B} \to X$ by $N(B) = x_B$.

Clearly, N is a net in X and hence must have a cluster point x. We show that $x \in \bigcap \{F \mid F \in \mathscr{A}\}$. Let B be any element of $\mathscr{B}$ and suppose that $B' \in \mathscr{B}$

and $B' \geq B$. Then $N(B') \in B' \subset B$. Therefore, the net N is eventually in B for each B, and since B is closed, we have that $x \in B$. It follows that $x \in \bigcap \{B \mid B \in \mathscr{B}\} \subset \bigcap \{F \mid F \in \mathscr{A}\}$, and thus X is compact.

Conversely, suppose that X is compact and that $N : D \to X$ is a net. We use the finite intersection property for compact sets to find a cluster point for N. For each $\alpha \in D$, let $F_\alpha = \{N(\beta) \mid \beta \geq \alpha\}$. It is easy to see that $\mathscr{F} = \{F_\alpha \mid \alpha \in D\}$ is a collection of sets with the finite intersection property; furthermore, $\overline{\mathscr{F}} = \{\overline{F_\alpha} \mid \alpha \in D\}$ is a family of closed sets that also has the finite intersection property and hence has nonempty intersection. Let $y \in \bigcap \{\overline{F_\alpha} \mid \alpha \in D\}$. We show that y is a cluster point for N. If y is not a cluster point for N, there is a neighborhood U of y and an $\alpha \in D$ such that if $\beta \geq \alpha$, then $N(\beta) \notin U$. This, however, is impossible, since $y \in U \cap \overline{F_\alpha}$.

(11.B.10) Exercise. Let $(D, \geq)$ be a directed set and suppose that $\{A_\alpha \mid \alpha \in D\}$ is a family of nonempty, compact, connected subsets of a T_2 space X such that $A_\alpha \subset A_\beta$ if and only if $\alpha \geq \beta$. Show that $\bigcap_{\alpha \in D} A_\alpha$ is nonempty, compact, and connected.

C. FILTERS

The concept of filter, while seemingly far removed from the idea of sequence, nevertheless provides an important and useful theory of convergence. We have seen that neighborhoods of a point are critical in determining convergence at the point. Filters may be considered as an abstraction of the following two properties of these neighborhoods:

(i) the intersection of two neighborhoods is again a neighborhood, and
(ii) if G is a neighborhood of x, then any set containing G is also a neighborhood.

(11.C.1) Definition. A *filter* $\mathscr{F}$ on a set X is a collection of nonempty subsets of X such that

(i) if $F_1, F_2 \in \mathscr{F}$, then $F_1 \cap F_2 \in \mathscr{F}$, and
(ii) if $F \in \mathscr{F}$, and $F \subset G$, then $G \in \mathscr{F}$.

If $\mathscr{F}$ is a filter then $\mathscr{B} \subset \mathscr{F}$ is called a *filter base* for $\mathscr{F}$ if an only if each member of $\mathscr{F}$ contains a member of $\mathscr{B}$.

(11.C.2) Exercise. Suppose that $\mathscr{B}$ is a collection of subsets of a set X. Show that if $\mathscr{B}$ has the finite intersection property, then $\mathscr{B}$ is a base for a unique filter on X.

(11.C.3)   Examples.
   1.   The neighborhood base at a point $x \in X$ is a filter, called the *neighborhood filter* at x.
   2.   Let $N : D \to X$ be a net in X. Then $\mathscr{F} = \{A \subset X \mid N$ is eventually in $A\}$ is a filter.
   3.   The family $\mathscr{B} = \{(a,\infty) \mid a \in \mathbf{R}^1\}$ is a base for a filter.
   4.   If X is an infinite set, then the family of all sets with finite complements is a filter in X.

(11.C.4)   Definition.   Suppose that $\mathscr{F}$ is a filter in a set X. Then $\mathscr{F}$ is *eventually in* a subset $A \subset X$ if and only if $A \in \mathscr{F}$. A filter $\mathscr{F}$ in a topological space X *converges* to a point $x \in X$ if and only if $\mathscr{F}$ is eventually in each neighborhood of x.

   Thus one might picture such a filter as funneling down toward x.
   Suppose that $\mathscr{F}$ is a filter in a set X and that $f : X \to Y$ is a function from X into an arbitrary set Y. Although $f(\mathscr{F}) = \{f(F) \mid F \in \mathscr{F}\}$ is not necessarly a filter (why?), it is at least a base for the filter $\{G \subset Y \mid f(F) \subset G$ for some $F \in \mathscr{F}\}$. Abusing notation slightly, we also denote this filter by $F(\mathscr{F})$.

(11.C.5)   Exercise.   Suppose that X and Y are topological spaces and that $f : X \to Y$. Show that f is continuous at a point x_0 if and only if for each filter $\mathscr{F}$ converging to x_0, the filter $f(\mathscr{F})$ converges to $f(x_0)$.

   Filters like nets may cluster at a point.

(11.C.6)   Definition.   A filter $\mathscr{F}$ *clusters* at a point x if and only if x belongs to the closure of each member of $\mathscr{F}$.

   Subnets, when translated into filter terminology, become "finer" filters (11.D.2). If $\mathscr{F}$ and $\mathscr{G}$ are filters on X, then $\mathscr{F}$ is *finer* than $\mathscr{G}$ provided $\mathscr{G} \subset \mathscr{F}$.

(11.C.7)   Theorem.   A filter $\mathscr{F}$ clusters at a point x if and only if there is a filter finer than $\mathscr{F}$ that converges to x.

   Proof.   Suppose that $\mathscr{F}$ clusters at x. Let $\mathscr{B}$ be the family of all sets of the form $F \cap U$, where $F \in \mathscr{F}$ and U is a neighborhood of x. If $\mathscr{G}$ is the filter with base $\mathscr{B}$, then $\mathscr{G}$ is finer than $\mathscr{F}$ and converges to x (why does $U \in \mathscr{G}$?).
   Suppose now that $\mathscr{F} \subset \mathscr{G}$ and that $\mathscr{G}$ converges to x. Since the neighborhood base at x is contained in $\mathscr{G}$, every member of the base must intersect each $F \in \mathscr{F}$ (since $\varnothing \notin \mathscr{G}$). Hence, $\mathscr{F}$ clusters at x.

(11.C.8) Definition. A filter $\mathscr{U}$ in a set X is an *ultrafilter* if and only if whenever $\mathscr{V}$ is a filter such that $\mathscr{U} \subset \mathscr{V}$, then $\mathscr{U} = \mathscr{V}$, i.e., there does not exist a filter in X that is finer than $\mathscr{U}$.

(11.C.9) Theorem. Every filter $\mathscr{F}$ in a set X is contained in some ultrafilter.

Proof. Zorn's lemma or some variation of it is unavoidable. Let $\mathscr{F}$ be a filter in X and let $\mathbf{G} = \{\mathscr{G} \mid \mathscr{G}$ is a filter finer than $\mathscr{F}\}$. Order $\mathbf{G}$ by inclusion ($\mathscr{G}_\alpha \leq \mathscr{G}_\beta$ if and only if $\mathscr{G}_\alpha \subset \mathscr{G}_\beta$), and let $\{\mathscr{G}_\alpha \mid \alpha \in \Lambda\}$ be a nest in $\mathbf{G}$. It is easily checked that $\bigcup \{\mathscr{G}_\alpha \mid \alpha \in \Lambda\}$ is a filter finer than each filter in the nest, and hence each nest has an upper bound. Therefore, $\mathbf{G}$ has a maximal element, the desired ultrafilter.

The next theorem gives us one of the more peculiar properties of ultrafilters.

(11.C.10) Theorem. A filter $\mathscr{U}$ in a set X is an ultrafilter if and only if for each $A \subset X$, either A or its complement in X belongs to $\mathscr{U}$.

Proof. Suppose that $\mathscr{U}$ is an ultrafilter and let $\mathscr{U}' = \mathscr{U} \cup \{A\}$. If $A \cap U \neq \varnothing$ for each $U \in \mathscr{U}$, then $\mathscr{U}'$ is a base for a filter $\mathscr{U}''$. However, since $\mathscr{U}''$ is finer than $\mathscr{U}$, $\mathscr{U}''$ must be equal to $\mathscr{U}$, and hence $A \in \mathscr{U}$. On the other hand, if there is a $U \in \mathscr{U}$ such that $A \cap U = \varnothing$, then $U \subset X \setminus A \in \mathscr{U}$.

(11.C.11) Exercise. Prove the converse of (11.C.10).

(11.C.12) Exercise. Show that if $\mathscr{U}$ is an ultrafilter on X and $f : X \to Y$ is onto, then $f(\mathscr{U})$ is an ultrafilter.

(11.C.13) Exercise. Show that if each filter in X has a cluster point, then each ultrafilter in X converges.

Compactness may be characterized in terms of filters.

(11.C.14) Theorem. A space X is compact if and only if each filter in X has a cluster point.

Proof. Since X is compact, it has the finite intersection property. Suppose that $\mathscr{F}$ is a filter on X. Then $\{\bar{F} \mid F \in \mathscr{F}\}$ has the finite intersection property, and hence the intersection of all these sets will be nonempty. Any point in this intersection is obviously a cluster point of $\mathscr{F}$.

Suppose now that each filter clusters in X, but that there exists an open cover $\mathscr{U}$ with no finite subcover. We exhibit an ultrafilter that does not converge, thus contradicting (11.C.13). Let $\mathscr{B}$ be the collection of all complements of finite unions of members of $\mathscr{U}$, i.e., a typical $B \in \mathscr{B}$ has the form $X \setminus \bigcup_{i=1}^{n} U_i$, where $U_i \in \mathscr{U}$. Then $\mathscr{B}$ is a base for a filter $\mathscr{G}$. By (11.C.9), $\mathscr{G}$ is contained in an ultrafilter $\mathscr{H}$, which by (11.C.13) converges to a point x. There is a $U \in \mathscr{U}$ such that $x \in U$, and the definition of convergence implies that $U \in \mathscr{H}$. On the other hand, it is immediate from the construction of $\mathscr{G}$ that $X \setminus U \in \mathscr{G} \subset \mathscr{H}$. Thus, we have that $U \in \mathscr{H}$ and $X \setminus U \in \mathscr{H}$, which is impossible, since $U \cap (X \setminus U) = \varnothing \notin \mathscr{H}$.

Problems involving compactness are often easily solved with the aid of filters. For instance, the next two exercises yield a very elegant proof of the Tihonov theorem.

(11.C.15)   Exercise.   Suppose that $\{X_\alpha \mid \alpha \in \Lambda\}$ is a collection of topological spaces and that $\{x_\alpha\}$ is a point in $\prod_{\alpha \in \Lambda} X_\alpha$. Show that a filter $\mathscr{F}$ converges to $\{x_\alpha\}$ in $\prod_{\alpha \in \Lambda} X_\alpha$ if and only if for each $\alpha \in \Lambda$ the filter $p_\alpha(\mathscr{F})$ converges to $p_\alpha(\{x_\alpha\})$ in X_α, where p_α is the α-th projection.

(11.C.16)   Exercise (Tihonov Theorem).   Show that if $\{X_\alpha \mid \alpha \in \Lambda\}$ is a family of compact spaces, then $\prod_{\alpha \in \Lambda} X_\alpha$ is compact. [Hint: Use Ultrafilters.]

D. NETS AND FILTERS

Nets were first introduced by E. H. Moore and H. L. Smith [1922], and refinements of the theory were added by Kelley [1950a]. The French, exhibiting their customary enthusiasm for American inspirations, have demonstrated a marked predilection for filters. The concept of the filter found its origin in the work of the noted French mathematician Cartan [1937]. The purpose of this section is to underscore the futility of any filter vs. net contest by showing that nets arise from filters in a most natural way, and, no less naturally, filters may be derived from nets.

Suppose that $N = \{x_\alpha\}_{\alpha \in D}$ is a net in a set X. For each $\alpha \in D$, let $T_\alpha = \{x_\beta \mid \beta \geq \alpha\}$. Then it is easily checked that $\{T_\alpha \mid \alpha \in D\}$ has the finite intersection property and therefore is a base for a unique filter, which we denote $\mathscr{F}_N$.

(11.D.1)   Theorem.   A net N is eventually in a set $A \subset X$, if and only

if $\mathscr{F}_N$ is eventually in A; consequently, if X is a topological space, and N converges to $x \in X$, so must $\mathscr{F}_N$.

Proof. If N is eventually in A, there is an $\alpha \in D$ such that for $\beta \geq \alpha$, $x_\beta \in A$. Thus, $T_\alpha \subset A$, and hence $A \in \mathscr{F}_N$.

If $\mathscr{F}_N$ is eventually in A, then $A \in \mathscr{F}_N$, which implies that there is an α such that $T_\alpha \subset A$. Hence, for all $\beta \geq \alpha$, we have that $x_\beta \in T_\alpha \subset A$, i.e., $\{x_\alpha\}$ is eventually in A.

We previously alluded to a possible link between subnets and filter refinements. This relationship is clarified in the next theorem.

(11.D.2) *Theorem.* Suppose that $N : D \to X$ is a net in X with associated filter $\mathscr{F}_N$. Suppose further that $P : E \to D$, where E is a directed set and $NP : E \to X$ is a subnet of N. Then the filter $\mathscr{F}_{NP}$ refines $\mathscr{F}_N$.

Proof. Suppose that $T_\alpha \in \mathscr{F}_N$, where $\alpha \in D$. There is a $\beta \in E$ such that $P(\gamma) \geq \alpha$ for each $\gamma \geq \beta$. Let $S_\beta = \{x_{P(\gamma)} \mid \gamma \geq \beta\}$. Then $S_\beta \in \mathscr{F}_{NP}$ and $S_\beta \subset T_\alpha$. Therefore, $T_\alpha \in \mathscr{F}_{NP}$.

Now let us see how filters give rise to nets. Suppose $\mathscr{F} = \{F_\alpha \mid \alpha \in D\}$ is a filter in a set X. Order D so that $\alpha_1 \leq \alpha_2$ if and only if $F_{\alpha_2} \subset F_{\alpha_1}$. Then clearly, D is a directed set. For each $F_\alpha \in \mathscr{F}$, select a point $x_\alpha \in F_\alpha$. Then $N : D \to X$ defined by $N(\alpha) = x_\alpha$ is clearly a net.

The proof of the next theorem is trivial.

(11.D.3) *Theorem.* A filter $\mathscr{F}$ is eventually in a set $A \subset X$ if and only if the net $N_\mathscr{F}$ is eventually in A; consequently, if X is a topological space and $\mathscr{F}$ converges to $x \in X$, so must $N_\mathscr{F}$.

One aesthetic mishap occurs, however.

(11.D.4) *Exercise.* Show that a refinement of the filter $\mathscr{F}$ associated with a net N is not necessarily a subnet of N.

PROBLEMS

Sections A and B

1. Suppose that X and Y are topological spaces. Show that a function $f : X \to Y$ is continuous if and only if whenever a net $\{x_\alpha\}$ converges to a point x^*, $\{f(x_\alpha)\}$ converges to $f(x^*)$.

2. Show that (11.A.3) is valid when sequences are replaced by nets.
3. A net $\{x_\alpha\}$ in X is a *universal net* if and only if for each $A \subset X$, $\{x_\alpha\}$ is eventually either in A or in $X \setminus A$. Show that if a universal net is frequently in a set, then it is eventually in the set.
4. Show that a universal net converges to each of its cluster points.
5. Show that a space X is compact if and only if every universal net in X converges to an element in X.
6. Show that every subnet of a universal net is a universal net.
7. Suppose that $\{X_\alpha \mid \alpha \in \Lambda\}$ is a family of topological spaces and that $X = \prod_{\alpha \in \Lambda} X_\alpha$. Show that a net $\{z_\beta \mid \beta \in \psi\}$ in X converges to a point $x^* \in X$ if and only if $\{p_\alpha(z_\beta)\}_{\beta \in \psi}$ converges to $p_\alpha(x^*)$ for each projection map p_α.
8. Suppose that (N,D) is a net in a space X, and that A is the set of cluster points of (N,D). Show that A is closed.

Section C

1. If $\mathcal{F}$ is a filter, then the *adherence of* $\mathcal{F}$, $\mathrm{Adh}(\mathcal{F})$ is defined to be $\bigcap_{F \in \mathcal{F}} \bar{F}$.

   Show that if a filter $\mathcal{F}$ in X converges to a point x, then $x \in \mathrm{Adh}(\mathcal{F})$.
2. Show that a T_2 space is compact if and only if for each filter $\mathcal{F}$ in X, $\mathrm{Adh}(\mathcal{F}) \neq \varnothing$.
3. Show that the collection of subsets of $\mathbf{R}^1$, each of which contain an interval of the form $(0,\infty)$, is a filter.
4. Show that the family of all subsets of $\mathbf{R}^1$ whose complements are finite is a filter.
5. Suppose that $\mathcal{F}$ is a filter in a set X. A subset $\mathcal{S} \subset \mathcal{F}$ is a *subbase* for $\mathcal{F}$ if each set in $\mathcal{F}$ contains the intersection of a finite number of sets in $\mathcal{S}$. Show that if $\mathcal{G}$ is a collection of subsets of X with the finite intersection property, then $\mathcal{G}$ is a subbase for some filter $\mathcal{F}$ in X.
6. Suppose that $\mathcal{F}$ is an ultrafilter in a set X.
   (a) Show that $\bigcap_{F \in \mathcal{F}} F$ consists of at most one point.
   (b) Show that if $\bigcap_{F \in \mathcal{F}} F = \{x\}$, then $\mathcal{F} = \{A \subset X \mid x \in A\}$.
7. Show that a space is T_2 if and only if each filter in X converges to at most one point. Find an example of a filter that has no cluster points.
8. Suppose that (X,d) is a pseudometric space and $\mathcal{F}$ is a filter in X. Show that $\mathcal{F}$ converges to some point $x \in X$ if and only if, for each $\varepsilon > 0$, there is an $F \in \mathcal{F}$ such that diam $F < \varepsilon$.
9. If $\mathcal{F}$ is a filter in a space X, denote the set of points of convergence of $\mathcal{F}$ by $\mathrm{Lim}\ \mathcal{F}$. Show that if $\mathcal{F}$ is an ultrafilter in a space X, then $\mathrm{Adh}(\mathcal{F}) = \mathrm{Lim}(\mathcal{F})$.

10. Show that a neighborhood filter of a point x is an ultrafilter if and only if x is an isolated point.

11. Prove or disprove: Suppose that X is a topological space and that $A \subset X$. Then a point $x \in X$ is an accumulation point of A if and only if there is a sequence $\{x_n\}$ in $A \cap (X \setminus \{x\})$ that converges to x.

12. Show that a filter $\mathscr{F}$ in a set X is an ultrafilter in X if and only if whenever $A \cap F \neq \varnothing$ for each $F \in \mathscr{F}$, then $A \in \mathscr{F}$.

13. Suppose that $\mathscr{F}$ is a filter in a set X such that whenever $A \cup B \in \mathscr{F}$, then either $A \in \mathscr{F}$ or $B \in \mathscr{F}$. Show that $\mathscr{F}$ is an ultrafilter.

14. Find an example of a filter that is not an ultrafilter.

15. Show that the intersection of a family of filters on a set X is a filter on X.

16. Determine whether or not a neighborhood filter is an ultrafilter.

17. Show that if $(X, \mathscr{T})$ is a completely regular space, then there is a uniform structure $\mathscr{U}$, for X such that $\mathscr{T} = T_{\mathscr{U}}$ (see Chapter 4, Problem D-18). [Hint: Consider the filter $\mathscr{U}$ with subbase $\mathscr{S} = \{U_{f,\varepsilon} \,|\, f : X \to [0,1],\ \varepsilon > 0\}$ where f is continuous and $U_{f,\varepsilon} = \{(x,y) \,|\, |f(x) - f(y)| < \varepsilon\}$.]

Section D

1. Show that a net associated with an ultrafilter is a universal net.

2. Show that the filter generated by a universal net is an ultrafilter.

3.* Show that every net has a universal subnet.

4. Show that a point x is a cluster point of a net $\{x_\alpha\}$ if and only if x is a cluster point of the filter associated with x.

5. Suppose that $N = \{x_\alpha\}_{\alpha \in \Lambda}$ is a net in X and $\mathscr{F}$ is the associated filter base. Let $\mathscr{G} = \{G_\beta\}_{\beta \in \Psi}$ be a refinement of $\mathscr{F}$. Show that the following construction leads to a subnet of N. Order Ψ by declaring $\beta_1 \leq \beta_2$ if and only if $G_{\beta_2} \subset G_{\beta_1}$. Rename the points x_α as y_β's, requiring only that x_α cannot be named y_β unless it belongs to G_β. Show that $\{y_\beta\}_{\beta \in \Psi}$ is a subnet of N.

Chapter 12

THE ALGEBRAIZATION OF TOPOLOGY

In this chapter, we approach the interstices between the sublime realm of topology and the nether world of algebra. Hitherto, we have tackled topological problems by imposing one or more conditions on the topological structure. In algebraic topology, one tries to answer topological questions by translating them into algebraic ones (which may or may not be easier to resolve). A way will be found to assign groups to topological spaces and to assign group homomorphisms to continuous functions in such a manner that composition is preserved and that identity homomorphisms are associated with identity maps. Actually, there are a number of ways of doing this, but we restrict our attention to what is perhaps the most geometrically intuitive method.

Algebraic topology is one of the more active and difficult areas of mathematics, and our presentation here only gives some indication of the profundity of the subject matter. Once adequate topological and algebraic machinery is set up, an extraordinary variety of deep and interesting results may be obtained.

A. THE FUNDAMENTAL GROUP (THE FIRST HOMOTOPY GROUP)

We have tried to stress throughout the book that a central problem in topology is that of determining whether or not two spaces are homeomorphic. Invariant properties such as connectedness, compactness, metrizability, etc.

are a definite aid in attacking this problem, but it will soon be apparent that
there is a need for far more powerful tools than these. For instance, how
might one distinguish topologically between an open disk and an open disk
with the center removed?

 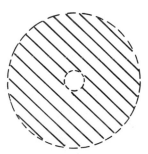

Although both spaces are connected, locally connected, locally compact,
metrizable, etc., they do not appear to be homeomorphic. At this juncture,
the reader might find it edifying to attempt to prove for himself that the
spaces are indeed topologically distinct. As we shall soon discover, the resolution of this problem becomes trivial when appropriate tools are available.

(12.A.1) Definition.   Suppose that X is a topological space and that x_0 is a
point in X. The family $\mathscr{L}$ of *loops* in X based at x_0 is defined to be the set of
all paths in X that start and end at x_0, i.e., $\mathscr{L} = \{f \mid f : \mathbf{I} \to X, f$ is continuous
and $f(0) = x_0 = f(1)\}$. If $x_0 \in X$, then the constant loop $c_{x_0} : \mathbf{I} \to X$ is
defined by $c_{x_0}(t) = x_0$ for all $t \in \mathbf{I}$.

From $\mathscr{L}$ it is frequently possible to extract essential information about
the space X. For example, suppose that X is the open disk with the center
removed and Y is the open disk with the center intact. Let x_0 and y_0 be points
in X and Y, respectively. In X consider the loops ℓ_1 and ℓ_2 in the figure that
follows. Note that ℓ_1 can be "shrunk" (or pulled back) to x_0 without any
trouble; however, ℓ_2 becomes hung up on the hole if any such shrinking is
attempted. In Y, any loop based at y_0 can easily be shrunk to y_0. Thus, it
would appear that a distinction between these two spaces might be based on
the behavior of certain loops.

What does all this have to do with groups? Suppose that for each space
X, the loops based at a point x_0 could in some manner be considered as elements of a group. Then in the preceding examples, the group element associated with the loop ℓ_2 would apparently have no counterpart in the group
derived from the loops in Y based at y_0. Consequently, the group associated
with X would differ from that associated with Y, and if it could be established

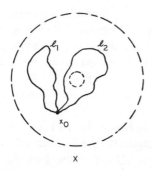

that homeomorphic spaces yield isomorphic groups, then one could conclude that X and Y are not homeomorphic. Thus, the primary goal of this section is to find a reasonable way of assigning groups to topological spaces.

It turns out that $\mathscr{L}$ has too many loops for our purposes. For example, in the next figure, the loop ℓ_3 provides little information about X that could not have been obtained from ℓ_2. Consequently, we will want to consider these loops as being essentially the same, or equivalent. Observe that ℓ_3 could be pushed or "deformed continuously" into ℓ_2 without moving the base point x_0. Such "deformations" are given mathematical significance in the following definitions.

(12.A.2) Definition. A *topological pair* is an ordered pair (X,A), where X is a topological space and A is a subspace of X. If (X,A) and (Y,B) are topological pairs and $f : X \to Y$ is a function such that $f(A) \subset B$, then we write $f : (X,A) \to (Y,B)$.

(12.A.3) Definition. Suppose that (X,A) and (Y,B) are topological pairs and that f and g are continuous functions from (X,A) into (Y,B). Then f *is*

homotopic to g relative to A if and only if there is a continuous function $F : (X \times I, A \times I) \to (Y,B)$ such that for each $x \in X$, $F(x,0) = f(x)$, $F(x,1) = g(x)$, and for each $a \in A$, $F(a,t) = f(a) = g(a)$. In this case, we write $f \sim g(\text{mod } A)$ or $f \underset{A}{\sim} g$. The map F is called a *relative homotopy* between f and g. If $A = \varnothing$, then we say that *f is homotopic to g* (denoted $f \sim g$).

If $f : X \to Y$ is homotopic to a constant map, then f is *null homotopic*.

(12.A.4) Notation. Suppose that $F : (X \times I, A \times I) \to (Y,B)$ is a relative homotopy between functions f and g and that $t \in I$. Then for each $t \in I$, define a map $F_t : (X,A) \to (Y,B)$ by $F_t(x) = F(x,t)$. Note that $F_0 = f$ and $F_1 = g$.

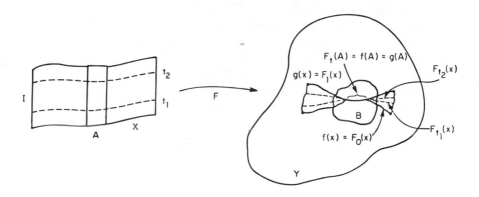

(12.A.5) Notation. If (X,A) and (Y,B) are topological pairs, then $\mathbf{C}((X,A), (Y,B))$ will denote $\{f : (X,A) \to (Y,B) \mid f \text{ is continuous}\}$.

(12.A.6) Theorem. Relative homotopy (mod A) is an equivalence relation in $\mathbf{C}((X,A), (Y,B))$.

Proof. Suppose that $f,g,h \in \mathbf{C}((X,A), (Y,B))$.

Reflexivity. Define $F : (X \times I, A \times I) \to (Y,B)$ by $F(x,t) = f(x)$ for all x and t.

Symmetry. Suppose that $f \underset{A}{\sim} g$, and let F be the homotopy between f and g. Define $G : (X \times I, A \times I) \to (Y,B)$ by $G(x,t) = F(x, 1 - t)$. Then $G_0 = g$, $G_1 = f$, and $G_t(a) = f(a)$ for all $t \in I$ and $a \in A$; therefore, $g \underset{A}{\sim} f$.

Transitivity. Suppose that $f \underset{A}{\sim} g$ and $g \underset{A}{\sim} h$ where the homotopies are given by F and G, respectively. Define a homotopy $H(X \times I, A \times I) \to (Y,B)$ by

$$H(x,t) = \begin{cases} F(x,2t) & \text{for} \quad 0 \le t \le \frac{1}{2} \\ G(x, 2t - 1) & \text{for} \quad \frac{1}{2} \le t \le 1 \end{cases}$$

Then H is the desired homotopy between f and h. The continuity of H follows from the map gluing theorem.

(12.A.7) *Exercise.* Suppose that (X,A), (Y,B), and (Z,C) are topological pairs and that $f \underset{A}{\simeq} g$, where $f,g \in C((X,A), (Y,B))$ and $h \in C((Y,B), (Z,C))$. Show that $hf \underset{A}{\simeq} hg$.

We now return to loops by restricting our attention to the following special situation. Suppose $X = \mathbf{I}$, $A = \{0,1\}$, Y is an arbitrary space, and $B = \{y_0\}$, where $y_0 \in Y$. The preceding theorem yields an equivalence relation on the family $\mathscr{L}$ of loops based at y_0. (Note that if $f \underset{\{0,1\}}{\simeq} g$, then geometrically one may think of the homotopy as moving the loop f continuously onto g without ever disturbing the base point.) If $f \in \mathscr{L}$, then $[f]$ will denote the corresponding equivalence class and $\mathscr{G}$ will denote the family of these equivalence classes. We now define a binary operation on $\mathscr{G}$, and then show that this operation converts $\mathscr{G}$ into a group.

Suppose that $[f],[g] \in \mathscr{G}$. A way must be found of "multiplying" $[f]$ and $[g]$ to obtain an element $[f] \cdot [g]$ also in $\mathscr{G}$. This is easily done. The loop f starts at y_0, runs around a circuit as the parameter $t \in \mathbf{I}$, journeys from 0 to 1, and then returns to y_0; the same, of course, holds for g. A loop that represents $[f] \cdot [g]$ will look like f on the first half of $\mathbf{I}$ (even though it must race with the parameter twice as fast as f did to arrive back at y_0 when $t = \frac{1}{2}$), and then becomes an accelerated version of g on the second half of $\mathbf{I}$. This new loop is denoted by $f * g$. More precisely, we have the following definition.

(12.A.8) *Definition.* If f and $g \in \mathscr{L}$, then $f * g$ is the loop defined by

$$(f * g)(t) = \begin{cases} f(2t) & \text{for} \quad 0 \le t \le \frac{1}{2} \\ g(2t - 1) & \text{for} \quad \frac{1}{2} \le t \le 1 \end{cases}$$

Note that at $t = \frac{1}{2}$, $f(2t) = g(2t - 1) = y_0$, and hence $f * g$ is continuous by the map gluing theorem. The next theorem shows that this definition induces a well defined operation in $\mathscr{G}$, namely $[f] \cdot [g] = [f * g]$. We will use the customary abbreviation and denote $[f] \cdot [g]$ by $[f][g]$.

(12.A.9) *Theorem.* Suppose that $f,g,f',g' \in \mathscr{L}, f \underset{\{0,1\}}{\simeq} f'$, and $g \underset{\{0,1\}}{\simeq} g'$. Then $[f][g] = [f'][g']$.

Proof. It suffices to show that $f * g \underset{\{0,1\}}{\simeq} f' * g'$. By hypothesis, there are homotopies $F : I \times I \to Y$ and $G : I \times I \to Y$ such that $F_0 = f$, $F_1 = f'$, $G_0 = g$, and $G_1 = g'$. Furthermore, $F(0,t) = F(1,t) = G(1,t) = y_0$. Define $H : I \times I \to Y$ by setting

$$H(s,t) = \begin{cases} F(2s,t) & \text{if} \quad 0 \le s \le \tfrac{1}{2} \\ G(2s - 1, t) & \text{if} \quad \tfrac{1}{2} \le s \le 1 \end{cases}$$

It is easily checked that H is the desired relative homotopy between $f * g$ and $f' * g'$.

Frequently, it is necessary to "multiply" arbitrary paths together. This can be done, provided that the paths match up properly.

(12.A.10) Definition. Suppose that $f : I \to X$ and $g : I \to X$ are continuous functions satisfying $f(1) = g(0)$. Then the product $f * g : I \to X$ is defined by

$$f * g(t) = \begin{cases} f(2t) & \text{if} \quad 0 \le t \le \tfrac{1}{2} \\ g(2t - 1) & \text{if} \quad \tfrac{1}{2} \le t \le 1 \end{cases}$$

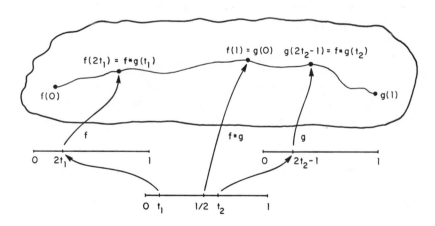

(12.A.11) Exercise. State and prove the analog of (12.A.9) for matching paths.

(12.A.12) Exercise. Suppose that $f,g : I \to X$ are continuous functions such that $f(1) = g(0)$. Let $h : X \to Y$ be continuous. Show that $[h(f * g)] = [hf * hg]$ (paths $p_1, p_2 : I \to Y$ are defined to be equivalent if and only if $p_1 \underset{\{0,1\}}{\sim} p_2$).

Theorem (12.A.9) assures a well-defined operation for $\mathscr{G}$. The rest of this section will be devoted to finding inverses, identity elements, etc., in order to show that $(\mathscr{G},\cdot)$ is indeed a group.

The homotopies to follow are rather tedious to untangle analytically. However, contemplation of the accompanying figures should enable the reader to penetrate some of the mysteries behind the rather arcane formulae that appear in the next few theorems. In any case, the reader is advised to check the given homotopies for their relevance and their accuracy.

It is not difficult to guess which equivalence class will act as an identity element. Throughout the remainder of this section, the reader should remember that $\mathscr{L}$ is the set of loops in Y based at y_0.

(12.A.13) Theorem. Let $e = [c_{y_0}]$ be the equivalence class represented by the constant map $c_{y_0} : \mathbf{I} \to Y$. Suppose that $f \in \mathscr{L}$. Then $(c_{y_0} * f)_{\{\widetilde{0,1}\}} f$ and $(f * c_{y_0})_{\{\widetilde{0,1}\}} f$, and hence $e[f] = f$ and $[f]e = [f]$.

Proof. Define a map $H : \mathbf{I} \times \mathbf{I} \to Y$ by

$$H(s,t) = \begin{cases} c_{y_0}(s) & \text{if} \quad 0 \le t \le -2s + 1 \\ f\left(\dfrac{2s + t - 1}{t + 1}\right) & \text{if} \quad -2s + 1 \le t \le 1 \end{cases}$$

As usual, the map gluing theorem rescues continuity.

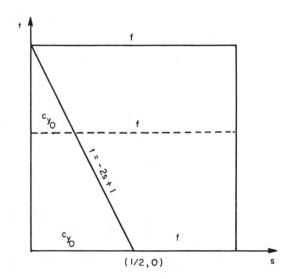

Note that $H_0 = c_{y_0} * f$, $H_1 = f$, and $H(0,t) = H(1,t) = y_0$ for all $t \in \mathbf{I}$.

A similar homotopy may be displayed to demonstrate that $f \underset{\{0,1\}}{\simeq} f * c_{y_0}$.

(12.A.14) Theorem. Suppose that $f, g, h \in \mathcal{L}$. Then $([f][g])[h] = [f]([g][h])$, and thus $(\mathcal{G}, \cdot)$ is associative.

Proof. It suffices to show that $(f * g) * h \underset{\{0,1\}}{\simeq} f * (g * h)$.
Define

$$H(s,t) = \begin{cases} f\left(\dfrac{4s}{2-t}\right) & \text{for} \quad 0 \le s \le \dfrac{2-t}{4} \\[2ex] g(4s - 2 + t) & \text{for} \quad \dfrac{2-t}{4} \le s \le \dfrac{3-t}{4} \\[2ex] h\left(\dfrac{4s - 3 + t}{1 + t}\right) & \text{for} \quad \dfrac{3-t}{4} \le s \le 1 \end{cases}$$

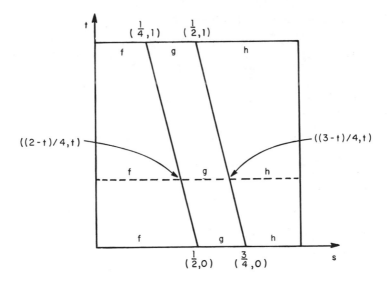

Note that $H_0 = f * (g * h)$ and $H_1 = (f * g) * h$.

(12.A.15) Exercise. State and prove analogs of the previous two theorems when f, g, and h are paths that match properly.

Finally, inverses for elements of $(\mathcal{G}, \cdot)$ must be found. This is done with the aid of the well-known principle that if you walk down a path in one direction and then retrace your steps, then you might as well not have left in the first place.

(12.A.16) **Definition.** Suppose that $f : I \to Y$ is a path in Y. Then the *reverse* of f, f^r, is defined by $f^r(t) = f(1 - t)$.

(12.A.17) **Theorem.** If $f \in \mathcal{L}$, then $[f][f^r] = [f^r][f] = [c_{y_0}] = e$.

Proof. It is sufficient to show that $(f * f^r) \underset{\{0,1\}}{\sim} c_{y_0}$ and $(f^r * f) \underset{\{0,1\}}{\sim} c_{y_0}$. A glance at the next figure reveals the following homotopy:

$$H(s,t) = \begin{cases} c_{y_0}(s) & \text{for} \quad 0 \le s \le \dfrac{t}{2} \\[2mm] f(2s - t) & \text{for} \quad \dfrac{t}{2} \le s \le \dfrac{1}{2} \\[2mm] f^r(2s + t - 1) & \text{for} \quad \dfrac{1}{2} \le s \le \dfrac{2 - t}{2} \\[2mm] c_{y_0}(s) & \text{for} \quad \dfrac{2 - t}{2} \le s \le 1 \end{cases}$$

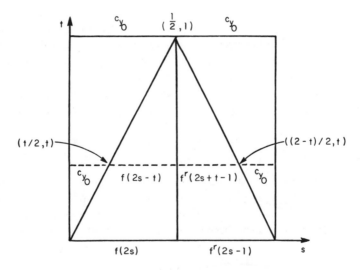

Note that $H_0 = (f * f^r)$ and $H_1 = c_{y_0}$. A similar figure and homotopy can be given to establish that $f^r * f \underset{\{0,1\}}{\sim} c_{y_0}$.

The previous theorems may be combined to yield the principal result.

(12.A.18) **Theorem.** Suppose that Y is a topological space and that $y_0 \in Y$. Let $\mathcal{L}$ be the collection of loops based at y_0 and let $\mathcal{G}$ be the set of equivalence classes obtained from $\mathcal{L}$ by declaring two loops equivalent if

and only if they are homotopic mod$\{0,1\}$. Then $\mathscr{G}$ with the binary operation $\cdot$ defined by $[f] \cdot [g] = [f * g]$ whenever $f,g \in \mathscr{L}$ is a group.

The group $\mathscr{G}$ is called the *fundamental group* of Y based at y_0 (or alternately the *first homotopy group* of Y based at y_0); we follow the usual custom of denoting $\mathscr{G}$ by $\pi_1(Y,y_0)$.

The elegance of forthcoming proofs that involve the fundamental group is not at all presaged by the somewhat tawdry details used to show that $\pi_1(Y,y_0)$ is a group.

B. ELEMENTARY PROPERTIES OF THE FUNDAMENTAL GROUP

(12.B.1)   Definition.   Suppose that X is a space and that $x_0 \in X$. Then the pair (X,x_0) is called a *pointed space*. The point x_0 is called either the *base point* or the *preferred point* of (X,x_0).

As we have just seen, it is a relatively easy matter to associate a group with each pointed topological space. However, it is quite another problem to actually recognize one of these groups; in fact, except for a number of special cases, this has not been successfully done.

Before an attempt is made to calculate a fundamental group, we first show that in path connected spaces, the role of the base point is minimal.

(12.B.2)   Theorem.   Suppose that X is a path connected space and that $x_0,x_1 \in X$. Then $\pi_1(X,x_0)$ and $\pi_1(X,x_1)$ are isomorphic.

Proof.   The proof is straightforward. Suppose that x_0 and x_1 are in X, and let $k : \mathbf{I} \to X$ be a path from x_0 to x_1. For each loop based at x_0, we want to associate a loop based at x_1, and vice versa. Let $f : \mathbf{I} \to X$ be an arbitrary loop at x_0, and observe that there is a natural choice for a corresponding loop

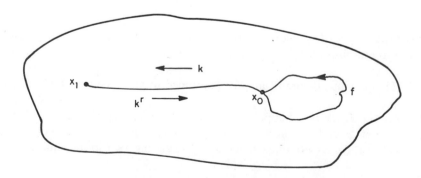

based at x_1, namely $k^r * f * k$ (walk backwards along k to go from x_1 to x_0, then proceed along f, and finally return to x_1 via k).

Define functions $K_\# : \pi_1(X,x_0) \to \pi_1(X,x_1)$ by $K_\#([f]) = [k^r * f * k]$ and $K_\#^r : \pi_1(X,x_1) \to \pi_1(X,x_0)$ by $K_\#^r([g]) = [k * g * k^r]$. The reader should verify that these are well-defined homomorphisms and are inverse to each other (use (0.C.11)).

Frequently, if X is path connected, we will denote the fundamental group of (X,x_0) by $\pi_1(X)$.

In the next theorem, we see that the assignment of fundamental groups to topological spaces is well behaved with respect to finite products.

(12.B.3) Theorem. Suppose that (X,x_0) and (Y,y_0) are pointed topological spaces. Then $\pi_1(X \times Y, (x_0,y_0))$ is isomorphic to $\pi_1(X,x_0) \times \pi_1(Y,y_0)$, where the symbol $\times$ represents the direct product of groups.

Proof (Outline). For each loop in $X \times Y$ with base point (x_0,y_0), corresponding loops must be located in X and Y. This is accomplished via the projection maps p_X and p_Y, which send $X \times Y$ onto X and Y, respectively. If $f : I \to X \times Y$ is a loop in $X \times Y$, then $p_X f$ and $p_Y f$ are loops in X and Y. Define a function $\psi : \pi_1(X \times Y, (x_0,y_0)) \to \pi_1(X,x_0) \times \pi_1(Y,y_0)$ by $\psi([f]) = ([p_X f],[p_Y f])$. It is not difficult to verify that this map is well defined.

To see that ψ is a homomorphism, observe that $\psi([f][g]) = \psi([f * g]) = ([p_X(f * g)], [p_Y(f * g)]) = ([p_X f * p_X g], [p_Y f * p_Y g]) = ([p_X f][p_X g], [p_Y f][p_Y g]) = \psi([f])\psi([g])$. The third quality follows from (12.A.12).

To see that ψ is one to one, suppose that $\psi([f]) = \psi([g])$. Then $([p_X f],[p_Y f]) = ([p_X g],[p_Y g])$, which implies that $[p_X f] = [p_X g]$ and $[p_Y f] = [p_Y g]$. Let H and K be the corresponding homotopies. Define $F : I \times I \to X \times Y$ by $F(s,t) = (H(s,t), K(s,t))$. It is easily checked that F is a suitable homotopy between f and g, and hence $[f] = [g]$.

The map ψ is onto, since if $([f],[g]) \in \pi_1(X,x_0) \times \pi_1(Y,y_0)$, then the equivalence class determined by the following loop in $X \times Y$

$$k(t) = \begin{cases} (f(2t), y_0) & 0 \le t \le \tfrac{1}{2} \\ (x_0, g(2t - 1)) & \tfrac{1}{2} \le t \le 1 \end{cases}$$

is mapped onto $([f],[g])$.

Rigorous calculations of fundamental groups of some rather simple topological spaces are given in the next section. Before succumbing to such formalities, however, we attempt to discover in an intuitive fashion the fundamental group of $\mathscr{S}^1$. (Two complete but distinct proofs of the fact that $\pi_1(\mathscr{S}^1,s_0)$ is infinite cyclic are given in Chapter 13 and 14.)

A fairly obvious candidate for the generator of $\pi_1(\mathscr{S}^1,s_0)$ is the equivalence class [α] determined by the loop shown in the next figure, where α wraps once around $\mathscr{S}^1$ in the direction indicated.

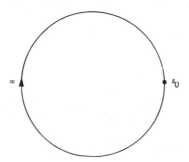

If a loop β is wrapped n times around $\mathscr{S}^1$ in the direction of α, it is reasonable to expect that the corresponding equivalence class in $\pi_1(\mathscr{S}^1, s_0)$ is $[α]^n$. The inverse of [β] can be derived from the loop that runs around the circle n times in the opposite direction. Of course, not all loops are so simple, since many maps any reverse themselves back and forth repeatedly. Nevertheless, the reader perhaps can convince himself that no matter how involved a loop is, it is probably homotopic to a loop that wraps around $\mathscr{S}^1$ a finite number of times in a single direction, and thus is equal to $[α]^n$ for some integer n. Consequently, it seems reasonable that $\pi_1(\mathscr{S}^1, s)$ is isomorphic to the group of integers under addition.

(12.B.4) Exercise. Assume that $\pi_1(\mathscr{S}^1, s) = \mathbf{Z}$. Find the fundamental group of the torus.

C. CONTINUOUS FUNCTIONS AND HOMOMORPHISMS

One of the most exciting and beautiful links between topology and algebra is the simple way that continuous functions may be converted into homomorphisms. Suppose that $f : (X, x_0) \to (Y, y_0)$ is continuous. If α is a loop in X based at x_0, then $fα$ is a loop in Y based at y_0. Furthermore, it is easily seen that if $α \sim α'$, then $fα \sim fα'$ (12.A.7). Thus, f induces a well-defined map $f_* : \pi_1(X, x_0) \to \pi_1(Y, y_0)$ defined by $f_*([α]) = [fα]$. We say that f_* is the homomorphism induced by f. The ease of establishing the following assertions is inversely proportional to their importance.

(12.C.1) Exercise. Suppose that $f : (X, x_0) \to (Y, y_0)$, $g : (X, x_0) \to (Y, y_0)$

and $h : (Y,y_0) \to (Z,z_0)$ are continuous maps. Show that:

(a) f_* is a homomorphism;

(b) $f_*([\alpha]^{-1}) = (f_*([\alpha]))^{-1}$ for $[\alpha] \in \pi_1(X,x_0)$;

(c) $(hf)_* = h_*f_*$;

(d) if $f \underset{\{x_0\}}{\simeq} g$, then $f_* = g_*$;

(e) if $id : (X,x_0) \to (X,x_0)$ is the identity map, then $id_* : \pi_1(X,x_0) \to \pi_1(X,x_0)$ is the identity homomorphism.

A key concept of this section is the following.

(12.C.2) Definition. Two topological spaces X and Y are *homotopically equivalent* if and only if there are maps $f : X \to Y$ and $g : Y \to X$ such that $fg \sim id_Y$ and $gf \sim id_X$. Topological pairs (X,A) and (Y,B) are *homotopically equivalent* if and only if there are maps $f : (X,A) \to (Y,B)$ and $g : (Y,B) \to (X,A)$ such that $fg \sim id_Y(\text{mod } B)$ and $gf \sim id_X(\text{mod } A)$. We call f (or g) a *homotopy equivalence* and say that f is a *homotopy inverse* of g (or g is a homotopy inverse of f).

(12.C.3) Examples.

1. Homeomorphic spaces are clearly homotopically equivalent.

2. Suppose that D is the unit disk minus its center and that $E = \mathscr{S}^1$. Let $f : E \to D$ be the inclusion map, and $g : D \to E$ be the obvious projection map. Then it is easy to see that $gf = id_E$ and $fg \sim id_D$; hence, D and E are homotopically equivalent, and f and g are homotopy equivalences.

We are now able to prove a result that implies that homeomorphic spaces possess isomorphic fundamental groups, or equivalently, if the fundamental groups of two spaces differ, the spaces cannot be homeomorphic.

(12.C.4) Theorem. If pointed spaces (X,x_0) and (Y,y_0) are homotopically equivalent, then $\pi_1(X,x_0)$ and $\pi_1(Y,y_0)$ are isomorphic (cf. Problem C-17).

Proof. Let $f : (X,x_0) \to (Y,y_0)$ and $g : (Y,y_0) \to (X,x_0)$ be homotopy inverses. Then $f_*g_* = (fg)_* = id_{\pi_1(Y,y_0)}$, $g_*f_* = (gf)_* = id_{\pi_1(X,x_0)}$, and consequently each of f_* and g_* is an isomorphism.

(12.C.5) Corollary. If $h : (X,x_0) \to (Y,y_0)$ is a homeomorphism, then $h_* : \pi_1(X,x_0) \to \pi_1(Y,y_0)$ is an isomorphism.

A large number of spaces have trivial fundamental groups; included among them are the contractible spaces,

(12.C.6) Definition. A space X is *contractible* if and only if there is a point x_0 in X such that the identity map $id : X \to X$ is homotopic to the constant map $c_{x_0} : X \to X$. In this case, X is said to be contractible to x_0.

(12.C.7) Example. Convex subsets of $\mathscr{E}^n$ are contractible. Suppose that $A \subset \mathscr{E}^n$ is convex. Let $x_0 \in A$. Then $F : A \times \mathbf{I} \to A$, defined by $F(a,t) = at + (1 - t)x_0$ is the desired homotopy.

In (12.C.7), the contraction homotopy does not move the point x_0. That this is not always the case may be seen in the following example.

(12.C.8) Example. For each $n \in \mathbf{Z}^+$, let $A_n = \{((1/n),y) \in \mathscr{E}^2 \mid 0 \le y \le 1\}$ and set $A_0 = \{(0,y) \in \mathscr{E}^2 \mid 0 \le y \le 1\}$. Then $X = (\mathbf{I} \times \{0\}) \cup (\bigcup_{n=0}^{\infty} A_n)$ is contractible to $p = (0,1) \in \mathscr{E}^2$, although p must move during the contraction.
Some basic properties of contractible spaces are given next.

(12.C.9) Theorem.

(i)   Contractible spaces are path connected.
(ii) If X is contractible, Y is a topological space, and $f,g : Y \to X$ are continuous, then f and g are homotopic.
(iii) A space X is contractible if and only if X is homotopically equivalent to a point.

Proof.
(i)   Suppose that X is contractible to x_0 and that $x_1 \in X$. Let $H : X \times \mathbf{I} \to X$ be a homotopy such that $H_0 = id$ and $H_1 = c_{x_0}$ where c_{x_0} is defined by $c_{x_0}(x) = x_0$. Then $f : \mathbf{I} \to X$ defined by $f(t) = H(x_1,t)$ is a path from x_1 to x_0. It now follows easily from (2.C.3) that X is path connected.
(ii) Let $H : X \times \mathbf{I} \to X$ be the homotopy given in (i). Define $\psi : X \times \mathbf{I} \to Y$ by $\psi(x,t) = H(f(x),t)$. Then ψ is a homotopy between f and c_{x_0}. Hence, all maps from Y into X are homotopic to the constant function c_{x_0}.
(iii) Suppose that X is contractible to $x_0 \in X$. Let $g : \{x_0\} \to X$ be the inclusion map and let $f : X \to \{x_0\}$ be the constant function. Then $gf \sim id_X$ and $fg = id_{\{x_0\}}$. Thus, X and $\{x_0\}$ are homotopically equivalent.
Conversely, if X and $\{p\}$ are homotopically equivalent, where $p \in X$, there are continuous maps $f : X \to \{p\}$ and $g : \{p\} \to X$ such that $fg \sim id_{\{p\}}$ and $gf \sim id_X$. Then the constant map gf is homotopic to id_X, and hence X is contractible.

(12.C.10) Definition. A path connected space X is *simply connected* if and only if $\pi_1(X,x_0) = \{e\}$.

It is not completely obvious that contractible spaces are simply connected (especially in view of Example (12.C.8) above). Suppose that X is contractible to $x_0 \in X$. Then certainly by the previous theorem, every loop in X based at x_0 is homotopic to $c_{x_0} : I \to X$; however, there is no guarantee that the base point will not be moved during the homotopy. The simple connectedness of contractible spaces follows from the next theorem (it also follows from problem C.17).

(12.C.11)   Theorem. Suppose that X is a path connected space with the property that each continuous function $f : \mathscr{S}^1 \to X$ can be extended to the disk $\mathscr{B}^2$. Then X is simply connected. Conversely, if X is simply connected, then each continuous function $f : \mathscr{S}^1 \to X$ can be extended to the disk $\mathscr{B}^2$.

Proof. First note (or proceed to (12.C.13)) that if three sides of $I \times I$ are identified to a point then the resulting quotient space is homeomorphic to $\mathscr{B}^2$ (and the fourth side becomes $\mathscr{S}^1$). Let $x_0 \in X$ and let $g : I \to X$ be a loop based at x_0. Define $F : I \times \{0\} \to X$ by $F(s,0) = g(s)$. Identify in $I \times I$ the sides $I \times \{1\}$, $\{0\} \times I$, $\{1\} \times I$ to a point, and let B be the resulting quotient space with quotient map $P : I \times I \to B$. Define $f : \mathscr{S}^1 \to X$ as follows. Suppose that $\hat{s} \in \mathscr{S}^1$. Then there is an $s \in I$ such that $P(s,0) = \hat{s}$. Now set $f(\hat{s}) = g(s)$. Then by hypothesis, f may be extended to a map $G : B \to X$. Extend F to $I \times I$ by setting $F(s,t) = G(P(s,t))$. Then $F_0 = g$, $F_1 = c_{x_0}$, and $F(0,t) = F(1,t) = x_0$ for all $t \in T$. Therefore, $[g] = e$ and X is simply connected.

Conversely, suppose that X is simply connected. Let $f : \mathscr{S}^1 \to X$ be a continuous map. Define $\hat{f} : I \to X$ by $\hat{f}(t) = f((\cos 2\pi t, \sin 2\pi t))$. Since X is simply connected, there is a homotopy $H : I \times I \to X$ such that $H_0 = \hat{f}$, $H_1 = c_{x_0}$ where $x_0 = \hat{f}(0) = \hat{f}(1)$, and $H(0,t) = H(1,t) = x_0$ for all $t \in I$. It follows from (12.C.13) below that there is a continuous function $P : I \times I \to \mathscr{B}^2$ such that $P(t,0) = (\cos 2\pi t, \sin 2\pi t)$ for each $t \in I$, $P(I \times \{1\} \cup \{0\} \times I \cup \{1\} \times I) = (1,0)$, and furthermore, $P_{|\text{int}(I \times I)}$ is 1–1 and maps $\text{int}(I \times I)$ onto $\text{int } \mathscr{B}^2$. Define $F : \mathscr{B}^2 \to X$ by $F(z) = HP^{-1}(z)$. Then F is well defined and extends f continuously.

(12.C.12)   Corollary. If X is contractible to x_0, then $\pi_1(X,x_0) = \{e\}$.

Proof. Let $H : X \times I \to X$ be the contraction homotopy, where $H_0 = \text{id}$ and $H_1 = c_{x_0}$. Suppose that $f : \mathscr{S}^1 \to X$ is continuous. It suffices to find an extension of f to $\mathscr{B}^2$. Define $G : \mathscr{S}^1 \times I \to X$ by $G(x,t) = H(f(x),t)$. Let B be the space obtained by identifying the subset $\mathscr{S}^1 \times \{1\}$ of $\mathscr{S}^1 \times I$ to a point, and note that B is homeomorphic with $\mathscr{B}^2$. Let $h : \mathscr{B}^2 \to B$ be a homeomorphism which "is the identity" on $\mathscr{S}^1$. Let $P : \mathscr{S}^1 \times I \to B$ be the quotient map. Extend f to $F : \mathscr{B}^2 \to X$ by defining $F(x) = GP^{-1}h(x)$. Observe that G

is well defined, since H collapses $X \times \{1\}$ to x_0. Thus, by (12.C.11), X is simply connected.

(12.C.13)   Exercise. Show that if three sides of $\mathbf{I} \times \mathbf{I}$ are identified to a point, the the resulting quotient space is homeomorphic to $\mathscr{B}^2$.

We now give an algebraic proof of the no retraction theorem (5.C.21).

(12.C.14)   Theorem. Suppose that S is a simple closed curve lying in a contractible subset X of $\mathscr{E}^2$. Then there is no retraction of X onto S.

Proof. Let $s \in S$ and suppose that $r : (X,s) \to (S,s)$ is a retraction. Let $i : (S,s) \to (X,s)$ be the inclusion map. Then $ri = id_S$, and consequently $r_* i_* = id_{\pi_1(S,s)}$. However, since X is contractible, $\pi_1(X,s) = \{e\}$. Thus, r_* maps the trivial group onto the infinite cyclic group $\pi_1(S,s)$, an obvious contradiction (see (13.B.3)).

(12.C.15)   Corollary. There is no retraction of a disk onto its boundary

(12.C.16)   Corollary (Brouwer Fixed Point Theorem). Suppose that $f : \mathscr{B}^2 \to \mathscr{B}^2$ is a continuous map of $\mathscr{B}^2$ onto itself. Then f has a fixed point.

Proof. Suppose that f has no fixed point; hence, $f(x) \neq x$ for each $x \in \mathscr{B}^2$. A retraction r of $\mathscr{B}^2$ onto its boundary is constructed as indicated in the figure. Each point x in $\mathscr{B}^2$ is projected onto the point in $Bd \ \mathscr{B}^2$ obtained by intersecting $Bd \ \mathscr{B}^2$ and the ray from $f(x)$ through x (why is r continuous?).

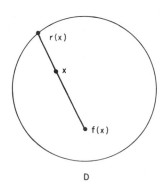

This of course contradicts (12.C.15).

Suppose that $f : (X,x_0) \to (Y,y_0)$ is a continuous function. Although the homomorphism $f_* : \pi_1(X,x_0) \to \pi_1(Y,y_0)$ is induced by f, f_* does not necessarily retain all of f's characteristics. For instance, even though the inclusion map $f : \mathcal{S}^1 \to \mathcal{E}^2$ is one to one, f_* is the trivial map. Furthermore, for any onto map $f : I \to \mathcal{S}^1$, the corresponding homomorphism f_* fails to be surjective. Nevertheless, in the following situation we see that f and f_* are more closely related.

(12.C.17) Theorem. Suppose that A is a retract of a space X and that $a \in A$. Let $r : (X,a) \to (A,a)$ be a retraction and $i : (A,a) \to (X,a)$ be the inclusion map. Then r_* is onto and i_* is one to one.

 Proof. Note that $ri = id_A$. Consequently, by (12.C.1), we have that $r_* i_* = id_{\pi_1(A,a)}$, and hence i_* must be one to one and r_* onto.

(12.C.18) Definition. A subspace A of a space X is a *deformation retract* of X if and only if there is a retraction $r : X \to A$ and a homotopy $H : X \times I \to X$ such that $H_0 = id$, $H_1 = ir$, and $H(a,t) = a$ for each $a \in A$ and $t \in I$.

 Note that a point x_0 in a space X is a deformation retract of X if and only if X is contractible to $x_0 (\mathrm{mod}\ x_0)$.

 An important property of deformation retracts is the following.

(12.C.19) Theorem. Suppose that A is a deformation retract of X and $x_0 \in A$. Then $\pi_1(X,x_0)$ is isomorphic to $\pi_1(A,x_0)$.

 Proof. Let H be the deformation retraction homotopy given in (12.C.18), and set $r = H_1$. Then by (12.C.17), r_* is a homomorphism onto. Since r is homotopic to the identity (mod A), we have $i_* r_* = id_{\pi_1(X,x_0)}$ and hence r_* is also one to one.

 Deformation retracts are used frequently in computing fundamental groups. For instance, it is easy to describe the necessary deformation retractions in the following examples.

(12.C.20) Examples.
 1. The fundamental group of an annulus is isomorphic to that of a circle.
 2. The fundamental group of a Moebius strip is **Z**.

(12.C.21) Exercise. Show that a disk D (open or closed) is not homeomorphic to D minus its center.

D. CATEGORIES AND FUNCTORS

Homotopy theory represents an example *par excellence* of how one can pass from one class of mathematical "things" (topological spaces and continuous functions) to another (groups and homomorphisms). As we shall see presently, this is merely a special case of a far broader theory involving categories (certain objects together with certain maps) and functors (ways of moving from category to category).

The reader in his previous studies has most likely encountered a variety of mathematical systems: vector spaces and linear transformations, groups and homomorphisms, topological spaces and continuous functions, etc. A category (to be defined below) may be viewed as a very general system with just enough properties to encompass all of the above systems (and many more) under its aegis.

(12.D.1)   Definition.   A *category* $(\mathfrak{O},\mathfrak{F})$ consists of a collection $\mathfrak{O}$ of sets called the *objects* of the category together with a collection $\mathfrak{F}$ of sets whose elements are called *morphisms* (or *mappings*) with the property that for each $A,B \in \mathfrak{O}$, there is a set $\mathrm{Hom}_{\mathfrak{O}}(A,B) \in \mathfrak{F}$, satisfying the following properties:

(i)   for each triple $A,B,C \in \mathfrak{O}$, there is a composite function $\circ$ which assigns to every pair of morphisms $f \in \mathrm{Hom}_{\mathfrak{O}}(A,B)$ and $g \in \mathrm{Hom}_{\mathfrak{O}}(B,C)$ a morphism $g \circ f \in \mathrm{Hom}_{\mathfrak{O}}(A,C)$;

(ii)   if $f \in \mathrm{Hom}_{\mathfrak{O}}(A,B)$, $g \in \mathrm{Hom}_{\mathfrak{O}}(B,C)$, $h \in \mathrm{Hom}_{\mathfrak{O}}(C,D)$, then $(f \circ g) \circ h = f \circ (g \circ h) \in \mathrm{Hom}_{\mathfrak{O}}(A,C)$;

(iii)   if $A \in \mathfrak{O}$, there is a morphism $1_A \in \mathrm{Hom}_{\mathfrak{O}}(A,A)$ such that if $f \in \mathrm{Hom}_{\mathfrak{O}}(A,B)$ and $g \in \mathrm{Hom}_{\mathfrak{O}}(B,A)$, then $f \circ 1_A = f$ and $1_A \circ g = g$.

(12.D.2)   Examples.
1.   Let $\mathfrak{T}$ be the class of all topological spaces. For each $A,B \in \mathfrak{T}$, set $\mathrm{Hom}_{\mathfrak{T}}(A,B) = \{f : A \to B \mid f \text{ is continuous}\}$. If $\mathfrak{F}$ is the class of all $\mathrm{Hom}_{\mathfrak{T}}(A,B)$ such that $A,B \in \mathfrak{T}$, then $(\mathfrak{T},\mathfrak{F})$ is a category. This is the category of topological spaces and continuous maps.

2.   Let $\mathfrak{T}_P$ be the class of all pointed topological spaces. For each $(A,a),(B,b) \in \mathfrak{T}_P$, let $\mathrm{Hom}_{\mathfrak{T}_P}((A,a),(B,b)) = \{f : (A,a) \to (B,b) \mid f \text{ is continuous}\}$. This is the category of pointed topological spaces and pointed, continuous maps.

3.   Let $\mathfrak{T}_R$ be the class of all topological pairs, and for each two such pairs, define $\mathrm{Hom}_{\mathfrak{T}_R}$ as in the previous two examples. This is the category of pairs of topological spaces and continuous mappings of pairs.

4.   Let $\mathfrak{G}$ be the class of all groups, and for each $G,H \in \mathfrak{G}$, define $\mathrm{Hom}_{\mathfrak{G}}(G,H)$ to be the set of all homomorphisms from G into H.

5. Let $\mathfrak{S}$ be the class of all topological spaces, and for each $A,B \in \mathfrak{S}$, let $\text{Hom}_{\mathfrak{S}}(A,B)$ be the set of homeomorphisms from A onto B. Compare with example 1, and note that in example 5 the map population is so severely restricted that there results a large number of empty Hom's.

6. Let Π be the class of all topological spaces, and $\text{Hom}_{\Pi}(A,B)$ be the set of equivalence classes under Homotopy. Then Π is called the homotopy category.

Functors are mappings between categories. The mapping is done in such a way that the structure is preserved, i.e., due respects are paid to the composition of functions.

(12.D.3) Definition. Suppose that $(\mathfrak{C},\mathfrak{F})$ and $(\mathfrak{D},\mathfrak{H})$ are categories. A *covariant functor* $F : (\mathfrak{C},\mathfrak{F}) \to (\mathfrak{D},\mathfrak{H})$ consists of a map F from the class $\mathfrak{C}$ into the class $\mathfrak{D}$ and a map (also denoted by F) from the class $\bigcup (\bigcup \mathfrak{F})$ where $\bigcup (\bigcup \mathfrak{F})$ is $\bigcup \{\text{Hom}_{\mathfrak{C}}(A,B) \mid A \in \mathfrak{C} \text{ and } B \in \mathfrak{C}\}$ to the class $\bigcup (\bigcup \mathfrak{H})$ such that

(i)   if $f \in \text{Hom}_{\mathfrak{C}}(A,B)$, then $F(f) \in \text{Hom}_{\mathfrak{D}}(F(A),F(B))$;
(ii) $F(1_A) = 1_{F(A)}$;
(iii) $F(f \circ g) = F(f) \circ F(g)$.

If conditions (i), (ii), and (iii) are replaced by

(i')   if $f \in \text{Hom}_{\mathfrak{C}}(A,B)$, then $F(f) \in \text{Hom}_{\mathfrak{D}}(F(B),F(A))$;
(ii') $F(1_A) = 1_{F(A)}$;
(iii') $F(f \circ g) = F(g) \circ F(f)$,

then F is called a *contravariant functor*.

(12.D.4) Example. Let $(\mathfrak{T},\mathfrak{F})$ be the category of topological spaces and continuous functions, and let $(\mathfrak{R},\mathfrak{M})$ denote the category of sets and set functions. Define a covariant functor $F(\mathfrak{T},\mathfrak{F}) \to (\mathfrak{R},\mathfrak{M})$ as follows. For each topological space $(X,\mathcal{U})$, let $F((X,\mathcal{U})) = X$, and for each continuous function $f : (X,\mathcal{U}) \to (Y,\mathcal{V})$, define $F(f)$ to be the set function $f : X \to Y$. F is called the *forgetful functor*.

(12.D.5) Exercise. Express the relationship between pointed topological spaces and their fundamental groups in terms of categories and functors.

Categories and functors form the essential vocabulary for any thorough study of homology theory (which is treated in courses in algebraic topology), and we make use of this terminology in Section F as well as in Chapter 14.

In Section B, we saw that the homotopy functor sends finite products of topological spaces to direct products of groups. In Section F, we shall see that direct limits are also fairly well behaved. It will also be shown that wedge products of topological spaces (disjoint unions of pointed spaces with the base points identified) are sent to free products of groups.

One of the advantages of studying categories and functors is that one can recognize the similarity of structures in the different categories, e.g., products of topological spaces vs. direct products of groups, wedge products of topological spaces vs. free products of groups, etc. Once the corresponding structures are identified, then it becomes natural to ask which structures are preserved by a given functor. Thus, category theory may be employed to help bring into focus a broad range of natural questions concerning functors between arbitrary pairs of categories.

E. THE SEIFERT-VAN KAMPEN THEOREM

Prerequisite reading for this section is the Appendix, Sections A and B. Difficulties encountered in calculating fundamental groups soon converge on the insurmountable. However, some relief is afforded by the Seifert-van Kampen theorem, two versions of which are given below. These results were established independently by Seifert [1931] and van Kampen [1933]. The main emphasis of this section is to make plausible the statement of the Seifert-van Kampen theorem and to show how the theorem may be applied. Basic proofs are omitted.

We begin with a nonrigorous calculation of the fundamental group of the figure eight shown in the next figure.

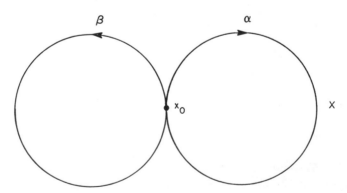

Loops based at x_0 are evidently eventually reducible to the form $\cdots *$ $\alpha^m * \beta^n * \alpha^p * \beta^q * \cdots$, where one alternately journeys a finite number of times around α and β ($\alpha^m = \alpha * \alpha * \cdots * \alpha$). Consequently, it is reasonable to conclude that $\pi_1(X, x_0)$ is a free group on two generators (or, equivalently, the free product of two infinite cyclic groups), and such is indeed the case.

To attack problems involving the fundamental groups of more complicated spaces such as the Klein bottle, the projective plane, complements of knots, etc., rather sophisticated results are needed. A favorite tactic used in obtaining the fundamental group of a space X is to split X into appropriate pieces whose fundamental groups are known, and then to exploit these groups to find the fundamental group of the original space X. Suppose, then, that $X = X_1 \cup X_2$, where X_1 and X_2 are open subsets of X. Suppose further that X, X_1, X_2, and $X_0 = X_1 \cap X_2 \neq \emptyset$ are path connected and $x^* \in X_0$. Let $G_0 = \pi_1(X_0,x^*)$, $G_1 = \pi_1(X_1,x^*)$, $G_2 = \pi_1(X_2,x^*)$, and $G = \pi_1(X,x^*)$. The homomorphisms given in the following commutative diagram are those induced by the inclusions maps between the appropriate spaces.

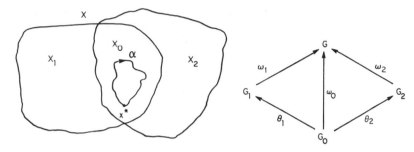

To achieve a first approximation of $\pi_1(X,x^*)$, one might imitate the argument used in predicting the fundamental group of the figure eight and conclude that G is the free product of G_1 and G_2. Complications arise, however, since the loop α in X_0 (in the preceding figure) is essentially counted twice in the free product, once as $g_1 = \theta_1([\alpha])$ and again as $g_2 = \theta_2([\alpha])$. This problem is overcome by introducing relations in $G_1 * G_2$ of the form $\theta_1([\alpha])\theta_2([\alpha])^{-1}$ for each $[\alpha] \in \pi_1(X_0,x^*)$.

How are loops in X accounted for that do not lie completely in either X_1 or X_2: for instance, the loop α in the next figure?

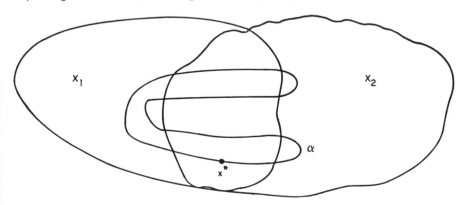

The basic procedure followed is to break α up into paths α_i, each of which lies entirely either in X_0, X_1, or X_2.

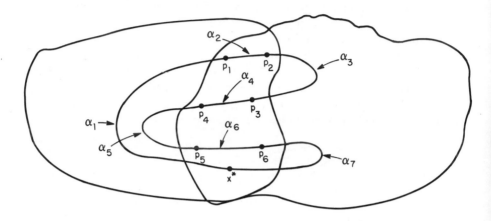

Since X_0 is path connected, for each i a path β_i may be constructed in X_0 from x^* to p_i.

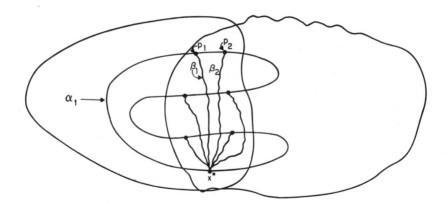

Note that $\alpha = (\alpha_1 * \beta_1^r) * (\beta_1 * \alpha_2 * \beta_2^r) * (\beta_2 * \alpha_3 * \beta_3^r) * \cdots * (\beta_6 * \alpha_7)$. Thus, G may be viewed as being generated by loops, each of which lies entirely in either X_1 or X_2. With all this in mind, it should seem plausible that G is the "amalgamated" product of G_1 and G_2, i.e., the free product of G_1 and G_2 modulo the relations mentioned previously. More precisely, suppose that $\{A;R_A\}$ and $\{B;R_B\}$ are presentations of G_1 and G_2, respectively. Then $G = \{A,B;R_A,R_B,\theta_1([\alpha])\theta_2([\alpha])^{-1}$ where $[\alpha] \in G_0\}$. The following is the classic formulation of the Seifert-van Kampen theorem.

(12.E.1)   Theorem (Seifert-van Kampen).   Suppose that $X = X_1 \cup X_2$ is a path connected space, where X_1 and X_2 are open, path connected subsets such that $X_0 = X_1 \cap X_2$ is nonempty and path connected. Let $x^* \in X_0$. Let $\theta_1 : \pi_1(X_0,x^*) \to \pi_1(X_1,x^*)$ and $\theta_2 : \pi_1(X_0,x^*) \to \pi_1(X_2,x^*)$ be the homomorphisms induced by the respective inclusion maps. Suppose that $\{A;R_A\}$ and $\{B;R_B\}$ are presentations of $\pi_1(X_1,x^*)$ and $\pi(X_2,x^*)$, respectively. Then $\{A,B;R_A,R_B,\theta_1([\alpha])\theta_2([\alpha])^{-1}$, where $[\alpha] \in \pi_1(X_0,x^*)\}$ is a presentation for $\pi_1(X,x^*)$.

Note that in the special case where X_0 is simply connected, G is simply the free product of the groups G_1 and G_2. Suppose that X is the figure eight described earlier. To see that the fundamental group of X is in fact a free group on two generators, first write X as the union of the path connected open sets

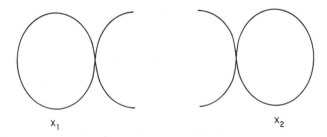

X_1 X_2

and then observe that $\pi_1(X_1,x^*) = \mathbf{Z}$ each $\pi_1(X_2,x^*) = \mathbf{Z}$, and that by (12.C.19), $\pi_1(X_1 \cap X_2, x^*) = \{e\}$. Consequently, $\pi_1(X_1,x^*)$ and $\pi_1(X_2,x^*)$ each have presentations of the form $\{a; \}$. Thus, by the Seifert-van Kampen theorem, $\pi_1(X)$ is a free group on two generators.

The fundamental group of $\mathscr{S}^n$ $(n > 1)$ is also easily calculated. Note that $\mathscr{S}^n$ can be written as the union of two overlapping (open) "hemispheres," each of which has trivial fundamental groups (the hemispheres are contractible), and hence $\pi_1(\mathscr{S}^n) = \{e\}$.

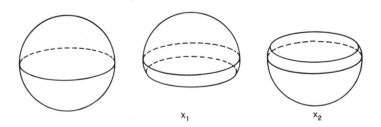

X_1 X_2

There is a more modern, but less transparent statement of the Seifert-van Kampen Theorem that has the virtue of being frequently easier to apply than its classical counterpart. A proof is not given here, but the interested readers may find detailed presentations in either Massey [1967] or Crowell and Fox [1963].

(12.E.2)   Theorem.   Suppose that X_1 and X_2 are path connected open subsets of a path connected space X such that $X = X_1 \cup X_2$ and $X_0 = X_1 \cap X_2 \neq \varnothing$ is path connected. Let $x^* \in X_0$. As before, there is a commutative diagram

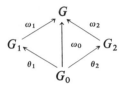

where $G = \pi_1(X,x^*)$, the G_i's are the respective fundamental groups of the X_i's, and the indicated homomorphisms are those induced by the inclusion mappings.
    Then

   (i)   the groups $\omega_1(G_1)$ and $\omega_2(G_2)$ generate G, and
   (ii)   if H is an arbitrary group and $\psi_i : G_i \to H$, $i = 0, 1, 2$ are homomorphisms satisfying $\psi_0 = \psi_1\theta_1 = \psi_2\theta_2$, then there is a unique homomorphism $\lambda : G \to H$ such that $\psi_{\cdot} = \lambda\omega_i$, $i = 0, 1, 2$.

There are a number of corollaries that follow rather easily from the main result.

(12.E.3)   Corollary.   If X_1 and X_2 are simply connected, then so is X.

   Proof.   This is immediate from (i).

(12.E.4)   Corollary.   If X_0 is simply connected and G_1 and G_2 are free groups with free bases $A = \{x_1, x_2, \ldots\}$ and $B = \{y_1, y_2, \ldots\}$, respectively, then G is a free group with basis $\{\omega_1(x_1), \omega_1(x_2), \ldots, \omega_2(y_1), \omega_2(y_2), \ldots\}$.

   Proof.   Let H be a free group with basis $\{c_1, c_2, \ldots, d_1, d_2, \ldots\}$. Define $\psi_1 : A \to H$ by $\psi_1(x_i) = c_i$, $i = 1, 2, \ldots$, and $\psi_2 : B \to H$ by $\psi_2(y_j) = d_j$, $j = 1, 2, \cdots$. Extend ψ_1 homomorphically to G_1 and ψ_2 to G_2.
   Let $\psi_0 : G_0 \to H$ be the only homomorphism possible (the trivial one) and note that $\psi_0 = \psi_1\theta_1 = \psi_2\theta_2$. By (12.E.2), there is a homomorphism

$\lambda : G \rightarrow H$ such that $\psi_i = \lambda \omega_i$, $i = 0, 1, 2$. We show that λ is an isomorphism by exhibiting an inverse. It follows from the "freeness" of H that there is a homomorphism $\rho : H \rightarrow G$ such that $\rho(c_i) = \omega_1(x_i)$, $i = 1, 2, \ldots$ and $\rho(d_j) = \omega_2(y_j)$, $j = 1, 2, \cdots$. Then $\rho\lambda(\omega_1(x_i)) = \rho\psi_1(x_i) = \rho(c_i) = \omega_1(x_i)$ and $\lambda\rho(c_i) = \lambda(\omega_1(x_i)) = \psi_1(x_i) = c_i$. Since the same sort of thing occurs with ω_2 and y_i and d_j, we have that $\lambda\rho$ and $\rho\lambda$ are identity maps.

(12.E.5)   Corollary.   If X_2 is simply connected, then

    (i) ω_1 is onto, and
    (ii) if $\{x_1, x_2, \ldots\}$ generates G_0, then ker ω_1 is the consequence of $R = \{\theta_1(x_1), \theta_1(x_2), \ldots\}$.

Therefore, $\pi_1(X)$ is the quotient group $\pi_1(X_1)/\langle R \rangle$, where $\langle R \rangle$ is the normal subgroup generated by R.

    Proof.
    (i)   This is trivial by part (i) of (12.E.2).
    (ii) Since $\omega_1(\theta_1(x_i)) = \omega_0(x_i) = \omega_2(\theta_2(x_i)) = e$, we have that R is contained in ker ω_1. Suppose that $z \in$ ker ω_1. Let $H = G_1/\langle R \rangle$, let $\psi_1 : G_1 \rightarrow H$ be the natural projection, and let $\psi_2 : G_2 \rightarrow H$ be the trivial homomorphism. Then if $\psi_0 = \psi_1\theta_1$, we have that $\psi_0 = \psi_1\theta_1 = \psi_2\theta_2$ (all these maps are trivial). Consequently, by (12.E.2), there is a homomorphism $\lambda : G \rightarrow H$ such that $\psi_1 = \lambda\omega_1$. However, this implies that $\psi_1(z) = \lambda(\omega_1(z)) = e$, and therefore $z \in \langle R \rangle$.

We apply (12.E.5) to calculate the fundamental group of the projective plane P_2. The projective plane is the space obtained by identifying diametrically opposite points on the edge of the disk, as indicated in the next figure.

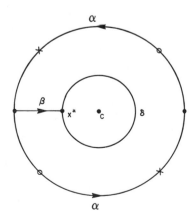

As usual, the first step consists in finding suitable open subsets. Let c denote the center of the disk and let x^* be an arbitrary point in the interior of the disk. Let $X_1 = P_2 \setminus \{c\}$ and let X_2 be the interior of the disk (considered as a subset of P_2). Note that after identification has taken place, α is a circle, and furthermore there is a deformation retract from X_1 onto α. Therefore, $G_1 = \pi_1(X_1, x^*) = \mathbf{Z}$, and G_1 is generated by the equivalence class represented by $\gamma = \beta^r * \alpha * \beta$. The group $\pi_1(X_2, x^*)$ is trivial and $\pi_1(X_1 \cap X_2, x^*)$ is infinite cyclic with generator $[\delta]$. Note that $\theta_1([\delta]) = [\gamma]^2$. Therefore, by (12.E.5), $\pi_1(P_2, x^*) = \pi_1(X_1, x^*)/K$, where K is the subgroup generated by $[\gamma]^2$. Thus, $\pi_1(P_2, x^*)$ is cyclic of order 2.

F.   DIRECT LIMITS

In this section, direct systems and direct limits of topological spaces are defined and characterized by a "universal mapping property." The homotopy functor will transform a direct system of pointed topological spaces into a direct system of groups (see Section D of the Appendix). The amazing thing that happens is that (with some reasonable conditions on the topological spaces) the fundamental group of the direct limit of the topological spaces is isomorphic to the direct limit of the fundamental groups. This result will be indispensable in constructing the examples in Chapter 15, Section E.

(12.F.1)   Notation.   If $\{X_\alpha \mid \alpha \in \Lambda\}$ is a family of topological spaces, then $\bigcup_{\alpha \in \Lambda} X_\alpha$ will denote the disjoint union of the X_α with the disjoint union topology.

(12.F.2)   Definition.   A *direct system of topological spaces* $(X_\alpha, f_{\alpha\beta}, D)$ consists of a directed set D, a family of topological spaces $\{X_\alpha\}_{\alpha \in D}$, and a family of continuous functions $\{f_{\alpha\beta} \mid \alpha, \beta \in D$ and $\alpha \leq \beta\}$ such that

   (i) $f_{\alpha\beta} : X_\alpha \to X_\beta$, whenever $\alpha \leq \beta$,
   (ii)   for each $\alpha \in D, f_{\alpha\alpha} = id$, and
   (iii)   if $\alpha \leq \beta \leq \gamma$, then $f_{\alpha\gamma} = f_{\beta\gamma} f_{\alpha\beta}$.

Let $X = \bigcup_{\alpha \in D} X_\alpha$. If $x_\alpha \in X_\alpha$ and $x_\beta \in X_\beta$, then declare x_α and x_β to be equivalent $(x_\alpha \sim x_\beta)$ if and only if there is a $\gamma \in D$ such that $\alpha \leq \gamma$, $\beta \leq \gamma$, and $f_{\alpha\gamma}(x_\alpha) = f_{\beta\gamma}(x_\beta)$.

   Denote $X/\sim$ by X^∞ or $\varinjlim_{\alpha \in D} X_\alpha$. Then X^∞ is called the *direct limit* of the direct system $(X_\alpha, f_{\alpha\beta}, D)$. If $x_\alpha \in X_\alpha$, we denote $P(x_\alpha)$ by $[x_\alpha]$, where P is the quotient map from X onto $X/\sim$.

(12.F.3) **Exercise.** Show that $\sim$ is an equivalence relation on X.

It should be noted that unlike inverse limits, the direct limit cannot be empty provided that any one of the X_α's is nonempty.

(12.F.4) **Notation.** If $\{X_n\}_{n\in N}$ is a family of topological spaces and $\{f_i : X_i \to X_{i+1}\}_{i\in N}$ is a family of continuous functions, then the direct system (X_i, f_{nm}, N) will often be denoted by (X_i, f_i, N), where $f_{nm} = f_{m-1} \cdots f_{n+1}f_n$.

(12.F.5) **Examples.**
1. Let $X_n = \mathscr{E}^1$ for each $n \in N$ and define $f_n : \mathscr{E}^1 \to \mathscr{E}^1$ by $f(x) = 0$ for all $x \in \mathscr{E}^1$. Then X^∞ is a point with the usual topology.
2. Suppose that for each $n \in N$, $X_n \subset X_{n+1}$ and $f_n : X_n \to X_{n+1}$ is the inclusion map. Then X^∞ is homeomorphic to $Y = \bigcup\limits_{n=0}^{\infty} X_n$, where Y is given the strong topology.
3. Let $X_n = \mathscr{E}^n$ and $f_i : \mathscr{E}^i \to \mathscr{E}^{i+1}$ be defined by $f(x_1, \ldots, x_i) = (x_1, \ldots, x_i, 0)$ (see problem 12.F.1).

(12.F.6) **Theorem.** Suppose that X^∞ is the direct limit of the direct system $(X_\alpha, f_{\alpha\beta}, D)$. For each $\alpha \in D$, define $f_\alpha : X_\alpha \to X^\infty$ by $f_\alpha = Pi_\alpha$, where $i_\alpha : X_\alpha \to \bigcup\limits_{\alpha \in D} X_\alpha$ is the inclusion map and $P : \bigcup\limits_{\alpha \in D} X_\alpha \to X^\infty$ is the quotient map.

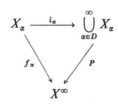

Then

(i) f_α is continuous,
(ii) for each $\alpha, \beta \in D$ with $\alpha \le \beta$ the following diagram is commutative,

(iii) $X^\infty \subset \bigcup_{\alpha \in D} f_\alpha(X_\alpha)$.

Proof.
(i)   This is trivial.
(ii)   Suppose that $x_\alpha \in X_\alpha$ and let $x_\beta = f_{\alpha\beta}(x_\alpha)$. Then $f_\beta f_{\alpha\beta}(x_\alpha) = Pi_\beta f_{\alpha\beta}(x_\alpha)$
$= P(x_\beta) = P(x_\alpha) = f_\alpha(x_\alpha)$.
(iii)   This is clear, since $Pi_\alpha(x_\alpha) = f_\alpha(x_\alpha)$.

(12.F.7)   *Theorem.*   Let $(X_\alpha, f_{\alpha\beta}, D)$ be a direct system of topological spaces. Suppose that Y is a topological space and the $\{g_\alpha : X_\alpha \to Y \mid \alpha \in D\}$ is a family of continuous functions such that $Y = \bigcup_{\alpha \in D} g_\alpha(Y_\alpha)$ and the following diagram is commutative whenever $\alpha\beta \in D$ and $\alpha \le \beta$.

Then there is a unique continuous onto function $h : X^\infty \to Y$ such that the tetrahedron is commutative for all $\alpha \le \beta$.

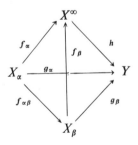

Proof.   It is clear how h must be defined, but it is less clear that what arises is a map, or that it has the requisite properties. Suppose that $x \in$ $(\bigcup_{\alpha \in D} X_\alpha)/\sim \;= X^\infty$. Let x_α be an element in $\bigcup_{\alpha \in D} X_\alpha$ such that $[x_\alpha] = x$. If the following diagram is to be commutative there is no choice but to set $h(x) = g_\alpha(x_\alpha)$. Hence, h (if it is a function) is unique. Furthermore, it is easy to show that if h is a map, then the tetrahedron is commutative.

To see that h is well defined, suppose $x_\alpha \in X_\alpha$ and $x_\beta \in X_\beta$ have the property that $[x_\alpha] = [x_\beta]$. Then there is a γ such that $\alpha \leq \gamma$, $\beta \leq \gamma$, and $f_{\alpha\gamma}(x_\alpha) = f_{\beta\gamma}(x_\beta) = x_\gamma$. We show that $g_\alpha(x_\alpha) = g_\beta(x_\beta) = g_\gamma(x_\gamma)$. By the commutativity of the following diagram, we have that $g_\alpha(x_\alpha) = g_\gamma(f_{\alpha\gamma}(x_\alpha)) = g_\gamma(x_\gamma)$, and similarly $g_\beta(x_\beta) = g_\gamma(x_\gamma)$.

That h is onto may be seen as follows. Suppose that $y \in Y$. By hypothesis, there is an $\alpha \in D$ and an $x_\alpha \in X_\alpha$ such that $y = g_\alpha(x_\alpha)$. Then $h([x_\alpha]) = y$.

It remains to show that h is continuous. Suppose that U is open in Y. Since X^∞ has the quotient topology, $h^{-1}(U)$ will be open in X^∞ if $P^{-1}h^{-1}(U)$ is open in $\bigcup_{\alpha \in D} X_\alpha$. However, $P^{-1}h^{-1}(U) = \bigcup_{\alpha \in D} g_\alpha^{-1}(U)$ which is clearly open in $\bigcup_{\alpha \in D} X_\alpha$.

The next theorem gives a characterization of direct limits that is trivial to establish in spite of the rather awesome hypothesis.

(12.F.8) Theorem (Universal Mapping Characterization). Let $(X_\alpha, f_{\alpha\beta}, D)$ be a direct system of topological spaces. Suppose that Y is a topological space and that $\{g_\alpha : X_\alpha \to Y \mid \alpha \in D\}$ is a family of continuous functions such that the following diagram is commutative and that $Y = \bigcup_{\alpha \in D} g_\alpha(X_\alpha)$.

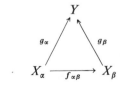

Suppose further that for each topological space Z and for each family of continuous functions $\{h_\alpha : X_\alpha \to Z \mid \alpha \in D\}$ with the properties that the following diagram is commutative and that $Z = \bigcup_{\alpha \in D} h_\alpha(X_\alpha)$, there is a unique continuous surjection $\phi_Z : Y \to Z$ such that the following diagram is commutative.

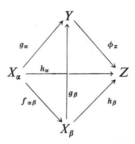

Then Y is homeomorphic to X^∞.

Proof. Consider the following diagram:

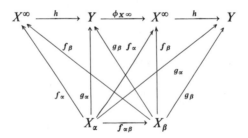

By the uniqueness of the map in (12.F.7), we have that $\phi_{X^\infty} h = id_{X^\infty}$. By the uniqueness in the hypothesis of the present theorem, it follows that $h\phi_{X^\infty} = id_Y$. Thus, h is a homeomorphism.

The reader will want to reread the entire section in order to observe that nothing goes awry if one assumes that all topological spaces are endowed with a base point, and that all continuous functions preserve the base point. To nudge the less than diligent reader, we present the following exercise.

(12.F.9) Exercise. Restate and prove the analog of (12.F.7) for the category of pointed topological spaces.

The reader should now check the Appendix and observe that the analog of each of the preceding theorems also holds in the category of groups and homomorphisms.

Next we examine what is produced when the homotopy functor operates on a direct system.

(12.F.10) *Theorem.* If $((X_\alpha, x_\alpha), f_{\alpha\beta}, D)$ is a direct system of pointed topological spaces, then $(\pi_1(X_\alpha, x_\alpha), (f_{\alpha\beta})_*, D)$ is a direct system of groups.

Proof. By (12.C.1) (or since π_1 is a covariant functor), if $\alpha \le \beta \le \gamma$, then $(f_{\alpha\gamma})_* = (f_{\beta\gamma} f_{\alpha\beta})_* = (f_{\beta\gamma})_* (f_{\alpha\beta})_*$. Furthermore, if $\alpha \le \beta$, then $(f_{\alpha\beta})_*$: $\pi_1(X_\alpha, x_\alpha) \to \pi_1(X_\alpha, x_\alpha)$ is a homomorphism.

Now for the jewel in the crown.

(12.F.11) *Theorem.* Let $((X_\alpha, x_0), f_{\alpha\beta}, D)$ be a direct system of topological spaces. Suppose that $f_\alpha(X_\alpha, x_0)$ is open in (X^∞, x_0) for each α, and that all of the $f_{\alpha\beta}$'s are inclusion maps. Then there is an isomorphism $h : \varinjlim(\pi_1(X_\alpha, x_0) \to \pi_1(X^\infty, x_0)$. Furthermore, the following diagram is commutative.

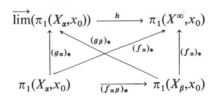

Proof. By the previous theorem, $(\pi_1(X_\alpha, x_0), (f_{\alpha\beta})_*, D)$ is a direct system. Furthermore, for each $\alpha \le \beta$, the following diagram is commutative.

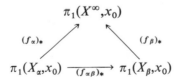

Thus, if it can be shown that $\pi_1(X^\infty, x_0) \subset \bigcup_{\alpha \in D} (f_\alpha)_* (\pi_1(X_\alpha, x_0))$, then by Theorem (Ap.D.2), there will be an onto homomorphism $h : \varinjlim(\pi_1(X_\alpha, x_0)) \to \pi_1(X^\infty, x_0)$.

Let $[\ell] \in \pi_1(X^\infty, x_0)$. Since $\{f_\alpha(X_\alpha) \mid \alpha \in D\}$ forms an open cover of $\ell(\mathrm{I})$, there are a finite number of $f_\alpha(X_\alpha)$'s that cover $\ell(\mathrm{I})$. Let β be an index larger than all of the α's in the finite cover. Then we have that $\ell(\mathrm{I}) \subset f_\beta(X_\beta)$.

Since the $f_{\alpha\beta}$'s are inclusion maps, f_β is one to one. Hence, we may define a path $m = f_\beta^{-1}\ell$. Then $(f_\beta)_*[m]$ equals $[\ell]$.

All that remains is to show that h is one to one. Suppose that $x \in \varprojlim(\pi_1(X_\alpha,x_0))$ and that $h(x) = e$; we show that $x = e$. Since $\varprojlim(\pi_1(X_\alpha,x_0)) \subset \bigcup_{\alpha\in D} (g_\alpha)_*(\pi_1(X_\alpha,x_0))$, there is an $\alpha \in D$ and a $y \in \pi_1(X_\alpha,x_0)$ such that $(g_\alpha)_*(y) = x$.

The diagram

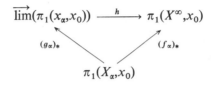

is commutative, and therefore $(f_\alpha)_*(y) = e$. Now let $\psi : I \to X_\alpha$ be a loop based at x_0 such that $[\psi] = y$. Since $(f_\alpha)_*(y) = e$, there is a homotopy $H : I \times I \to (X^\infty,x_0)$ such that $H_0 = f_\alpha\psi$ and $H_1 = c_{x_0}$. Let β be sufficiently large so that $\operatorname{Im} H \subset f_\beta(X_\beta)$ (and $\alpha \leq \beta$). As before, f_β is one to one, and hence $f_\beta^{-1}H$ is a homotopy between $f_{\alpha\beta}\psi$ and c_{x_0}. Therefore, we have that $(f_{\alpha\beta})_*(\psi) = e$. By the commutativity of the following diagram,

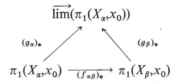

we have that $x = (g_\alpha)_*(y) = (g_\alpha)_*([\psi]) = (g_\beta)_*(f_{\alpha\beta})_*[\psi] = (g_\beta)_*[c_{x_0}] = (g_\beta)_*(e) = e$, which completes the proof.

PROBLEMS

Sections A and B

1. Show that if $f,g : X \to Y$ are homotopic and $A \subset X$, then $f_{|A}$ and $g_{|A}$ are homotopic.
2. Find an example of a space X containing points x_0 and x_1 such that $\pi_1(X,x_0)$ and $\pi_1(X,x_1)$ are not isomorphic.
3. Show that a continuous function $f : X \to Y$ is null homotopic if and only if f can be extended to the cone on X.

4. Suppose that $x_0 \in X$ and C is the path component of X containing x_0. Show that $\pi_1(X,x_0)$ and $\pi_1(C,x_0)$ are isomorphic.

5. Suppose that X_1 and X_2 are disjoint topological spaces and that $X = X_1 \cup X_2$ is given the disjoint union topology. Show that if $x_1 \in X_1$ and $x_2 \in X_2$, then $\pi_1(X_1,x_1) = \pi_1(X,x_1)$ and $\pi_1(X,x_2) = \pi_1(X_2,x_2)$.

6. Suppose that $\{Y_\alpha \mid \alpha \in \Lambda\}$ is a family of topological spaces and let X be an arbitrary topological space. Suppose further that $f,g : X \to \prod_{\alpha \in \Lambda} Y_\alpha$ are continuous functions and let $p_\beta : \prod_{\alpha \in \Lambda} Y_\alpha \to Y_\beta$ be the projection map. Show that f is homotopic to g if and only if $p_\alpha f$ is homotopic to $p_\alpha g$ for all α.

7. Suppose that X and Y are topological spaces and that $f,g : X \to Y$ are continuous. Show that Y may be embedded in a space Z by a function h so that hf is homotopic to hg.

8.* Let X and Y be topological spaces and suppose that $C(X,Y)$ has the compact-open topology.
   (a) Show that if X is locally compact, then $f \sim g$ if and only if f and g belong to the same path component of $C(X,Y)$,
   (b) If X is arbitrary, does (a) hold?

9. A path connected space X is *1-simple* if and only if every two paths p and q in X with $p(0) = q(0)$ and $p(1) = q(1)$ induce the same homomorphism from $\pi_1(X,\alpha(0)) \to \pi_1(X,\alpha(1))$, i.e., $\phi([\alpha]) = [p][\alpha][p'] = [q][\alpha][q']$. Show that X is 1-simple if and only if $\pi_1(X)$ is abelian.

10. Prove that a space X is simply connected if and only if, for each pair of paths α and β in X such that $\alpha(0) = \beta(0)$ and $\alpha(1) = \beta(1)$, then $\alpha \underset{\{0,1\}}{\simeq} \beta$.

11. Find the fundamental group of an open tin can.

12. Find the fundamental group of the Moebius strip.

13. Prove or disprove: if $f,g : X \to Y$ and $f \sim c_1$ and $g \sim c_2$, where c_1 and c_2 are constant maps, then $f \sim g$.

14.* Investigate the relationships between infinite products of topological spaces and the corresponding products of their fundamental groups.

Section C

1. Give an example of an embedding $f : (X,x_0) \to (Y,y_0)$ such that f_* is neither 1-1 nor onto.

2. Give an example of an onto map $f : (X,x_0) \to (Y,y_0)$ such that f_* is neither 1-1 nor onto.

3. Show that any retract of a contractible space is contractible.

4. Suppose that A is a deformation retract of a space X and that B is a deformation retract of a space Y. Show that $A \times B$ is a deformation retract of $X \times Y$.

5. Suppose that X and Y are topological spaces and that $f : X \to Y$ is con-

tinuous. The *mapping cylinder of f*, M_f is $((X \times I) \bigcup Y)/\sim$, where $\sim$ is the equivalent relation generated by $(x,1) \sim f(x)$ for $x \in X$ and $(X \times I) \bigcup Y$ is the disjoint union of $X \times I$ and Y with the disjoint union topology. Show that

(a) Y and M_f are homotopically equivalent, and

(b) Y is a deformation retract of M_f.

6.   Prove that the Hilbert cube is contractible. Generalize.

7.   What is the fundamental group of a starfish?

8.   Show that if a space X is contractible Y is path connected and $f,g: X \to Y$ are continuous functions, then f and g are homotopic.

9.* Suppose F is a homotopy between maps $f,g : X \to Y$, where X and Y are path connected spaces. Let $x_0 \in X$ and let $y_0 = f(x_0)$ and $\hat{y}_0 = g(x_0)$. Show that the following diagram is commutative, where $k(t) = F(x_0,t)$ and that f_* is an isomorphism if and only if g_* is.

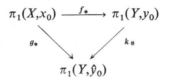

(See the proof of (12.B.2) for the definition of $k_\#$.)

10. Show that if A is a deformation retract of X and B is a deformation retract of A, then B is a deformation retract of X.

11.* Show that if A, B, and $A \cap B$ are nonempty, path connected subsets of $A \cup B$, and if A and B are simply connected, then so is $A \cup B$.

12. Use problem 11 to show that $\pi_1(\mathcal{S}^2) = \{e\}$.

13. Let n and s be the north and south poles of an n-sphere. Prove that $\{n,s\}$ is not a retract of $\mathcal{S}^n$.

14. Suppose that $x_0 \in \mathcal{S}^1$. Show that $\mathcal{S}^1 \times \{x_0\}$ is not a deformation retract of $\mathcal{S}^1 \times \mathcal{S}^1$.

15. Two continuous functions $f,g : I \to Y \subset \mathcal{E}^2$ are *adjacent in Y* if and only if there is a partition $0 = x_0 < x_1 < x_2 < \cdots < x_n = 1$ of I such that $f([x_{i-1},x_i]) \cup g([x_{i-1},x_i]) \subset D_i$ for each i, where D_i is a disk in Y. If f and g are adjacent in Y, are f and g homotopic?

16. Do there exist continuous maps f and g such that the following hold?

(i) $f : (\mathcal{S}^1 \times \{0\}) \cup (\{x_0\} \times I) \cup (\mathcal{S}^1 \times \{1\}) \to \mathcal{S}^1 \times I$,

(ii) $g : (\mathcal{S}^1 \times I) \to (\mathcal{S}^1 \times \{0\}) \cup (\{x_0\} \times I) \cup (\mathcal{S}^1 \times \{1\})$, and

(iii) $gf = id$.

17. Suppose that X and Y are topological spaces and that $f : X \to Y$ is a

homotopy equivalence with homotopy inverse $g : Y \to X$. Let $x \in X$. Show that $\pi_1(X,x_0)$ is isomorphic with $\pi_1(Y,f(y_0))$. [Warning: While $gf \sim id$, the homotopy need not keep x_0 fixed.]

18. Suppose that T is a solid triangle in $\mathscr{E}^2$ with boundary S. Show that there is no continuous function $f : T \to S$ which maps each edge of T into itself.

Section D

1. Find (somewhere in mathematics) an example of a contravariant functor.

2.* (*The Higher Homotopy Groups*) Let X be a topological space. Let $\partial I^n = \{(x_1, \ldots, x_n) \in I^n \mid$ for some i, $x_i = 0$ or $1\}$, and let $\Omega_n(X,x_0) = C((I^n,\partial I^n),(X,x_0))$. If $f,g \in \Omega_n(X,x_0)$, define

$$(f * g)(x_1, \ldots, x_n) = \begin{cases} f(2x_1, x_2, \ldots, x_n) & 0 \le x_1 \le \tfrac{1}{2} \\ g(2x - 1, x_2, \ldots, x_n) & \tfrac{1}{2} \le x_1 \le 1. \end{cases}$$

Partition $\Omega_n(X,x_0)$ into equivalence classes via homotopies relative to ∂I^n. Denote the set of equivalence classes by $\pi_n(X,x_0)$. Finally, if $[f]$ and $[g] \in \pi_n(X,x_0)$, define $[f] \cdot [g] = [f * g]$.

(a) Show that $\cdot$ is well defined and that $\pi_n(X,x_0)$ with this operation is a group.

(b) Prove analogs of (12.C.1).

(c) Show that for each $n > 1$, the construction above gives rise to a functor from $\mathfrak{T}_P$ to the category of abelian groups and homomorphisms. (It should be no surprise that this is called the n-th homotopy functor.)

(d) Show that if $n > 1$, then $\pi_n(X \times Y, (x_0,y_0)) \simeq \pi_n(X,x_0) \times \pi_n(Y,y_0)$.

(e) Show that if (X,x_0) and (Y,y_0) are homotopically equivalent, then $\pi_n(X,x_0)$ and $\pi_n(Y,y_0)$ are isomorphic.

Remark: In general, it is exceedingly difficult to compute the higher homotopy groups. In fact, their computation even for n-spheres is still not fully done. Some of the known results leave one gasping, e.g., $\pi_{15}(\mathscr{S}^8) = \mathbf{Z} \times \mathbf{Z}_{120}$ and $\pi_{16}(\mathscr{S}^8) = \mathbf{Z}_2 \times \mathbf{Z}_2 \times \mathbf{Z}_2 \times \mathbf{Z}_2$ ($\mathbf{Z}_n$ is the group of integers mod n).

3. Suppose that $(\mathfrak{C},\mathfrak{F})$ is a category and that $f \in \text{Hom}_{\mathfrak{C}}(A,B)$ and $g \in \text{Hom}_{\mathfrak{C}}(B,A)$. Then f is a *right inverse* of g if and only if $g \circ f = i_B$, in which case g is called a *left inverse* of f.

(a) Show that in the category of sets (hence also in the category of topological spaces and continuous functions), if f has a left inverse, then f is 1–1, and if f has a right inverse, then f is onto.

A map f is called a $\mathfrak{C}$-*equivalence* if and only if f has both a left and a right inverse.

(b) Show that if f has a left inverse g and a right inverse h, then $g = h$.

(c) Show that a homotopy equivalence is a $\mathfrak{C}$-equivalence, where $\mathfrak{C}$ is the homotopy category.

4. Define a category as follows: Let $\mathfrak{D} = \mathbf{Z}^+$ be the class of objects. If $m,n \in \mathfrak{D}$, define $\mathrm{Hom}_{\mathfrak{D}}(m,n) = \{(m,a,n) \mid a$ is a common divisor of m and $n\}$. Define $(m,a,n) \circ (p,b,q) = (m,d,q)$, where d is the greatest common divisor of a and b. Show that $(\mathfrak{D},\mathfrak{H})$ is a category, where $\mathfrak{H} = \{\mathrm{Hom}_{\mathfrak{D}}(m,n) \mid m,n \in \mathfrak{D}\}$.

Section E

1. Calculate the fundamental group of the five-petaled rose: ✳ .
2. Calculate the fundamental group of $\infty\!\!\infty$.
3. Calculate the fundamental group of a 2-sphere with three holes in it.
4. Calculate the fundamental group of the torus (using the Seifert-van Kampen theorem).
5. Calculate the fundamental group of the Klein bottle.
6. Calculate the fundamental group of $\mathscr{E}^3 \setminus \mathscr{S}^1$.
7. Calculate the fundamental group of a torus with a hole in it.
8. For each $n > 2$, show how to construct a space having a cyclic fundamental group of order n.

Section F

1. Show that X^∞ in example 3 of (12.F.5) is not Hilbert space.
2. Show that $\mathscr{S}^1$ cannot be written as a direct limit of subspaces, unless one of the X_α's is $\mathscr{S}^1$.
3. Compute $\pi_1(X)$, where $X = \infty\!\!\infty\!\!\infty\!\!\infty \cdots$.
4. For $n \in \mathbf{Z}^+$, let $X_n = [-1,1]$, and for $m \geq n$, let $\phi_{nm} : X_n \to X_m$ be defined by $\phi_{nm}(x_n) = x_n/2^{m-n}$. Show that $\overrightarrow{\lim}(X_n,\phi_{nm},\mathbf{Z}^+)$ is homeomorphic to $\mathscr{E}^1$.
5. Show that any CW-complex can be considered as a direct limit in a nontrivial way.
6.* Show that any T_2 k-space can be considered as a direct limit in a nontrivial way.

Chapter 13

COVERING SPACES

The beautiful and surprisingly close interplay between algebra and topology is well illustrated in the theory of covering spaces. This theory is very reminiscent of Galois theory and has numerous applications in both topology and geometry. In particular, some of the results may be used to obtain an elegant computation of the fundamental group of $\mathscr{S}^1$. Throughout this chapter, the reader is especially advised to construct numerous figures that illustrate both the definitions as well as the hypotheses and proofs of theorems.

A. THE LIFTING THEOREMS

(13.A.1) Definition. Suppose that X and $\hat{X}$ are connected, locally path connected topological spaces and that $p : \hat{X} \to X$ is a continuous map. The pair $(\hat{X},p)$ is a *covering space* of X if and only if for each $x \in X$, there is a connected open set U containing x such that

  (i) $p^{-1}(U) = \bigcup\limits_{\alpha \in \Lambda_x} S_\alpha$, where the S_α's form a collection of mutually dis-
      joint open sets in $\hat{X}$, and
  (ii) $p_{|S_\alpha} : S_\alpha \to U$ is a homeomorphism (onto) for each $\alpha \in \Lambda_x$.

The space X is called the *base space*, and the open sets U are called *canonical neighborhoods*; p is referred to either as the *projection* or the *covering map*.

(13.A.2) Exercise. A map $f : X \to Y$ is a *local homeomorphism* if and only if for each $x \in X$, there is an open set V_x containing x such that $f(V_x)$ is open in Y and $f_{|V_x} : V_x \to f(V_x)$ is a homeomorphism. Show that covering maps are local homeomorphisms.

The decision to define covering spaces only in the context of connected, locally path connected spaces is made with some reluctance. Less restrictive definitions are often given (Spanier, [1966]) and a number of the basic theorems are still valid in a more general setting. However, in practice, much work is done with spaces that satisfy these connectedness conditions, and this fact together with blatant expediency is our justification for confining the presentation to such spaces.

(13.A.3) Example. Let $X = \mathscr{S}^1$ and $\hat{X} = \mathscr{E}^1$. Define $p : \mathscr{E}^1 \to \mathscr{S}^1$ by $p(t) = (\cos 2\pi t, \sin 2\pi t)$. In this case, p wraps $\mathscr{E}^1$ around $\mathscr{S}^1$ an infinite number of times. In fact for each n, the interval $[n, n + 1]$ goes once around $\mathscr{S}^1$. Any connected open proper subset of $\mathscr{S}^1$ serves as a canonical neighborhood.

(13.A.4) Example. The projective plane P_2 may be defined as the quotient space obtained by identifying antipodal points of $\mathscr{S}^2$. Then $(\mathscr{S}^2, P)$ is a covering space of P_2, where $P : \mathscr{S}^2 \to P_2$ is the quotient map.

(13.A.5) Example. Any connected, locally path connected space is a covering space of itself, with the identity map serving as the projection.

(13.A.6) Exercise. Suppose that $(\hat{X}, p)$ is a covering space of X and $(\hat{Y}, p')$ is a covering space of Y. Find a covering space of $X \times Y$.

(13.A.7) Exercise. Find three distinct covering spaces of the torus (one of which is the plane).

One of the most useful results in covering space theory is the following theorem. It states that a path α lying in the base space X may be "lifted in a unique fashion" to a path in the covering space. Here, "lifted" means that the projection of the lifted path coincides with α, and "unique" means that if $p\hat{\alpha} = p\hat{\alpha}$ and $\hat{\alpha}(0) = \hat{\alpha}(0)$, then $\hat{\alpha} = \hat{\alpha}$.

(13.A.8) Theorem (Unique path Lifting Theorem). Suppppose that $(\hat{X}, p)$ is a covering space of X and that $\alpha : \mathbf{I} \to X$ is a path in X with initial point x_0. Let $\hat{x}_0 \in p^{-1}(x_0)$. Then there is a unique path $\hat{\alpha} : \mathbf{I} \to \hat{X}$ with initial point $\hat{x}_0$ and such that $p\hat{\alpha} = \alpha$.

Proof. The idea of the proof is quite simple. We subdivide $\mathbf{I}$ finely enough so that each piece of the subdivision is mapped by α into a canonical neighborhood, and then the local homeomorphism p is used to lift the pieces. Some care is taken in order to ensure that the appropriate endpoint of the i-th piece matches up properly with a suitable endpoint of the $(i - 1)$-st piece.

Cover X with a family $\mathcal{U} = \{U_x \mid x \in X\}$ of canonical neighborhoods. Then $\alpha^{-1}(\mathcal{U})$ is an open cover of $\mathbf{I}$, and since $\mathbf{I}$ is compact, a finite subcover $\{\alpha^{-1}(U_{x_1}), \dots, \alpha^{-1}(U_{x_m})\}$ may be extracted from $\alpha^{-1}(\mathcal{U})$. Let γ be a Lebesgue number for this cover and let $0 = t_0 < t_1 < \cdots < t_n = 1$ be a partition of $\mathbf{I}$ of mesh less than γ. The following assertion is proven by induction: For each i, $0 \le i < n$, there is a unique continuous map $\alpha_i' : [0, t_i] \to \hat{X}$ such that $\alpha_i'(0) = \hat{x}_0$ and $p\alpha_i' = \alpha_{|[0, t_i]}$. Obviously, for $i = 0$, the map α_0' exists. Assume, then, that we have constructed such a map for some i, where $i < n$. By the choice of the partition, we have that $\alpha([t_i, t_{i+1}])$ is contained in some $U \in \mathcal{U}$. Let S_j be the component of $p^{-1}(U)$ that contains $\alpha_i'(t_i)$. Since $p_j = p_{|S_j}$ is an embedding, it follows that $p_j^{-1} : U \to S_j$ is a continuous function. Consequently the map $\alpha_{i+1}' : [0, t_{i+1}] \to X$ given by

$$\alpha_{i+1}'(t) = \begin{cases} \alpha_i'(t) & \text{if} \quad 0 \le t \le t_i \\ p_j^{-1}\alpha(t) & \text{if} \quad t_i \le t \le t_{i+1} \end{cases}$$

is well defined. Continuity of $\alpha_{i+1}'(t)$ follows from the map gluing theorem. It should be clear that α_{i+1}' has all the requisite properties. The function α_n' in the desired lifting.

Not only may paths be lifted, but so may be entire homotopies.

(13.A.9) Theorem (Homotopy Lifting Theorem). Suppose that $(\hat{X}, p)$ is a covering space of a space X, Y is a locally connected space, and $H : Y \times \mathbf{I} \to X$ a homotopy. Suppose, furthermore, that there is a continuous map $f : Y \to \hat{X}$ such that $pf = H_0$. Then there is a unique homotopy $H' : Y \times \mathbf{I} \to \hat{X}$ such that $pH' = H$ and $H_0' = f$.

Proof. The proof is based on the observation that for each $y \in Y$, $H_{|\{y\} \times \mathbf{I}}$ is essentially a path in X, and, hence, by the previous theorem may be lifted uniquely to a path in X starting at $f(y)$. Suppose that $y \in Y$. Let $\sigma_y : \mathbf{I} \to \hat{X}$ be the unique path with the property that $\sigma_y(0) = f(y)$ and, for each $t \in \mathbf{I}$, $p\sigma_y(t) = H(y, t)$. Define $H' : Y \times \mathbf{I} \to X$ by $H'(y, t) = \sigma_y(t)$. Clearly $pH' = H$; the problem is to show that H' is continuous.

Suppose that $y_* \in Y$. We construct a "tube" in $Y \times \mathbf{I}$ that contains $\{y_*\} \times \mathbf{I}$ and on which H' is continuous. Let $\mathcal{U} = \{U_\alpha \mid \alpha \in \Lambda\}$ be a cover of X by canonical open sets. For each $t \in \mathbf{I}$, let I_t be an open interval about t and let N_t be a connected open neighborhood of y_* such that $H(N_t \times I_t) \subset U_\alpha$ for some $U_\alpha \in \mathcal{U}$. Let $\{I_{t_1}, \dots, I_{t_n}\}$ be a finite subcover of the open cover

$\{I_t \mid t \in \mathbf{I}\}$, and let $N_{t_1}, \ldots, N_{t_n}$ represent the corresponding neighborhoods of y_*. Choose a connected open set N containing y_* such that $N \subset \bigcap_{i=1}^n N_{t_i}$.

Let γ be a Lebesgue number for $\{I_{t_1}, \ldots, I_{t_n}\}$ and let $0 < a_1 < a_2 < \cdots < a_n = 1$ be a partition of $\mathbf{I}$, where, for each i, $a_{i+1} - a_i < \gamma$. Let $A_i = [a_i, a_{i+1}]$ and note that for each i, $H(A_i \times N)$ is contained in some canonical neighborhood U_α.

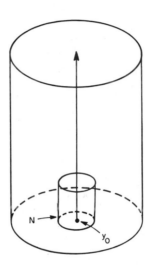

Consider $f(N \times \{0\})$. Since $pf = H_0$ and $N \times \{0\}$ is connected, there is an $\alpha_1 \in \Lambda$ such that $f(N \times \{0\})$ is contained in a component of $p^{-1}(U_{\alpha_1})$ and $H(N \times A_1) \subset U_{\alpha_1}$. Denote the restriction of p to this component by p_{α_1}. Then p_{α_1} is an embedding, and hence the map $F : N \times A_1 \to \hat{X}$, defined by $F(y,t) = p_{\alpha_1}^{-1} H(y,t)$, is well defined and continuous.

Select $\alpha_2 \in \Lambda$ such that

(i) $H(N \times A_2) \subset U_{\alpha_1}$, and
(ii) $F(N \times \{a_1\})$ is contained in a component of $p^{-1}(U_{\alpha_2})$.

If p_{α_2} denotes the restriction of p to this component, then F may be extended to $N \times A_2$ by setting $F(y,t) = p_{\alpha_2}^{-1} H(y,t)$ for $y \in N$, $t \in A_2$. Note that F is well defined at $t = a_1$ (why?). This process is repeated for each A_i; the resulting map $F : N \times \mathbf{I} \to \hat{X}$ clearly has the property that $F(y,t) = \sigma_y(t)$ for each $y \in N$ and $t \in \mathbf{I}$. Thus, we have that $F = H'_{|N \times \mathbf{I}}$, and since F is continuous so must be H'.

(13.A.10) Exercise. Show that H' is unique.

(13.A.11) Exercise. Suppose that $(\hat{X},p)$ is a covering space of X. Show that p_* is 1–1, where $p_* : \pi_1(\hat{X},\hat{x}) \to \pi_1(X,x)$ is the homomorphism induced by the projection map p.

The next exercise yields the amazing result that the same number of points always lie above any point in the base space. This result makes possible many of the beautiful theorems that follow.

(13.A.12) Exercise. Suppose that $(\hat{X},p)$ is a covering space of X. Show that card $p^{-1}(x) = $ card $p^{-1}(y)$ for each $x,y \in X$ and that $p^{-1}(x)$ with the relative topology is discrete; card $p^{-1}(x)$ is the *number of sheets* of the covering space $(\hat{X},p)$.

B. $\pi_1(\mathscr{S}^1, s) = Z$ (A DIVERSION)

The previous two theorems can be employed to confirm our intuitive result that $\pi_1(\mathscr{S}^1,s) = Z$. In the earlier discussion of this matter, it was claimed that the integer assigned to a member of $\pi_1(\mathscr{S}^1,s)$ would somehow indicate the number of times a path was wrapped around $\mathscr{S}^1$. Let $s = (1,0)$. Consider a typical loop σ based at s. By (13.A.3), $(\mathscr{E}^1,p)$ is a covering space of $\mathscr{S}^1$, where $p(t) = (\cos 2\pi t, \sin 2\pi t)$. By (13.A.8), σ may be lifted uniquely to a path σ' starting at 0 such that $p\sigma'(0) = \sigma(0) = s$. Furthermore, since $p\sigma'(1) = \sigma(1) = s$, it follows that $\sigma'(1)$ must be an integer. The reader should recognize that this integer reflects the true number of circuits that the loop σ makes around $\mathscr{S}^1$ as σ runs from 0 to 1. A loop σ that starts around $\mathscr{S}^1$ in one direction, but from time to time reverses itself, is characterized by zigzags traced out by σ' in $\mathscr{E}^1$. This leads to the following definition.

(13.B.1) Definition. Let $\phi : \pi_1(\mathscr{S}^1,s) \to Z$ be defined by $\phi([\sigma]) = \sigma'(1)$, where σ' is the unique lifting of σ that starts at 0. The function ϕ is called the *degree map*.

(13.B.2) Lemma. The degree map ϕ is well defined.

Proof. Suppose that σ and τ are homotopic loops (mod $\{0,1\}$) based at s, with liftings σ' and τ', respectively. We show that $\sigma'(1) = \tau'(1)$.

Let $F : I \times I \to \mathscr{S}^1$ be the homotopy satisfying

(i) $F_0 = \sigma$,
(ii) $F_1 = \tau$, and
(iii) $F(0,t) = F(1,t) = s$ for each $t \in I$.

By (13.A.9), there is a unique lifting homotopy $H : I \times I \to \mathscr{E}^1$ with the property that $pH = F$ and $H_0 = \sigma'$. By part (iii), $H(0,t)$ and $H(1,t) \in p^{-1}(s)$ for each t. However, $p^{-1}(s)$ consists of the integers which is a discrete set of points. Therefore, since $H(1,0) = \sigma'(1)$, it follows that $H(1,t) = \sigma'(1)$ for all t ($H(\{1\} \times I)$ is connected and $H(0,t) = \sigma'(0) = \tau'(0) = 0$). Furthermore, since $pH_0 = F_0 = \sigma$ and $pH_1 = F_1 = \tau$, it follows from (13.A.9) and (13.A.8) that $H_0 = \sigma'$ and $H_1 = \tau'$. Finally, we have that $\sigma'(1) = H(1,0) = H(1,1) = \tau'(1)$, and hence the function ϕ is well defined.

(13.B.3) Theorem. The degree map $\phi : \pi_1(\mathscr{E}^1,s) \to \mathbf{Z}$ is an isomorphism.

Proof. First, we show that ϕ is a homomorphism. Suppose that $[\sigma]$ and $[\tau]$ are elements of $\pi_1(\mathscr{E}^1,s)$ and let $m = \sigma'(1)$ and $n = \tau'(1)$. Define a map $\tau'' : I \to \mathscr{E}^1$ by $\tau''(s) = \tau'(s) + m$. Then $\tau''(0) = m$ and $\tau''(1) = m + n$. Furthermore, it is clear that $p\tau''(s) = p(\tau'(s) + m) = \tau(s)$, and only slightly less obvious that $p(\sigma' * \tau'') = \sigma * \tau$ (this can be verified directly). Hence, $\phi([\sigma][\tau]) = \phi([\sigma * \tau]) = (\sigma' * \tau'')(1) = \tau''(1) = m + n = \phi([\sigma]) + \phi([\tau])$.

Next, we show that ϕ is one to one. Suppose that $\phi([\sigma]) = 0$. This means of course that $\sigma'(1) = 0$, and hence σ' is actually a loop in $\mathscr{E}^1$ based at 0. Define a homotopy $H : I \times I \to \mathscr{E}^1$ by $H(s,t) = \sigma'(s) \cdot (1 - t)$. It may be routinely verified that pH is a homotopy (mod $\{0,1\}$) between σ and the constant function at s. Hence, $[\sigma] = e$ and ϕ is one to one.

Finally, we show that ϕ is onto. Suppose that n is an integer. Let $\sigma'(s) = ns$ for each $s \in I$ and set $\sigma = p\sigma'$. Clearly, $\phi([\sigma]) = \sigma'(1) = n$.

C. REGULAR COVERING SPACES

In (13.A.11), it was observed that if $(\hat{X},p)$ is a covering space of X and $x_0 \in X$, then the induced homomorphism p_* embeds $\pi_1(\hat{X},\hat{x})$ as a subgroup of $\pi_1(X,x_0)$, where $\hat{x} \in p^{-1}(x_0)$. In this section and in the corresponding problems we investigate in some detail the nature of $p_*(\pi_1(\hat{X},\hat{x}))$ and the topological information it provides.

(13.C.1) Theorem. Suppose that $(\hat{X},p)$ is a covering space of X and $x_0 \in X$. Then $\mathscr{C} = \{p_*(\pi_1(\hat{X},\hat{x})) \mid \hat{x} \in p^{-1}(x_0)\}$ forms a complete conjugate class of subgroups of $\pi_1(X,x_0)$ (i.e., if $\hat{x},\hat{x}' \in p^{-1}(x_0)$, then there is a $g \in \pi_1(X,x_0)$ such that $p_*(\pi_1(\hat{X},\hat{x})) = g^{-1}p_*(\pi_1(\hat{X},\hat{x}'))g$, and furthermore, if $C \in \mathscr{C}$ and $g \in \pi_1(X,x_0)$, then $g^{-1}Cg \in \mathscr{C}$).

Proof. We first establish that $\mathscr{C}$ is a conjugate class of subgroups. Suppose that $p_*(\pi_1(\hat{X},\hat{x}))$ and $p_*(\pi_1(\hat{X},\hat{x}')) \in \mathscr{C}$. Let $\tau : I \to \hat{X}$ be a path such that $\tau(0) = \hat{x}'$ and $\tau(1) = \hat{x}$. Then $p\tau$ is a loop in X based at x_0; furthermore,

$[p\tau]^{-1} = [p\tau']$. We show that $p_*(\pi_1(\hat{X},\hat{x})) = g^{-1}p_*(\pi_1(\hat{X},\hat{x}'))g$, where $g = [p\tau]$. To see this, observe that if $[\alpha] \in \pi_1(\hat{X},\hat{x})$, then $p_*[\alpha] = g^{-1}p_*([\beta])g$, where $[\beta] = [\tau * \alpha * \tau'] \in \pi_1(\hat{X},\hat{x}')$. This follows, since $g^{-1}p_*([\beta])g = [p\tau]^{-1}[p\beta][p\tau]$ $= [p\tau]^{-1}[p\tau][p\alpha][p\tau]^{-1}[p\tau] = p_*([\alpha])$. Thus, $p_*([\alpha]) \in g^{-1}p_*(\pi_1(\hat{X},\hat{x}'))g$, and consequently we have that $p_*(\pi_1(\hat{X},\hat{x})) \subset g^{-1}p_*(\sigma_1(\hat{X},\hat{x}'))g$. If τ is replaced by τ', then the same argument may be used to show that $p_*(\pi_1(\hat{X},\hat{x}')) \subset gp_*(\pi_1(\hat{X},\hat{x}))g^{-1}$, and hence it follows that $g^{-1}p_*(\pi_1(\hat{X},\hat{x}'))g \subset p_*(\pi_1(\hat{X},\hat{x}))$.

Now suppose that $p_*(\pi_1(\hat{X},\hat{x})) \in \mathscr{C}$ and $g = [\beta] \in \pi_1(X,x_0)$. The path β has a unique lifting τ that starts at $\hat{x}$ and ends at some point $\hat{x}'$ in $p^{-1}(x_0)$ (of course, $\hat{x}'$ may be $\hat{x}$). We show that $g^{-1}p_*(\pi_1(\hat{X},\hat{x}))g = p_*(\pi_1(\hat{X},\hat{x}'))$, and consequently that $g^{-1}p_*(\pi_1(\hat{X},\hat{x}))g \in \mathscr{C}$. Suppose that $[\alpha] \in \pi_1(\hat{X},\hat{x})$. Then $g^{-1}p_*([\alpha])g = p_*([\tau' * \alpha * \tau]) \in p_*(\pi_1(\hat{X},\hat{x}'))$. Conversely, if $[\gamma] \in \pi_1(\hat{X},\hat{x}')$, let $[\alpha] = [\tau * \gamma * \tau']$. It is easy to verify that $p_*([\gamma]) = g^{-1}p_*([\alpha])g$.

(13.C.2) **Definition.** A covering space $(\hat{X},p)$ of X is *regular* if and only if there is a point $x_0 \in X$ such that for some $\hat{x} \in p^{-1}(x_0)$, $p_*(\pi_1(\hat{X},\hat{x}))$ is a normal subgroup of $\pi_1(X,x_0)$.

(13.C.3) **Theorem.** Suppose that $(\hat{X},p)$ is a regular covering space of X with $x_0 \in X$ the special point of definition (13.C.2). If $\hat{x},\hat{x}' \in p^{-1}(x_0)$, then $p_*(\pi_1(\hat{X},\hat{x})) = p_*(\pi_1(\hat{X},\hat{x}'))$.

Proof. Without loss of generality, we may assume that $\hat{x}$ is the point of $p^{-1}(x_0)$ such that $p_*(\pi_1(\hat{X},\hat{x}))$ is normal. By the previous theorem, there is a $g \in \pi_1(X,x_0)$ such that $g^{-1}p_*(\pi_1(\hat{X},\hat{x}))g = p_*(\pi_1(\hat{X},\hat{x}'))$. Since $p_*(\pi_1(\hat{X},\hat{x}))$ is normal, the result follows.

The two previous theorems together with (12.B.2) may be used to obtain the following result.

(13.C.4) **Exercise.** If $(\hat{X},p)$ is a regular covering space of X, then for any $x_0 \in X$ and any $\hat{x} \in p^{-1}(x_0)$, $p_*(\pi_1(\hat{X},\hat{x}))$ is normal.

The following exercise and theorem are frequently useful.

(13.C.5) **Exercise.** Suppose that H is a normal subgroup of a group G and that $a,b \in G$. Then $ab \in H$ if and only if $ba \in H$, in which case, $aH = b^{-1}H$.

(13.C.6) **Theorem.** Suppose that $(\hat{X},p)$ is a regular covering space of X. Suppose further that $x_0 \in X$, $\hat{x}_0,\hat{x},\hat{\hat{x}} \in p^{-1}(x_0)$, and that α and β are paths from $\hat{x}_0$ to $\hat{x}$ and from $\hat{x}_0$ to $\hat{\hat{x}}$, respectively. If $[p\alpha]p_*(\pi_1(\hat{X},\hat{x}_0)) = [p\beta]p_*(\pi_1(\hat{X},\hat{x}_0))$, then $\hat{x} = \hat{\hat{x}}$,

Proof.   Since $[p\beta]^{-1}[p\alpha] \in p_*(\pi_1(\hat{X},\hat{x}_0))$, it follows from the previous exercise that $[p\alpha][p\beta]^{-1} = [p\alpha][p\beta'] = [(p\alpha) * (p\beta')] \in p_*(\pi_1(\hat{X},\hat{x}_0))$. Hence, $(p\alpha) * (p\beta')$ lifts to a loop. However, by the unique path lifting theorem, this is impossible unless $\hat{x} = \hat{\hat{x}}$.

D.   MAP LIFTINGS

Suppose that $(\hat{X},p)$ is a covering space of X. We have already seen that paths in X can be lifted uniquely to paths in $\hat{X}$. Can this property be generalized? That is, given a map $f : (Y,y_0) \to (X,x_0)$ does there exist a lifting $f' : (Y,y_0) \to (\hat{X},\hat{x})$ such that $pf' = f$? Remarkably enough, the answer to this query can be formulated in purely algebraic terms. This is expressed in the following extraordinary theorem.

(13.D.1)   Theorem.   Suppose that $(\hat{X},p)$ is a covering space of X, $x_0 \in X$, $\hat{x}_0 \in p^{-1}(x_0)$, and Y is a connected and locally path connected space. Then if $f : (Y,y_0) \to (X,x_0)$ is continuous, there is a unique continuous function $f' : (Y,y_0) \to (\hat{X},\hat{x}_0)$ such that $pf' = f$ if and only if $f_*(\pi_1(Y,y_0)) \subset p_*(\pi_1(\hat{X},\hat{x}_0))$.

Proof.   Half of the proof is trivial, for if such an f' exists, then $f = pf'$ and hence $f_* = p_* f_*'$. This latter equation yields $f_*(\pi_1(Y,y_0)) \subset p_*(\pi_1(\hat{X},\hat{x}_0))$.

Now the remarkable part: the creation of a function f' mapping (Y,y_0) into $(\hat{X},\hat{x}_0)$. Suppose that $y \in Y$, and let α be any path from y_0 to y. Then $f\alpha$ is a path from x_0 to $f(y)$, and by (13.A.8), $f\alpha$ may be lifted uniquely to a path τ_α in $\hat{X}$ with initial point $\hat{x}_0$. Define $f'(y) = \tau_\alpha(1)$. We first establish that f' is well defined, i.e., f' is not dependent on α.

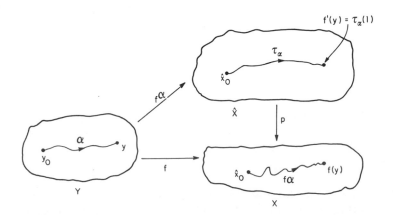

Suppose that β is another path from y_0 to y. We wish to show that $\tau_\alpha(1) = \tau_\beta(1)$, where τ_β is the lifting of the path $f\beta$ with initial point $\hat{x}_0$.

Observe that $\alpha * \beta^r$ is a loop in Y based at y_0 and consequently, $f(\alpha * \beta^r) = \sigma$ is a loop in X based at x_0.

By (13.A.8) there is a path $\tau_{\alpha*\beta^r}$ that lifts $f(\alpha * \beta^r)$. We first show that $\tau_{\alpha*\beta^r}$ is a loop. Since $f(\alpha * \beta^r)$ is a loop, by hypothesis there is a loop h based at $\hat{x}_0$ such that $[ph] = [f(\alpha * \beta^r)]$. Let $H : \mathbf{I} \times \mathbf{I} \to X$ be a homotopy such that $H_0 = ph$, $H_1 = f(\alpha * \beta^r)$, and $H(\{1\} \times \mathbf{I}) = H(\{0\} \times \mathbf{I}) = x_0$. Observe that $ph = H_0$ and hence by the Homotopy Lifting Theorem, H can be lifted to a homotopy $\hat{H} : \mathbf{I} \times \mathbf{I} \to \hat{X}$ such that $\hat{H}_0 = h$ and $\hat{H}_1 = \tau_{\alpha*\beta^r}$ ($p\hat{H}_1 = f(\alpha * \beta^r)$). Furthermore, we have that $\hat{H}(1,t) \in p^{-1}(x_0)$, $\{1\} \times \mathbf{I}$ is connected, and $\hat{H}(1,0) = \hat{x}_0$. Since $p^{-1}(x_0)$ is discrete, it follows that $\hat{H}(1,1) = \hat{x}_0$, and consequently $\tau_{\alpha*\beta^r}$ is a loop based at x_0.

Next we observe that for $0 \le t \le 1$, $p\tau_{(\alpha*\beta^r)}((2-t)/2) = f(\alpha * \beta^r)((2-t)/2)$ $= (f\alpha * f\beta^r)((2-t)/2) = f\beta^r(2((2-t)/2) - 1) = f\beta^r(1-t) = f\beta(t) = p\tau_\beta(t)$. Thus by the uniqueness of path lifting, we have that if $0 \le t \le 1$, then $\tau_\beta(t) = \tau_{\alpha*\beta^r}((2-t)/2)$. Similarly, it is easily shown that for $0 \le t \le 1$, $\tau_\alpha(t) = \tau_{\alpha*\beta^r}(t/2)$. Hence, $\tau_\beta(1) = \tau_{\alpha*\beta^r}(1/2) = \tau_\alpha(1)$ and it follows that f' is well defined.

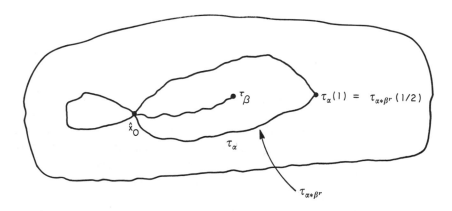

Now we establish the continuity of f'. Suppose that $y \in Y$ and that U is a neighborhood of $f'(y)$. Let V be a canonical open neighborhood of $pf'(y) = f(y)$, and let $\hat{V}$ be the component of $p^{-1}(V)$ that contains $f'(y)$. If $W = U \cap \hat{V}$, then $p_{|W}$ is an embedding and $p(W)$ is open in X. We show that if N is a path connected neighborhood of y such that $f(N) \subset p(W)$, then $f'(N) \subset W \subset U$.

Suppose that $z \in N$. Let α be a path from y_0 to y. The path $f\alpha$ may be lifted to a path τ_α with initial point $\hat{x}_0$; of course, $f'(y) = \tau_\alpha(1)$ by the definition of f'. Let β be a path from y to z contained in N. Then $\alpha * \beta$ is a path from y_0 to z and $f(\alpha * \beta)$ may be uniquely lifted to a path $\tau_{\alpha*\beta}$ with initial point $\hat{x}_0$. Since $(p_{|W})^{-1}(f\beta(t)) \subset W$ for each $t \in \mathbf{I}$, the lifting of $f(\alpha * \beta)$ is given by

$$\tau_{\alpha*\beta}(t) = \begin{cases} \tau_{\alpha}(2t) & 0 \le t \le \tfrac{1}{2} \\ (p_{|W})^{-1}f\beta(2t-1) & \tfrac{1}{2} \le t \le 1. \end{cases}$$

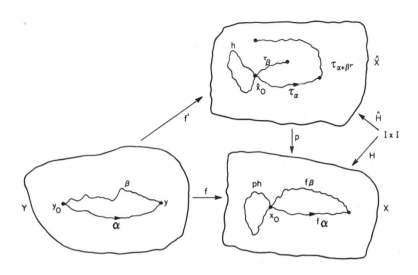

The terminal point of $\tau_{\alpha*\beta}$ is in W, and since $f'(z) = \tau_{\alpha*\beta}(1)$, the result follows.

E. UNIVERSAL COVERING SPACES

A given space may have a number of distinct coverings (13.A.7). In this section, we investigate the possibility of whether or not there is a connection between these covering spaces, and, in particular, whether it is possible to find a "super" covering space: one that covers all the other coverings.

(13.E.1) Definition. Suppose that $(\hat{X}_1, p_1)$ and $(\hat{X}_2, p_2)$ are covering spaces of X. A *covering morphism* is a continuous function $\phi : \hat{X}_1 \to \hat{X}_2$ that yields the following commutative diagram.

(13.E.2) Exercise. Show that each covering morphism ϕ is onto [Hint: Use (13.A.8) and the fact that all the spaces involved are path connected.]

(13.E.3) **Theorem.** Suppose that $(\hat{X}_1, p_1)$ and $(\hat{X}_2, p_2)$ are covering spaces of X and $\phi : \hat{X}_1 \to \hat{X}_2$ is a covering morphism. Then $(\hat{X}_1, \phi)$ is a covering space of $\hat{X}_2$.

Proof. Canonical neighborhoods in $\hat{X}_2$ are constructed as follows. Suppose that $z \in \hat{X}_2$. Let U_1 be an open subset of X that contains $p_2(z)$ and is canonical with respect to p_1. Let U_2 be an open subset of X containing $p_2(z)$ that is canonical with respect to p_2. Finally, let V be the path component of $U_1 \cap U_2$ that contains $p_2(z)$. We show that the path component W of $p_2^{-1}(V)$ that contains z is a canonical open set with respect to ϕ.

Let $\{S_\alpha \mid \alpha \in \Lambda\}$ be the set of path components of $\phi^{-1}(W)$. It must be shown that $\phi_{|S_\alpha} : S_\alpha \to W$ is a homeomorphism onto. However, this follows immediately, since $p_1^{-1}(V) = \phi^{-1} p_2^{-1}(V)$ and V is canonical with respect to both p_1 and p_2.

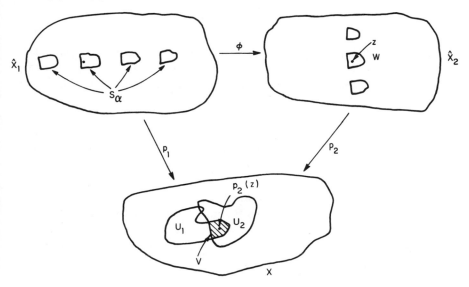

(13.E.4) **Definition.** A covering space $(\hat{X}, p)$ of X is a *universal covering space* if and only if for each covering space $(\hat{X}', p')$ of X, there is a covering morphism ϕ such that the following diagram is commutative.

Note that by (13.E.3), $(\hat{X}, \phi)$ is a covering space for $\hat{X}'$.

To define a universal covering space scarcely constitutes a guarantee that any actually exist. The following theorem gives a simple criterion for establishing that a given covering space is universal. This theorem, together with (13.E.8), will ensure the existence of universal covering spaces under fairly general circumstances.

(13.E.5) Theorem.   If $(\hat{X},p)$ is a covering space of X and $\pi_1(\hat{X},\hat{x}_0) = \{e\}$, then $(\hat{X},p)$ is a universal covering space for X.

Proof.   Suppose that $(\hat{X}',p')$ is any covering space of X. Consider the following diagram.

Since $\pi_1(\hat{X},\hat{x}_0) = \{e\}$, we have that $p_*(\pi_1(\hat{X},\hat{x}_0)) \subset p'_*(\pi_1(\hat{X}',\hat{x}'_0))$ and hence, by (13.D.1), there is a lifting ϕ of p such that the diagram is commutative. Thus, ϕ is a covering morphism, and therefore $(\hat{X},p)$ is a universal covering space.

It follows from the previous theorem that any simply connected space serves as its own universal covering space (where the projection map is the identity). However, we will presently obtain a more substantial result.

(13.E.6) Definition.   A path connected space X is *semilocally simply connected* if and only if X has a basis **B** consisting of path connected sets with the property that if $x \in B \in \mathbf{B}$ and α is a loop in B based at x, then α may be shrunk to x in X, i.e., there is a homotopy $F : \mathbf{I} \times \mathbf{I} \to X$ such that $F_0 = \alpha$, $F_1 = c_x$, and $F(0,t) = F(1,t) = x$ for all $t \in \mathbf{I}$.

(13.E.7) Example.   If M is an n-manifold, then M is semi-locally simply connected.

The next result again illustrates how intricately interwoven are the algebraic and topological aspects of covering space theory.

(13.E.8) Theorem.   Suppose that X is a semilocally simply connected (and, hence, a locally path connected and connected) space and that $x_0 \in X$. Then, corresponding to each subgroup G of $\pi_1(X,x_0)$, there is a covering $(\hat{X}_G,p)$ of X such that $p_* : \pi_1(\hat{X}_G,\hat{x}_0) \to \pi_1(X,x_0)$ is an isomorphism onto G.

Before establishing this theorem, we observe that it implies the existence of a universal covering space for X. Simply let $G = \{e\}$, and apply the preceding theorem.

Proof of (13.E.8). For convenience we subdivide the proof into the following parts:

- (A) Definition of the set $\hat{X}_G$ and the function p.
- (B) Construction of a family **B** of subsets of $\hat{X}_G$.
- (C) Properties of **B**.
- (D) Demonstration that $(\hat{X}_G, p)$ is a covering space (with basis **B**).
- (E) Demonstration that $p_*(\pi_1(\hat{X}_G, \hat{x})) = G$.

(A) Let $\mathscr{P}$ be the collection of all paths in X with initial point x_0. Define an equivalence relation in $\mathscr{P}$ by declaring paths α and β equivalent $(\alpha \underset{G}{\sim} \beta)$ if and only if (i) $\alpha(1) = \beta(1)$ and (ii) $[\alpha * \beta^r] \in G$. Denote the set of equivalence classes $\{\langle \alpha \rangle \mid \alpha \in \mathscr{P}\}$ by $\hat{X}_G$ and observe that the map $p : \hat{X}_G \to X$ defined by $p(\langle \alpha \rangle) = \alpha(1)$ is in fact well defined.

(B) For each $\hat{x} = \langle \alpha \rangle \in \hat{X}_G$ and each open set U in X containing $p(\hat{x})$, define $U_{\hat{x}}$ to be $\{\langle \alpha * \beta \rangle \in \hat{X}_G \mid \beta : \mathbf{I} \to U, \beta(0) = \alpha(1)\}$. The set $U_{\hat{x}}$ is well defined since if $\alpha \underset{G}{\sim} \alpha'$, then $\langle \alpha * \beta \rangle = \langle \alpha' * \beta \rangle$ (this follows from the fact that $[(\alpha * \beta) * (\alpha' * \beta)^r] = [(\alpha * \beta) * (\beta^r * (\alpha')^r)] = [\alpha * (\alpha')^r] \in G$. Since X is semilocally simply connected, X has a basis $\mathscr{S}$ consisting of path connected sets with the property that if ℓ is a loop in $U \in \mathscr{S}$ that is based at x, then $\ell \underset{(0,1)}{\simeq} c_x$ where the homotopy may wander outside of U. Let $\mathbf{B} = \{U_{\hat{x}} \mid U \in \mathscr{S}, \hat{x} \in \hat{X}_G$, and $p(\hat{x}) \in U\}$.

(C) First we show that if $\hat{x}_2 \in U_{\hat{x}_1}$, then $U_{\hat{x}_1} = U_{\hat{x}_2}$. Let $\hat{x}_1 = \langle \alpha \rangle$ and $\hat{x}_2 = \langle \alpha * \beta \rangle$ where $\beta : \mathbf{I} \to U$ and $\beta(0) = \alpha(1)$. Suppose that $\langle \alpha * \sigma \rangle \in U_{\hat{x}_1}$. Then $\langle \alpha * \sigma \rangle = \langle (\alpha * \beta) * (\beta^r * \sigma) \rangle \in U_{\hat{x}_2}$. Hence we have that $U_{\hat{x}_1} \subset U_{\hat{x}_2}$. On the other hand, any member $\langle (\alpha * \beta) * \gamma \rangle \in U_{\hat{x}_2}$ may be written $\langle \alpha * (\beta * \gamma) \rangle$, which of course places it in $U_{\hat{x}_1}$. Note that it follows trivially that if $\hat{x}_1, \hat{x}_2 \in p^{-1}(U)$, then either $U_{\hat{x}_1} = U_{\hat{x}_2}$ or $U_{\hat{x}_1} \cap U_{\hat{x}_2} = \varnothing$.

Next we show that if U is open in X, then $p^{-1}(U) = \bigcup \{U_{\hat{x}} \mid p(\hat{x}) \in U\}$. It is immediate from the definition of $U_{\hat{x}}$ that $\bigcup \{U_{\hat{x}} \mid p(\hat{x}) \in U\} \subset p^{-1}(U)$. On the other hand, if $\hat{z} = \langle \alpha \rangle \in p^{-1}(U)$ then $p(\hat{z}) = \alpha(1) \in U$ and hence $\hat{z} = \langle \alpha \rangle = \langle \alpha * \beta \rangle \in U_{\hat{z}} \in \{U_{\hat{x}} \mid p(\hat{x}) \in U\}$ where β is the constant path at $\alpha(1)$.

Finally we see that **B** is a basis for a topology for $\hat{X}_G$. Suppose that $\hat{z} \in U_{\hat{x}_1} \cap V_{\hat{x}_2}$ where $\hat{z} = \langle \gamma \rangle$, $\hat{x}_1 = \langle \gamma_1 \rangle$, and $\hat{x}_2 = \langle \gamma_2 \rangle$. Then $\hat{z} = \langle \gamma_1 * \beta_1 \rangle$ and $\hat{z} = \langle \gamma_2 * \beta_2 \rangle$ where $\beta_1(\mathbf{I}) \subset U$ and $\beta_2(\mathbf{I}) \subset V$. Consequently, we have that $\gamma \underset{G}{\sim} \gamma_1 * \beta_1$ and $\gamma \underset{G}{\sim} \gamma_2 * \beta_2$. Let $W = U \cap V$. We show that $\hat{z} \in W_{\hat{z}} \subset U_{\hat{x}_1} \cap V_{\hat{x}_2}$. If $\hat{y} \in W_{\hat{z}}$, we have that $\hat{y} = \langle \gamma * \beta \rangle$ where $\beta(\mathbf{I}) \subset W \subset U \cap V$. Therefore, $\hat{y} = \langle \gamma_1 * (\beta_1 * \beta) \rangle = \langle \gamma_2 * (\beta_2 * \beta) \rangle \in U_{\hat{x}_1} \cap V_{\hat{x}_2}$. Thus, **B** is a basis.

(D) Members of the family $\{U \mid U \in \mathscr{S}\}$ will serve as canonical neighborhoods. The continuity of p follows from the properties of $\mathbf{B}$, which may be easily verified by the reader. We prove the following assertion: if $U \in \mathscr{S}$, $\hat{x} = \langle \alpha \rangle$, and $U_{\hat{x}} \subset \hat{X}_G$, then $p_{|U_{\hat{x}}} : U_{\hat{x}} \to U$ is a homeomorphism.

It is easy to show that $p_{|U_{\hat{x}}}$ is onto. Suppose that $y \in U$, and let β be a path in U from $p(\hat{x}) = \alpha(1)$ to y. Then $p(\langle \alpha * \beta \rangle) = y$.

That $p_{|U_{\hat{x}}}$ is one to one may be seen as follows. Suppose that $p(\langle \alpha * \beta \rangle) = p(\langle \alpha * \tau \rangle)$; we show that $\langle \alpha * \beta \rangle = \langle \alpha * \tau \rangle$. Since $p(\langle \alpha * \beta \rangle) = p(\langle \alpha * \tau \rangle)$, we have that $\beta(1) = \tau(1)$, and consequently $\beta * \tau^r$ is a loop in U. It remains to establish that $[(\alpha * \beta) * (\alpha * \tau)^r] \in G$. This is done by showing that $(\alpha * \beta) * (\alpha * \tau)^r = (\alpha * \beta) * (\tau^r * \alpha^r)$ is homotopically equivalent to the constant path $c_{x_0} : I \to X$, which implies that $[(\alpha * \beta) * (\alpha * \tau)^r] = e \in G$.

Since $\beta * \tau^r$ is a loop in U, there is a homotopy $H : I \times I \to X$ such that:

(i) $H(s,0) = \beta * \tau^r(s)$ for each $s \in I$
(ii) $H(s,1) = x_0$ for each $s \in I$
(iii) $H(0,t) = H(1,t) = x_0$ for each $t \in I$

By repeatedly applying the technique used in gluing together the homotopies H and K in (12.B.3), we may "stack" homotopies as indicated in the figure below to obtain the desired equivalence between $\alpha * \beta * \tau^r * \alpha^r$ and c_{x_0}. Thus p is an injection.

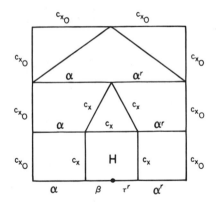

Therefore, we have a covering space of X.

(E)   Finally, we show that $p_*(\pi_1(\hat{X}_G, \hat{x}_0)) = G$, where $\hat{x}_0 = \langle c_{x_0} \rangle$. Before proceeding further, we take advantage of our previous work to glean more information about certain path liftings. Suppose that σ is a path in X with initial point x_0. Then of course σ may be lifted uniquely to a path in X_G that starts at $\langle c_{x_0} \rangle$. Under ordinary circumstances, nothing more could be said about this path; however, if $\sigma' : I \to \hat{X}_G$ is defined by setting $\sigma'(t) = \langle \sigma_t \rangle$, where $\sigma_t(s) = \sigma(ts)$, then it is clear that σ' must be the unique lifting of σ.

Suppose now that $[\alpha] \in \pi_1(\hat{X}_G, \hat{x}_0)$. Then $\tau = p\alpha$ is a loop based at x_0 and can be lifted as indicated above to a map τ'. However, since α also represents a lifting of $p\alpha$, we have that $\alpha = \tau'$. Therefore, $\langle c_{x_0} \rangle = \alpha(1) = \tau'(1) = \langle \tau_1 \rangle = \langle \tau \rangle = \langle p\alpha \rangle$. Since c_{x_0} and $p\alpha$ are equivalent, it follows that $[(p\alpha) * c_{x_0}^r] \in G$. However, $p_*([\alpha]) = [(p\alpha) * c_{x_0}^r]$, and thus $p_*(\pi_1(\hat{X}_G, \hat{x}_0)) \subset G$.

If $[\tau] \in G$, then τ and c_{x_0} belong to the same equivalence class $([\tau * c_{x_0}^r] \in G)$, and hence $\langle \tau \rangle = \hat{x}_0$. Therefore, τ' must be a loop based at $\hat{x}_0$, and since $p_*([\tau']) = [\tau]$, we conclude that $G \subset p_*(\pi_1(\hat{X}_G, \hat{x}_0))$.

(13.E.9) Exercise. Write out the details of the homotopy needed in part (D).

(13.E.10) Definition. Suppose that $(\hat{X}, p)$ is a covering space of X and that $h : (\hat{X}, p) \to (\hat{X}, p)$ is a covering morphism. Then if h is a homeomorphism, h is called a *covering translation* (or a *Deckbewegung*, or a *sheet shifting map*).

It is clear that the set of all sheet shifting maps form a group, where the group operation is composition. This group, which we denote $D(\hat{X}, p)$, is called the *Deckbewegungs group*.

Given a regular covering space $(\hat{X}, p)$, there are a variety of groups that may be associated with it. Two such groups include $\pi_1(X, x_0)/p_*(\pi_1(\hat{X}, \hat{x}))$ (which superficially does not seem to say very much geometrically) and $D(\hat{X}, p)$. This latter group appears to be somewhat removed from any homotopy considerations. The next theorem asserts that an amazing coincidence happens: these two seemingly unrelated groups are isomorphic. This is one of the most aesthetically pleasing results in mathematics.

(13.E.11) Theorem. If $(\hat{X}, p)$ is a regular covering space of X, then $D(\hat{X}, p)$ is isomorphic to $\pi_1(X, x_0)/p_*(\pi_1(\hat{X}, \hat{x}))$, where $\hat{x} \in p^{-1}(x_0)$.

Proof. Suppose that $h \in D(\hat{X}, p)$. Then we have that $h(\hat{x}) = x' \in p^{-1}(x_0)$. Let α be a path from $\hat{x}$ to x' and define $\phi : D(\hat{X}, p) \to \pi_1(X, x_0)/p_*(\pi_1(\hat{X}, \hat{x}))$ by $\phi(h) = [p\alpha]p_*(\pi_1(\hat{X}, \hat{x}))$.

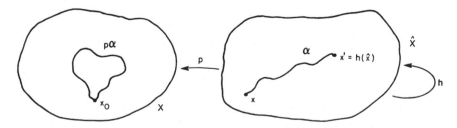

That ϕ is well defined is left to the reader as an easy exercise (recall (13.C.5)).

We now show that ϕ is a homomorphism. Let $x' = h(\hat{x})$, $x'' = g(\hat{x})$, and $x''' = gh(\hat{x}) = g(x')$. Suppose that α is a path from $\hat{x}$ to x' and that β is a path from $\hat{x}$ to x''. Observe that $g\alpha$ is a path from x'' to x''', and hence $\beta * g\alpha$ is a path from $\hat{x}$ to x'''.

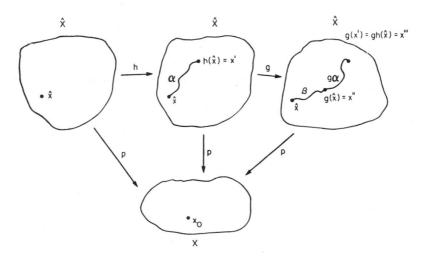

Hence,

$$\phi(gh) = [p(\beta * g\alpha)]p_*(\pi_1(\hat{X},\hat{x})) = [p\beta * pg\alpha]p_*(\pi_1(\hat{X},\hat{x}))$$
$$= [p\beta][p\alpha]p_*(\pi_1(\hat{X},\hat{x})) \quad (\text{since } pg = p)$$
$$= [p\beta]p_*(\pi_1(\hat{X},\hat{x}))[p\alpha]p_*(\pi_1(\hat{X},\hat{x}))$$
$$= \phi(g)\phi(h).$$

That ϕ is one to one may be seen as follows. Suppose that $g,h \in D(\hat{X},p)$ and $\phi(g) = \phi(h)$. Let $g(\hat{x}) = x'$ and $h(\hat{x}) = x''$. Let α and β be paths from $\hat{x}$ to x' and $\hat{x}$ to x'', respectively. Since $[p\alpha]p_*(\pi_1(\hat{X},\hat{x})) = [p\beta]p_*(\pi_1(\hat{X},\hat{x}))$, it follows from (13.C.6) that $x' = x''$. We finish the proof that ϕ is one to one by showing the following: whenever $g,h \in D(\hat{X},p)$ and there is a point $\hat{x} \in \hat{X}$ such that $g(\hat{x}) = h(\hat{x})$, then $g = h$.

Let $p(\hat{x}) = x_0$ and note that both $\hat{x}$ and $g(\hat{x}) = h(\hat{x}) = x'$ are elements of $p^{-1}(x_0)$. Since $(\hat{X},p)$ is regular, we may apply (13.C.4) and (13.C.3) to obtain $p_*(\pi_1(\hat{X},\hat{x})) = p_*(\pi_1(\hat{X},x'))$. Hence, by (13.D.1) there is a unique lifting f of p such that $f(\hat{x}) = x'$. Since both g and h are liftings of p with this property, we have that $g = h$.

Finally we show that ϕ is onto. Consider $[\alpha]p_*(\pi_1(\hat{X},\hat{x}))$ where $[\alpha] \in \pi_1(X,x_0)$. Let τ be the lifting of α which starts at $\hat{x}$, and set $x' = \tau(1)$. We wish to find an $h \in D(\hat{X},p)$ such that $h(\hat{x}) = x'$. (Actually, we show that whenever $x' \in p^{-1}(x_0)$, there is an $h \in D(\hat{X},p)$ such that $h(\hat{x}) = x'$).

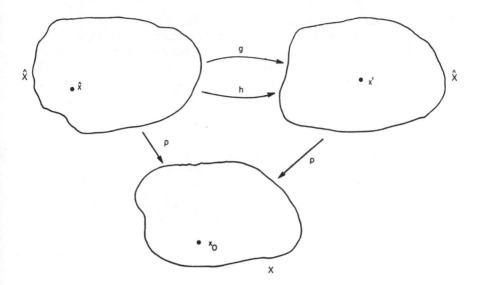

Since $(\hat{X},p)$ is regular, it follows that $p_*(\pi_1(\hat{X},\hat{x})) = p_*(\pi_1(\hat{X},x'))$. By (13.D.1) there are covering morphisms such that the following diagram is commutative.

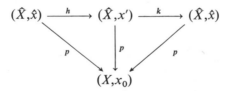

By uniqueness of map liftings, we have that $kh = hk = id_{\hat{x}}$. Hence, $h \in D(\hat{X},p)$.

(13.E.12) Corollary. If $(\hat{X},p)$ is a simply connected covering space of X, then $D(\hat{X},p)$ is isomorphic to $\pi_1(X)$ and the order of $\pi_1(X)$ is the sheet number of the covering space.

(13.E.13) Corollary of Proof. If $(\hat{X},p)$ is a regular covering space with sheet number α, then $D(\hat{X},p)$ and $\pi_1(X,x_0)/p_*(\pi_1(\hat{X},\hat{x}))$ are isomorphic with a transitive subgroup of the symmetric group S_α.

F. TWO EXAMPLES

It follows from (13.E.8) that universal covering spaces are guaranteed for a

fairly wide range of spaces, and may be identified as universal coverings with
the aid of (13.E.5). We begin this section by building in a completely geometric fashion the universal covering space for the figure eight. Suppose that
X is a figure eight,. The universal cover is built up inductively from open
line segments in $\mathscr{E}^2$, as indicated in the following figure. The space $\hat{X}$ will
be the (infinite) union of the A_n, where it is understood that each A_{i+1}
contains A_i.

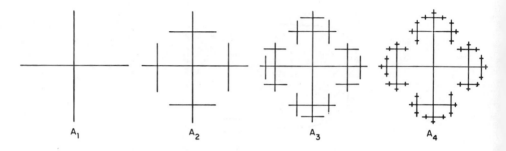

A_1 $\qquad$ A_2 $\qquad$ A_3 $\qquad$ A_4

To obtain a mapping p from $\hat{X}$ onto X, orient X and the segments of $\hat{X}$
in the manner shown in the next figure. Each segment marked with one arrow
is sent by p onto the left half of X in the appropriate manner, and segments
with two arrows are mapped by p onto the right side of X.

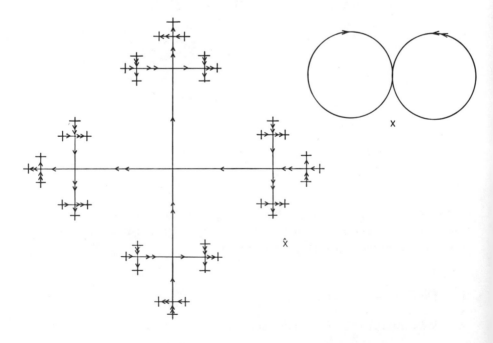

(13.F.1) *Exercise.* Show that $(\hat{X}, p)$ is a universal covering space of X.

As a second example of a universal cover, we demonstrate that the plane is a universal covering space of the Klein bottle K. This result follows immediately from the next exercise and (13.A.7), once it is established that the torus is a covering space for K.

(13.F.2) *Exercise.* Suppose that $(\hat{X}, p_1)$ is a covering space of X, $(\hat{\hat{X}}, p_2)$ is a covering space of $\hat{X}$, and $\pi_1(\hat{\hat{X}}) = \{e\}$. Show that $(\hat{\hat{X}}, p_1 p_2)$ is a universal covering space of X.

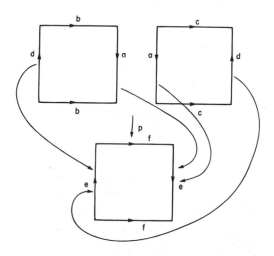

The top two squares together yield one torus after identification.
The bottom square yields a Klein bottle.

That the torus T is a two-sheeted covering space of the Klein bottle K may be seen pictorially above.
Here, b and c are sent to f, while d and a are sent to e in the manner indicated.
Analytically, this may be described as follows. Consider the torus T as $\mathscr{S}^1 \times \mathscr{S}^1$. Let $q : \mathbf{I} \times \mathbf{I} \to K$ be the identification map that identifies sides of $\mathbf{I} \times \mathbf{I}$ as shown in the next figure.
Then $p : T \to K$ defined by

$$p((\cos 2\pi x, \sin 2\pi x), (\cos 2\pi y, \sin 2\pi y))$$

$$= \begin{cases} q(2x, y) & \text{if} \quad 0 \le x \le \tfrac{1}{2} \\ q(2x - 1, 1 - y) & \text{if} \quad \tfrac{1}{2} \le x \le 1 \end{cases}$$

yields a covering space (T, p) of K.

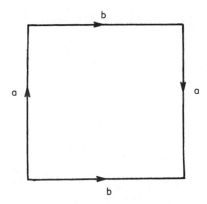

(13.F.3) Exercise. Show that (T,p) is a covering space for K, and consequently that the plane is a universal covering space for K.

PROBLEMS

Sections A and B

1. Prove or disprove the converse of (13.A.2), where f is onto.
2. Suppose that $(\hat{X},p)$ is a covering space of X and that Y is connected. Show that if $f,g : (Y,y_0) \to (\hat{X},\hat{x}_0)$ and $pf = pg$, then $f(y) = g(y)$ for each $y \in Y$.
3. Suppose that $(\hat{X},p)$ is a covering space of X and $f,g : \mathbf{I} \to \hat{X}$ are continuous functions with the same initial point. Show that if $pf \sim pg$ (mod $\{0,1\}$), then $f \sim g$ (mod $\{0,1\}$).
4. Suppose that X and Y are connected and path connected. Show that if X is compact, Y is T_2, and $f : X \to Y$ is a local homeomorphism, then (X,f) is a covering space of Y.
5. Show that if

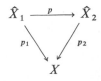

is commutative and $(\hat{X}_1,p_1)$ and $(\hat{X}_2,p_2)$ are covering spaces of X, then $(\hat{X}_1,p)$ is a covering space of $\hat{X}_2$.

6. Show that if X is a compact space and $f : X \to Y$ is a local homeomorphism, then for any point $y \in Y$, card $f^{-1}(y) < \infty$. Show that if it is also assumed that Y is connected and T_2, then f is onto.

7. Suppose that X is a connected and locally path-connected space, and that U and V are path-connected open subsets of X such that $X = U \cup V$. Suppose further that $U \cap V$ consists of two disjoint open sets A and B. Let $a \in A$ and $b \in B$. Show that if a and b can be joined by paths in U and in V, then $\pi_1(X,a)$ is not trivial. [Hint: Assign $Y = \bigcup \{(U \times \{2n\}) \cup (V \times \{2n + 1\}) \mid n \in \mathbf{Z}\}$ the relative topology inherited from $X \times \mathbf{Z}$. Construct a space $\hat{X}$ by identifying in Y:

$$(x,2n) \text{ and } (x, 2n - 1) \text{ for each } x \in A$$

$$(y,2n) \text{ and } (y, 2n + 1) \text{ for each } y \in B.$$

Define now $p : \hat{X} \to X$ by $p((x,n)) = x$. Then $(\hat{X},p)$ is a covering space of X. If α is a path from a to b in U and β is a path from b to a in U, There is a lifting of $\alpha * \beta$ that is not a loop. Conclude that $\alpha * \beta$ is not null homotopic.] Munkres [1975] has used this and another similar result to obtain a very clever and readable proof of the Jordan Curve Theorem.

Section C

1. Find a two sheeted covering space of the figure eight.
2. Show that the covering space found in problem 1 is regular (or else find another one that is).
3. Show that $(\hat{X},p)$ is a regular covering space of X if and only if whenever ℓ is a loop in X based at x_0 that lifts to a loop at some $\hat{x}_0 \in p^{-1}(x_0)$, then ℓ lifts to a loop at each $\hat{x} \in p^{-1}(x_0)$.
4. Suppose that $(\hat{X},p)$ is a covering space of X and that $x_0 \in X$. Let α and β be loops based at x_0. Show that if α and β can be lifted to paths based at $\hat{x}_0$ with a common end point, then α and β determine the same coset of $\pi_1(X,x_0)/p_*(\pi_1(\hat{X},\hat{x}_0))$.
5. Suppose that $(\hat{X},p)$ is a regular covering space of X and that $x_0 \in X$. Show that $(\hat{X},p)$ is an n-sheeted cover of X if and only if the index of $p_*(\pi_1(\hat{X},\hat{x}_0))$ in $\pi_1(X,x_0)$ is n. [Hint: Use the preceding two problems.]

Section D

1. Show that if $(\hat{X},p)$ is a covering space of X and Y is contractible to $y_0 \in Y$, then any map $f : (Y,y_0) \to (X,x_0)$ can be lifted to a map $f' : (Y,y_0) \to (\hat{X},\hat{x}_0)$.

2. A map $p : Y \to B$ is called a *fibration* if p has the homotopy lifting property described in Section B (therefore, any covering map is a fibration). Show that if $p : Y \to B$ is a fibration, and $\sigma : I \to B$ is a path such that $\sigma(0) \in p(Y)$, then σ can be lifted to $\sigma' : I \to Y$.

3. Show that if a fibration $p : Y \to B$ has the *unique* path lifting property, then for each $b \in B$ and for each path α in $p^{-1}(b)$, α is constant.

4.* Prove the converse of problem 3.

5. Suppose that for each $\alpha \in \Lambda$, $f_\alpha : X_\alpha \to B_\alpha$ is a fibration. Show that $f : \prod_{\alpha \in \Lambda} X_\alpha \to \prod_{\alpha \in \Lambda} B_\alpha$ defined by $f(\{x\}_\alpha)_{\alpha \in \Lambda} = \{f_\alpha(x_\alpha)\}_{\alpha \in \Lambda}$ is a fibration.

6. Show that if $f : Y \to B$ is a fibration, then $f(Y)$ is a union of path components of B.

7. Suppose that $(\hat{X}, p)$ is a covering space of X. A *section* of p is a map $s : X \to \hat{X}$ such that $ps = id_X$. Show that p has a section if and only if $p_*(\pi_1(\hat{X}, \hat{x}_0)) = \pi_1(X, x_0)$.

8. Suppose that $(\hat{X}, p)$ is a covering space of X. Show that p has a section if and only if there is a $g : X \to \hat{X}$ such that $pg \sim id$.

Section E

1. Show that if $\hat{X}$ and $\tilde{X}$ are universal covering spaces of a space X, then $\hat{X}$ and $\tilde{X}$ are homeomorphic. [Hint: Use (13.D.1).]

2. Suppose that X is simply connected. Show that if $(\hat{X}, p)$ is a covering space of X, then p is a homeomorphism.

3. Suppose that $(\hat{X}, p)$ is a covering space of X and that $\pi_1(\hat{X}) = \{e\}$. Show that every point $x \in X$ has a neighborhood U with the property that $i_* : \pi_1(U, x) \to \pi_1(X, x)$ is the trivial map (i_* is the homomorphism induced by the inclusion map).

4. Suppose that $(\hat{X}, p)$ is a covering space of X and $\hat{x}_1, \hat{x}_2 \in p^{-1}(x_0)$. Show that there is an $h \in D(\hat{X}, p)$ such that $h(\hat{x}_1) = \hat{x}_2$ if and only if $p_*(\pi_1(\hat{X}, \hat{x}_1)) = p_*(\pi_1(\hat{X}, \hat{x}_2))$.

Problems 5–10 yield the Borsuk-Ulam theorem (problem 8) and some of its consequences.

5. Suppose that $q : \mathscr{S}^1 \to \mathscr{S}^1$ identifies antipodal points. Show that the quotient space derived from q is $\mathscr{S}^1$. Let $p_1 : \mathscr{S}^1 \to \mathscr{S}^1$ be the quotient map. Show that if $p_2 : \mathscr{S}^2 \to P_2$ is the map described in (13.A.4) and $f : \mathscr{S}^2 \to \mathscr{S}^1$ is antipodal (i.e., $f(-x) = -f(x)$ for all $x \in \mathscr{S}^2$), then there is a map $g : P^2 \to \mathscr{S}^1$ such that the following diagram is commutative.

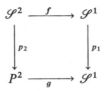

6. Show that $g_* : \pi_1(P_2) \to \pi_1(\mathscr{S}^1)$ is trivial.

7. Suppose that $\alpha : \mathbf{I} \to \mathscr{S}^1 \subset \mathscr{S}^2$ is a path and that $\alpha(0) = -\alpha(1)$. Show that
    (a) $(p_2)_*([\alpha]) \neq e$ and $(p_1)_* f_*([\alpha]) \neq e$, and
    (b) $(p_2)_*([\alpha]) \in \pi_1(P_2,a_0)$ and $(p_1)_* f_*([\alpha]) \in \pi_1(\mathscr{S}^1,b_0)$, where $p_2\alpha(0) = a_0$ and $p_1 f\alpha(a) = b_0$.

8. Show that the function f defined in problem 5 cannot exist (i.e., there does not exist an antipodal function from $\mathscr{S}^1$ onto $\mathscr{S}^1$).

9. Show that if $f : \mathscr{S}^2 \to \mathscr{B}^2$ is continuous and $f(-x) = -f(x)$ for each $x \in \mathscr{S}^2$, then for some $x \in \mathscr{S}^2$, $f(x) = \mathbf{O}$.

10. Show that no subspace of $\mathscr{E}^2$ is homeomorphic to $\mathscr{S}^2$.

11.* Suppose that $(\hat{X},p)$ is an n-sheeted regular covering space of X. Let $x_0 \in X$ be given. Let S_n denote the permutation group on n objects. Suppose that $h \in D(\hat{X},p)$. Note that $h(p^{-1}(x_0)) = p^{-1}(x_0)$, and hence this yields a permutation of n objects. Define $\phi : D(\hat{X},p) \to S_n$, where $\phi(h)$ is the permutation indicated above. Show that ϕ is a 1–1 homomorphism.

12.* Find an upper bound for the number of four-sheeted regular covering spaces of the figure eight. Find as many of them as possible.

13. Suppose that X is a semilocally simply connected space such that $\pi_1(X,x_0)$ is isomorphic to $\mathbf{Z}$. How many distinct covering spaces does X have, and how does one know they are distinct?

14. Find an example of a semilocally simply connected space that is not locally simply connected.

15.* Use (13.E.6) to show that if X is a semilocally simply connected compact space, then $\pi_1(X)$ is finitely generated.

16. Find $D(\mathscr{E}^2,p)$, where $(\mathscr{E}^2,p)$ is a universal covering space of the torus $\mathscr{S}^1 \times \mathscr{S}^1$.

17. Show that if $(\hat{X},p)$ is a covering space of X and $\pi_1(\hat{X}) = \{e\}$, then X is semilocally simply connected.

Chapter 14

SOME ELEMENTS OF
SIMPLICIAL THEORY

When we established the Jordan curve theorem in Chapter 5, the result was first proven for simple closed curves composed of straight line segments, and then generalized to arbitrary simple closed curves. This illustrates one of the basic techniques in topology; theorems are proven for spaces with some kind of linear structure, and then efforts are made to extend these results to more general spaces. Frequently, however, this technique leads to more questions than answers. In this chapter, we shall work almost exclusively with spaces that have or may be given a local linear structure. The advantages inherent in such spaces will become apparent as the theory is developed.

A. THE POLYHEDRAL CATEGORY

(14.A.1) Definition. A set of $k + 1$ points $\{v_0, v_1, \ldots, v_k\}$ in $\mathscr{E}^n$ are *independent* if and only if the vectors $v_1 - v_0, v_2 - v_0, \ldots, v_k - v_0$ are linearly independent (in the algebraic sense).

(14.A.2) Exercise. Show that points $\{v_0, v_1, \ldots, v_k\}$ in $\mathscr{E}^n$ are independent if and only if whenever $\sum_{i=0}^{k} \alpha_i v_i = \mathbf{O}$ and $\sum_{i=0}^{k} \alpha_i = 0$, then $\alpha_i = 0$ for each i.

(14.A.3)   Example.   The points (1,4), (3,0), (2,2) in $\mathscr{E}^2$ are independent but the points (1,4), (2,8), (3,12) are not.

(14.A.4)   Definition.   Suppose that $\{v_0, v_1, \ldots, v_k\}$ is a set of independent points in $\mathscr{E}^n$. Then the set $\sigma = \{\sum_{i=0}^{k} \lambda_i v_i \in \mathscr{E}^n \mid \lambda_i > 0, \sum_{i=0}^{k} \lambda_i = 1\}$ is the *Euclidean k-simplex with vertices* $v_0, v_1, \ldots, v_k$. In this context, σ is said to be *spanned* by $\{v_0, v_1, \ldots, v_k\}$. Such simplices will often be denoted by $\sigma = \langle v_0 v_1 \cdots v_k \rangle$. The *dimension* of σ, dim σ, is k.

If $\{v_0, v_1, \ldots, v_k\}$ is a set of independent points in $\mathscr{E}^n$, then the set

$$\sigma = \left\{ \sum_{i=0}^{k} \lambda_i v_i \in \mathscr{E}^n \mid \lambda_i \geq 0, \sum_{i=0}^{k} \lambda_i = 1 \right\}$$

is called a *closed k-simplex*.

An *r-face* of $\sigma = \langle v_0 v_1 \cdots v_k \rangle$ is the Euclidean *r*-simplex spanned by any subcollection of $\{v_0, v_1, \ldots, v_k\}$ consisting of $r + 1$ vertices. If τ is a face of σ, we write $\tau < \sigma$. Note that every simplex is a face of itself. Quite often, we shall drop the term Euclidean in referring to (Euclidean) *k*-simplices. It is convenient to define the (-1)-simplex as the empty set and to decree that it is a face of all simplices. The *closure* of $\langle v_0 \cdots v_k \rangle$, denoted $\overline{\langle v_0 v_1 \cdots v_k \rangle}$, is defined to be $\{\sigma \mid \sigma$ in a face of $\langle v_0 \cdots v_k \rangle\}$.

In the problem set for this section, another kind of simplex, the abstract simplex, will be introduced.

Analytic geometry is all that is needed to verify that:

(i) $\langle v_0 v_1 \rangle$ is an open line segment with end points v_0 and v_1;

(ii) $\langle v_0 v_1 v_2 \rangle$ is the interior of a triangle with vertices v_0, v_1, and v_2;

(iii) $\langle v_0 v_1 v_2 v_3 \rangle$ is the interior of a tetrahedron;

(iv)   the closed *k*-simplex with vertices $\{v_0, v_1, \ldots, v_k\}$ is equal to $\bigcup \{\sigma \mid \sigma \in \overline{\langle v_0 v_1 \cdots v_k \rangle}\}$;

(v)   if $\langle v_0 v_1 \cdots v_k \rangle = \langle w_0 w_1 \cdots w_k \rangle$, then each v_i is equal to some w_j, and conversely;

(vi)   diam$\langle v_0 v_1 \cdots v_k \rangle = \max\{d(v_i, v_j) \mid 0 \leq i, j \leq k\}$.

Simplices may be used to form a variety of both commonplace and exotic spaces. The precise manner in which this is done is specified in the following definition.

(14.A.5)   Definition.   A *simplicial complex K* (often referred to simply as a *complex K*) is a finite collection of simplices lying in some Euclidean space $\mathscr{E}^n$ that satisfies the following properties:

(i)   if $\sigma \in K$ and $\tau < \sigma$, then $\tau \in K$, and

(ii)   if $\sigma, \tau \in K$, and $\sigma \cap \tau \neq \varnothing$, then $\sigma = \tau$.

If a complex K contains an n-simplex, and for all $k > n$ K contains no k-simplex, then we say that the *dimension* of K is n. We write dim $K = n$.

(14.A.6) Example. In the figures below we illustrate a 2-complex and a 3-complex.

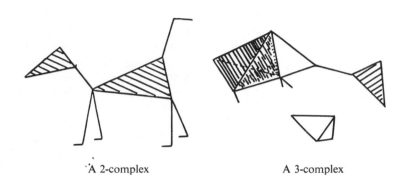

A 2-complex   A 3-complex

Although our primary concern is with complexes defined as above, the reader should be aware that there is a more general definition (to be given shortly) involving "infinite complexes." First, however, let us see what is permitted under the present definition and what is not. The figure on the left represents a complex, while the other two figures fail to represent complexes.

(14.A.7) Definition. A *subcomplex* of a complex K is any subset L of K that is a complex. If L is a subcomplex of K, then the pair (K,L) is called a *simplicial pair*.

(14.A.8) Definition. If K is a complex and L is a subset of K, then the *complex generated by* L (denoted by L^g) is $\{\sigma \in K \mid$ there is a $\tau \in L$ and $\sigma < \tau\}$.

(14.A.9) Exercise. Suppose that L is a subset of a complex K. Show that L^g is a subcomplex of K.

(14.A.10) Definition. Let K be an n-dimensional complex. The *m-skeleton of K* is the set $K_m = \{\sigma \in K \mid \dim \sigma \leq m\}$.

(14.A.11) Exercise. Show that the m-skeleton of a simplicial complex K is a subcomplex.

It should be emphasized that a simplicial complex is merely a collection of simplices, and, as such, is not a topological space. Nevertheless, a given complex K in $\mathscr{E}^n$ determines a subset of $\mathscr{E}^n$, namely the union of all simplices contained in K. This subset is called the *carrier of K* and will be denoted $|K|$. Carriers of complexes are frequently referred to as *Euclidean polyhedra* or *geometric complexes*. We will also on occasion consider $|K|$ when K is not a simplicial complex, but just a set of simplices in $\mathscr{E}^n$. The topology given to a carrier is the relative topology with respect to $\mathscr{E}^n$. If σ is a simplex, the *boundary* of σ, denoted by $\dot{\sigma}$, is the carrier of the union of the proper faces of σ.

(14.A.12) Exercise. Suppose that K is a simplicial complex. Show that if $A \subset |K|$, then A is closed if and only if $A \cap |\sigma^g|$ is closed in $|\sigma^g|$ for each $\sigma \in K$.

The next concept is one of the most basic in topology.

(14.A.13) Definition. A topological space X is *triangulable* if and only if X is homeomorphic to the carrier of some polyhedron. In this case, the homeomorphic image of a simplex is called a *curvilinear simplex*.

(14.A.14) Example.

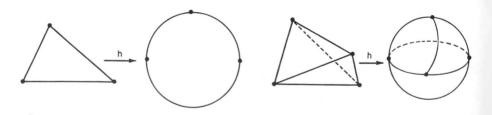

In Chapter 16, a more general notion of simplicial complex and triangulation is employed. The following definitions show in what sense noncompact spaces may be triangulated.

(14.A.15) Definition. An *infinite simplicial complex K* is a collection of

simplices lying in some Euclidean space $\mathscr{E}^n$ (or in Hilbert space) that satisfies the following properties:

(i)   if $\sigma \in K$ and $\tau < \sigma$, then $\tau \in K$;
(ii) if $\sigma, \tau \in K$ and $\sigma \cap \tau \neq \varnothing$, then $\sigma = \tau$.
(iii) if v is a vertex of K, then v is a face of only finitely many simplices of K (*locally finite property*).

If K is an infinite simplicial complex, then the carrier of K, $|K|$, is defined to be $\bigcup \{\sigma \mid \sigma \in K\}$. The carrier $|K|$ is assigned the relative topology.

Condition (iii) ensures that the relative topology and the topology obtained by gluing simplices together will coincide. Interiors, boundaries, etc. are defined as before.

(14.A.16) Definition. A topological space X is *(infinitely) triangulable* if and only if there is a homeomorphism from $|K|$ to X, where K is an *(infinite) complex.*

Throughout the remainder of this chapter, we work with finite complexes; however, the reader may wish to check whether or not the definitions and theorems that appear can be extended to infinite complexes.

(14.A.17) Definition. Suppose that K and L are complexes. A map $f : |K| \rightarrow |L|$ (also denoted $f : K \rightarrow L$) is *simplicial* if and only if

(i)   whenever $\langle v_0 v_1 \cdots v_k \rangle$ is a simplex in K, then $f(v_0), f(v_1), \ldots,$ $f(v_k)$ are the vertices (not necessarily distinct) of a simplex in L, and

(ii) f is linear on each simplex, i.e., if $x = \sum_{i=0}^{k} \lambda_i v_i \in \langle v_0 v_1 \cdots v_k \rangle$, then

$$f(x) = \sum_{i=0}^{k} \lambda_i f(v_i).$$

If (K,L) and (M,N) are simplicial pairs, then a map $f : (K,L) \rightarrow (M,N)$ is *simplicial* if and only if $f : K \rightarrow M$ is simplicial and $f_{|L} : L \rightarrow N$ is simplicial.

(14.A.18) Example. In the figure below

$$K = \{v_1, v_2, v_3, v_4, \langle v_1 v_2 \rangle, \langle v_1 v_3 \rangle, \langle v_2 v_3 \rangle, \langle v_1 v_4 \rangle, \langle v_2 v_4 \rangle, \langle v_1 v_2 v_3 \rangle\},$$

$$L = \{v_1, v_2, v_3, \langle v_1 v_2 \rangle, \langle v_1 v_3 \rangle\},$$

$$M = \{w_1, w_2, w_3, \langle w_1 w_2 \rangle, \langle w_1 w_3 \rangle, \langle w_2 w_3 \rangle\},$$

$$N = \{w_1, w_2, \langle w_1 w_2 \rangle\}, \text{ and if } f \text{ is determined by}$$

$$f(v_1) = w_1$$

$$f(v_2) = f(v_3) = w_2$$
$$f(v_4) = w_3.$$

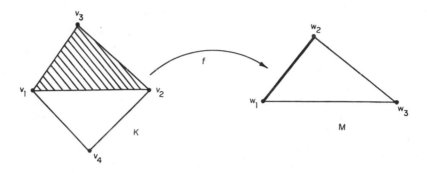

K

M

(14.A.19)   Exercise. Show that simplicial maps are continuous, and that if K and L are complexes, then every map of the vertices of K to the vertices of L determines *at most* one simplicial map.

We are now in a position to define the (finite) polyhedral category (PC). The objects in the category are Euclidean polyhedra and the morphisms are obtained as follows. If X and Y are polyhedra, then a map f belongs to $\mathrm{Hom}_{\mathrm{PC}}(X, Y)$ if and only if there are simplicial complexes K and M such that $|K| = X$, $|M| = Y$, and $f : |K| \to |M|$ is simplicial. Maps in the polyhedral category are called *piecewise linear*.

The difficulty in establishing that the polyhedral category is a category lies in showing that the composition of two morphisms is a morphism. The next section will be devoted to building the machinery needed to sketch a solution to this nontrivial problem.

It should be noted that virtually every question that may be asked in the topological category may also be asked in the polyhedral category, and although polyhedra are fairly manageable spaces, this is offset somewhat by the fact that the allowable mappings are more restricted. Hence, a theorem is not automatically true in the polyhedral category just because it is true in the category of topological spaces and continuous maps.

Perhaps the most fundamental illustration of this problem is the *Hauptvermutung*: if X and Y are homeomorphic Euclidean polyhedra, then there is a piecewise linear homeomorphism between X and Y. It is known, for example, that this conjecture holds for 2-complexes (Papakyriapoulos [1943]) and 3-manifolds (Bing [1954] or Moise [1954]), but fails for 7-manifolds (Milnor [1961]). Although a great deal of work has been done in connection with the Hauptvermutung (Kirby and Siebenmann [1969]), there still remain a number of unanswered questions.

We conclude this section with a theorem whose proof follows immediately from (4.B.28).

(14.A.20)   Theorem.   Any finite simplicial complex is an ANR_M.

B. BASIC CONSTRUCTIONS

In this section, we develop a substantial amount of the structure used in the study of simplicial theory and then indicate how one can show that the polyhedral category is in fact a category.

The reader should be warned that the whole area of simplicial topology is permeated with what euphemistically might be termed quasi-proofs. Often, it is quite easy to visualize pictorially or geometrically what is transpiring, but very difficult to write out the details. We shall occasionally succumb to the rationalization that "little is gained, and often a great deal is obscured by detailed proofs" which has earned a certain degree of respectability in simplicial theory.

(14.B.1)   Definition.   Suppose that K is a simplicial complex and that $\sigma \in K$. Then the *star of σ in K*, star (σ, K), is defined to be $\{\hat{\sigma} \in K \mid \sigma < \hat{\sigma}\}$. The *link of σ in K*, link(σ, K), is defined to be $(\text{star}(\sigma, K))^g \setminus L$, where L is the set of simplices in K that share at least one vertex with σ, and $(\text{star}(\sigma, K))^g$ is the subcomplex of K generated by star(σ, K). When the complex K in clear from the context, we shall often write star(σ), link(σ) and $(\text{star}(\sigma))^g$.

(14.B.2)   Examples.

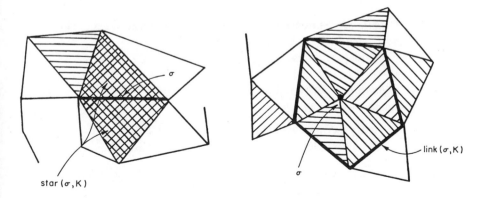

star (σ, K)

link (σ, K)

(14.B.3)   Exercise.   Show that if K is a simplicial complex and v_0 is a vertex of K, then $|\text{star}(v_0, K)|$ is open in $|K|$ and v_0 is the only vertex of K in

$|\mathrm{star}(v_0,K)|$. Furthermore, show that $\{|\mathrm{star}(v,K)| \mid v \in K_0\}$ forms an irreducible open cover of $|K|$.

(14.B.4) Definition. Suppose that σ and τ are simplices in $\mathscr{E}^n$. Then σ and τ are *joinable* if and only if the vertices of σ together with the vertices of σ form an independent set of points. If σ and τ are joinable, their *join* $\sigma \cdot \tau$ is the simplex whose vertices are those of σ and τ.

(14.B.5) Remark. If τ is the (-1)-dimension simplex, then $\sigma \cdot \tau = \sigma = \tau \cdot \sigma$.

(14.B.6) Example.

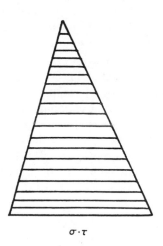

(14.B.7) Definition. Suppose that K and L are complexes in $\mathscr{E}^n$. Then the *join of K and L*, denoted by $K \cdot L$, is defined to be $\{\sigma \cdot \tau \mid \sigma \in K, \tau \in L,$ and σ and τ are joinable$\}$.

(14.B.8) Definition. Suppose that K and L are complexes in $\mathscr{E}^n$. Then K and L are *joinable* if and only if

    (i)   for each $\sigma \in K$ and each $\tau \in L$, σ and τ are joinable, and
    (ii) $K \cdot L$ is a complex.

(14.B.9) Theorem. Suppose that K is a complex and $\sigma \in K$. Then

    (i) $\mathrm{link}(\sigma,K) = \{\tau \in K \mid \sigma$ and τ are joinable and $\sigma \cdot \tau \in K\}$, and

(ii) $(\text{star}(\sigma,K))^g = (\sigma)^g \cdot \text{link}(\sigma,K)$.

Proof.
(i) Recall that $\text{link}(\sigma,K) = (\text{star}(\sigma,K))^g \setminus L$, where L is the set of simplices in K that share at least one vertex with σ. Let $A = \{\tau \in K \mid \sigma \text{ and } \tau \text{ are joinable and } \sigma \cdot \tau \in K\}$. If $\tau \in A$, then τ is a proper face of a simplex $\sigma \cdot \tau$. Since $\sigma \cdot \tau$ lies in $\text{star}(\sigma,K)$, it follows that $\tau \in (\text{star}(\sigma,K))^g$. Furthermore, $\tau \notin L$ (if τ shares a vertex with σ, then τ and σ are not joinable). Thus we have that $A \subset \text{link}(\sigma,K)$.

Now suppose that $\tau \in \text{link}(\sigma,K)$, i.e., $\tau < s \in K$, where $\sigma < s$ and σ and τ do not share any vertices. Then, since σ and τ are both faces of s, it follows that σ and τ are joinable, and furthermore, $\sigma \cdot \tau < s$; therefore, $\sigma \cdot \tau \in K$.

(ii) Suppose that $\tau \in (\text{star}(\sigma,K))^g$. We consider two cases.

Case 1. Suppose that $\tau \in \text{star}(\sigma,K)$, i.e., $\sigma < \tau$. Let γ be the simplex spanned by the vertices of τ that are not vertices of σ. Then by part (i), $\gamma \in \text{link}(\sigma,K)$ and $\sigma \cdot \gamma = \tau$; hence, $\sigma \cdot \gamma \in \sigma^g \cdot \text{link}(\sigma,K)$.

Case 2. Suppose that $\tau \in (\text{star}(\sigma,K))^g \setminus \text{star}(\sigma,K)$. Then there is a simplex $\gamma \in \text{star}(\sigma,K)$ such that $\tau < \gamma$ and $\sigma \not< \tau$. Let $\hat\sigma$ be the simplex spanned by the vertices that σ shares with τ, and let $\hat\tau$ be the simplex spanned by the vertices of τ that are not shared with σ. Then we have that $\hat\sigma \cdot \hat\tau = \tau$, $\hat\sigma \in \sigma^g$, and $\hat\tau \in \text{link}(\sigma,K)$. Therefore, $\tau \in \sigma^g \cdot \text{link}(\sigma,K)$.

Suppose that $\hat\sigma \in \sigma^g$ and $\hat\tau \in \text{link}(\sigma,K) = \{\tau \in K \mid \sigma \text{ and } \tau \text{ are joinable and } \sigma \cdot \tau \in K\}$. Then $\hat\sigma$ and $\hat\tau$ are joinable, $\hat\sigma \cdot \hat\tau < \sigma \cdot \hat\tau$, and $\sigma \cdot \hat\tau \in \text{star}(\sigma,K)$ by definition. Consequently, $\hat\sigma \cdot \hat\tau \in (\text{star}(\sigma,K))^g$.

(14.B.10) Definition. Suppose that K and L are simplicial complexes such that $|K| = |L|$. Then K is a *subdivision* of L if and only if $\hat\sigma \in K$ implies that there is an $\sigma \in L$ such that $\hat\sigma \subset \sigma$.

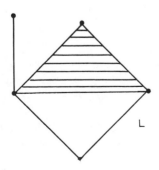

L

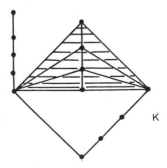

K

Note that K is a subdivision of L if and only if $|K| = |L|$ and each simplex of L is the union of simplices of K.

Given any complex with positive dimension, there are an uncountable number of possible subdivisions. We shall describe two particular types of subdivision that are of frequent use.

(14.B.11)   *Definition.*   Suppose that K is a simplicial complex and that $\sigma = \langle v_0 v_1 \cdots v_m \rangle \in K$. The *barycenter* of σ is the point $b_\sigma = (1/(m+1)) \sum_{i=0}^{m} v_i$.

(14.B.12)   *Definition.*   If σ is a simplex in a complex K, then we can barycentrically subdivide σ^g as follows. For each face $s < \sigma$, let b_s denote its barycenter. The complex consisting of all simplices of the form $\langle b_{s_1}, b_{s_2}, \ldots, b_{s_r} \rangle$, where $s_1 < s_2 < \cdots < s_r$ and each $s_i < \sigma$, is called the *barycentric subdivision* of σ^g. The *barycentric subdivision* of a complex K is the complex $K^{(1)}$ resulting from barycentrically subdividing each simplex in K.

(14.B.13)   *Example.*

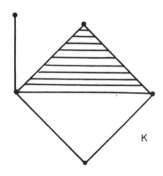
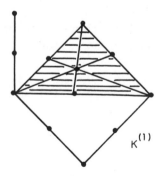

K $K^{(1)}$

(14.B.14)   *Definition.*   If K is a complex, then the *second barycentric subdivision* of K, denoted by $K^{(2)}$, is the complex obtained by barycentrically subdividing $K^{(1)}$. In general, the *n-th barycentric subdivision* of K is $(K^{(n-1)})^{(1)}$ (see 14.B.15).

More generally, if one performs the construction employed in building a barycentric subdivision, but uses, in place of each b_s, an arbitrary point in s, then a *first derived complex*, K', is obtained (see 14.B.16).

(14.B.15)   Example.

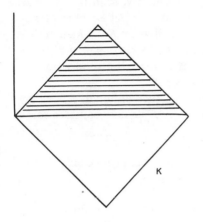
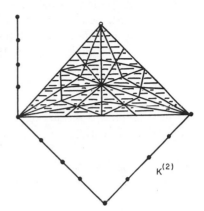

(14.B.16)   Example.

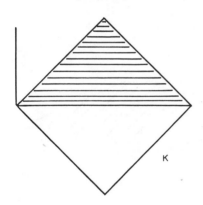
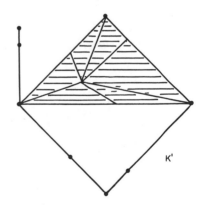

(14.B.17)   Exercise.   Suppose that K is a simplicial complex. Let K' be a derived complex of K and $K^{(1)}$ be the barycentric subdivision of K. Show that there is a simplicial map $\phi : |K'| \to |K^{(1)}|$ that is a homeomorphism.

We now outline one way of showing that the composition of two piecewise linear maps is piecewise linear. Suppose that K, L, M, and N are complexes such that $|L| = |M|$. Let $f : |K| \to |L|$ and $g : |M| \to |N|$ be simplicial maps. We want to find simplicial complexes $\hat{K}$ and $\hat{N}$ such that $|\hat{K}| = |K|$, $|\hat{N}| = |N|$ and $gf : |\hat{K}| \to |\hat{N}|$ is simplicial.

We proceed as follows. First, we shall show that if $|L| = |M|$, then there are subdivisions $\hat{L}$ and $\hat{M}$ of L and M, respectively, such that $\hat{L} = \hat{M}$ (14.B.30). Next, we indicate how one can show that if $f : |K| \to |L|$ is simplicial and L' is a subdivision of L, then there is a subdivision K' of K such that $f : |K'| \to |L'|$ is simplicial (14.B.31). Finally, it is proven that if $g : |M| \to |N|$ is simplicial and M' is a subdivision of M, then there is a subdivision N' of N such that $g : |M'| \to |N'|$ is simplicial (14.B.34).

First, notice that if σ and τ are simplices in $\mathscr{E}^n$, then although $\sigma \cap \tau$ may not be a simplex, it is a convex set (possibly empty). We now consider some special convex sets.

(14.B.18)   Definition.   Suppose that $A \subset \mathscr{E}^n$. Then the *convex hull* of A is the intersection of all convex subsets in $\mathscr{E}^n$ that contain A.

(14.B.19)   Definition.   A *linear cell* is the convex hull C of a finite set of points S in $\mathscr{E}^n$. The minimal subset V of S with convex hull C is called the *set of vertices* of C. The *dimension* of C is the largest number m such that V contains $m + 1$ independent points. In this case, we say that C is a *linear m-cell*.

(14.B.20)   Example.

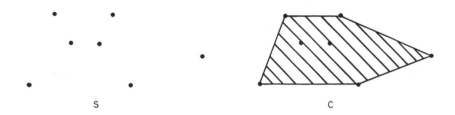

S   C

Note that C has five vertices and is of dimension 2.

(14.B.21)   Exercise.   Show that a linear n-cell C is homeomorphic with $|\sigma^g|$ where σ is an n-simplex.

(14.B.22)   Definition.   Suppose that C is a linear n-cell with vertex set V. A subset F of C is an $(n - 1)$-*face of* C if and only if

(i) F is a maximal linear $(n - 1)$-cell with vertex set a subset of V, and
(ii) $C \setminus F$ is connected.

An $(n - 2)$-face of C is an $(n - 2)$-face of an $(n - 1)$-face of C, and in general, an $(n - k)$-face of C is an $(n - k)$-face of an $(n - k + 1)$-face of C.

We assume that $\varnothing$ is a (-1)-face of any n-cell C, and that C is the only n-face of C.

The *boundary of* C, denoted by $\dot{C}$, is the union of the $(n-1)$-faces of C. If C is an n-cell, $n \geq 1$, then the *interior of* C, denoted by $\overset{\circ}{C}$, is $C \setminus \dot{C}$. If C is a 0-cell, then $\overset{\circ}{C} = C$.

(14.B.23)   Definition.   A *linear cell complex* K is a finite set of linear cells such that

    (i)   if $C \in K$ and D is a face of C, then $D \in K$, and
    (ii) if $C, D \in K$, then $C \cap D \in K$.

Since in this chapter we be deal only with linear cells and linear cell complexes, the modifier "linear" is dropped with the hope that no confusion will arise later in the book when more general types of cells are considered.

(14.B.24)   Example.

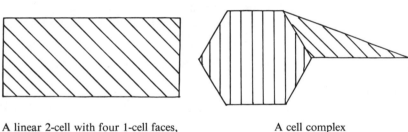

A linear 2-cell with four 1-cell faces, four 0-cell faces, and one (-1)-cell face

A cell complex

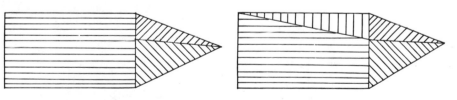

Not a cell complex

A cell complex

(14.B.25)   Definition.   The *carrier* $|K|$ of a cell complex K in $\mathscr{E}^n$ is defined by $|K| = \bigcup_{C \in K} C$. It is assigned the relative topology in $\mathscr{E}^n$.

A corollary of our first theorem will be that the carrier of a linear cell complex is a Euclidean polyhedron.

(14.B.26) Definition. Suppose that K and L are cell complexes. Then L is a subdivision of K if and only if

   (i) $|K| = |L|$, and
   (ii)   for each cell $C \in K$, there are cells $D_1, \ldots, D_m \in L$ such that $|C| = \bigcup_{i=1}^{m} |D_j|$.

(14.B.27) Theorem. Suppose that K is a cell complex. Then K has a subdivision L that is a collection of closed simplices whose interiors form a simplicial complex with the same set of vertices as K.

   Proof. Let $\hat{L} = \{C \in K \,|\, C$ is a vertex or C is a linear 1-cell$\}$. Let $v_0, v_1, \ldots, v_m$ be the vertices of K. For each cell C_i of K, let v_{C_i} be the vertex in C_i with smallest index.
   For each 2-cell C, we have that $|C| = |v_C \cdot D_1|$, where D_1 is the collection of 0- and 1-faces of C that do not contain v_C. Add $v_C \cdot D$ to $\hat{L}$. For each 3-cell $C \in K$, we have that $|C| = |v_C \cdot D_2|$, where D_2 is the collection of members of $\hat{L}$ that lie in C and do not contain v_C. Add $v_C \cdot D_2$ to $\hat{L}$.
   We continue the construction until all the cells of K have been subdivided into closed simplices.

(14.B.28) Example.

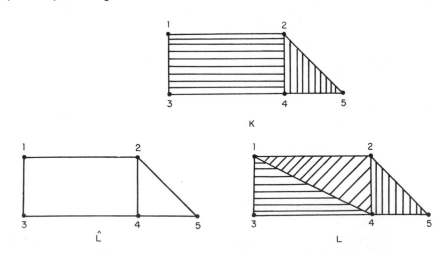

(14.B.29) Corollary. If K is a cell complex, then $|K|$ is a Euclidean polyhedron.

(14.B.30) Corollary. Suppose that K and L are simplicial complexes such

that $|K| = |L|$. Then there is a simplicial complex M that is a common subdivision of K and L.

Proof. Observe that the intersection of two closed simplices is a linear cell. Let $N = \{|\sigma^g| \cap |\tau^g| \mid \sigma \in K, \tau \in L\}$. Then N is a cell complex. Furthermore, N is a cell subdivision of the cell complexes that corresponds in a natural way to the simplicial complexes K and L. We apply (14.B.27) to N to obtain a collection of closed simplices whose interiors form a simplicial complex M that will be a subdivision of K and L.

(14.B.31) **Theorem.** Suppose that K and L are simplicial complexes and that $f : K \to L$ is simplicial. Let M be a simplicial subdivision of L. Then there is a simplicial subdivision N of K such that $f : N \to M$ is simplicial.

Sketch of Proof. Since f is simplicial, it follows that if $\sigma \in K$, then $f(\sigma) \in L$. Let $\hat{K} = \{|\sigma^g| \cap f^{-1}(|\tau^g|) \mid \sigma \in K, \tau \in M\}$. If $\hat{K}$ is a cell complex, then it is clearly a subdivision of the cell complex associated with K, and an application of (14.B.27) yields a simplicial subdivision of K on which f is simplicial.

That $\hat{K}$ is a cell complex follows from the three following assertions, which are more or less visually clear, but technically gruesome:

(i) $|\sigma^g| \cap f^{-1}(|\tau^g|)$ is a linear cell;
(ii) faces of $|\sigma^g| \cap f^{-1}(|\tau^g|)$ are obtained from $|s^g| \cap f^{-1}(|\mu^g|)$, where $s < \sigma$ and $\mu < \tau$;
(iii) $(|\sigma^g|) \cap f^{-1}(|\tau^g|) \cap (|\mu^g| \cap f^{-1}(|\gamma^g|)) = (|\sigma \cap \mu|^g) \cap f^{-1}(|\tau \cap \gamma|^g)$ is a face of both $|\sigma^g| \cap f^{-1}(|\tau^g|)$ and $|\mu^g| \cap f^{-1}(|\gamma^g|)$.

(14.B.32) **Exercise.** Suppose that X, Y, and Z are polyhedra. Show that if $f \in \mathrm{Hom}_{PC}(X,Y)$ and $g \in \mathrm{Hom}_{PC}(Y,Z)$ and g is onto, then $gf \in \mathrm{Hom}_{PC}(X,Z)$.

If g is not onto, there is still one complication in showing that the composition of piecewise linear maps is piecewise linear. This difficulty is overcome by showing that whenever K is a subcomplex of L and $\hat{K}$ is a subdivision of K, there is a subdivision of L that contains $\hat{K}$ as a subcomplex.

(14.B.33) **Theorem.** Suppose that K is a subcomplex of a simplicial complex L and $\hat{K}$ is a subdivision of K. Then there is a subdivision of L that contains $\hat{K}$ as a subcomplex.

Proof. Let $F_1 = \{\sigma \in L \mid \sigma \in K\}$, let $F_2 = \{\sigma \in L \mid \sigma^g \subset L \setminus K\}$, and let $F_3 = \{\sigma \in L \mid \sigma \notin K$ and there is a $\tau < \sigma$ such that $\tau \in K\}$. We construct the desired subdivision of L as follows.

(i)   For each $\sigma \in F_1$, use the subdivision induced by $\hat{K}$.

(ii) Leave each $\sigma \in F_2$ alone.

(iii) Order the simplices in F_3 so that if $i < j$, then dim $\sigma_i \leq$ dim σ_j.

For each σ_i, pick a point $v_i \in \sigma_i$. Subdivide σ_1 by taking $v_1 \cdot B_1$, where $B_1 = \{_1\tau_1, {}_1\tau_2, \ldots, {}_1\tau_{n_1}\}$ is the set of simplices obtained in (i) and (ii) such that $|_1\tau_i| \subset |\dot{\sigma}_1|$. Subdivide σ_2 by taking $v_2 \cdot B_2$, where $B_2 = \{_2\tau_1, \ldots, {}_2\tau_{n_2}\}$, where each $|_2\tau_i|$ is a subset of $|\sigma_2^q|$ and $_2\tau_i$ is a simplex obtained from either (i) or (ii) or is a simplex obtained from the previous subdivision of σ_1. Subdivide σ_2 by taking $v_2 \cdot B_2$. This process is repeated inductively to obtain L.

(14.B.34) Corollary. Suppose that K and L are simplicial complexes, $f : K \to L$ is simplicial, and $\hat{K}$ is a subdivision of K. Then there is a subdivision $\hat{L}$ of L and a subdivision $\hat{\hat{K}}$ of $\hat{K}$ such that $f : \hat{\hat{K}} \to \hat{L}$ is simplicial.

Proof.   Note that $f(\hat{K})$ yields a subdivision of a subcomplex of L.

(14.B.35) Theorem. Suppose that X, Y, and Z are Euclidean polyhedra. If $f \in \text{Hom}_{PC}(X,Y)$ and $g \in \text{Hom}_{PC}(Y,Z)$, then $gf \in \text{Hom}_{PC}(X,Z)$.

Proof.   Let K, L, M, and N be simplicial complexes such that $|K| = X$, $|L| = |M| = Y, |N| = Z, f : K \to L$ is simplicial, and $g : M \to N$ is simplicial. By (14.B.30), L and M have a common subdivision P. Next observe that there is a subcomplex Q of N such that $|Q| = g(|M|)$. By (14.B.32) there is a subdivision $\hat{K}$ of K and a subdivision $\hat{Q}$ of Q such that $gf : \hat{K} \to \hat{Q}$ is simplicial. Finally apply (14.B.33) to obtain a subdivision $\hat{N}$ of N that contains $\hat{Q}$ as a subcomplex, and $gf : \hat{K} \to \hat{N}$ is still simplicial.

C.   SOME CONNECTIONS WITH THE TOPOLOGICAL CATEGORY: THE SIMPLICIAL APPROXIMATION THEOREM

The simplicial approximation theorem will show that when attention is restricted to homotopy relations, the failings of the Hauptvermutung, mentioned at the end of Section A, are not so disastrous. In fact, with the aid of this theorem, polyhedral structures may be exploited to yield a number of interesting results. First, however, further definitions and lemmas are in order.

(14.C.1) Definition. The *mesh* of a complex K is max $\{\text{diam } \sigma \mid \sigma \in K\}$.

If K is a complex, it would appear, at least geometrically, that the mesh of $K^{(n)}$ converges to 0 as n increases. This is a trivial corollary of the following theorem.

(14.C.2) *Theorem.* Suppose that K is an n-dimensional complex with mesh γ. Then mesh $K^{(1)} \le (n/(n+1))\gamma$.

Proof. Suppose that $\tau \in K^{(1)}$ and that $\sigma = \langle v_0, v_1 \cdots v_t \rangle \in K$ is a simplex such that $\tau \in (\sigma^g)^{(1)}$. We show that diam $\tau \le (n/(n+1))\gamma$. Let v and w be any two vertices of τ. Allowing for possible reordering, we may write $v = 1/r \sum_{i=0}^{r-1} v_i$ and $w = 1/s \sum_{j=0}^{s-1} v_j$, where $0 \le s < r \le t+1 \le n+1$. Then $d(v,w)$ is the length of the vector $v - w = (1/s \sum_{j=0}^{s-1} v_j) - (1/r \sum_{i=0}^{r-1} v_i) = (1/r)(r/s \sum_{j=0}^{s-1} v_j - \sum_{i=0}^{r-1} v_i) = 1/r (\sum_{i=0}^{r-1} 1/s \sum_{j=0}^{s-1} (v_j - v_i)) = 1/(rs) \sum_{i=0}^{r-1} \sum_{j=0}^{s-1} (v_j - v_i)$. Now recall from linear algebra that $|u + u'| \le |u| + |u'|$. Hence $d(v,w) \le 1/(sr) \sum_{i=0}^{r-1} \sum_{j=0}^{s-1} |v_j - v_i| \le ((rs - s)/rs)\gamma \le ((r-1)/r)\gamma \le (n/(n+1))\gamma$.

(14.C.3) *Theorem.* Suppose that $v_0, v_1, \ldots, v_k$ are vertices of a complex K. Then $v_0, v_1, \ldots, v_k$ are the vertices of a simplex in K if and only if $\bigcap_{i=0}^{k} \text{star}(v_i) \ne \varnothing$ (or equivalently, $\bigcap_{i=0}^{k} |\text{star}(v_i)| \ne \varnothing$).

Proof. If $v_0, v_1, \ldots, v_k$ are the vertices of a simplex σ, then clearly, $\sigma \in \text{star}(v_i)$ for each i, and hence, $\sigma \in \bigcap_{i=0}^{k} \text{star}(v_i)$ and furthermore, $|\sigma| \subset \bigcap_{i=0}^{k} |\text{star}(v_i)|$.

Now suppose that $z \in |\text{star}(v_i)|$ for $i = 0, 1, \ldots, k$. Let σ be the unique simplex of K that contains z. Since $\sigma \subset |\text{star}(v_i)|$ for each i, we have that $\sigma \in \text{star}(v_i)$, and hence for each i, $v_i \in \sigma$. Since any subset of the vertices of a simplex spans a face of the simplex, it follows that $v_0, v_1, \ldots, v_k$ determines a simplex of K.

(14.C.4) *Definition.* Suppose that $f : |K| \to |L|$ is an arbitrary map. A simplicial map $\phi : |K| \to |L|$ is a *simplicial approximation* to f if and only if for each vertex v in K, $f(|\text{star}(v)|) \subset |\text{star}(\phi(v))|$.

(14.C.5) *Remark.* Note that if $\phi : |K| \to |L|$ is a simplicial approximation to $f : |K| \to |L|$ and if mesh$\{\{\text{star}(v)\} \mid v \in L\} = \varepsilon$, then $d(f(x),\phi(x)) \le \varepsilon$ for each $x \in |K|$.

Before taking up the question of the existence of simplicial approximations, we exhibit a sequence of lemmas that give some connections between simplicial approximations and homotopy.

(14.C.6) Lemma. If $\phi : |K| \to |L|$ is a simplicial approximation to $f : |K| \to |L|$ (f need not be continuous), then for all $x \in |K|$, $f(x)$ and $\phi(x) \in |\tau^g|$ for some $\tau \in L$.

Proof. Suppose that $x \in |K|$. Then $x \in \sigma^n = \langle v_0 v_1 \cdots v_n \rangle$ for some $\sigma^n \in K$. For each $i = 0, 1, \ldots, n$, we have that $f(x) \in f(\sigma^n) \subset f(|\text{star}(v_i)|) \subset |\text{star}(\phi(v_i))|$. Thus, $f(x) \in \bigcap_{i=0}^{n} |\text{star}(\phi(v_i))| \neq \varnothing$, and it follows from (14.C.3) that $\{\phi(v_i)\}_{i=1}^{n}$ spans a simplex $v \in L$. Although $f(x)$ need not be an element of v, we do have that $f(x)$ is an element of some simplex $\tau \in L$. Furthermore, for each i, $\tau \cap |\text{star}(\phi(v_i))| \neq \varnothing$. Hence, it follows that $\tau \in \bigcap_{i=0}^{n} \text{star}(\phi(v_i))$, i.e., each $\phi(v_i)$ is a vertex of τ. Now if $x = \sum_{i=0}^{n} a_i v_i$, then we have that $\phi(x) = \sum_{i=0}^{n} a_i \phi(v_i) \in |\tau^g|$.

(14.C.7) Lemma. If $\phi : |K| \to |L|$ is a simplicial approximation to $f : |K| \to |L|$ and if for some vertex v of K, $f(v)$ is a vertex of L, then $f(v) = \phi(v)$.

Proof. Recall that $|\text{star}(f(v))|$ contains only one vertex (14.B.3).

(14.C.8) Lemma. If $\phi : |K| \to |L|$ is a simplicial approximation to $f : |K| \to |L|$ and if $f_{|\sigma}$ is simplicial for some $\sigma \in K$, then $\phi_{|\sigma} = f_{|\sigma}$.

Proof. By the preceding lemma, $\phi(v_i) = f(v_i)$ for each vertex of σ. Suppose that $x = \sum_{i=0}^{n} a_i v_i$, where each v_i is a vertex of σ. Since f is simplicial on σ, we have that $f(x) = \sum_{i=0}^{n} a_i f(v_i) = \sum_{i=0}^{n} a_i \phi(v_i)$.

(14.C.9) Lemma. If $\phi : |K| \to |L|$ is a simplicial approximation to a continuous function $f : |K| \to |L|$ and if $f|_{|M|}$ is simplicial where M is a subcomplex of K, then $f \sim \phi (\text{mod } |M|)$.

Proof. Let $F : |K| \times I \to |L|$ be defined by $F(x,t) = t\phi(x) + (1 - t)f(x)$. This makes sense, since if $x \in |K|$, then $\phi(x)$ and $f(x)$ lie in the same closed simplex of L. Furthermore, by (14.C.8), we have that $\phi(x) = f(x)$ wherever f is simplicial. Hence, if $x \in |M|$, then $F(x,t) = t\phi(x) + (1 - t)\phi(x) = \phi(x) = f(x)$.

This last lemma suggests why it is worthwhile to seek simplicial approximations. Maps that can be simplicially approximated may be represented in the homotopy category by simplicial maps. Some consequences of this will be given in the corollaries of the simplicial approximation theorem.

(14.C.10) Example. The following continuous function f cannot be simplicially approximated.

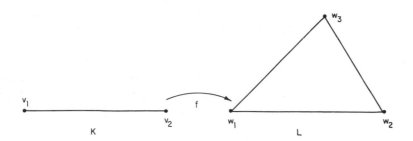

The function, f maps K onto the perimeter of the triangle $w_1 w_2 w_3$ and sends both v_1 and v_2 to w_1. The problem here is that K does not possess enough vertices for f to be simplicially approximated.

(14.C.11) Theorem (Simplicial Approximation Theorem I). Suppose that $f : |K| \to |L|$ is a continuous map. Then there is an integer n such that $f : |K^{(n)}| \to |L|$ may be simplicially approximated.

Proof. It follows from (14.B.3) that for each vertex $v \in L$, $|\text{star } (v)|$ is open in $|L|$. Thus, $\mathcal{U} = \{f^{-1}(|\text{star}(v)|) \mid v$ a vertex of $L\}$ is an open covering of the compact space $|K|$. Let δ be a Lebesgue number for $\mathcal{U}$ and choose an integer n large enough so that mesh $K^{(n)} < \delta/2$ (14.C.2). Then for each vertex $w \in K^{(n)}$, there is a vertex $v \in L$, such that $f(|\text{star}(w)|) \subset |\text{star}(v)|)|$.

We define a simplicial approximation ϕ to $f : |K^{(n)}| \to |L|$ as follows. If w is a vertex in $K^{(n)}$, let $\phi(w)$ be any vertex in L with the property that $f(|\text{star}(w)|) \subset |\text{star}(\phi(w))|$. The map ϕ then has the following property: If $w_0, w_1, \ldots, w_r$ are the vertices of a simplex in $K^{(n)}$, then the collection of vertices $V = \{\phi(w_0), \phi(w_1), \ldots, \phi(w_r)\}$ are the vertices of a simplex in L (of course, some of the vertices in V may be listed more than once). To see this, note that $\bigcap_{i=1}^{r} |\text{star}(w_i)| \neq \varnothing$, and that $f(|\text{star}(w_i)|) \subset |\text{star}(\phi(w_i))|$ for each i. Thus, $\bigcap_{i=1}^{r} \text{star}(\phi(w_i)) \neq \varnothing$, and hence, by (14.C.3), the vertices in V are vertices of some simplex in L.

The map ϕ may now be extended linearly to the interiors of simplices in $K^{(n)}$ (first to the 1-simplices, then the 2-simplices, etc.), which completes the proof.

The previous theorem and (14.C.9) may be combined to yield the following result.

(14.C.12) Theorem (Simplicial Approximation Theorem II). Suppose that K and L are complexes and $f : |K| \to |L|$ is continuous. Then there are an integer n and a simplicial approximation ϕ to $f : |K^{(n)}| \to |L|$ such that $\phi \sim f \pmod A$, where $A = \{x \in |K| \mid f(x) = \phi(x)\}$.

The next corollary follows easily from (14.C.12), (14.C.2), and (14.C.4).

(14.C.13) Corollary. Suppose that K and L are simplicial complexes and that $f : |K| \to |L|$ is continuous. Let $\varepsilon > 0$ be given. Then there are integers m and n and a simplicial map $\phi : |K^{(n)}| \to |L^{(m)}|$ that is homotopic to $f \pmod A$ and moves points a distance less than ε (where A is $\{x \in |K| \mid f(x) = \phi(x)\}$).

(14.C.14) Corollary. Suppose that K and L are complexes. Then every homotopy equivalence class of $\mathrm{Hom}(|K|,|L|)$ can be represented by a simplicial map. In particular, if $f : |K| \to |L|$ is a homeomorphism, then there is a simplicial approximation ϕ of f such that the isomorphism f_* is equal to ϕ_* (of course ϕ need not be a simplicial homeomorphism).

(14.C.15) Corollary. The fundamental group of $\mathscr{S}^n$ is trivial for $n \geq 2$ (cf. the discussion between (12.E.1) and (12.E.2)).

Proof. Let $\mathscr{S}^n$ be represented by the boundary of an $n + 1$ simplex σ. Let v_0 be a vertex, and let $f : \mathbf{I} \to |\dot\sigma|$ be a loop based at v_0. If $|\dot\sigma| \setminus f(\mathbf{I}) \neq \varnothing$, then it is an exercise in analytic geometry to show that $f \sim c_{v_0} \pmod{\{0,1\}}$. If $f(\mathbf{I})$ covers all of $|\dot\sigma|$, we proceed as follows. By (14.C.11), $\mathbf{I}$ may be subdivided so that there is a simplicial map $\gamma : |\mathbf{I}^{(n)}| \to |\dot\sigma|$ that is homotopic to f. Furthermore, since $f(0) = f(1) = v_0$, we have by (14.C.7) that $\gamma(0) = \gamma(1) = v_0$. Since $\gamma(|\mathbf{I}^{(n)}|)$ is in the 1-skeleton of $|\dot\sigma|$, it follows that $|\dot\sigma| \setminus \gamma(|\mathbf{I}^{(n)}|) \neq \varnothing$. Again, an exercise in analytic geometry yields that $\gamma \sim c_{v_0} \pmod{\{0,1\}}$, which concludes the proof.

(14.C.16) Corollary. For $n \geq 2$, every map $f : \mathscr{S}^n \to \mathscr{S}^1$ is null homotopic.

Proof. Since $\pi_1(\mathscr{S}^n) = \{e\}$, f can be lifted to the universal covering space $\mathscr{E}^1$ of $\mathscr{S}^1$ (13.D.1).

(14.C.17) Exercise. Show that every continuous map $f : \mathscr{S}^k \to \mathscr{S}^n$ $(k < n)$ is null homotopic.

(14.C.18) Exercise. Suppose that $f : \mathscr{S}^k \to X$ is null homotopic. Show that f may be extended to a continuous function $f : \mathscr{B}^{k+1} \to X$.

The advantages inherent in triangulable spaces are well illustrated in the proof of the following result.

(14.C.19)   Theorem. Suppose that A is a closed subset of $\mathscr{S}^n$ and that $f : A \to \mathscr{S}^n$ is continuous. Then f has a continuous extension $\hat{f} : \mathscr{S}^n \to \mathscr{S}^n$.

Proof. We assume that $\mathscr{S}^n = \dot{\sigma}^{n+1}$. By (4.B.21), there is an open neighborhood U of A and an extension $\hat{f} : U \to \mathscr{S}^n$ of f. Let $\delta = d(A, X \setminus U)$. Subdivide $\mathscr{S}^n$ so that the mesh of the triangulation is less than $\delta/2$, and call the corresponding complex K. Note that any simplex that intersects A is contained in U. Let $|L|$ be the union of all simplices that are in the closure of a simplex that intersects A. Then $\hat{f}|_{|L|} : |L| \to |\mathscr{S}^n|$ is an extension of f. Extend $\hat{f}|_{|L|}$ to $|L| \cup |K_0|$ by arbitrarily mapping the points of the 0-skeleton K_0 to vertices of $K \setminus L$. Assume that f has been extended to $\hat{f} : |L| \cup |K_m| \to \mathscr{S}^n$ and let $\sigma_{m+1} \in K \setminus L$. Then $\hat{f} : \dot{\sigma}_{m+1} \to \mathscr{S}^m$ and, since by (14.C.17) $\hat{f}|_{\dot{\sigma}_{m+1}}$ is null homotopic, it follows that f can be extended over $|(\sigma_{m+1})^g|$ (14.C.18). Thus, there is an extension of $\hat{f}$ to $|L| \cup |K_{m+1}|$. Inductively, we extend f to a map $\hat{f} : |L| \cup |K_n| \to \mathscr{S}^n$, which completes the proof, since $|L| \cup |K_n| = \mathscr{S}^n$.

A slightly stronger version of the simplicial approximation theorem II is frequently useful. Suppose that K and L are complexes and that $f : |K| \to |L|$. If f is already simplicial on some subcomplex M of K, it would be desirable to obtain a simplicial approximation ϕ such that $\phi|_{|M|} = f|_{|M|}$. The problem, however, is that when K is subdivided as in the proof of the simplicial approximation theorem I, so is M, and f is no longer simplicial on the corresponding subdivision of M. This problem may be avoided by carefully subdividing K (almost barycentrically) in such a way that (i) M is not disturbed, and (ii) outside of M, the subdivision is fine enough so that the construction in

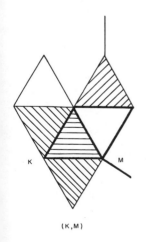
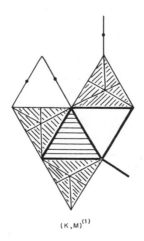
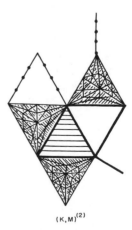

(K,M) $(K,M)^{(1)}$ $(K,M)^{(2)}$

the proof of the approximation theorem still works. We indicated in the previous figure how this may be done. Successive subdivisions are denoted by $(K,M)^{(1)}, (K,M)^{(2)}, \cdots$.

The following simplicial approximation theorem is due to Zeeman; we do not prove it, but instead refer the interested reader to Maunder [1970] for a clear exposition.

(14.C.20)   Theorem (Zeeman [1964]).   Suppose that (K,M) is a simplicial pair and L is a complex. Let $f : |K| \to |L|$ be a continuous function with the property that $f|_{|M|}$ is simplicial. Then there are an integer n and a simplicial map $g : |(K,M)^{(n)}| \to |L|$ such that $g|_{|M|} = f|_{|M|}$ and $g \sim f(\mathrm{mod}\ |M|)$.

Although g is not actually a simplicial approximation of f, it is at least "close," i.e., it is homotopic to f and agrees with f on M.

Zeeman's theorem may be used to establish the no retraction theorem for all n, which in turn yields the general Brouwer fixed point theorem.

(14.C.21)   Theorem.   There is no retraction from $\mathscr{B}^n$ onto its boundary $\mathscr{S}^{n-1}$.

Proof.   To apply Zeeman's theorem, a simplicial pair is needed; consider $\mathscr{B}^n$ to be the closure of an n-simplex, σ_n, and $\mathscr{S}^{n-1}$ to be the boundary, $\dot{\sigma}_n$, of σ_n. Suppose that there is a continuous map $\phi : |\sigma_n^g| \to |\dot{\sigma}_n|$ that is the identity on $|\dot{\sigma}_n|$. By Zeeman's result, there are an integer n and a simplicial map $g : |(\sigma_n^g, \dot{\sigma}_n)^{(n)}| \to |\dot{\sigma}_n|$ that also leaves $\dot{\sigma}_n$ fixed. Let b_s be the barycenter of an $(n-1)$-simplex s lying in $\dot{\sigma}_n$. We shall show that $g^{-1}(b_s)$ is a polygonal arc with one end at b_s and the other end at a distinct point of $|\dot{\sigma}_n|$; this will contradict the fact that g acts as the identity on $|\dot{\sigma}_n|$. Let σ be any n-simplex in $(\sigma_n^g, \dot{\sigma}_n)^{(n)}$, and suppose $g^{-1}(b_s) \cap |\sigma^g| \neq \varnothing$. Then of course, σ must be mapped onto s. Since it can be shown that exactly two faces of σ map onto s, it follows that $|\sigma^g| \cap g^{-1}(b_s)$ is the line segment in σ joining the barycenters of the two faces. Thus, if σ is an n-simplex, then $g^{-1}(b_s) \cap |\sigma^g|$ is either a line segment or empty, and consequently, there is a sequence of line segments that cannot cross themselves. Since only a finite number of simplices are involved, eventually the string must come to an end at a point in $|\dot{\sigma}_n|$ from whence it did not start (each $(n-1)$-simplex in $\mathscr{S}^{n-1}$ is the face of only one n-simplex in $\mathscr{B}^n$).

(14.C.22)   Corollary (Brouwer Fixed Point Theorem).   Suppose that $f : \mathscr{B}^n \to \mathscr{B}^n$ is continuous. Then f has a fixed point.

Proof.   The proof is the same, mutatis mutandis, as that used in (12.C.16).

D. EDGE PATH GROUPS: HOW TO COMPUTE $\pi_1(|K|, v_0)$ FOR ANY COMPLEX

To demonstrate a particularly striking use of the simplicial approximation theorem II, we renew our attack on the calculation of fundamental groups. The spaces to be considered will all be triangulable, which is not an unduly restrictive condition, since many common and/or interesting spaces may be triangulated.

First, a new group, the edge path group, is defined. Then it is shown that the edge path group is isomorphic to the fundamental group. By the very nature of its definition, the edge path group is (in a very illusory way) more computable than the fundamental group.

Although the details are somewhat involved, the basic ideas are quite simple. The fundamental group of a space X was based on the notion of loops, i.e., maps of $\mathbf{I}$ into X that send the end points to a base point x_0. Loops were defined to be equivalent if they were homotopic relative to $\{0,1\}$. Suppose now that X is triangulable and v_0 is a vertex of the triangulation. From what we have already seen, a loop in X based at v_0 may be simplicially approximated by a map ϕ that is homotopic to f relative to $\{0,1\}$, provided that the triangulation of X is sufficiently fine. Thus, rather vaguely speaking, it seems reasonable to consider "linear" maps when trying to associate a group with the given space. This leads in a fairly natural way to the following definition, which gives the "simplicial" counterparts to paths and loops.

(14.D.1) *Definition.* Suppose that K is a simplicial complex and that v and v' are vertices of K. Then an *edge path in K from v to v'* is a sequence of vertices, $w_0, w_1, \ldots, w_n$, where for each i, w_i and w_{i+1} span a simplex in K (either 0- or 1-dimensional), $w_0 = v$, and $w_n = v'$. If $v = v'$, then the path is called an *edge loop*.

Multiplication of edge paths is effected in the obvious manner. If $\alpha = w_0 w_1 \cdots w_n$ and $\beta = v_0 v_1 \cdots v_m$ are edge paths with $w_n = v_0$, then $\alpha\beta$ is defined to be the edge path $w_0 w_1 \cdots w_n v_1 v_2 \cdots v_m$. The definition of the reverse of a path is equally straightforward. If $\alpha = v_0 v_1 :\cdots v_n$ is a path, then α^r is defined to be $v_n v_{n-1} \cdots v_0$.

An equivalence relation is introduced on the set of edge paths as follows. Two edge paths with identical end points are *equivalent* if and only if one may be obtained from the other by a finite sequence of the following operations.

I. Suppose that $\alpha = v_0 v_1 \cdots v_i v_{i+1} v_{i+2} \cdots v_n$, and that $v_i = v_{i+1}$. Then α may be replaced by $v_0 v_1 \cdots v_i v_{i+2} \cdots v_n$.

I^{-1}.   Suppose that $\alpha = v_0v_1 \cdots v_iv_{i+1} \cdots v_n$ and $w = v_i$. Then α may be replaced by $v_0v_1 \cdots v_iwv_{i+1} \cdots v_n$.

II.   Suppose that $\alpha = v_0v_1 \cdots v_{i-1}v_iv_{i+1} \cdots v_n$ and that $\{v_{i-1},v_i,v_{i+1}\}$ spans a simplex in K. Then α may be replaced by $v_0v_1 \cdots v_{i-1}v_{i+1} \cdots v_n$.

II^{-1}.   Suppose that $\alpha = v_0v_1 \cdots v_{i-1}v_{i+1} \cdots v_n$ and that $\{v_{i-1},v_i,v_{i+1}\}$ spans a simplex in K. Then α may be replaced by $v_0v_1 \cdots v_{i-1}v_iv_{i+1} \cdots v_n$.

It is easy to verify that these four operations lead to an equivalence relation on the edge paths of K.

(14.D.2)   Examples.

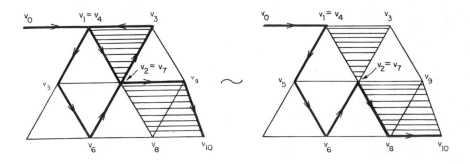

That the edge path $v_0v_1v_2v_3v_4v_5v_6v_7v_9v_{10}$ is equivalent to the edge path $v_0v_1v_5v_6v_7v_8v_{10}$ is shown by the following chain of equivalences: $v_0v_1v_2v_3v_4v_5v_6v_7v_9v_{10} \sim v_0v_1v_3v_4v_5v_6v_7v_9v_{10} \sim v_0v_1v_4v_5v_6v_7v_9v_{10} \sim v_0v_1v_5v_6v_7v_9v_{10} \sim v_0v_1v_5v_6v_7v_8v_9v_{10} \sim v_0v_1v_5v_6v_7v_8v_{10}$.

(14.D.3)   Exercise.   Suppose that α and β are equivalent edge paths with end points v and v', and that α' and β' are equivalent edges with end points v' and u. Show that $\alpha\alpha' \sim \beta\beta'$.

(14.D.4)   Exercise.   Show that if $\alpha \sim \beta$, then $\alpha^r \sim \beta^r$.

(14.D.5)   Exercise.   Show that if α is an edge path with end points v and w, then $v\alpha \sim \alpha \sim \alpha w$, $\alpha\alpha^r \sim v$ and $\alpha^r\alpha \sim w$.

Trivial as the preceding exercises are, they are nevertheless indispensable in allowing one to define a group structure based on edge loops.

(14.D.6)   Definition.   Suppose that v_0 is a vertex of a simplicial complex K. For each edge loop α based at v_0, let $[\alpha]$ denote the corresponding equiva-

lence class. Define the product of equivalence classes $[\alpha][\beta]$ to be $[\alpha\beta]$. If $[\alpha]$ is an equivalence class, define $[\alpha]^{-1}$ to be $[\alpha^r]$. These operations yield a group with identity element $[v_0]$. This group is the *edge path group*, and is denoted by $e(K,v_0)$.

The reader may establish easily that $e(K,v_0)$ is actually a group. It will eventually be shown that $e(K,v_0)$ is isomorphic to $\pi_1(|K|,v_0)$ and hence is a topological invariant. First, however, we see how these groups are calculated using the edge paths. Groups may be determined by listing their generators and a sufficient number of relations between these generators (see Appendix). However, the student should realize that finding generators and relations does not always satisfy one's curiosity about or increase ones intuition for the nature of the group.

To obtain generators and relations for the edge path group, the following concept is useful.

(14.D.7) Definition.   If K is a simplicial complex, then a 1-dimensional subcomplex L is a *tree* if and only if $|L|$ is contractible.

The existence and some properties of trees may be deduced from the following set of exercises.

(14.D.8) Exercise.   Suppose that K is a simplical complex. Show that K contains a maximal tree (i.e., a tree not properly contained in any other tree).

(14.D.9) Exercise.   Show that a tree L in a connected simplicial complex K is maximal if and only if L contains all the vertices of K.

(14.D.10) Exercise.   Suppose that L_1 and L_2 are subcomplexes of a simplicial complex K such that $L_1 \cup L_2 = K$. Suppose that $M = L_1 \cap L_2$ is connected and that T is a maximal tree in M. Show that there is a maximal tree T' in K such that $T' \cap L_i$ is a maximal tree in L_i for $i = 1,2$ and $T' \cap M = T$.

(14.D.11) Definition.   Suppose that K is a simplicial complex. Order the set of vertices of K, $v_0 \prec v_1 \prec v_2 \prec \cdots \prec v_n$. Then a 1-simplex $\langle v_i v_j \rangle$ is *ordered* if and only if $v_i \prec v_j$. The ordered 1-simplex $\langle v_i v_j \rangle$ is denoted by s_{ij} or s_{ji}^{-1}. A 2-simplex $\langle v_i v_j v_k \rangle$ is *ordered* if and only if $v_i \prec v_j \prec v_k$.

Now suppose that K is a simplicial complex, $|K|$ is path connected and L is a subcomplex of K with contractible carrier and containing all of the vertices of K. A group G is formed that has as generators the symbols s_{ij}, one for each ordered 1-simplex s_{ij} in $K \setminus L$, and having relations of the form $s_{ij}s_{jk}s_{ik}^{-1}$, one for each ordered 2-simplex $\langle v_i v_j v_k \rangle$ in $K \setminus L$. (If $\langle v_i v_j \rangle \in L$, interpret s_{ij} as e.)

(14.D.12) Theorem. The group G is isomorphic to $e(K, v_0)$.

Proof. Homomorphisms $\phi : G \to e(K, v_0)$ and $\psi : e(K, v_0) \to G$ are defined that are inverses of each other. For each vertex v_i of K, pick an edge path e_i that goes from v_0 to v_i. If s_{ij} is an ordered 1-simplex in $K \setminus L$, define $\phi(s_{ij}) = [e_i v_i v_j e_j^r]$. If $\langle v_i v_j v_k \rangle$ is an ordered 2-simplex of $K \setminus L$, it may be easily verified that $\phi(s_{ij})\phi(s_{jk})(\phi(s_{ik}))^{-1} = e$. Consequently, ϕ may be extended to a homomorphism (Ap.C.3) from G into $e(K, v_0)$.

To define the map ψ, we will use the following notation. Suppose that v_i and v_j are vertices in K that span a simplex in K. Let

$$t_{ij} = \begin{cases} s_{ij} & \text{if } \langle v_i v_j \rangle \text{ is an ordered 1-simplex in } K \setminus L, \\ s_{ji}^{-1} & \text{if } \langle v_j v_i \rangle \text{ is an ordered 1-simplex in } K \setminus L, \\ e & \text{otherwise.} \end{cases}$$

Suppose that $\alpha = v_0 v_{i_1} v_{i_2} \cdots v_{i_k} v_0$ is an edge loop in K based at v_0. Define $\psi([\alpha]) = t_{0 i_1} t_{i_1 i_2} \cdots t_{i_k 0}$. The rest of the proof is straightforward, but the reader should verify that (i) ψ is well defined, and (ii) $\phi\psi$ and $\psi\phi$ are the identity isomorphisms on $e(K, v_0)$ and G, respectively.

(14.D.13) Notation. With the notation as in the foregoing proof, we shall henceforth denote the generators $\phi(s_{ij})$ of $e(K, v_0)$ by g_{ij}.

To illustrate the relative ease of calculating edge path groups with the aid of the preceding theorem, we start with the unit circle $\mathscr{S}^1$. Since $\mathscr{S}^1$ is topologically the same as a triangle, a contractible subcomplex L containing all vertices may be obtained as indicated in the next figure.

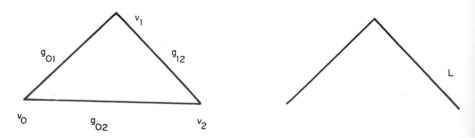

Thus, G is a group with one generator, g_{02}, and no relations, i.e., the infinite cyclic group $\mathbf{Z}$. Before proceeding further, let us establish that the foregoing result is no coincidence, i.e., for any simplicial complex K, $e(K,v_0)$ is isomorphic to $\pi_1(|K|,v_0)$.

(14.D.14) Theorem.   Suppose that K is a simplicial complex with connected carrier. Then $e(K,v_0)$ is isomorphic with $\pi_1(|K|,v_0)$.

Proof.   We associate with each edge loop, $\alpha = v_0 v_{i_1} v_{i_2} \cdots v_{i_k} v_0$ based at v_0, a map $f_\alpha : \mathbf{I} \to |K|$ as follows. Let $0 = a_0 < a_1 < \cdots < a_{k-1} < a_k < a_{k+1} = 1$ be a partition of $\mathbf{I}$. Then f_α is defined to be the simplicial map that for each j maps the simplex $\langle a_j a_{j+1} \rangle$ onto $\langle v_{i_j} v_{i_{j+1}} \rangle$. Define $\phi : e(K,v_0) \to \pi_1(|K|,v_0)$ by setting $\phi([\alpha]) = [f_\alpha]$. The reader may verify that ϕ is well defined and is a homomorphism.

The simplicial approximation theorem is used to show that ϕ is onto. Suppose that $[f] \in \pi_1(|K|,v_0)$. Then $f : \mathbf{I} \to |K|$ and $f(0) = v_0 = f(1)$. By the simplicial approximation theorem II and (14.C.7), there is a triangulation of $\mathbf{I}$ whose vertices form a partition of $\mathbf{I}$, $0 = a_0 < a_1 < \cdots < a_k < a_{k+1} = 1$, · and a simplicial map $\hat{f}$ (relative to this triangulation) such that $\hat{f} : \mathbf{I} \to |K|$, and $f \sim \hat{f}(\mathrm{mod}\ \{0,1\})$. If $\alpha = f(a_0)f(a_1) \cdots f(a_k)f(a_{k+1})$, then clearly, $\phi([\alpha]) = [f]$.

To show that ϕ is 1–1 requires somewhat more effort. Suppose that $\phi([\alpha]) = e$, where $\alpha = v_0 v_{i_1} v_{i_2} \cdots v_{i_k} v_0$, i.e., $f_\alpha : \mathbf{I} \to |K|$ is homotopic via H to $c_{v_0}(\mathrm{mod}\ \{0,1\})$.

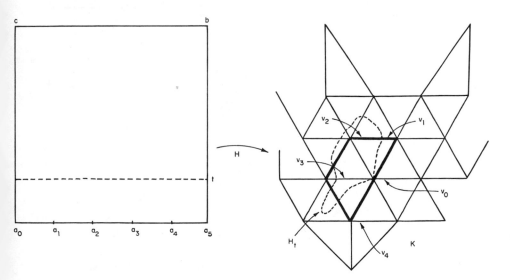

Triangulate $\mathbf{I} \times \mathbf{I}$ as indicated in (a) of the next figure, and note that H restricted to the boundary B of $\mathbf{I} \times \mathbf{I}$ is simplicial. Hence, by Zeeman's version of the simplicial approximation theorem, there is a subdivision $(\mathbf{I} \times \mathbf{I}, B)^{(n)}$ (which leaves B alone) (see (b)) and a simplicial map $G : |L| \to |K|$ such that $G_{|B} = F$ and G is homotopic to $H(\mathrm{mod}\ B)$.

In L however, the edge path $a_0 a_1 \cdots a_{k+1} b c a_0$ is equivalent to a_0. (Figures (b), (c), and (d) give a few of the steps that may be taken when $L = (\mathbf{I} \times \mathbf{I}, B)^{(1)}$ to show that these two paths are indeed equivalent. For $n \geq 2$, similar, albeit many more, equivalence operations may be performed to yield the edge path equivalence.)

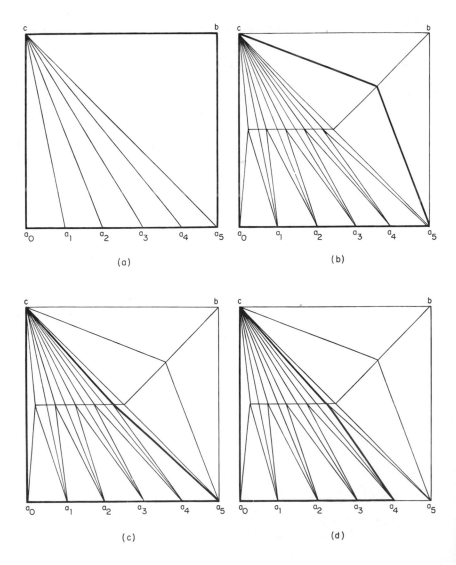

To see that $[\alpha]$ is equal to the identity element in $e(K,v_0)$, observe that $\alpha = v_0 v_{i_1} \cdots v_{i_k} v_0 \sim v_0 v_{i_1} \cdots v_{i_k} v_0 v_0 v_0 v_0 = G(a_0)G(a_1) \cdots G(a_{k+1})G(b)G(c) G(a_0) \sim G(a_0) = v_0$. The last equivalence is due to the fact that $a_0 a_1 \cdots a_{k+1} b c a_0$ is equivalent to a_0 and G is simplicial. Therefore, ϕ is 1–1.

Observe that it follows from the proof of the foregoing theorem that the fundamental group $\pi_1(|K|,v_0)$ depends only on the carrier of the 2-skeleton $|K_2|$ of K.

Despite the apparent ease of calculating fundamental groups via the edge path group, the reader should realize that all is not completely well. Many triangulable spaces require an inordinate number of 2-simplexes (the torus, for example requires at least twenty one 1-simplexes, fourteen 2-simplexes, and seven 0-simplexes). Thus, the number of generators and relations may become astronomical, even for relatively simple spaces. Computations of the edge path group are often greatly facilitated with the aid of the following theorem, which is due to Seifert and van Kampen. It is the polyhedral analog of the Seifert-van Kampen theorem stated in Chapter 12.

Recall from group theory that if G and H are groups with presentations $\{g_1, \ldots, g_n; r_1, \ldots, r_m\}$ and $\{g'_1, \ldots, g'_p; r'_1, \ldots, r'_j\}$, respectively, then the free product of G and H is isomorphic to the group with presentation $\{g_1, \ldots, g_m, g'_1, \ldots, g'_p; r_1, \ldots, r_m, r'_1, \ldots, r'_j\}$ (Ap.B.10).

(14.D.15) Theorem (Seifert-van Kampen Theorem). Suppose that K is a simplicial complex and that L_1 and L_2 are subcomplexes of K whose union is K. Let $N = L_1 \cap L_2$ and suppose that $|L_1|$, $|L_2|$, and $|N|$ are connected. Let $i_1 : |L_1| \to |K|$ and $i_2 : |L_2| \to |K|$ be inclusion maps, and let v_0 be a vertex of N. Then $e(K,v_0)$ is the free product of $e(L_1,v_0)$ and $e(L_2,v_0)$ with additional relations of the form $(i_1)_*(g)((i_2)_*(g))^{-1}$, one for each element g in a finite set of generators for $e(N,v_0)$.

Proof. By (14.D.10), there are trees T_N, T_{L_1}, and T_{L_2} such that

   (i) T_N is maximal in N,
   (ii) $T_{L_1} \cap N = T_N = T_{L_2} \cap N$,
   (iii) $T_{L_1} \cup T_{L_2} = T_K$ is a maximal tree in K, and
   (iv) T_{L_1} and T_{L_2} are maximal in L_1 and L_2, respectively.

Order the vertices of K; this induces an order on the vertices of each subcomplex of K. As before, $e(K,v_0)$ is the group generated by symbols of the form g_{ij} subject to relations $g_{ij} g_{jk} g_{ik}^{-1}$, where each g_{ij} corresponds to an ordered 1-simplex in $K \setminus T_K$. However, any 1-simplex in $K \setminus T_K$ is certainly a 1-simplex in either $L_1 \setminus T_{L_1}$ or $L_2 \setminus T_{L_2}$, and vice versa. Hence if a typical ordered 1-simplex in $L_1 \setminus T_{L_1}$ gives a generator denoted by g_{ij}^1 and a 1-simplex in $L_2 \setminus T_{L_2}$ gives a generator denoted by g_{ij}^2, then $e(K,v_0)$ is the group generated by all of the g^1_{ij} and g^2_{ij} with the same relations as before (but relabeled to agree with the new names of the generators). There is a complication, however. It is possible that $g^1_{ij} = g^2_{ij}$ (in which case, g^1_{ij} and g^2_{ij} correspond to the same ordered simplex in $N \setminus T_N$). This problem is eliminated by adding relations $g^1_{ij}(g^2_{ij})^{-1}$ whenever $g^1_{ij} = g^2_{ij}$. Thus, $e(K,v_0)$ may be described as the free product of $e(L_1,v_0)$ and $e(L_2,v_0)$ with additional relations $(i_1)_*(g_{ij})((i_2)_*(g_{ij}))^{-1} = e$ for each generator $g_{ij} \in e(N,v_0)$.

(14.D.16) Corollary. Suppose that $K = L_1 \cup L_2$, where $|K|$, $|L_1|$, and $|L_2|$ are connected. If $|N| = |L_1 \cap L_2|$ is simply connected, then $e(K,v_0)$ is the free product of $e(L_1,v_0)$ and $e(L_2,v_0)$.

(14.D.17) Corollary. Suppose that $K = L_1 \cup L_2$, and that $|K|$, $|L_1|$, $|L_2|$, and $|N| = |L_1 \cap L_2|$ are connected. If $|L_2|$ is simply connected, then $e(K,v_0)$ may be obtained from $e(L_1,v_0)$ by adding the relation $(i_1)_*(g_{ij}) = e$ for each $g_{ij} \in N \setminus T_N$.

Proof. Obviously, the free product of $e(L_1,v_0)$ and $e(L_2,v_0)$ is isomorphic to $e(L_2,v_0)$. By (14.D.15), there are additional relations of the form $(i_1)_*(g_{ij})((i_2)_*(g_{ij}))^{-1} = e$ for each symbol $g_{ij} \in e(N,v_0)$, where g_{ij} corresponds to an ordered 1-simplex in N. Since $(i_2)_*(g_{ij}) = e$, we have that $((i_2)_*(g_{ij}))^{-1} = e$, and hence the relations added are of the form $(i_1)_*(g_{ij}) = e$.

This latter corollary proves to be especially useful in the calculation of fundamental groups of 2-manifolds. Many spaces (including all compact connected 2-manifolds) may be constructed by attaching 2-cells to appropriate complexes. In fact, it will follow from results in Chapter 16 that any compact connected 2-manifold can be formed by gluing a 2-cell in a suitable manner to the reiterated wedge product of $\mathscr{S}^1$ with itself. (The *wedge product* of pointed spaces (X,x_0) and (Y,y_0) denoted by $X \vee Y$ is the space obtained from the disjoint union $X \cup Y$ by identifying the points x_0 and y_0.) Consider once more the torus as a quotient space obtained by identifying sides of the unit square.

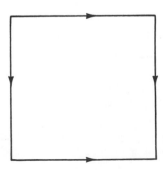

Note that under this identification, the sides of the square form the wedge product of $\mathscr{S}^1$ with itself, and a 2-cell (the original square) has been attached to the product. Alternatively, this may be described as indicated in the next figure. A 2-cell D is attached to $\mathscr{S}^1 \vee \mathscr{S}^1$ by gluing ab onto $vstv$, bc onto $vs't'v$, cd onto $vtsv$, and da onto $vt's'v$.

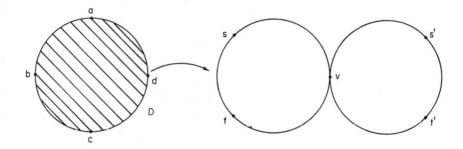

The end result is the torus.

In a similar manner, the projective plane is obtained by attaching a 2-cell to $\mathscr{S}^1$ as shown in the next figure.

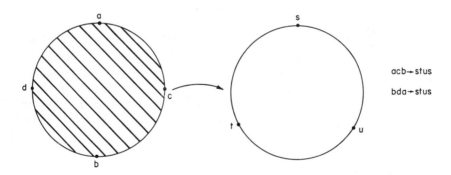

(14.D.18) **Exercise.** Show that the two-holed torus may be obtained by an appropriate attachment of a 2-cell to the wedge product of four circles.

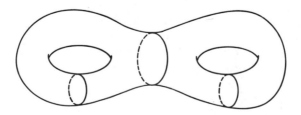

More generally, if K is any complex and $\alpha = v_0 v_1 \cdots v_n v_0$ is an edge loop in K that has no identical adjacent vertices, then a 2-cell may be attached to K along α in the following manner. Let P be a regular polygon in $\mathscr{E}^2$ with $n + 1$ sides. Triangulate P as indicated below and label the vertices b, v_0', $v_1', \ldots, v_n'$. Let f be a simplicial map that sends the boundary of P onto the carrier of the loop α by mapping vertices v_i' onto v_i for each i. Then $|\hat{K}| = |K| \cup_f |P|$ is the carrier of a simplicial complex $\hat{K}$.

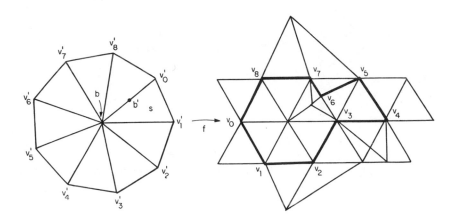

Let s be the 2-simplex with vertices v_0', v_1', and b, and let $|s^g|$ be the closure of s. Let b' be the barycenter of the 1-simplex $\langle bv_0' \rangle$. Radial projection from b' yields a deformation retraction of $|\hat{K}| \setminus |s^g|$ onto $|K|$ that leaves points of $|K|$ pointwise fixed. Of course, the projection is originally done in $|P|$ and then transferred over to $|\hat{K}|$ by f; furthermore, the projection is not piecewise linear. By (12.C.19), $\pi_1(|\hat{K}| \setminus |s^g|, v_0)$ is isomorphic to $\pi_1(|\hat{K}|, v_0)$. Since $\pi_1(|s^g|) = \{e\}$, (14.D.17) may be applied, where $L_2 = s^g$ and $L_1 = \hat{K} \setminus s$, to see that $e(\hat{K}, v_0)$ may be obtained from $e(\hat{K} \setminus s, v_0)$ by adding the relations arising from generators in $L_1 \cap L_2$. Note that any one of the ordered 1-simplices in $L_1 \cap L_2$ serves as a generator of $e(L_1 \cap L_2, v_0)$.   •

In essence, what has occured is that if a cell is glued to a loop in K, then the group element determined by this loop is obliterated (reduced to the identity).

(14.D.19) Example. Consider the torus T as a square with sides identified. The cell C in the torus indicated by the dashed lines is deformable to $\mathscr{S}^1 \vee \mathscr{S}^1$. Hence, since $e(C) = \{e\}$ and $e(\mathscr{S}^1 \vee \mathscr{S}^1) = \{g_1, g_2; \}$, we have that $e(T) = \{g_1, g_2; g_1 g_2 g_1^{-1} g_2^{-1}\}$.

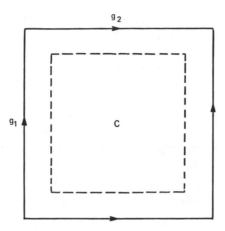

There are two rather startling results that may with some imagination be extracted from our work thus far. The enterprising reader might wish to consider the following two assertions, the first of which is not difficult to establish.

(i) Suppose that G is a finitely presented group. Then there is a two-dimensional polyhedron whose fundamental group is isomorphic to G.

(ii) Suppose that G is a finitely presented group. Then there is a 4-manifold that has G as its fundamental group.

The latter result coupled with the work of Rabin [1958] leads one to conclude that there is no countable algorithm that classifies 4-manifolds.

PROBLEMS

Section A

1. Show that if K is a finite collection of simplices lying in $\mathscr{E}^n$, then K is a simplicial complex if and only if

    (i) if $\sigma \in K$ and $\tau < \sigma$, then $\tau \in K$, and
    (ii) if $\sigma, \tau \in K$, then $\sigma^g \cap \tau^g$ is the closure of a face of both σ and τ.

2. Suppose that K is an infinite simplicial complex lying in $\mathscr{E}^n$ (or in Hilbert space). Show that the relative topology on $|K|$ coincides with $\mathscr{U} = \{ U \subset |K| \mid U \cap \sigma \text{ is open in } \sigma \text{ for each } \sigma \in K \}$.

3. Show that the result of problem 2 fails to hold if the local finiteness condition is removed from (14.A.15).

4. Let V be a set of elements (called *abstract vertices*) and let K be a collection of finite subsets of V with the property that any subset of a set in K is also in K. Then K is called an *abstract complex*. Any set of $n + 1$ vertices $a_0, a_1, \ldots, a_n$ in K is called an *abstract n-simplex*. Show how an infinite simplicial complex may determine an abstract complex.

5. Two complexes K and L (simplicial or abstract) are *isomorphic* if and only if there is a simplicial bijective map $\phi : K \to L$ such that ϕ^{-1} is simplicial. Show that any finite abstract complex with $n + 1$ vertices is isomorphic to a subcomplex of the closure of an n-dimensional simplex.

6. Suppose that $(X, \mathcal{U})$ is a topological space. Let $A = \prod_{U \in \mathcal{U}} \mathscr{E}_U^1$, where $\mathscr{E}_U^1 = \mathscr{E}^1$ for each $U \in \mathcal{U}$. For each $V \in \mathcal{U}$, let $x_V \in \prod_{U \in \mathcal{U}} \mathscr{E}_U^1$ be the point in A whose V-th coordinate is 1, and whose remaining coordinates are 0. Let $\mathscr{V}$ be a finite open cover of $(X, \mathcal{U})$. A complex K associated with $\mathscr{V}$ is defined as follows. The vertices of K consist of the points x_V, where $V \in \mathscr{V}$. Vertices $x_{V_1}, \ldots, x_{V_n}$ determine a simplex in K if and only if $\bigcap_{i=1}^{n} V_i \neq \varnothing$. Show that $K_{\mathscr{V}}$ is an abstract complex and is isomorphic to to a simplicial complex lying in a Euclidean space $\mathscr{E}^m$. The complex $K_{\mathscr{V}}$ is called the *nerve* of the cover $\mathscr{V}$.

7. Find an open cover $\mathscr{V}$ of $\mathscr{S}^1$ such that the carrier of the simplicial complex isomorphic to the nerve of $\mathscr{V}$ (see problem 6) is homeomorphic to $\mathscr{S}^1$.

8.* Suppose that K is a simplicial complex. Let $\mathcal{U} = \{star(v) \mid v$ is a vertex of $K\}$. Let N be the simplicial complex associated with the nerve of $\mathcal{U}$ (see problem 6). Show that there is a homeomorphism from $|K|$ onto $|N|$ such that h maps each simplex of K onto a simplex of N.

9. Show that every open set in $\mathscr{E}^n$ can be triangulated by a countably infinite complex.

10. Show that a triangulation of a compact manifold must be finite.

11.* Show that every countable, locally finite, n-dimensional simplicial complex can be embedded in $\mathscr{E}^{2n+1}$.

Section B

1. Show that if K is a simplicial complex, then $|K| \times \mathbf{I}$ is triangulable.

2. Show that the product of two simplices is a linear cell.

3.* Suppose that K is a simplicial complex in $\mathscr{E}^m$ and that L is a simplicial complex in $\mathscr{E}^n$. Define $K \times L$ to be $\{\tau \times \sigma \mid \tau \in K, \sigma \in L\}$. Show that

$K \times L$ is a cell complex and that with a suitable subdivision, $K \times L$ can be made into a simplicial complex (without the introduction of any new vertices).

4. Show that the intersection and union of two polyhedra is a Euclidean polyhedron.

5. Let s be an n-simplex, and suppose that σ and τ are opposite faces of s, where σ is a p-simplex and τ is a q-simplex and $p + q = n - 1$. Show that every point x of s is of the form $x = (1 - \alpha)y + \alpha z$, where $y \in \sigma$ and $z \in \tau$. Find a homotopy that retracts $|s^g| \setminus |\tau^g|$ onto $|\sigma^g|$.

6. Suppose that (K,L) is a simplicial pair. Show that the simplices of $K^{(1)}$ that have no vertices in common with $L^{(1)}$ form a subcomplex M of $K^{(1)}$. Show that $|L^{(1)}|$ is a deformation retract of $|K^{(1)} \setminus M|$ and that $|K^{(1)} \setminus M|$ is a neighborhood of $|L^{(1)}|$.

7.* Use problems A.4 and B.6 to show that every finite simplicial complex is an ANR_M.

8.* Suppose that X and Y are Euclidean polyhedra and that $f : X \to Y$ is a continuous function. Show that f is simplicial if and only if the graph of f, $\{(x,f(x)) \in X \times Y \mid x \in X\}$, is a polyhedron.

9. Suppose that X and Y are polyhedra lying in $\mathcal{E}^n$. For each $x \in X$ and $y \in Y$, let ℓ_{xy} be the line segment in $\mathcal{E}^n$ with end points x and y. Show that X and Y are joinable if and only if (i) $\ell_{xy} \cap X = x$ and $\ell_{xy} \cap Y = y$ for each $x \in X$ and $y \in Y$, and (ii) if $x,x' \in X$ and $y,y' \in Y$ and $\ell_{xy} \neq \ell_{x'y'}$, then $\ell_{xy} \cap \ell_{x'y'}$ is empty or is an end point.

10. An n-simplex σ of a simplicial complex K has a *free face* τ if and only if τ is an $n - 1$ face of σ but is a face of no other n-simplex of K. An *elementary collapse* is the process of replacing K by $K \setminus \{\sigma \cup \tau\}$. If L is a subcomplex of K, we say K *collapses* to L, denoted by $K \searrow L$ if and only if L can be obtained from K by a sequence of elementary collapses. Show that if K collapses to L, then L is a deformation retract of K.

11. The house with two rooms consists of the top labeled T; the partition, P; the bottom, B; and the walls, W_1, W_2, W_3, and W_4; the two curtains, K_1, K_2; and the two tunnels, C_1, C_2. One enters the bottom room by crawling down from the roof through tunnel C_1. Similarly, one enters the top room by crawling up from under the floor B through tunnel C_2 (see next figure). Show that the house with two rooms (which is triangulable) can be obtained by collapsing a 3-cell; hence, the house with two rooms is a retract of a 3-cell, and thus is contractible.

Show that the house with two room is not collapsible to a point, but if curtain K_1 is thickened into a 3-cell as shown in the second figure, then the resulting complex is collapsible.

Show that the product of the house with two rooms and I is collapsible.

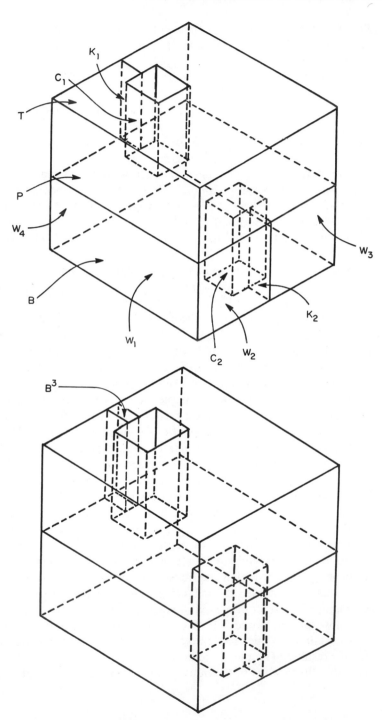

12. The *dunce hat* D is constructed by identifying the sides of a closed 2-simplex as indicated in the figure.

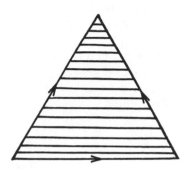

It can be shown that D is contractible (Zeeman [1963]). Show that D is triangulable, but that there is no triangulation of D for which D is collapsible to a point.

Section C

1. Suppose that K and L are simplicial complexes and that $f_1, f_2 : K \to L$ are simplicial maps. The maps f_1 and f_2 are *contiguous* if and only if for each $\langle v_0 v_1 \cdots v_k \rangle \in K$, there is a simplex $\tau \in L$ such that $f_1(v_0), \ldots, f_1(v_k)$ and $f_2(v_0), \ldots, f_2(v_k)$ are vertices of τ.
    (a) Suppose that K and L are simplicial complexes and that $f_1, f_2 : K \to L$ are simplicial approximations to a continuous map $\phi : |K| \to |L|$. Show that f_1 and f_2 are contiguous.
    (b) Show that if K and L are simplicial complexes, and $f_1, f_2 : K \to L$ are contiguous simplicial maps, then f_1 and f_2 are homotopic.
2. Show that if K and L are finite simplicial complexes, then there is only a countable number of homotopy classes of mappings from $|K|$ into $|L|$.
3. Suppose that A is a closed subset of $\mathscr{S}^n$ and that $f : A \to \mathscr{S}^{n-1}$ is continuous. Let $\{C_\alpha \mid \alpha \in \Lambda\}$ be the family of components of $\mathscr{S}^n \setminus A$. For each $\alpha \in \Lambda$, remove a point c_α from C_α. Show that there is a continuous extension of f, $F : \mathscr{S}^n \setminus \bigcup \{c_\alpha \mid \alpha \in \Lambda\} \to \mathscr{S}^{n-1}$.
4. Suppose that K, L, and M are simplicial complexes. Suppose that $f : |K| \to |L|$ and $g : |L| \to |M|$ are maps and that ϕ_f and ϕ_g are simplicial approximations of f and g, respectively. Show that $\phi_g \phi_f$ is a simplicial approximation of gf.
5. Suppose that K and L are simplicial complexes and that $f : |K| \to |L|$. Show that there is a simplicial approximation to f if and only if the

carrier of each simplex $\sigma \in K$ is contained in $f^{-1}(\text{star}(v))$ for some vertex $v \in L$.

6. Suppose that A is a compact subset of $\mathscr{E}^n$ and that $h : A \to \mathscr{E}^n$ is an embedding such that $h_{|\text{Fr } A}$ is the identity. Show that $h(A) = A$.

7. Suppose that U is a bounded open subset of $\mathscr{E}^n$. Show that Fr U is not a retract of $\overline{U}$.

8. Show that $id : \mathscr{S}^n \to \mathscr{S}^n$ is not null homotopic.

9.* Suppose that A is a compact subset of $\mathscr{E}^n$ and B is a subset of A homeomorphic to $\mathscr{B}^n$. Show that if $f : A \to \mathscr{E}^n$ is continuous and maps Fr A into B, then f has a fixed point.

Section D

1. Suppose that T is a tree and α_0 is the number of vertices of T and α_1 the number of 1-simplices. Show that $\alpha_0 - \alpha_1 = 1$.

2. Use the edge path group to calculate the fundamental group of the projective plane P_2.

3. Show that if K is a polyhedron, then the suspension of $|K|$ is a polyhedron.

4. Suppose that K is a simplicial complex and that K is a path connected polyhedron. Use the edge path group to calculate the fundamental group of the suspension of $|K|$.

5. The double torus T_2 may be considered as the quotient space obtained by identifying edges of the octagon as indicated in the figure.

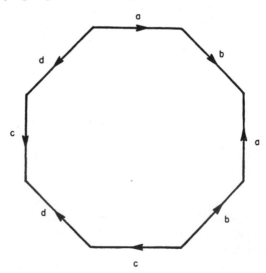

Find the fundamental group of T_2.

Chapter 15

FURTHER APPLICATIONS OF HOMOTOPY

Thus far, we have studied two major areas of topology where the concept of homotopy has proven to be indispensable: the creation of the fundamental group, and the the development of covering spaces. These represent, however, by no means the only contexts in which the notion of homotopy is relevant. This chapter is devoted to exploring some additional problems where homotopy theory is crucial.

A. THE EXTENSION PROBLEM (REVISITED)

We have already discussed a number of problems involving extensions of functions (Tietze, etc.). Proofs of many theorems hinge on whether or not a given map may be extended to a larger domain. Homotopy theory is often invaluable in resolving such problems.

A rather sophisticated result along these lines is the following theorem due to Borsuk.

(15.A.1) Theorem (Borsuk's Top Hat Theorem). Suppose that A is a closed subset of a normal space X and that Y is an ANR. Let f and g be homotopic maps from A into Y. If there is an extension F of f to all of X, then there is also an extension G of g to all of X, and furthermore, G may be chosen to be homotopic to F.

Proof. Let $H : A \times I \to Y$ be the homotopy between f and g with

391

$H_0 = f$ and $H_1 = g$. Extend H to $T = (A \times \mathbf{I}) \cup (X \times \{0\})$ by defining $\hat{H}(x,t) = H(x,t)$ if $(x,t) \in A \times \mathbf{I}$, and $\hat{H}(x,0) = F(x)$, otherwise. This function is continuous by the map gluing theorem.

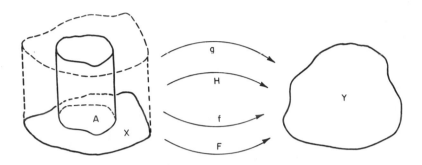

Since T is a closed subset of $X \times \mathbf{I}$, and Y is an ANR, $\hat{H}$ may be extended to an open set U containing T. Denote this extension by $\hat{\hat{H}}$.

By (2.G.14), there is an open set V in X such that $A \subset V$ and $V \times \mathbf{I} \subset U$. Since A and $X \setminus V$ are disjoint closed subsets of a normal space X, we can apply Uryson's lemma to define a function $\Psi : X \to \mathbf{I}$ that maps all of A to 1 and $X \setminus V$ to 0. Define $\hat{G} : X \times \mathbf{I} \to Y$ by setting $\hat{G}(x,t) = \hat{\hat{H}}(x, t \cdot \Psi(x))$. Then $\hat{G}_0 = F$, and $\hat{G}_1$ is the desired extension of g that is homotopic to F.

(15.A.2) Definition.   A map $f : X \to Y$ is *inessential* (*null homotopic*) if and only if f is homotopic to a constant map. Otherwise, f is *essential*.

With this terminology we have the following corollary.

(15.A.3) Corollary.   Suppose that A is a closed subset of a normal space X and $f : A \to Y$ is a continuous function from A into an ANR Y. If f is inessential, then f may be extended to X.

Proof.   Since f is inessential, it is homotopic to a constant map $c : A \to Y$. Clearly, c may be extended, and, consequently, so may be f (in fact, to an inessential map).

(15.A.4) Exercise.   Suppose that $f : (X,x_0) \to (\mathscr{S}^1,s)$ is inessential. Show that $f \sim c_s (\mathrm{mod}\ x_0)$, and hence $f_* : \pi_1(X,x_0) \to \pi_1(\mathscr{S}^1,s)$ is trivial.

(15.A.5) Exercise.   Show that if A is a closed subset of $\mathscr{E}^n$ and $f : A \to \mathscr{S}^m$ is extendable to $\mathscr{E}^n$, then f is inessential.

The restriction that Y be an ANR in the top hat theorem may be relaxed if one is willing to place additional conditions on A and X. Essentially, we define away the problem.

(15.A.6) Definition. Suppose that (X,A) is a topological pair. Then A has the *absolute homotopy extension property* $(AHEP)$ if and only if for each continuous map $F : X \to Z$ where Z is arbitrary, and for each homotopy $H : A \times I \to Z$ with $H_0 = F_{|A}$, there is a homotopy $\hat{H} : X \times I \to Z$ such that $\hat{H}$ extends H and $\hat{H}_0 = F$.

The reader may feel that the foregoing definition is a typical mathematical ruse to get around a problem, i.e., avoid it by restricting attention only to those spaces that behave themselves properly. While this may indeed be the case, there are nevertheless a number of topological pairs that yield the $AHEP$, e.g., polyhedral pairs.

(15.A.7) Theorem. Finite simplicial pairs have the $AHEP$.

Proof. Suppose that (K,L) is a finite simplicial pair and X is an arbitrary space. Let $H : |L| \times I \to X$ and $F : |K| \to X$ be continuous maps such that $H_0 = F|_{|L|}$. Suppose that σ is a simplex in $K \setminus L$. Then there is a projection p_σ that maps $|\sigma^g| \times I$ onto $(|\dot{\sigma}| \times I) \cup (\sigma \times \{0\})$ as indicated in the following figure.

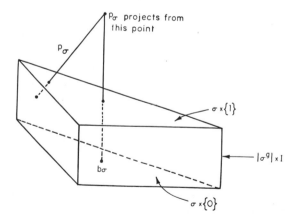

Note that p_σ is actually a retraction. Similar projections exist for each $\tau \in K$. If proper care is exercised in matching up the projections (start with the highest dimensional simplices first), then the map gluing theorem yields a retract $p : |K| \times I \to (|L| \times I) \cup (|K| \times \{0\})$.

Let $\hat{H} : ((|L| \times I) \cup (|K| \times \{0\})) \rightarrow X$ be defined by

$$\hat{H}(x,t) = \begin{cases} H(x,t) & \text{if} & x \in |L| & \text{and} & t \in I \\ F(x) & \text{if} & x \in |K| & \text{and} & t = 0 \end{cases}$$

Then $\hat{H}p : |K| \times I \rightarrow X$ is the required homotopy.

In general, when does the *AHEP* infect a space? One important criterion is given in the next theorem.

(15.A.8)   Theorem.   Suppose that (X,A) is a topological pair. Then A has the *AHEP* with respect to X if and only if $(X \times \{0\}) \cup (A \times I)$ is a retract of $X \times I$.

Proof.   Suppose that A has the *AHEP*. Define F to be the natural embedding of X into $(X \times \{0\}) \cup (A \times I)$, i.e., $F(x) = (x,0)$, and let H be the similar embedding of $A \times I$. Then $F_{|A} = H_0$, and by the *AHEP*, H may be extended to a homotopy $\hat{H} : X \times I \rightarrow (X \times \{0\}) \cup (A \times I)$ with $\hat{H}_0 = F$; $\hat{H}$ is obviously the retraction that was sought.

Now suppose that $(X \times \{0\}) \cup (A \times I)$ is a retract of $X \times I$. Let $F : X \rightarrow Y$ be a continuous function and suppose that $H : A \times I \rightarrow Y$ is a homotopy with $H_0 = F_{|A}$. The problem is to extend H to $X \times I$. The map H may first be extended to $(X \times \{0\}) \cup (A \times I)$ by setting $H(x,0) = F(x)$, and since this set is a retraction of $X \times I$, H may be further extended to all of $X \times I$ (4.B.15).

A second criterion for the *AHEP* is due to Young [1964].

(15.A.9)   Theorem.   Suppose that X is a normal space and that C is a closed subset of X. Then C has the *AHEP* if C is a G_δ subset, and there is an open set U containing C and a map $h : U \times I \rightarrow X$ such that

(i) $h(u,0) = u$ and $h(u,1) \in C$ for each $u \in U$, and
(ii) $h(c,t) = c$ for each $c \in C$, $t \in I$, i.e., C is a deformation retract of U in X (mod C).

Proof.   By (15.A.8), it suffices to show that $T = (X \times \{0\}) \cup (C \times I)$ is a retract of $X \times I$. Since C is a G_δ set, it follows from the proof of (4.B.6) that there is a map $\Psi : X \rightarrow I$ such that $\Psi^{-1}(1) = C$, and $\Psi(X \setminus U) = 0$, where U is the open set given in the hypothesis. The retraction $r : X \times I \rightarrow T$ is defined as follows:

(i) $r(x,t) = (x,0)$ if $\Psi(x) = 0$ and $t \in I$;
(ii) $r(x,t) = h(x,2\Psi(x)t)$ if $0 < \Psi(x) \leq \frac{1}{2}$ and $t \in I$;
(iii) $r(x,t) = h(x,t/2(1 - \Psi(x)))$ if $\frac{1}{2} \leq \Psi(x) < 1$ and $t \leq 2(1 - \Psi(x))$;

(iv) $r(x,t) = (h(x,1), t - 2(1 - \Psi(x)))$ if $\frac{1}{2} < \Psi(x) \le 1$ and $1 \ge t \ge 2(1 - \Psi(x))$;

(v) $r(x,t) = (x,t)$ if $\Psi(x) = 1$ and $t \in I$.

The reader will enjoy verifying that r is the promised retraction (actually, continuity of r as $\Psi(x)$ approaches 1 is the only troublesome detail to be checked).

B. THE SEPARATION PROBLEM

The Jordan curve theorem states in part that a simple closed curve lying in $\mathscr{E}^2$ separates the plane. Obvious as this result seemed from a visual standpoint, it was no trivial matter to verify. Hence, one might begin to suspect that in general it could prove difficult to determine whether or not a given subset of $\mathscr{E}^n$ separates the whole space. The goal in this section is to establish one important criterion for such a separation to take place. Homotopy theory will once more prove to be an unwitting accomplice.

We shall make use of certain maps first described by Borsuk, which commonly bear his name. These maps are constructed as follows. For each $p \in \mathscr{E}^n$, define a continuous function $\Psi_p : (\mathscr{E}^n \setminus \{p\}) \to \mathscr{S}^{n-1}$ by setting $\Psi_p(x) = (x - p)/\|x - p\|$, where x and p are considered as vectors, and $\|x - p\|$ represents the usual norm (distance of $x - p$ to the origin). If $A \subset \mathscr{E}^n \setminus \{p\}$, then $\Psi_{p|A}$ is called a *Borsuk map* for A based at p.

Borsuk maps are given a prime role in the proof of the following theorem due to Borsuk.

(15.B.1)   Theorem. Suppose that A is a compact subset $\mathscr{E}^n$. Then A separates $\mathscr{E}^n$ if and only if there is a map $f : A \to \mathscr{S}^{n-1}$ that is essential.

Proof. Suppose that A separates $\mathscr{E}^n$. Let C be a bounded component of $\mathscr{E}^n \setminus A$ (why does one exist?), and pick any point $p \in C$. The Borsuk map for A based at p will prove to be essential. If Ψ_p is inessential, then by (15.A.3), Ψ_p can be extended to all of $\mathscr{E}^n$. In particular, Ψ_p can be extended to a ball B containing A with center p and radius $r > 0$. We show that this is an impossibility by constructing a retraction of B onto its boundary. Let Φ denote the extension of Ψ_p to B, and note that, in particular, Φ is defined at p. Then $f : B \to \mathrm{Fr}\, B$ defined by

$$f(x) = \begin{cases} p + r \cdot \dfrac{x - p}{\|x - p\|} & \text{for} \quad x \in B \setminus C \\ p + r \cdot \Phi(x) & \text{otherwise} \end{cases}$$

is a retraction, which leads to a contradiction of (14.C.21).

The converse is more difficult and is based on the following lemma.

(15.B.2) Lemma. Suppose that K is a compact subset of $\mathscr{E}^n$, C is a closed subset of K, and $f : C \to \mathscr{S}^{n-1}$ cannot be extended to K. Then there is an open subset V of $\mathscr{E}^n$ such that $V \subset (K \setminus C)$ and Fr $V \subset C$.

Proof. Since f cannot be extended to K, there is, by (4.B.23), a closed subset C^* of K such that

(i) f can not be extended to $C \cup C^*$, and
(ii) f can be extended to the union of C with any proper closed subset of C^*.

We first show that $C^* \setminus C$ is open in $\mathscr{E}^n$. Suppose that $C^* \setminus C$ is not open and that $x \in (C^* \setminus C) \cap \text{Fr}\,(C^* \setminus C)$. Let U be the interior of a small n-ball that contains x. Since $C^* \setminus U$ is properly contained in C^*, there is an extension of $f, \hat{f}$, that maps $C \cup (C^* \setminus U)$ into $\mathscr{S}^{n-1}$. In particular, $\hat{f}$ maps $C^* \cap (\text{Fr } U)$ into $\mathscr{S}^{n-1}$, and since $C^* \cap (\text{Fr } U)$ is a closed subset of the $(n-1)$-sphere Fr U, we may apply (14.C.19) to obtain an extension of $\hat{f}_{|C^* \cap (\text{Fr } U)}$ to all of Fr U. Thus, we now have a map $\tilde{\hat{f}} : ((C^* \setminus U) \cup ((\text{Fr } U) \cup C) \to S^{n-1}$ that extends f. Let $p \in U \cap \text{ext } C^*$, and from p project points of $C^* \cap U$ into Fr U by a map P. Finally we extend f to $F : (C \cup C^*) \to \mathscr{S}^{n-1}$ by setting

$$F(x) = \begin{cases} \tilde{\hat{f}}P(x) & \text{if} \quad x \in U \cap (C \cup C^*) \\ \hat{f}(x) & \text{if} \quad x \in (C^* \setminus U) \cap \text{Fr } U \end{cases}$$

Since this extension contradicts the nature of C^*, we have that $C^* \setminus C$ is open in $\mathscr{E}^n$. It now follows easily that $\text{Fr}(C^* \setminus C) \subset C$, since C^* is closed and $C^* \setminus C$ is open. This completes the proof of the lemma.

We now complete the proof of (15.B.1). Suppose that $f : A \to \mathscr{S}^{n-1}$ is essential. Let B be a large ball in $\mathscr{E}^n$ that contains A. It follows from (15.A.5) that it is impossible to extend f to B, and hence by the foregoing lemma, there is an open set U in $B \setminus A$ whose frontier lies completely in A. Therefore, U is both open and closed in $\mathscr{E}^n \setminus A$, which of course implies that A separates $\mathscr{E}^n$.

(15.B.3) Corollary. Suppose that A is a compact subset of $\mathscr{E}^n$ and A' is another subspace of $\mathscr{E}^n$ homeomorphic to A. Then A separates $\mathscr{E}^n$ if and only if A' separates $\mathscr{E}^n$.

Note that this corollary yields a portion of the Jordan curve theorem, although it should be observed that neither (15.B.1) nor the corollary gives the number of components that a separation may cause, nor if this number is a topological invariant. Results of this nature may be found in Dugundji [1966].

A very important consequence of (15.B.1) is contained in the next result, which identifies one of the most powerful characteristics of $\mathscr{E}^n$: invariance of domain. The theorem states that embeddings map open subsets of $\mathscr{E}^n$ into open subsets of $\mathscr{E}^n$. This property also holds for n-manifolds (see problem B.10).

(15.B.4)   Theorem (Invariance of Domain).   Suppose that U is an open subset of $\mathscr{E}^n$ and that $h : U \to \mathscr{E}^n$ is an embedding. Then $h(U)$ is open in $\mathscr{E}^n$.

Proof.   Suppose that $x = h(u) \in h(U)$. Since U is open, there is an $\varepsilon > 0$ such that if $V = S_\varepsilon(u)$, then $\overline{V} \subset U$. Note that since $\overline{V}$ does not separate $\mathscr{E}^n$, it follows from the previous theorem that $h(\overline{V})$ also fails to separate $\mathscr{E}^n$. Furthermore, $h(\overline{V}) \setminus h(\mathrm{Fr}\ V)$ is connected, since $h(\overline{V} \setminus \mathrm{Fr}\ V) = h(\overline{V}) \setminus h(\mathrm{Fr}\ V)$ and $\overline{V} \setminus \mathrm{Fr}\ V$ is clearly connected. Hence, $\mathscr{E}^n \setminus h(\mathrm{Fr}\ V)$ may be written as the disjoint union of the connected sets $\mathscr{E}^n \setminus h(\overline{V})$ and $h(\overline{V}) \setminus h(\mathrm{Fr}\ V)$.

Since these two sets are connected, they must be components of the open set $\mathscr{E}^n \setminus h(\mathrm{Fr}\ V)$. However, components of open subsets of a locally connected space are open; hence, in particular, $W = h(\overline{V} \setminus \mathrm{Fr}\ V)$ is open, and since $x \in W \subset h(U)$, it follows that $h(U)$ is an open subset of $\mathscr{E}^n$.

C.   UNICOHERENCE

As a third illustration of the versatility of homotopy theory, we reconsider briefly the concept of unicoherence. Recall that a connected space X is *unicoherent* if and only if whenever $X = C_1 \cup C_2$ where C_1 and C_2 are closed and connected, then $C_1 \cap C_2$ is connected. The unit circle $\mathscr{S}^1$ is an example of a space that is not unicoherent, since overlapping semicircles do not have connected intersection. Loosely speaking, unicoherence reflects a lack of "holes" in a space. The following theorem is considerably more precise.

(15.C.1)   Theorem.   Suppose that X is a connected, locally path connected metric space with the property that every continuous map $f : X \to \mathscr{S}^1$ is inessential. Then X is unicoherent.

In order to prove the theorem, a pair of lemmas are needed. The proof of the first lemma follows readily from covering space theory, and the proof of the second is left as an easy exercise.

(15.C.2)   Lemma.   Suppose that X is a connected, locally path connected space and that $f : X \to \mathscr{S}^1$ is inessential. Then there is a map $g : X \to \mathscr{E}^1$ such that $f(x) = (\cos g(x), \sin g(x))$ for each $x \in X$.

Proof.   Consider the following diagram.

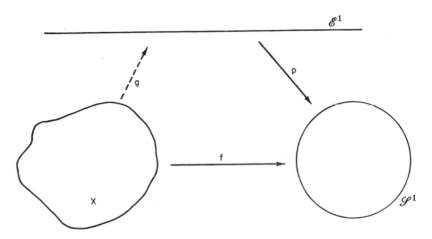

Here, $p : \mathscr{E}^1 \to \mathscr{S}^1$ is defined by $p(x) = (\cos x, \sin x)$. Then $(\mathscr{E}^1, p)$ is a covering space for $\mathscr{S}^1$. Since $f_*(\pi_1(X, x_0)) = \{e\} \subset p_*(\pi_1(\mathscr{E}^1, 0))$, it follows from (13.D.1) that f may be lifted to a map g such that $pg = f$.

(15.C.3)   Lemma.   Suppose that f and g are maps from X into $\mathscr{E}^1$ and that $\hat{f}$ and $\hat{g}$ are defined by $\hat{f}(x) = (\cos f(x), \sin f(x))$ and $\hat{g}(x) = (\cos g(x),$ $\sin g(x))$. If $\hat{f}(x) = \hat{g}(x)$ for each $x \in X$, then there is an integer n such that $f(x) = g(x) + 2n\pi$ for each $x \in X$.

Proof of (15.C.1).   Suppose that X is not unicoherent. Then there are closed connected subsets A and B such that $X = A \cup B$ and $A \cap B$ is not connected. Write $A \cap B$ as the disjoint union of nonempty closed subsets C_1 and C_2. For each $x \in X$, let $q_x = d(x, C_1)/[d(x, C_1) + d(x, C_2)]$. Define functions $f_A : A \to \mathscr{S}^1$ and $f_B : B \to \mathscr{S}^1$ by letting $f_A(x) = (\cos \pi q_x, \sin \pi q_x)$ and $f_B(x) = (\cos \pi q_x, -\sin \pi q_x) = (\cos (-\pi q_x), \sin (-\pi q_x))$. Observe that $f_{A|A\cap B} = f_{B|A\cap B}$, and consequently the map gluing theorem ensures that the function

$$f(x) = \begin{cases} f_A(x) & \text{if} \quad x \in A \\ f_B(x) & \text{if} \quad x \in B \end{cases}$$

is both well defined and continuous. By hypothesis, f must be inessential; thus, by (15.C.2), there is a map $g : X \to \mathscr{E}^1$ such that $f(x) = (\cos g(x),$ $\sin g(x))$. Lemma (15.C.3) asserts that there are integers m and n such that $g(x) = \pi q_x + 2m\pi$ if $x \in A$ and $g(x) = -\pi q_x + 2n\pi$ if $x \in B$.

However, if $x \in C_1 \subset A \cap B$, then $g(x) = 2m\pi = 2n\pi$, and hence $m = n$. On the other hand, if $x \in C_2 \subset A \cap B$, it follows that $g(x) = \pi +$

$2mn\pi = -\pi + 2n\pi$, and we have shown that $\pi = -\pi$. Thus, X must be unicoherent.

(15.C.4) Corollary. The spaces $\mathscr{E}^n$, $\mathbf{I}^n$, $\mathscr{S}^{n+2}$ are unicoherent for each $n \in \mathbf{N}$.

D. $\pi_1(\mathscr{E}^3 \setminus G)$, WHERE G IS A POLYGONAL GRAPH IN $\mathscr{E}^3$

The material in this section is largely an adaptation of Chapter 6 of Crowell and Fox [1963]. The present goal is to convince the reader of the existence of an algorithm that may be used to compute the fundamental group of the complement of a knot, or more generally of the complement of a finite graph lying in $\mathscr{E}^3$. A careful blend of mathematics and sleight of hand is employed. It is hoped that the normal reader will be able to convince himself that the hand waves can be done mathematically, and it is assumed that only the more naive or masochistic readers will actually attempt to carry out some of the constructions.

(15.D.1) Exercise. Show that $\pi_1(\mathscr{E}^3 \setminus \{p\}) = \{e\}$.

(15.D.2) Definition. A *graph* G is the carrier of a one dimensional simplicial complex in $\mathscr{E}^3$.

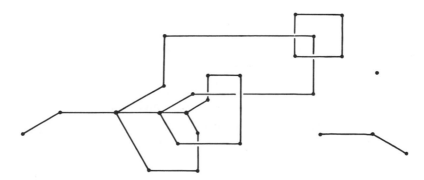

(15.D.3) Remark. If G is a graph and v_0 is a vertex of G that is the face of only one 1-simplex $\langle v_0 v_1 \rangle$, then there is a homeomorphism of $\mathscr{E}^3 \setminus (G \setminus \langle v_0 v_1 \rangle)$ into $\mathscr{E}^3 \setminus G$. Actually, there is an *isotopy*, i.e., a homotopy $H : (\mathscr{E}^3 \setminus G) \times \mathbf{I} \to \mathscr{E}^3 \setminus G$ such that $H_0 = id$, H_1 is onto $\mathscr{E}^3 \setminus (G \setminus \langle v_0 v_1 \rangle)$, and H_t is a homeomorphism for all t.

Thus, the fundamental group is unchanged if G is replaced by a graph

$\hat{G}$ in which no vertex is the face of just one 1-simplex (simply apply the fore-going remark a finite number of times). Furthermore, the Seifert-van Kampen theorem may be used in conjunction with (15.D.1) to enable us to eliminate components of G that consist of single vertices.

Hence, the fundamental group of the complement of the graph in the preceding figure is isomorphic to the fundamental group of the complement of the following graph.

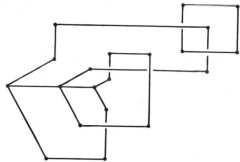

In the remainder of this section, it will be assumed that all graphs have been so simplified.

(15.D.4)   Definition.   A *branch point v* of a graph G is a vertex of G that is the face of three or more 1-simplices of G. The graph just pictured has three branch points.

(15.D.5)   Definition.   Let $p : \mathscr{E}^3 \to \mathscr{E}^2$ be the usual projection map onto the *xy* plane, i.e., $p(x,y,z) = (x,y,0)$, and let $\rho = p|_G$.

The map ρ projects G *normally* onto the *xy* plane (or $\rho(G)$ is a *normal projection*) if and only if

   (i) ρ is 1–1 except at a finite number of points of G;
   (ii)   for each $g \in G$, card($\rho^{-1}\rho(g)$) $\leq$ 2;
   (iii) if v is a vertex of G, then card($\rho^{-1}\rho(v)$) $= 1$.

Observe that (iii) rules out the following configurations.

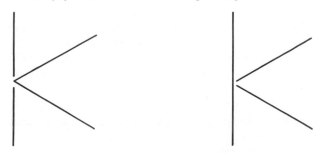

(15.D.6) Definition. Suppose that $\rho(G)$ is a normal projection and that $g,g' \in G$, $g \neq g'$, and $\rho(g) = \rho(g')$. Then whichever of g and g' has the larger z coordinate is called an *overcrossing*, and the other point is called an *undercrossing*.

It will be convenient·to indicate the projection of an overcrossing by a solid line and the image of an undercrossing by a broken line.

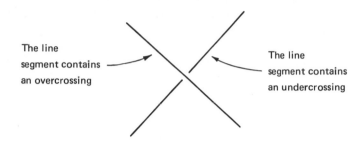

The line segment contains an overcrossing

The line segment contains an undercrossing

There is no reason to believe that the projection of a given graph onto the xy plane will be normal. The projection might, among other things, collapse a whole 1-simplex to a point. In the next theorem we see that if a projection is not normal, then it may be "normalized" by a slight rotation of the given coordinate system.

(15.D.7) Theorem. Suppose that G is a graph in $\mathcal{E}^3$. Then either $\rho(G)$ is a normal projection, or a small rotation of the coordinate system of $\mathcal{E}^3$ will yield a projection $\hat{\rho}$ that projects G normally onto the new xy plane.

Sketch of proof. We shall stand at the origin and keep track of the directions in which projections are to be proscribed.

1. Consider the vertices of G three at a time. A number of such triples span planes. Translate these planes to the origin; the z axis to be chosen will be selected in such a manner that it misses each of these planes (except at the origin, of course). Any z axis not contained in one of these planes yields a projection that satisfies parts (i) and (iii) of the definition of normal projection (15.D.5). However, there still may be points $g \in G$ such that $2 < \text{card}(\rho^{-1}\rho(g)) = n < \infty$. To eliminate this problem, the following is done.

2. For each triple of 1-simplexes of G, which when extended form a triple of skew lines, find the unique direction with the property that projection in this direction yields a triple point. Draw a line through the origin in each direction that is determined in this manner. Now there are a finite number of "unwanted" planes and lines passing through the origin. If the original z axis is in any one of the forbidden planes or lines, an arbitrarily small

rotation will move the z axis into an allowable position (which of course also changes the x and y axes, but that is of no concern). Projection relative to the new coordinate system is normal. Note that topologically we have used the obvious fact that the union of a finite number of planes and lines in $\mathscr{E}^3$ forms a nowhere dense subset of $\mathscr{E}^3$.

(15.D.8) Definition. Suppose that B is the set consisting of all under-crossings, overcrossings, and branch points of a graph G. Then $G \setminus B$ is a finite union of open arcs. Two points of B are *adjacent* if and only if they form the boundary of one of the arcs in $G \setminus B$.

In order to further position the graph G, we shall use a finite subset Y of G that satisfies the following properties:

(i)   each component of G contains at least two points of Y (hence, Y always contains at least two points);

(ii) two points of Y lie between adjacent branch points;

(iii) two points of Y lie between an adjacent branch point and an undercrossing;

(iv) one point of Y lies between an adjacent branch point and an over-crossing;

(v)   one point of Y lies between an adjacent undercrossing and an overcrossing;

(vi) two points of Y lie between two adjacent undercrossings;

(vii) two points of Y lie between two adjacent overcrossings.

The points of Y are marked by $\times$ in the following figure.

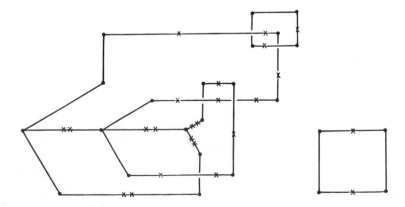

We construct a graph G' whose overcrossings lie above the xy plane, and whose undercrossings and branch points lie below. Furthermore, G' will have the following properties:

(i) $\rho(G) = \rho(G')$;

(ii) $G' \cap \mathscr{E}^2 = \rho(Y)$;

(iii)   there is a homeomorphism $h : (\mathscr{E}^3, G) \to (\mathscr{E}^3, G')$.

For each $y \in Y$, we map Y onto $\rho(Y)$. For each branch point $b_i \in G$, select a point $c_i \in p^{-1}\rho(b_i)$ with negative z coordinate. Let b_i be a branch point of G and suppose that $y \in Y$ has the property that there is a component α (an open arc) of $G \setminus B$ such that $b_i \in \bar{\alpha}$, and no other points of Y lie on the subarc β of $\bar{\alpha}$ that runs from b_i to y. Let β_{iy} be a polygonal arc in $p^{-1}\rho(\beta)$ between c_i and $\rho(y)$ such that β_{iy} lies below the xy plane (except at its end point $\rho(y)$).

Now suppose that γ is an arc of G between "two successive points of Y," and γ does not contain a branch point. Then one of the following must hold.

(i)   The arc γ contains an overcrossing.

(ii)   The arc γ contains an undercrossing.

(iii)   The arc γ contains no crossing and does not lie on a simple closed curve component of G. We have the following three subcases of (iii).

(a) γ lies between two consecutive overcrossings;

(b) γ lies between two consecutive undercrossings;

(c) γ lies between a branch point and an undercrossing.

(iv)   The arc γ contains no crossings and lies on a simple closed curve component of G.

In cases (i), and (b) and (c) of (iii), construct a polygonal arc U_i in $p^{-1}\rho(\gamma)$ which, except for the end points, lies above the plane $z = 0$. In cases (ii) and (a) of (iii), construct a polygonal arc $\hat{L}_i$ lying below the plane $z = 0$. In case (iv), Y cuts a component S of G into two components whose closures are arcs γ_1 and γ_2. Construct polygonal arcs in $p^{-1}(\rho(\gamma_1))$ and $p^{-1}(\rho(\gamma_2))$ one of which, U_i, lies above $\mathscr{E}^2$, and the other, $\hat{L}_i$, lies below $\mathscr{E}^2$ (except for the end points, of course). Let G' denote the resulting graph.

We do not define the homeomorphism h, but leave its construction as a well deserved exercise for the skeptical reader.

The net result of all this is that for computing $\pi_1(\mathscr{E}^3 \setminus G)$, we may assume that the graph G straddles the plane $z = 0$ with overcrossings lying above and undercrossings and branch points lying below the plane.

Let $U = \{U_1, U_2, \ldots, U_n\}$ be the collection of polygonal arcs constructed above $\mathscr{E}^2$ and let $L = \{L_1, \ldots, L_m\}$ be the components of $(\bigcup \hat{L}_i) \cup (\bigcup \beta_{iy})$. Note that the U_i's are all disjoint and the L_i's are either some $\hat{L}_i$ or the union of β_{iy}'s that share a common branch point. Henceforth, n and m will refer to the number of elements in U and L respectively.

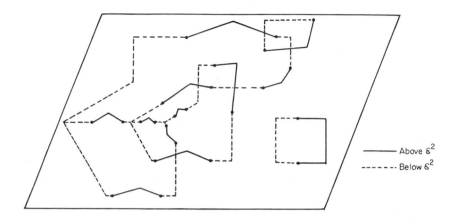

Let $X = \{x_1, x_2, \ldots, x_n\}$ be a set. We want to define a surjective homomorphism $\phi : F[X] \to \pi_1(\mathscr{E}^3 \setminus G)$, where $F[X]$ is the free group on X. We do this in what appears to be a rather oblique manner. This obliqueness will allow us eventually to obtain a geometric description of the kernel of ϕ. For bookkeeping purposes, an orientation is arbitrarily assigned to each $U_i \in U$.

(15.D.9) **Definition.** An oriented polygonal path α (not necessarily a loop) in $\mathscr{E}^3$ will be called *admissible* if and only if

    (i) $\alpha \subset E^2 \setminus \rho(L)$;
    (ii) $\operatorname{card}(\alpha \cap \rho(U))$ is finite;
    (iii) $\alpha \cap \rho(U)$ does not contain a vertex of either α or $\rho(U)$.

Let $\mathscr{R}$ be the set of admissible paths and define a map $\# : \mathscr{R} \to F(X)$ as follows. Suppose that $\alpha \in \mathscr{R}$. List in order the U_i's whose projections intersect α. Suppose that $U_{i_1}, U_{i_2}, U_{i_3}, \ldots, U_{i_k}$ is such a listing. Define $\#(\alpha) = x_{i_1}^{\varepsilon_1} x_{i_2}^{\varepsilon_2} \cdots x_{i_k}^{\varepsilon_k}$, where ε_i is 1 if α crosses $\rho(U_{i_j})$ from left to right and is -1 otherwise.

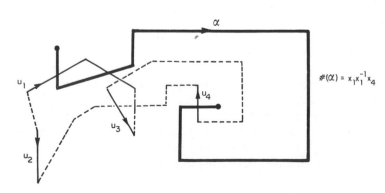

Observe that if $\alpha = \beta * \gamma$, then $\#(\alpha) = \#(\beta)\#(\gamma)$, and that $\#$ is almost never onto.

Now let $p_0 \in \mathscr{E}^3$ have z coordinate greater than the z coordinate of any point in G, and furthermore, pick p_0 such that $p(p_0) \cap p(G) = \varnothing$. For each point $u \in \mathscr{E}^2 \setminus p(G)$, let $\bar{u}$ be a path with initial point p_0, that follows along a line parallel to the xy plane to a point directly above u, and then drops straight down until it intersects u.

For each generator x_i of $F[X]$, let α_i be a polygonal path that crosses U_i once, from left to right. Now define $\phi : F[X] \to \pi_1(\mathscr{E}^3 \setminus G, p_0)$ by $\phi(x_0) = [\overline{\alpha_i(0)} * \alpha_i * \overline{\alpha_i(1)}{}^r]$.

It is clear geometrically that $\phi(x_i)$ is well defined. Since $F[X]$ is a free group on the alphabet X, ϕ may be extended uniquely to a homomorphism (that we also denote by ϕ) (Ap.A.13). Eventually, we show that ϕ is onto, but for now, we digress to describe some elements of $F[X]$ that turn out to generate the kernel of ϕ.

For each $L_i \in L$, let V_i be a neighborhood of $p(L_i)$ in the xy plane such that

    (i) V_i has an admissible simple closed curve v_i for a boundary,
    (ii) $p(p_0) \notin V_i$, and
    (iii) $V_i \cap V_j = \varnothing$.

See Example (15.D.12) below.

(15.D.10) Exercise. Show that it is possible to define such a family of neighborhoods. Assume that v_i traces Bd V_i in a counterclockwise direction and $v_i(0) \subset$ Bd $V_i \setminus p(U)$.

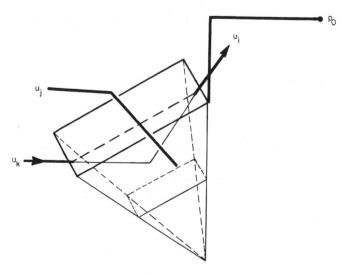

It is clear geometrically (see preceding figure) that $\#(v_i) \in \text{Ker } \phi$. We are now in a position to state the algorithm.

(15.D.11) Algorithm. With the notation developed above, $\{x_1, \ldots, x_n; r_1, \ldots, r_m\}_\phi$ is a presentation of $\pi_1(\mathscr{E}^3 \setminus G, p_0)$, where $r_i = \#(v_i)$.

Before continuing with a sketch of the proof of the algorithm, we give two examples illustrating its use. In the following examples, we indicate the U_i's with heavy lines (which is unnecessary) and label them by x_i. (If α_i is an admissible path that crosses only U_i and if α_i crosses U_i once from left to right, then $x_i = \#(\alpha_i)$.)

(15.D.12) Example.

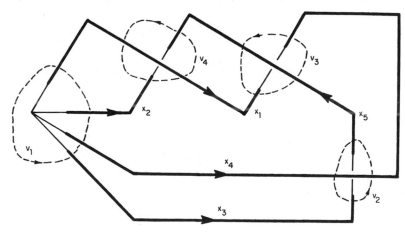

$$\pi_1(\mathscr{E}^3 \setminus G, p_0) \simeq \{x_1, x_2, x_3, x_4, x_5;\ x_3^{-1} x_4^{-1} x_2^{-1} x_1^{-1},\ x_4^{-1} x_5^{-1} x_4 x_3,\ x_4 x_5^{-1} x_1 x_5,\ x_5 x_1 x_2 x_1^{-1}\}$$

(15.D.13) Example.

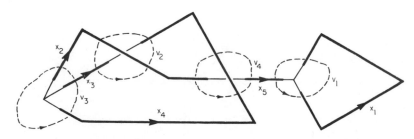

$$\pi_1(\mathscr{E}^3 \setminus G, p_0) \simeq \{x_1, x_2, x_3, x_4, x_5;\ x_1 x_5 x_1^{-1},\ x_4 x_2 x_3 x_2^{-1},\ x_2^{-1} x_4^{-1} x_3^{-1},\ x_5^{-1} x_4^{-1} x_2 x_4\}$$

We now outline the proof of the algorithm (readers interested only in the statement of the algorithm may proceed to Exercise (15.D.14)). We show that the map ϕ is onto, and that the kernel of ϕ is generated by $\{r_1, \ldots, r_m\}$. This will follow from several applications of the Seifert-van Kampen theorem.

Let S be a closed square satisfying the following properties:

(i) S lies in a plane parallel to the xy plane;

(ii)   the z coordinate of S is less than the z coordinate of each point of G;

(iii)   the projection of S in the xy plane contains $\rho(G)$, and $p(p_0) \in p(S)$.

It is easy to see that such a rectangle exists.

Let Z be any subset of G; Z^* will denote the space $S \cup Z \cup (\bigcup_{x \in Z} \ell(x))$, where $\ell(x)$ denotes the line segment between x and the point $s \in S$ such that $p(s) = \rho(x)$. Intuitively, Z^* is Z and S together with a curtain hanging from Z to S. The curtain can of course intersect itself.

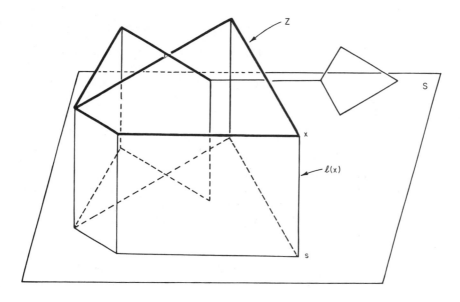

We compute in succession $\pi_1(\mathscr{E}^3 \setminus G^*)$, $\pi_1(\mathscr{E}^3 \setminus L^*)$, and $\pi_1(\mathscr{E}^3 \setminus G)$. It is geometrically clear that $\mathscr{E}^3 \setminus G^*$ is isotopic with $\mathscr{E}^3$ minus a point. (First, stretch out the point to uncover all of S, and then uncover the curtain starting at S to end up with $\mathscr{E}^3 \setminus G^*$.) Hence, $\pi_1(\mathscr{E}^3 \setminus G^*) = \{e\}$.

Next, observe that $\mathscr{E}^3 \setminus L^* = (\mathscr{E}^3 \setminus G^*) \cup (G^* \setminus L^*)$, and $G^* \setminus L^*$ is the disjoint union of n open disks $D_1, \ldots, D_n$, where D_i is the disk lying under the overpass U_i.

Let $\{\alpha_i \mid i = 1, 2, \ldots, n\}$ be the family of admissible paths previously chosen, i.e., α_i is a polygonal path in $\mathscr{E}^2 \setminus \rho(L)$ that passes through $\rho(U_i)$ once, and moves from left to right. Then the loop $\hat{\alpha}_i = \overline{\alpha_i(0)} * \alpha_i * \overline{\alpha_i(1)}^r$ intersects D_i once and otherwise lies in $\mathscr{E}^3 \setminus G^*$.

For each $i = 1, 2, \ldots, n$, let W_i be an open neighborhood of $D_i \cup$ Im $\hat{\alpha}_i$ such that

(i) W_i is path connected, and $\pi_1(W_i, p_0) \simeq \mathbf{Z}$ where $[\hat{\alpha}_i]$ is a generator;
(ii) $W_i \cap G^* = D_i$;
(iii) $\pi_1(W_i \setminus G^*, p_0) = \{e\}$.

The existence of the W_i is not analytically trivial, but the reader should be able to convince himself that such a construction is possible. (Take a small neighborhood of D_i that extends only in the xy directions and carefully thicken up $\hat{\alpha}_i$. The polygonality of $\hat{\alpha}_i$ is handy but not indispensable here.)

We now show with the aid of the Seifert-van Kampen theorem that $\pi_1(\mathscr{E}^3 \setminus L^*, p_0)$ is isomorphic to $F(X)$.

Let $Y_0 = \mathscr{E}^3 \setminus G^*$, and for $i = 1, 2, \ldots, n$, let $Y_i = Y_{i-1} \cup W_i$. Observe that for each i, $(\mathscr{E}^3 \setminus G^*) \cup W_i = \mathscr{E}^3 \setminus (G^* \setminus W_i)$, which by property (ii) equals $\mathscr{E}^3 \setminus (G^* \setminus D_i) = (\mathscr{E}^3 \setminus G^*) \cup D_i$. Hence, $Y_i = (\mathscr{E}^3 \setminus G^*) \cup (\bigcup_{k=1}^{i} D_k)$.

Thus, when $i = n$, we have that $Y_n = (\mathscr{E}^3 \setminus G^*) \cup (\bigcup_{i=1}^{n} D_i) = (\mathscr{E}^3 \setminus G^*) \cup (G^* \setminus L^*) = \mathscr{E}^3 \setminus L^*$. Finally, for $i \neq j$, $D_j \cap W_i = (G^* \cap W_j) \cap W_i = D_j \cap D_i = \varnothing$, and therefore, $Y_{i-1} \cap W_i = [(\mathscr{E}^3 \setminus G^*) \cup (\bigcup_{j=1}^{i-1} D_j)] \cap W_i = (\mathscr{E}^3 \setminus G^*) \cap W_i = W_i \setminus G^*$. Furthermore, by (iii) above, we have that $\pi_1((W_i \setminus G^*), p_0) = \{e\}$.

We now proceed inductively. We have that $\pi_1(\mathscr{E}^3 \setminus G^*, p_0) = \pi_1(Y_0, p_0) = \{e\}$. Assume that Y_{i-1} is path connected and that $\pi_1(Y_{i-1}, p_0)$ is a free group on the $i - 1$ generators $\{[\hat{\alpha}_1], \ldots, [\hat{\alpha}_{i-1}]\}$. Note that Y_{i-1}, W_i, and $Y_{i-1} \cap W_i$ are nonempty, path connected, and open. We apply the Seifert-van Kampen theorem to $Y_i = Y_{i-1} \cup W_i$ to obtain $\pi_1(Y_k, p_0)$ as a free group with generators $\{[\hat{\alpha}_1], \ldots, [\hat{\alpha}_{i-1}]\} \cup \{[\hat{\alpha}_i]\}$, since $\pi_1(W_i, p_0) = \{[\hat{\alpha}_i]\}$ and $\pi_1(Y_{i-1} \cap W_i) = \{e\}$. Hence, we have shown that $\pi_1(\mathscr{E}^3 \setminus L^*, p_0)$ is isomorphic to $F(X)$.

The Seifert-van Kampen theorem will be applied once more to obtain $\pi_1(\mathscr{E}^3 \setminus G, p_0)$, but first we make some geometric observations.

Let B_1 be the boundary of a rectangular prism with top face parallel to the xy plane and which contains S in its interior and G in its exterior, and let B_2 be the vertical line segment extending from the point p_0 to the top of the prism. Let $B = B_1 \cup B_2$. Note that B is of the same homotopy type as a 2-sphere. Let W be a neighborhood of $L^* \setminus G$ such that

(i) $\pi_1(W, p_0) = \{e\}$,

(ii) $W \cap G = \varnothing$,

(iii) $B \setminus L^*$ is a deformation retract of $W \setminus L^*$.

To obtain W, first fatten $L^* \setminus G$ with a small fringe, being careful that at the top, the fringe is added only in the xy direction. Then fatten B slightly.

We apply the Seifert-van Kampen theorem to $\mathscr{E}^3 \setminus L^*$, W, and $(\mathscr{E}^3 \setminus L^*) \cap W$. Note that

$$(\mathscr{E}^3 \setminus L^*) \cup W = \mathscr{E}^3 \setminus (L^* \setminus W) = \mathscr{E}^3 \setminus ((G \cup (L^* \setminus G)) \setminus W)$$

$$= \mathscr{E}^3 \setminus (\underbrace{(G \setminus W)}_{G} \cup \underbrace{((L^* \setminus G) \setminus W)}_{\varnothing})$$

$$= \mathscr{E}^3 \setminus G$$

It should also be noted that $(\mathscr{E}^3 \setminus L^*) \cap W = W \setminus L^*$ has the same homotopy type as $B \setminus L^*$. Furthermore, $B \setminus L^*$ is a box with m holes in it. The holes arising from underpasses are slits, the holes that come from branch points consist of three or more slits that come together.

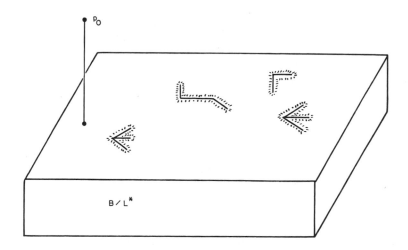

Hence, $\pi_1(B \setminus L^*, p_0)$ is isomorphic to $\pi_1(W \setminus L^*, p_0)$, and $\pi_1(B \setminus L^*, p_0)$ is a free group with $m - 1$ generators (what should be the m-th generator is a word in the other generators: think of slipping a loop around the back surface of B).

Since W is simply connected and ϕ is onto, the generators of $\pi_1(\mathscr{E}^3 \setminus G, p_0)$ will be just the generators of $\pi_1(\mathscr{E}^3 \setminus L^*, p_0)$, i.e. $\{[\hat{\alpha}_1], \ldots, [\hat{\alpha}_n]\}$. Furthermore, once we describe the generators of $\pi_1(B \setminus L^*, p_0)$ as words in the $[\hat{\alpha}_i]$, we will have the relations.

Let d be the projection of $\mathscr{E}^2$ down onto the plane containing the top of B. Then dv_i is a loop around the i-th hole. Now we alter the definition of $v_i(0)$ so that $v_i(0)$ becomes a vertical path from p_0 to $\mathscr{E}^2$ and is then an admissible path in $\mathscr{E}^2$ that misses all the V_j's except at its end point $v_i(0)$. With this altered definition of $\overline{v_i(0)}$, it is geometrically clear (see the next figure) that the path from p_0 down to the top of B, along $\overline{dv_i(0)}$, around dv_i, and back is homotopic to $\overline{v_i(0)} * v_i * \overline{v_i(0)}^r$.

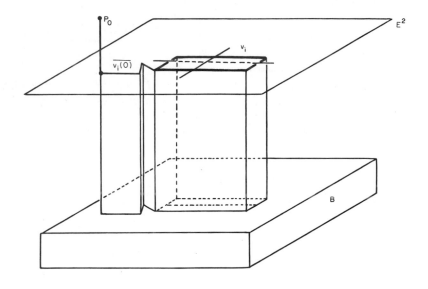

However, $\overline{v_i(0)} * v_i * \overline{v_i(0)}^r$ is associated with the word $\#(p(\overline{v_i(0)}))\#(v_i)$ $\#(p(\overline{v_i(0)}))^{-1})$ in $F[X]$. As a relation, this word is clearly equivalent to $\#(v_i)$.

Thus, we have obtained the desired result, with one additional benefit: any one of the m relations is a consequence of the others.

(15.D.14)   Exercise. Show that if G is a graph (or any compact set in $\mathscr{E}^3$ whose complement is path connected), then $\pi_1(\mathscr{E}^3 \setminus G)$ is isomorphic to $\pi_1(\mathscr{S}^3 \setminus G)$.

(15.D.15)   Definition. A *knot* is a graph homeomorphic to $\mathscr{S}^1$. The *group* of a knot K is $\pi_1(\mathscr{E}^3 \setminus K)$.

(15D.16)   Examples.
    1. The use of the algorithm for calculating fundamental groups of

knots and complements of graphs is illustrated in the following three examples.

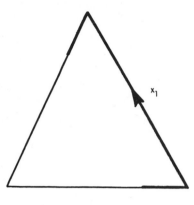

$$\pi_1(\mathscr{E}^3 \setminus G) = \{x_1; \ \} = \mathbf{Z}$$

2. *Trefoil knot.*

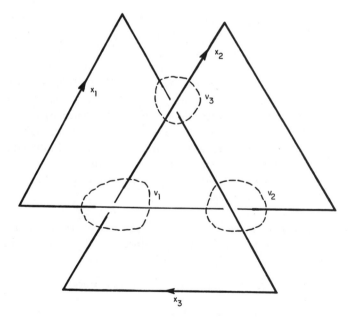

$$\pi_1(\mathscr{E}^3 \setminus G) = \{x_1, x_2, x_3; \ x_1^{-1}x_3x_1x_2^{-1}, \ x_1^{-1}x_3^{-1}x_2x_3, \ x_2x_3^{-1}x_2^{-1}x_1\}$$

3.

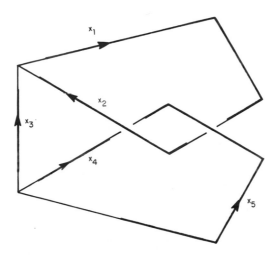

$$\pi_1(\mathscr{E}^3 \setminus G) = \{x_1, x_2, x_3, x_4, x_5; \; x_3x_2x_1^{-1}, \; x_5^{-1}x_4^{-1}x_3^{-1}, \; x_5x_1x_5^{-1}x_2^{-1}, \; x_4x_2x_5x_2^{-1}\}$$

(15.D.17)   Remark.   As we established in (12.C.5), one way of deciding that two spaces are not homeomorphic is to show that their fundamental groups are not isomorphic. For instance, one can show that $\mathscr{E}^3$ minus the boundary of a 2-simplex is not homeomorphic to $\mathscr{E}^3$ minus the trefoil knot, if it can be shown that the group in Example (15.D.16–2) is not isomorphic to **Z**. To show that these two groups are not isomorphic, it suffices to find a homomorphic image of one group that cannot be a homomorphic image of the other. Let $A = \pi_1(\mathscr{E}^3 \setminus G)$, where G is the trefoil knot. Observe that every homomorphic image of **Z** is abelian (in fact cyclic), and hence, if a nonabelian (or a noncyclic) homomorphic image of A can be found, or equivalently if a nonabelian (or a noncyclic) quotient group of A can be found, then A cannot be isomorphic to **Z**.

(15.D.18)   Reidemeister's Trick.   Suppose that G is the trefoil knot and that $A = \pi_1(\mathscr{E}^3 \setminus G)$. Consider $A/\langle x_1^2, x_2^2, x_3^2 \rangle$. Since $x_i = x_i^{-1}$, $A/\langle x_1^2, x_2^2, x_3^2 \rangle$ has a presentation

$$\{x_1, x_2, x_3; \; x_1^2, x_2^2, x_3^2, \; \underbrace{x_1x_3x_1x_2, \; x_1x_3x_2x_3, \; x_2x_3x_2x_1}\}$$
$$x_2 = x_1x_3x_1$$

Since $x_2 = x_1x_3x_1$, we may remove x_2 from the list of generators, and substitution of $x_1x_3x_1$ for x_2 in the relations yields the presentation

$$\{x_1,x_3; \, x_3^2,x_3^2, \, x_1x_3x_1x_1x_3x_1, \, x_1x_3x_1x_3x_1x_3, \, x_1x_3x_1x_3x_1x_3x_1x_1\}$$
$$= \{x_1,x_3; \, x_1^2,x_3^2, \, (x_1x_3)^3\}$$

It is easily seen that each coset can be represented by exactly one of the following words, x_1, x_1x_3, $(x_1x_3)x_1$, $(x_1x_3)^2$, $(x_1x_3)^2x_1$, e. (Note, for example, that $x_3x_1 = (x_1x_3)^{-1} = (x_1x_3)^2$.) That the group is not cyclic can now be checked by brute force. The quotient group $A/\langle x_1^2,x_2^2,x_3^2 \rangle$ that we have obtained is the dihedral group of order 6, and is neither cyclic nor abelian. Hence, A is not isomorphic to **Z**.

E. WILD SETS DO EXIST

(15.E.1) *Definition.* An n-cell B is any space homeomorphic to

$$\mathscr{B}^n = \{(x_1, x_2, \ldots, x_n) \in \mathscr{E}^n \mid x_1^2 + x_2^2 + \cdots + x_n^2 \le 1\}$$

An n-sphere S is any space homeomorphic to

$$\mathscr{S}^n = \{(x_1, x_2, \ldots, x_{n+1}) \in \mathscr{E}^{n+1} \mid x_1^2 + \cdots + x_{n+1}^2 = 1\}$$

(15.E.2) *Definition.* An n-cell C in $\mathscr{E}^n$ is *tame* if and only if there is a homeomorphism $h : \mathscr{E}^n \to \mathscr{E}^n$ such that $h(C) = \mathscr{B}^n$. If C is not tame, then C is *wild*. An n-sphere S in $\mathscr{E}^{n+1}$ is tame if and only if there is a homeomorphism $h : \mathscr{E}^{n+1} \to \mathscr{E}^{n+1}$ such that $h(S) = \mathscr{S}^n$.

(15.E.3) *Definition.* A compact subset X of $\mathscr{S}^n$ ($\mathscr{E}^n$) is *tame* if and only if there is a homeomorphism $h : \mathscr{S}^n \to \mathscr{S}^n$ ($h : \mathscr{E}^n \to \mathscr{E}^n$) such that $h(X)$ is a Euclidean polyhedron. If X is not tame, then X is *wild*.

(15.E.4) *Remark.* It is easy to show that an n-cell or an n-sphere that is tame in the sense of (15.E.2) is also tame in the sense of (15.E.3). For n-cells and for 2-spheres, the converse can also be shown (Rushing [1973]). For n-spheres, $n > 2$, it is unknown whether or not the converse holds.

(15.E.5) *Exercise.* Suppose that A is a polygonal arc in $\mathscr{E}^3$. Show that $\pi_1(\mathscr{E}^3 \setminus A)$ is trivial. Give a reasonable interpretation of a "polygonal arc" in $\mathscr{S}^3$ and prove an analogous result.

(15.E.6) *Definition.* Suppose that X is a locally connected topological space and that A is a closed subset of X. Then a *complementary domain* of A in X is a component of $X \setminus A$.

In this section, we give examples of

(i)   a wild arc in $\mathcal{E}^3$ ($\mathcal{S}^3$),

(ii) a 2-sphere in $\mathcal{S}^3$, the closure of one of whose complementary domains is not a cell,

(iii) a 2-sphere in $\mathcal{S}^3$ for which the closure of neither complementary domain is a cell, and

(iv) a wild cell in $\mathcal{E}^3$ ($\mathcal{S}^3$).

The latter three examples will be modifications of the first one.

In all of these examples, one proves wildness by showing that the fundamental group of a complementary domain is nontrivial. There are, however, examples of wild arcs and spheres whose complementary domains have trivial fundamental groups (see Fox and Artin [1948]). The proofs of wildness in these latter cases involve far more subtle properties of 3-cells than we shall consider.

*(15.E.7) **Example A Wild Arc** (Fox and Artin [1948]).* The construction of this arc involves a certain amount of tedious description. The reader is cautioned to read the next paragraphs with the figures below in sight.

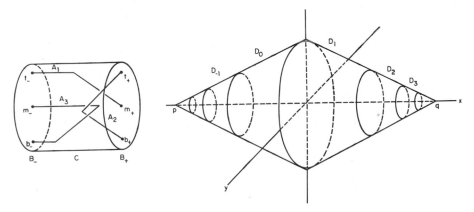

Let C be the right circular cylinder in the figure. The basic idea is to embed C in each of the regions D_i in such a manner that the end points of the arcs on the left-hand boundary of each D_i match up with the appropriate points on the right-hand side. On the base B_-, mark three collinear points $t_-, m_-,$ and b_-. On the other base B_+, mark collinear points $t_+, m_+,$ and b_+. Construct in C three disjoint polygonal arcs $\alpha_1, \alpha_2,$ and α_3 whose images are $A_1, A_2,$ and A_3 with the properties that A_1 joints t_- to m_+, A_2 joins b_+ to t_+, and A_3 joins m_- and b_-. These arcs are arranged as indicated in the figure (it is possible to describe the arcs analytically, but by now the reader should be convinced of the sterility of such exercises). Let $A = A_1 \cup A_2 \cup A_3$.

Let S be the solid, double spear point obtained by rotating about the

x axis the area bounded by $y = 0$, $y = (x/2) + 1$, and $y = (-x/2) + 1$. Subdivide S into a countable number of pieces by the family of planes $x = \pm(2 - 1/2^{m-1})$, where $m = 0, 1, \ldots$. Let D_n be the segment of S with x coordinates satisfying $(2 - 2^{2-n}) \le x \le (2 - 2^{1-n})$ for $n > 0$ or $(-2 + 2^{-n}) \le x \le (-2 + 2^{1-n})$ if $n \le 0$. Let $p = (-2,0,0)$ and $q = (2,0,0)$.

For each n, let f_n be a homeomorphism from C onto D_n with the following properties:

   (i)   The simple closed curves bounding the bases of C are mapped to the simple closed curves bounding the bases of D_n. Furthermore, the indicated orientation of the simple closed curves is to be preserved.
   (ii) The bases B_- and B_+ are mapped respectively to the left and right faces of D_n.
   (iii) The points $f_n(t_+)$, $f_n(m_+)$, and $f_n(b_+)$ lie in descending order on a vertical line through the x axis and coincide with the points $f_{n+1}(t_-)$, $f_{n+1}(m_-)$, and $f_{n+1}(b_-)$ respectively.
   (iv) The set $f_n(A)$ projects into the xz plane as

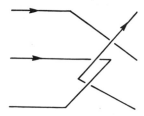

         where

         indicates that the point on the solid line has greater y coordinate than the corresponding point on the broken line.

(15.E.8) *Exercise.* Convince yourself that the homeomorphism f_n exists and that $X = \{p\} \cup (\bigcup_{n \in \mathbf{Z}} f_n(A)) \cup \{q\}$ is the image of the arc α in the following sketch.

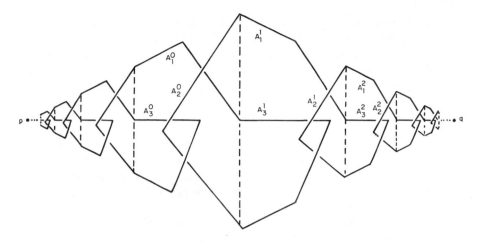

In order to facilitate the use of the algorithm that gives the description of the fundamental group of the complement of α, we redraw the figure and label subarcs that extend from undercrossing to undercrossing as follows:

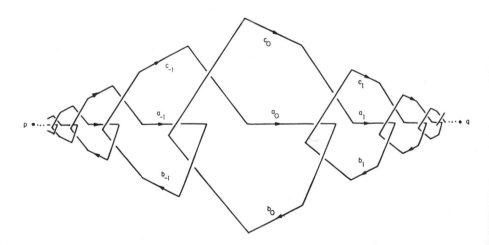

We show that α is a wild arc in $\mathscr{E}^3$. Suppose that α is tame. Then there is a homeomorphism $h : \mathscr{E}^3 \to \mathscr{E}^3$ that carries $\alpha(\mathbf{I}) = X$ onto a polygonal arc L. The map h also carries $\mathscr{E}^3 \setminus X$ homeomorphically onto $\mathscr{E}^3 \setminus L$, and consequently the fundamental group of $\mathscr{E}^3 \setminus X$ is isomorphic to $\pi_1(\mathscr{E}^3 \setminus L)$, which is the trivial group by (15.E.5). Thus, to show that α is wild, it suffices to prove that $\pi_1(\mathscr{E}^3 \setminus X)$ is nontrivial.

To compute $\pi_1(\mathscr{E}^3 \setminus X)$, we first note that $\mathscr{E}^3 \setminus X$ can be expressed as a direct limit. Let $X_n = X \cup (\overline{\bigcup_{|i| \geq n} D_i})$ and $Y_n = \mathscr{E}^3 \setminus X_n$, for each $n \in \mathbb{N}$. Define

the bonding maps $f_n : Y_n \to Y_{n+1}$ by letting f_n be the inclusion map for

each n. It is then clear that $\varinjlim(Y_n, f_n) = \mathscr{E}^3 \setminus X$.

With the aid of work in the previous section, the reader should be able to describe $\pi_1(\mathscr{E}^3 \setminus X_n)$ as $\{a_i, b_i, c_i (-n \leq i \leq n - 1); a_{i+1} = c_{i+1}c_i c_{i+1},$ $b_i = c_{i+1}^{-1} a_i c_{i+1}, c_{i+1} = b_i^{-1} b_{i+1} b_i, b_{-n} a_{-n}^{-1} c_{-n}^{-1} = e, c_{n-1} a_{n-1} b_{n-1}^{-1} = e \; (-n \leq i \leq n - 1)\}$. Next, we observe that $f_{n*} : \pi_1(\mathscr{E}^3 \setminus X_n) \to \pi_1(\mathscr{E}^3 \setminus X_{n+1})$ is an injection. This may be seen from the nature of the presentation and from the next figure, where it is clear that the loops A_{n-1} and A_n are homotopic. The same, of course, occurs for similar loops A_{-n} and $A_{-(n+1)}$ on the left-hand side of the spear point. These loops correspond to the relations $c_{n-1} a_{n-1} b_{n-1}^{-1} = e$, $c_n a_n b_n^{-1} = e$, and $b_{-n} a_{-n}^{-1} c_{-n}^{-1} = e$, $b_{-n-1} a_{-n-1}^{-1} c_{-n-1}^{-1} = e$. Furthermore, for $-n \leq i \leq n - 1$, f_{n*} maps a_i to a_i, b_i to b_i, and c_i to c_i, and of course preserves the relations.

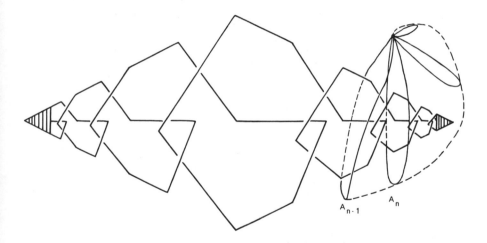

Since f_{n*} is the inclusion map, it now follows from the geometry above, (12.F.11), and (Ap.D.9) that $\pi_1(\mathscr{E}^3 \setminus X)$ is isomorphic to $\varinjlim \pi_1(\mathscr{E}^3 \setminus X_n) = \{a_i, b_i, c_i \; (i \in \mathbb{Z}); \quad a_{i+1} = c_{i+1}^{-1} c_i c_{i+1}, \quad b_i = c_{i+1}^{-1} a_i c_{i+1}, \quad c_{i+1} = b_i^{-1} b_{i+1} b_i, c_i a_i b_i^{-1} = e \; (i \in \mathbb{Z})\}$.

We shall show that $\pi_1(\mathscr{E}^3 \setminus X)$ is nontrivial by finding a nontrivial homomorphic image of $\pi_1(\mathscr{E}^3 \setminus X)$. First, however, note that the generators a_i and b_i may be removed since (*) $a_i = c_i^{-1} c_{i-1} c_i$ and (**) $b_i = c_{i+1}^{-1} a_i c_{i+1}$. This, of course, also leads to the disappearance of these two relations. The remaining two relations (***) $c_{i+1} = b_i^{-1} b_{i+1} b_i$ and (****) $c_i a_i b_i^{-1} = e$ may be rewritten as follows:

1. $c_{i+1} = b_i^{-1}b_{i+1}b_i \overset{(****)}{=} (a_i^{-1}c_i^{-1})(c_{i+1}a_{i+1})(c_ia_i)$

    $\overset{(*)}{=} (c_i^{-1}c_{i-1}^{-1})(c_ic_{i+1})(c_{i-1}c_i)$

    $= c_i^{-1}c_{i-1}^{-1}c_ic_{i+1}c_{i-1}c_i$

2. $c_i = b_ia_i^{-1} \overset{(**)}{=} c_{i+1}^{-1}a_ic_{i+1}a_i^{-1} \overset{(*)}{=} c_{i+1}^{-1}c_i^{-1}c_{i-1}c_ic_{i+1}c_i^{-1}c_{i-1}^{-1}c_i$

Note that in both cases 1 and 2 we have that $c_ic_{i+1}c_{i-1}c_i = c_{i-1}c_ic_{i+1}$. Thus, it follows from (Ap.C.9) and (Ap.C.13) that $\pi_1(\mathscr{E}^3 \setminus X)$ is isomorphic with $\{c_i \, (i \in \mathbf{Z}); c_ic_{i+1}c_{i-1}c_ic_{i+1}^{-1}c_i^{-1}c_{i-1}^{-1} = e \, (i \in \mathbf{Z})\}$. Now define (for no obvious reason) $h : \pi_1(\mathscr{E}^3 \setminus X) \to S_5$ (the symmetric group on five symbols) by

$$h(c_i) = \begin{cases} (1\ 2\ 3\ 4\ 5) & \text{if } i \text{ is odd} \\ (1\ 4\ 2\ 3\ 5) & \text{if } i \text{ is even} \end{cases}$$

Since $h(c_ic_{i+1}c_{i-1}c_ic_{i+1}^{-1}c_i^{-1}c_{i-1}^{-1}) = e$, h is a well-defined homomorphism by Exercise (Ap.C.3). Furthermore, the image of h is nontrivial which concludes the proof. It should be observed that if X is considered as a subset of $\mathscr{S}^3$, then by (15.D.14), $\pi_1(\mathscr{S}^3 \setminus X)$ is isomorphic to $\pi_1(\mathscr{E}^3 \setminus X)$.

(15.E.9) Example. A wild 2-sphere in $\mathscr{S}^3$, the closure of one of whose complementary domains is not a cell.

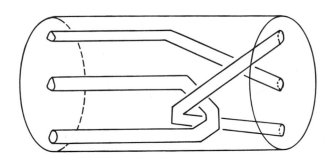

Consider the arc X in the preceding example as a subset of $\mathscr{S}^3$. In the previous construction, replace the arcs A_1, A_2, A_3 by slender hollow tubes T_1, T_2, T_3 similarly situated in C. If this is done carefully, the homeomorphisms constructed in (15.E.8) will properly match the ends of the tubes when C is mapped onto the D_i's. Then $Y = \{p\} \cup (\bigcup f_i(T_j)) \cup \{q\}$ is the desired 2-sphere.

(15.E.10) Exercise. Convince yourself that the closure of one of the complementary domains of Y is a 3-cell.

(15.E.11) Remark. It follows from (15.E.4) that both complementary domains of a tame 2-sphere S in $\mathscr{E}^3$ have trivial fundamental groups.

Let D be the complementary domain in the previous exercise whose closure is a cell. Since it is clear that $\mathscr{S}^3 \setminus \bar{D}$ is a deformation retract of $\mathscr{S}^3 \setminus X$, we have that $\pi_1(\mathscr{S}^3 \setminus \bar{D}) = \pi_1(\mathscr{S}^3 \setminus X) \neq \{e\}$; consequently, $\overline{\mathscr{S}^3 \setminus D}$ is not a cell, and by the Remark (15.E.11), Y is a wild 2-sphere.

(15.E.12) Example. A 2-sphere in $\mathscr{S}^3$ for which the closure of neither complementary domain is a cell.

In this example, push a bubble on the 2-sphere into the complementary domain that is a cell. Now cut off the bubble and graft in its place another copy of the wild sphere (minus a disk) to obtain the following:

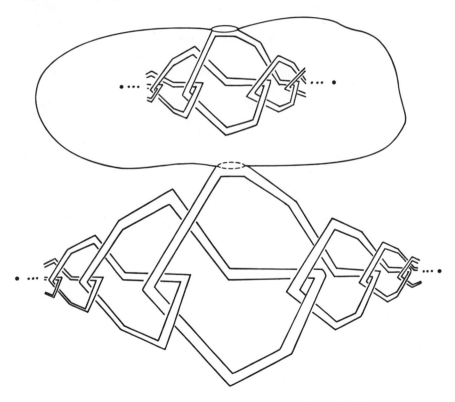

Here the closure of neither complementary domain is a 3-cell.

(15.E.13) Example (A Wild Cell). The cell given in (15.E.10) is wild (why?).

PROBLEMS

Section A

1. Show that $(\mathscr{S}^1, \mathscr{S}^1 \setminus \{(1,0)\})$ fails to have the *AHEP*.
2. A space X is *binormal* if and only if $X \times \mathbf{I}$ is normal. Suppose that A is a closed subset of a binormal space X such that $(A \times \mathbf{I}) \cup (X \times \{0\}) \cup (X \times \{1\})$ is an *ANR*. Show that (X,A) has the *AHEP*.
3. A pair (X,A) has the *homotopy extension property* (*HEP*) with respect to a space Y if and only if for each continuous map $F : X \to Y$ and for each homotopy $H : A \times \mathbf{I} \to Y$ with $H_0 = F_{|A}$, there is a homotopy $\hat{H} : X \times \mathbf{I} \to Y$ such that $\hat{H}$ extends H and $\hat{H}_0 = F$. Suppose that A is a closed subset of a space X and that B is a subset of a space Y. Suppose further that (X,A) has the *HEP* with respect to B and $(X \times \mathbf{I}, (X \times \{0,1\}) \cup (A \times \mathbf{I}))$ has the *HEP* with respect to Y. Show that if $f(X,A) \to (Y,B)$ is homotopic to $g : X \to B$, then $f \sim g(\mathrm{mod}\ A)$.
4. Find an example of a space Y such that the pair given in problem 1 has the *HEP* with respect to Y.
5. Suppose that A is a closed subset of a metric space X. Show that if $f,g : A \to \mathscr{S}^n$ are homotopic and if f can be extended to $F : X \to \mathscr{S}^n$, then so may be g.
6. Suppose that X is an *ANR* and that A is a closed subset of X. Show that $(X \times \{0\}) \cup (A \times \mathbf{I})$ is a retract of $X \times \mathbf{I}$ if and only if A is an *ANR*.
7.* Suppose that X is an *ANR* and that A is a closed subset of X. Show that (X,A) has the *AHEP* if and only if A is an *ANR*.
8. Suppose that X is a topological space. Show that (Cone X,X) has the *AHEP*.

Section B

1. Suppose that A is a compact subset of $\mathscr{E}^n$. Show that $\mathscr{E}^n \setminus A$ has precisely one unbounded component.
2. Show that $\mathscr{E}^m$ and $\mathscr{E}^n$ are not homeomorphic, and extend this result to manifolds.
3. Show that $\mathscr{S}^n$ is not homeomorphic to a proper subset of itself.
4. Suppose that $A \subset \mathscr{E}^n$ and that $h : A \to \mathscr{E}^n$ is an embedding. Show that if $x \in \mathrm{int}\ A$, then $h(x) \in \mathrm{int}\ h(A)$.
5. Suppose that $K \subset \mathscr{E}^n$ is compact. Show that K does not separate $\mathscr{E}^n$ if and only if $\mathbf{C}(A,\mathscr{S}^n)$ (with the compact open topology) is path connected.
6. Suppose that $K \subset \mathscr{E}^n$ is compact and that D is a retract of K. Show that $\mathscr{E}^n \setminus D$ does not have more components than $\mathscr{E}^n \setminus K$.

7. Suppose that S is an n-sphere lying in $\mathscr{E}^{n+1}$. Show that $\mathscr{E}^{n+1} \setminus S$ has finitely many components.
8. Suppose that K is a compact subset of $\mathscr{E}^n$ and that $x \in \mathscr{E}^n \setminus K$. Show that the component of $\mathscr{E}^n \setminus K$ that contains x is unbounded if and only if $\psi_{p|K}$ is null homotopic.
9. Suppose that K is a compact subset of $\mathscr{E}^n$ and that $D \subset K$ is a deformation retract of K. Show that there is a 1–1 correspondence between the components of $\mathscr{E}^n \setminus K$ and the components of $\mathscr{E}^n \setminus D$.
10. Suppose that M and N are n-manifolds, that U is an open set in M, and that $h : U \to N$ is an embedding. Show that $h(U)$ is open.
11. Suppose that U is an open subset of $\mathscr{E}^n$ and that $f : U \to \mathscr{E}^n$ is continuous and 1 to 1. Show that f is an embedding and that $f(U)$ is open.
12. Find an example of a function $f : \mathscr{E}^1 \to \mathscr{E}^2$ that is continuous and 1 to 1, but for which the image of some open subset of $\mathscr{E}^1$ is not open in $f(\mathscr{E}^1)$.
13. If $m < n$, show that there is no continuous 1 to 1 function mapping an open subset of $\mathscr{E}^n$ into $\mathscr{E}^m$.

Section C

1. Show that if the union of two continua C_1 and C_2 is $\mathscr{S}^n$, then $C_1 \cap C_2$ is connected.
2. Show that any contractible, locally path connected space is unicoherent.
3. Is the space described in (12.C.8) unicoherent?
4. Suppose that X is a compact metric space. Let $Z = \mathbf{C}(X, \mathscr{S}^1)$ be given the compact-open topology. Show that if Z is path connected, then X is unicoherent.
5. Show that if U_1 and U_2 are connected open subsets of $\mathscr{S}^2$ and Fr $U_1 \cap$ Fr U_2 is connected, then $U_1 \cap U_2$ is connected.

Section D

1. Find a presentation of group of the following knot, and show that the group is trivial.

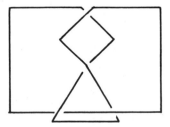

2. Find a presentation of the group of the following knot, and show that the group is nontrivial.

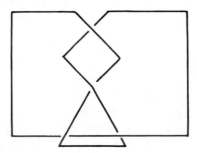

3. Find a presentation of the group of the following knot, and show that the knot group is nontrivial.

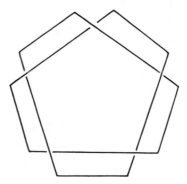

4.* Show that the groups obtained in problems 3 and 4 are not isomorphic.
5. Find a presentation of the group of the following knot, and show that the group is nontrivial.

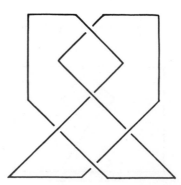

6. Show that the fundamental group of the complement of the following graph is {a,b; }.

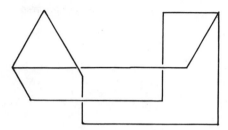

7.* Show that if K is a knot, G is the group of K, and $[G,G]$ is the commutator subgroup of G, then $G/[G,G]$ is infinite cyclic.

Chapter 16

2-MANIFOLDS

In this chapter, we focus our attention on some of the basic properties of 2-manifolds. Our investigation, while of interest in itself, also serves as an introduction to the techniques used and problems encountered in working with higher dimensional manifolds. Nevertheless, the reader should be warned that mastery of the 2-manifold does not presage any great competence for handling its higher dimensional counterparts. Just as we shall see that 2-manifolds are far richer than the 1-manifolds, the gap between 2-manifolds and 3-manifolds is even greater, and a 69-manifold is something to behold.

We shall soon be in a position to classify all compact 2-manifolds. However, we first establish that all 2-manifolds can be triangulated. A similar result holds for 3-manifolds, but its proof is far more difficult; the status of n-manifolds, $n \geq 4$, is not completely known.

A. THE TRIANGULATION OF 2-MANIFOLDS

The initial goal of this section is to establish a simple criterion that enables one to determine whether or not a connected open subset of $\mathscr{E}^2$ is homeomorphic to $\mathscr{E}^2$. Recall that a 2-cell is any space homeomorphic to $\mathscr{B}^2$, and that in the plane, every simple closed curve bounds such a cell.

(16.A.1) ***Theorem.*** If M is an open connected nonempty subset of $\mathscr{E}^2$

with the property that every simple closed curve in M bounds a 2-cell in M, then M is homeomorphic to $\mathscr{E}^2$.

Proof. Since M is σ-compact (connected manifolds are σ-compact by (10.A.18)) there is an increasing sequence of compact sets, K_1, K_2, ... that covers M (10.A.16). By (5.C.28), there is a simple closed curve γ_1 in M that encloses K_1 in its interior, $I(\gamma_1)$. The same exercise may be applied again to obtain a simple closed curve γ_2 such that $K_2 \cup \gamma_1 \subset I(\gamma_2)$, and inductively, for each integer n, there is a simple closed curve γ_n in M with the property that $K_n \cup \gamma_{n-1} \subset I(\gamma_n)$. Since the region between two nested simple closed curves is a 2-annulus (5.E.5), a homeomorphism between M and $\mathscr{E}^2$ can be constructed as follows: for each integer n, let $\mathscr{B}_n^2$ denote the disk of radius n lying in $\mathscr{E}^2$ with center at the origin. Let $H_1 : \overline{I(\gamma_1)} \to \mathscr{B}_1^2$ be a homeomorphism (5.D.10). Again one proceeds inductively. Suppose that $H_n : \overline{I(\gamma_n)} \to \mathscr{B}_n^2$ has been defined. Then there is a homeomorphism $\hat{H}_{n+1} : \overline{I(\gamma_{n+1}) \setminus I(\gamma_n)}$ $\to \mathscr{B}_{n+1}^2 \setminus \mathscr{B}_n^2$ such that $H_n \hat{H}_{n+1|Bd\ B_n^2}$ is a homeomorphism from $\mathscr{B}_n^2$ onto itself. The map $H_n \hat{H}_{n+1|Bd\ B_n^2}$ can be extended to a homeomorphism $g : \overline{\mathscr{B}_{n+1}^2 \setminus \mathscr{B}_n^2} \to \overline{\mathscr{B}_{n+1}^2 \setminus \mathscr{B}_n^2}$ (why?). Define $H_{n+1} : \overline{I(\gamma_{n+1})} \to \mathscr{B}_{n+1}^2$ by setting

$$H_{n+1}(x) = \begin{cases} H_n(x) & \text{if} \quad x \in I(\gamma_n) \\ g\hat{H}_{n+1}(x) & \text{if} \quad x \in \overline{I(\gamma_{n+1}) \setminus I(\gamma_n)} \end{cases}$$

Then $h : M \to \mathscr{E}^2$ defined by $h(x) = H_n(x)$ whenever $x \in \overline{I(\gamma_n)}$ is the desired homeomorphism.

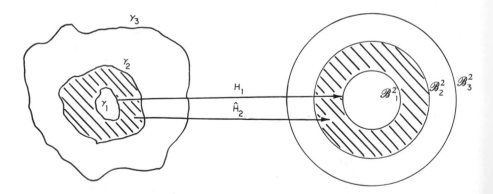

(16.A.2) *Exercise.* Show that h is a homeomorphism.

(16.A.3) *Corollary.* Suppose that U is a nonempty open simply connected subset of $\mathscr{E}^2$. Then U is homeomorphic with $\mathscr{E}^2$.

Proof. Suppose that γ is a simple closed curve in U. It must be shown that γ bounds a disk in U. Let $f : \mathscr{S}^1 \to \gamma$ be a homeomorphism. Since U is simply connected, by (12.C.11) f may be extended to a continuous map $\hat{f} : \mathscr{B}^2 \to U$. We show that $I(\gamma) \subset U$. Let p be an arbitrary point in $I(\gamma)$. Then there is a retraction r that maps $\mathscr{E}^2 \setminus \{p\}$ onto γ (why?). Suppose that $p \notin U$. Define $h : \mathscr{B}^2 \to \mathscr{S}^1$ by $h(x) = f^{-1}r\hat{f}(x)$. Then h retracts $\mathscr{B}^2$ onto its boundary, which contradicts (12.C.15). Thus, we conclude that every point in $I(\gamma)$ lies in U, and hence, each simple closed curve in U bounds a disk in U.

The next goal is to show that all 2-manifolds are triangulable. Since compactness will not be assumed here, the more general definition of triangulation (14.A.16) is applicable. A few additional lemmas are convenient.

(16.A.4) Lemma. Suppose that M is a 2-manifold. Then there is a sequence of disks $D_1, D_2, \ldots$ whose interiors cover M such that

(i) if $i \neq j$, then Bd $D_i \cap$ Bd D_j consists of at most a finite number of points, and

(ii) if $i \neq j \neq k \neq i$, then Bd $D_i \cap$ Bd $D_j \cap$ Bd $D_k = \varnothing$.

Proof. Since M is a separable metric space, M can be covered by a countable collection of open disks $E_1, E_2, \ldots$. For each i, choose a nested sequence of closed disks in E_i whose union is E_i. Sequentially order the collection of all these disks and denote then by $D_1^*, D_2^*, \cdots$.

The desired collection is defined inductively. Let $D_1 = D_1^*$. Suppose that

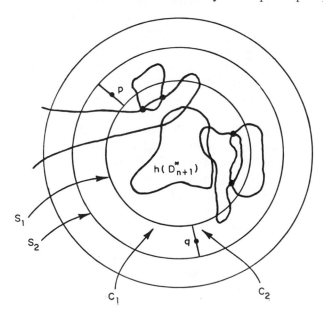

$D_1, \ldots, D_n$ have been chosen so that

    (i)   if $i \neq j$, then Bd $D_i \cap$ Bd D_j consists of a finite number of points,
    (ii) if $i \neq j \neq k \neq i$, then Bd $D_i \cap$ Bd $D_j \cap$ Bd $D_k = \varnothing$, and
    (iii) $D^* \subset D_i$ for each i.

Let F_n be $\bigcup_{i=1}^{n}$ Bd D_i. Of course, D_{n+1}^* is a subset of E_k for at least one
integer k. Let $h : E_k \to \text{int } \mathscr{B}^2$ be a homeomorphism and set $K = h(D_{n+1}^*) \cup$
$\{h(\text{Bd } D_i \cap \text{Bd } D_j) \cap E_k) \mid i \neq j$, and $i,j \leq n\}$. Since K is compact, we have
that $d(K, \text{Bd } \mathscr{B}^2) > 0$ and hence there are circles S_1 and S_2 in int $\mathscr{B}^2$ such
that $K \subset I(S_1) \subset \overline{I(S_1)} \subset I(S_2)$. Choose points p and q in $\overline{(I(S_2) \setminus I(S_1)}} \setminus$
$h(F_n \cap E_k)$ that do not lie on the same diameter, and draw radial lines
through p and q. These radial lines cut $(\overline{I(S_2) \setminus I(S_1)})$ into two cells C_1 and
C_2, which intersect along the radial lines.

    Claim.   There exist arcs α_1 and α_2 from p to q that lie in C_1 and C_2,
respectively, and which intersect $h(F_n \cap E_k)$ in at most a finite number of
points. Note that if such arcs can be found, then the union of these arcs forms
a simple closed curve and $D_{n+1} = h^{-1}(\overline{I(\alpha_1 \cup \alpha_2)})$ will satisfy the inductive
requirement.

    Proof of Claim.   We shall say that an arc in M is *admissible* if and only
if its intersection with F_n is finite. A point in M will be called a *crossing point*
if and only if it is on the boundary of more than one D_i. Observe that if
$p_1, p_2, \ldots, p_q$ is a collection of points in M and if for each $i = 1, 2, \ldots,$
$q - 1$, there is an admissible arc with end points p_i and p_{i+1}, then there is an
admissible arc with end points p_1 and p_q. Furthermore, it is easy to see that
every noncrossing point has a neighborhood in which any two points lie on
an admissible arc. (Use (5.D.9), for example.)
    In C_1, construct two small admissible arcs β_1 and β_2 with end points
$\{p, z_1\}$ and $\{q, z_2\}$, respectively, such that $(\beta_1 \setminus \{p\}) \cup (\beta_2 \setminus \{q\}) \subset \text{int } C_1$.
Since each point $x \in \text{int } C_1$ is a noncrossing point, there is an arc connected
neighborhood N_x of x lying in int C_1 such that any two points of N_x may be
joined by an admissible arc. By (2.F.2), there is a simple chain of the N_x's
between z_1 and z_2. It should now be obvious how to obtain the admissible
arc α_1 from p to q. Of course, an identical procedure may be used to join p
and q in C_2 with an admissible arc α_2. As indicated previously, this not only
completes the proof of the claim but also of the lemma.

(16.A.5) Definition. A *punctured disk* is any space homeomorphic to
$\mathscr{B}^2$ minus the interior of the union of a finite number of closed disjoint cells
lying in int $\mathscr{B}^2$. (A real disk is also assumed to be a punctured disk.)

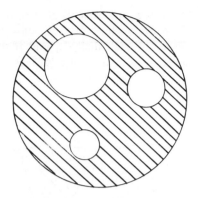

(16.A.6) *Exercise.* Suppose that P is a punctured disk. Show that P may be written as the union of a finite number of 2-cells with disjoint interiors.

(16.A.7) *Lemma.* Suppose that M is a 2-manifold and $\mathcal{D} = \{D_1, D_2, \ldots\}$ is a countable collection of punctured disks such that

(i) the intersection of the interiors of any two of these disks is disjoint, and
(ii) $\mathcal{D}$ is a locally finite cover of M (i.e., each point of M is contained in at most a finite number of punctured disks in $\mathcal{D}$).

Then M may be triangulated.

Proof. By (16.A.6), each punctured disk may be broken up into a finite number of disks with disjoint interiors. Let $C_1, C_2, \ldots$ be the family of all disks obtained in this fashion.

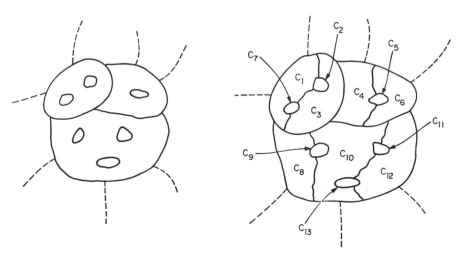

A point in M will be called a *vertex* in case it lies on the boundary of three or more of the C_i. More vertices are introduced as follows. A vertex v_i is selected in the interior of each disk C_i, and if necessary, vertices are added to Bd C_i to ensure that each Bd C_i has at least three vertices.

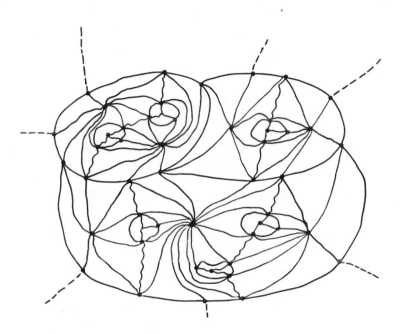

It is immediate from the construction and from (ii) of the hypothesis that C_i intersects at most a finite number of other disks, and consequently each disk contains only a finite number of vertices. In each C_i, connect v_i with each vertex on the boundary of C_i by arcs that are disjoint except at v_i. A curvilinear 2-simplex will be the region bounded by a 1-sphere composed of these arcs with one vertex at v_i and which does not contain any arcs in the interior of its closure. Thus, M has been triangulated.

(16.A.8)   Theorem.   Any 2-manifold M can be triangulated.

Proof.   It suffices to find a cover of M by punctured disks that satisfies the hypotheses of (16.A.7); this is accomplished via (16.A.4). Suppose that $D_1, D_2, \ldots$ is a sequence of disks that satisfies properties (i) and (ii) of (16.A.4). Let $\hat{D}_1 = D_1$. Note that Bd $\hat{D}_1$ "decomposes" D_2 into a finite number of punctured disks whose interiors are disjoint. Let $\hat{D}_2^1, \ldots, \hat{D}_2^{m_2}$ denote those punctured disks not in $\hat{D}_1$. We illustrate two of the possibilities in the next figure.

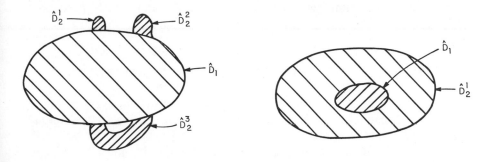

Similarly, D_3 is broken up by Bd $\hat{D}_1 \cup$ Bd $\hat{D}_2$ into a finite number of punctured disks.

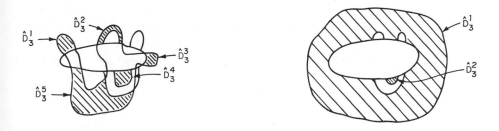

Inductively, at the n-th stage, D_n is decomposed by $\bigcup_{i=1}^{n-1}$ Bd D_i into a finite collection of punctured disks. These are labeled $\hat{D}_n^1, \hat{D}_n^2, \ldots, \hat{D}_n^{m_n}$. Then the family of punctured disks $\mathscr{D} = \{\hat{D}_n^j \mid n \in \mathbf{Z}^+, i \le j \le m_n\}$ covers M, and furthermore, any two elements of $\mathscr{D}$ have disjoint interiors. To see that $\mathscr{D}$ is locally finite, note that if $x \in M$, then $x \in \text{int } D_n$ for some n, and int D_n can intersect only those $\hat{D}_k^j$'s for which $1 \le k \le n$.

(16.A.9) Definition. A separable metric space M is a 2-*manifold with boundary* if and only if each point $x \in M$ has a neighborhood homeomorphic to $\mathscr{B}^2$. The *interior* of M, denoted by Int M, is $\{x \in M \mid x$ has neighborhood homeomorphic to $\mathscr{E}^2\}$, and the boundary of M, denoted by Bd M, is $M \setminus$ Int M.

(16.A.10) Exercise. Show that any compact 2-manifold with boundary can be triangulated.

It would be of some value for the reader to determine which parts of the foregoing construction generalize to 3-manifolds, and which parts are hopeless.

B.   THE CLASSIFICATION OF COMPACT 2-MANIFOLDS

We shall soon be able to provide an elegant description of every compact connected 2-manifold, as well as an algorithm that permits "easy" computation of their fundamental groups.

Suppose that M is a compact connected 2-manifold, and let T be any (finite) triangulation of M. Let S be a collection of disjoint 2-simplices lying in $\mathscr{E}^2$ that are in 1–1 correspondence with the 2-simplices of M. For each 2-simplex $\sigma^2 \in S$, let $h_{\sigma^2} : \sigma^2 \to M$ be an embedding that maps σ^2 onto the 2-simplex of M to which it corresponds in such a manner that vertices are sent to vertices.

We define a family ξ of equivalence relations. Assign $\hat{S} = \bigcup \{\sigma^2 \mid \sigma^2 \in S\}$ the relative topology. List the 1-simplices of S, $\sigma_1^1, \sigma_2^1, \ldots, \sigma_{3n}^1$. If σ_i^1 and σ_j^1 are faces of σ_p^2 and σ_q^2, then declare $x \in \sigma_i^1$ equivalent to $y \in \sigma_j^1$ if and only if $h_{\sigma_p^2}(x) = h_{\sigma_q^2}(y)$. This induces an equivalence relation that we denote by R_{ij}. Note that R_{ij} may be trivial. Let $\xi = \{R_{ij} \mid 1 \le i, j \le 3n$ and R_{ij} not trivial$\}$. Let R be the equivalence relation generated by $\bigcup \{R_{ij} \mid R_{ij} \in \xi\}$. It is easy to see that $\hat{S}/R$ is homeomorphic to M.

We select a subset ξ' of ξ with the property that if R' is the equivalence relation generated by $\bigcup \{R_{ij} \mid R_{ij} \in \xi'\}$, then $\hat{S}/R'$ is a polygon. We define ξ' inductively.

Suppose that $\sigma_1^2 \in S$. If $\sigma_{i_1}^1$ is a face of σ_1^2, then there is a unique 2-simplex σ_2^2, one of whose faces $\sigma_{j_1}^1$ is identified with $\sigma_{i_1}^1$ (why?).

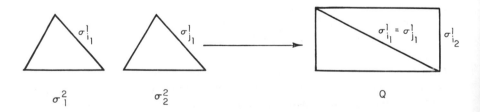

If the total number of 2-simplices in S is greater than 2, then the quadrilateral Q resulting from the identification of $\sigma_{i_1}^1$ with $\sigma_{j_1}^1$ (by $R_{i_1 j_1}$) will have at least one face $\sigma_{i_1}^1$ available for appropriate identification with a face $\sigma_{j_1}^1$ of a third simplex σ_3^2 in S, (otherwise M would not be connected!). Let $R_{i_2 j_2}$ be the relation responsible for this identification.

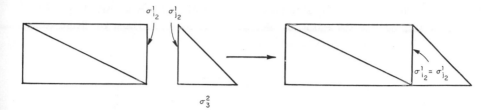

This procedure is continued ($n - 1$ times in all) until the simplices of S are exhausted. Let $\xi' = \{R_{i_k j_k} \mid k = 1, 2, \ldots, n - 1\}$ denote the set of relations used. If $\hat{\xi}'$ is the equivalence relation generated by $\bigcup \{R_{i_k j_k} \mid R_{i_k j_k} \in \xi'\}$, then $\hat{S}/\hat{\xi}'$ is homeomorphic to a regular n-sided polygon lying in the plane. Note that n must be even (why?).

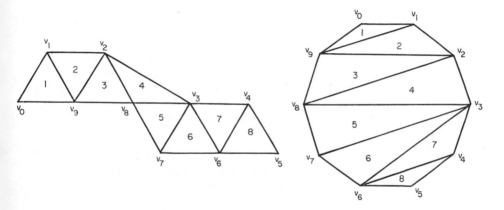

To obtain M once again, it is only necessary to identify the sides of the polygon by the equivalence relation generated by $\bigcup \{R_{ij} \mid R_{ij} \in \xi \setminus \xi'\}$.

Thus, we have obtained the rather spectacular result that every compact connected 2-manifold may be obtained by identifying appropriate pairs of sides of a regular polygon. For higher dimensional triangulable compact connected manifolds, a similar result holds, with slight adjustments in the proof. What is truly amazing is that there is an analogous theorem due to Brown for compact connected n-manifolds, which does not make use of a triangulation (17.E.1).

Suppose now that P is a regular $(2n)$-polygon obtained as above from a 2-manifold. Orient P, and consider two sides of P that are equivalent. Orient one of these sides arbitrarily and orient the other one so that the orientations agree when the sides are identified. If the orientation of the side agrees with the orientation induced by P denote it by a letter, a. If the orientation of the side is opposite to the induced one, denote it by a^{-1}. Thus, the following representation would be read $ab^{-1}cba^{-1}c$.

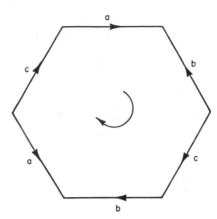

By use of this convention, we can interpret a collection of letters, e.g., $aba^{-1}cbd^{-1}cd^{-1}$, as specifying identification of sides of a polygon. We shall write $M = aba^{-1}c \cdots$ to mean that the polygon with sides identified via $aba^{-1}c \cdots$ is homeomorphic with M. Observe that each letter must appear exactly twice, and any compact connected 2-manifold will have at least one representation as a sequence of letters.

The key to the classification of compact 2-manifolds is that each sequence of letters can be put into a "normal form" and that different patterns of letters in normal form yield distinct 2-manifolds.

(16.B.1)   *Remark.*   The sequence of letters aa^{-1} will represent a 2-sphere,

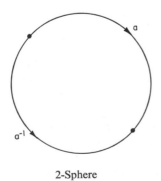

2-Sphere

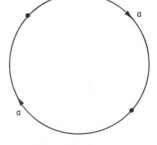

Projective plane

and the sequence *aa* will represent the projective plane. All other compact connected 2-manifolds will have at least two distinct letters in their representation.

(16.B.2)   Notation. In the following, $a, b, c, \ldots$ will denote sides of polygons, and Greek letters $\alpha, \beta, \gamma, \ldots$ will denote blocks of Roman letters, e.g., $\alpha = ab^{-1}c$. If α is such a block, then α^{-1} will denote the block obtained from α by reversing the letters and changing the exponents, e.g., if $\alpha = abb^{-1}cac^{-1}$, then $\alpha^{-1} = ca^{-1}c^{-1}bb^{-1}a^{-1}$.

That compact connected 2-manifolds may be represented by normal forms, is a consequence of the following sequence of Lemmas, whose proofs are sketched.

The normal forms are obtained through a series of cutting and pasting operations. For instance, the following figure

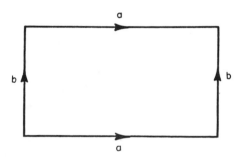

may be "cut in two"

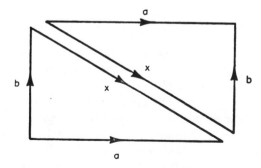

and then pasted together again along *b* to yield the next figure.

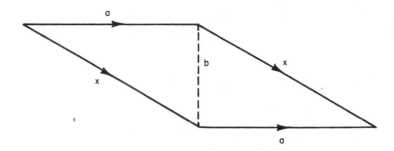

(16.B.3)   Lemma. If $M = aa^{-1}\beta$ (where $\beta \neq \varnothing$), then $M = \beta$.

Proof. A map ψ is constructed between the two polygons that is simplicial on some subdivision of $aa^{-1}\beta$ and on β. The description of such a map is illustrated in the next figure. Clearly, ψ induces a homeomorphism of M onto itself.

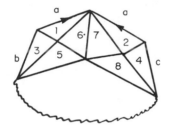 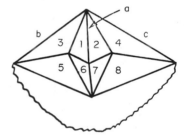

(16.B.4)   Lemma. Suppose that a compact connected 2-manifold M is represented by a polygon with more than two sides. Furthermore, suppose that P is not of the form $\alpha aa^{-1}\beta$. Then there is a polygon P' also representing M for which all of the vertices (on the boundary of P') are identified.

Proof. Suppose that the vertices of P fall into distinct equivalence classes $V_0, V_1, \ldots, V_q$. We obtain a polygon $\hat{P}$ representing M for which there is one more vertex in the class V_0 and one less in one of the remaining classes. Let $v_0 \in V_0$ be a vertex adjacent to a vertex $v_i \in V_i$, where $i \neq 0$. Let $a = v_i v$ be the 1-simplex on the other side of v_i, and let $\hat{a}$ be the side of P that is equivalent with a. Then $v_0 v_i \neq a^{-1}$, since by hypothesis it is assumed that all such configurations have been removed. Furthermore, $v_0 v_i \neq a$, for otherwise v_0 would be equivalent to v_i, which contradicts the way v_i was selected.

Depending on the orientation of a and $\hat{a}$, one of the following two situations must occur.

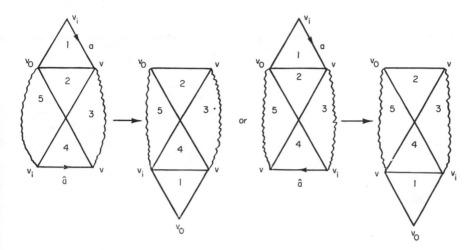

Note that in either case, there is now an additional vertex v_0 in V_0, and the two v_i's have become one.

Of course, under the maps indicated above, a configuration $\alpha c c^{-1} \beta$ may have been introduced; however, by (16.B.3), such configurations can be removed without increasing the number of elements in any of the equivalence classes. Repetition of the foregoing procedure will eventually place all of the vertices in the equivalence class containing v_0.

From this point on, it will be assumed that if P is a polygon representing a 2-manifold M, then all the vertices of P belong to the same equivalence class.

(16.B.5) Lemma. If $M = \alpha c \beta c$, then $M = a a \alpha \beta^{-1}$.

Proof.

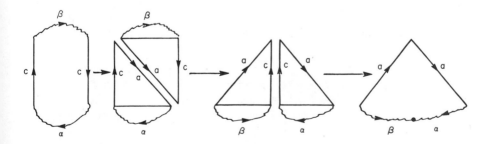

(16.B.6) Lemma. If $M = \alpha a \beta b \gamma a^{-1} \delta b^{-1}$, then $M = y x y^{-1} x^{-1} \delta \gamma \beta \alpha$.

Proof.

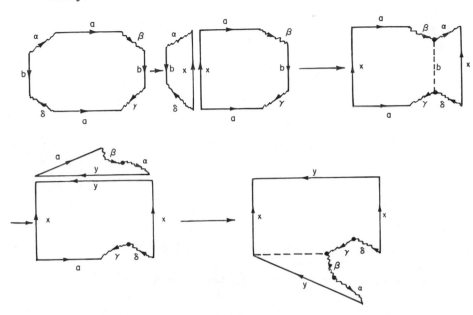

(16.B.7)   Exercise.   Show that if α and β have no letter in common and all vertices of P are identified, then such configurations as $a\alpha a^{-1}\beta$ cannot occur: the "single" vertex would not have a 2-cell neighborhood.

(16.B.8)   Lemma.   If $M = cc\alpha bab^{-1}a^{-1}\beta$, then $M = eeffgg\alpha^{-1}\beta^{-1}$.

Proof.

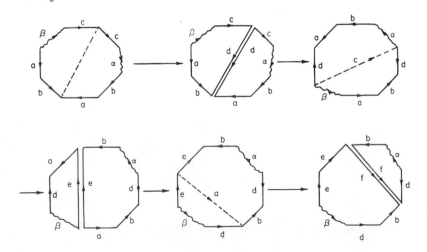

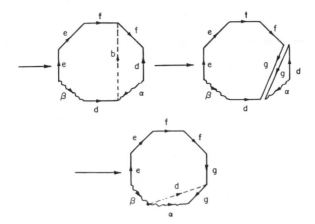

Integrating the preceding five lemmas with the exercise, we have the principal result.

(16.B.9)   Theorem. Every representation of a compact connected 2-manifold is either of the form

  (i) aa^{-1} or aa,
  (ii) $aabbccdd \ldots$, or
  (iii) $aba^{-1}b^{-1}cdc^{-1}d^{-1} \ldots$.

Thus, if a compact connected 2-manifold is not a 2-sphere or a projective plane, it may be described either as a collection of projective planes glued together or as a many holed torus. The forms just described are called normal forms. It remains to show that distinct normal forms are associated with distinct (nonhomeomorphic) manifolds. This is accomplished by demonstrating that manifolds with distinct normal forms have fundamental groups that are not isomorphic.

(16.B.10)   Exercise. Use the polyhedral Seifert-van Kampen theorem to establish that the fundamental group of a 2-manifold, represented by a $2n$ polygon with identifications via the word α, is $\{x_1, x_2, x_3, \ldots, x_n; \alpha\}$, where the x_i's are the n symbols in the word α (unless $\alpha = x_1 x_1^{-1}$).

If M and N are 2-manifolds with the same normal form, then it is easy to construct a homeomorphism between them. If M and N have distinct normal forms, we shall show that they are not homeomorphic, thus completing the classification of compact connected 2-manifolds. What we actually prove is that $\{x_1, \ldots, x_n; \alpha\}$ and $\{y_1, \ldots, y_n; \beta\}$ are not isomorphic whenever α and β are different in normal form.

We need the following theorem from group theory.

(16.B.11)   Theorem. If G and H are isomorphic groups, then $G/[G; G]$ and $H/[H; H]$ are isomorphic. ($[S; S]$ denotes the commutator subgroup of S.)

It follows that if $G/[G; G]$ and $H/[H;H]$ are not isomorphic, then G and H fail to be isomorphic.

(16.B.12)   Theorem. Compact connected 2-manifolds having distinct normal forms are not homeomorphic.

Proof. If $\alpha = x_1x_1x_2x_2x_3x_3 \cdots x_nx_n$, then

$$A = \{x_1x_2 \cdots x_n; \alpha\}$$

$$= \{x_1x_2x_3 \cdots x_n, x; x_1x_1x_2x_2x_3x_3 \cdots x_nx_n, x = x_1x_2x_3 \cdots x_n\}$$

by the Tiezte equivalence II (Ap.C.13). When we consider $A/[A;A]$, the first relation α is equivalent to x^2, and the second relation can be used to eliminate the generator x_n (Tiezte equivalence I (Ap.C.3)). Hence, abelianization yields $\mathbf{Z} \oplus \mathbf{Z} \oplus \cdots \oplus \mathbf{Z} \oplus \mathbf{Z}_2$, where there are $n - 1$ factors of $\mathbf{Z}$. Note that if $n = 1$, we obtain $\mathbf{Z}_2$.

If $\alpha = x_1x_2x_1^{-1}x_2^{-1}x_3x_4x_3^{-1}x_4^{-1} \cdots x_{n-1}x_nx_{n-1}^{-1}x_n^{-1}$, then $\{x_1, x_2, \ldots, x_n; \alpha\}$ when abelianized is just $\mathbf{Z} \oplus \mathbf{Z} \oplus \cdots \oplus \mathbf{Z}$ (since the relation becomes trivial), where there are n factors of $\mathbf{Z}$.

Recall that $\pi_1(\mathscr{S}^2) = \{e\}$, and thus $\mathscr{S}^2$ is not homeomorphic to any other compact 2-manifold. Hence, if $\alpha \neq \beta$ then $\{x_1, x_2, \ldots, x_m; \alpha\}$ and $\{x_1, x_2, \ldots, x_n; \beta\}$ cannot be isomorphic.

(16.B.13)   Remark.
One should note that it is now possible to determine whether or not any two compact 2-manifolds M and N are homeomorphic. Since M and N are compact, they have only a finite number of components, and each such component is a compact connected 2-manifold. The manifolds M and N will be homeomorphic if and only if there is a 1 to 1 correspondence between the components of M and the components of N in such a way that corresponding components are homeomorphic.

(16.B.14)   Remark. Success in classifying compact 2-manifolds has led to attempts to carry the work over to compact 3-manifolds. It is relatively easy to show that every compact connected 3-manifold is a 3-cell with triangles on the boundary identified. Unfortunately, no one has devised a neat scheme for deciding what a "normal form" would be. A special case that is of some interest yields the lens space (Brody [1960]; Reidemeister [1935]).

A somewhat different but related approach to 3-manifolds is to remove a maximal open 3-cell C from a 3-manifold M. This leaves the identified boundary of C, which is called the *spine* of the manifold. Unfortunately once again, since a manifold has many possible spines, the problem becomes one of finding a workable "normal form" (Zeeman [1963]).

C. A CHARACTERIZATION OF $\mathscr{E}^2$

In this section we employ the triangulability of 2-manifolds and our knowledge of compact 2-manifolds to obtain a characterization of $\mathscr{E}^2$. We show that any simply connected noncompact 2-manifold is homeomorphic to $\mathscr{E}^2$. The following two results will be needed.

(16.C.1) Exercise. Suppose that M is a compact 2-manifold with boundary. Show that each boundary component of M is a simple closed curve, and furthermore, that each boundary component can be collared, i.e., if C is a boundary component of M, then there is a simple closed curve $D \subset$ Int M and an embedding $h : \mathscr{S}^1 \times I \to M$ such that h maps $\mathscr{S}^1 \times \{0\}$ homeomorphically onto C, h maps $\mathscr{S}^1 \times \{1\}$ homeomorphically onto D, and $h(\mathscr{S}^1 \times I)$ is a neighborhood of C in M.

(16.C.2) Exercise. Suppose that M is a compact 2-manifold with boundary and that T is a triangulation of M. Let $T^{(1)}$ and $T^{(2)}$ denote the first and second barycentric subdivisions of T, and let $\mathscr{B} = \{b \in M \mid b$ is a barycenter of a 0, 1, or 2-dimension simplex in $T\}$. Let $\mathscr{S}$ be the family of closed stars of the points $b \in \mathscr{B}$ with respect to $T^{(2)}$. Show that if $\mathscr{S}'$ is any subfamily of $\mathscr{S}$, then $\bigcup \{S \mid S \in \mathscr{S}'\}$ is a 2-manifold with boundary.

(16.C.3) Theorem. If M is a simply connected noncompact 2-manifold, then M is homeomorphic to $\mathscr{E}^2$.

The proof is broken up into three lemmas.

(16.C.4) Lemma. Suppose that M is a triangulated simply connected noncompact 2-manifold with boundary and that $N \subset M$ is a compact connected 2-manifold with boundary. Then N can be embedded in $\mathscr{E}^2$.

Proof. Let $\mathscr{C} = \{C_1, C_2, \ldots, C_n\}$ be the boundary components of N. By (16.C.1), the C_i's are simple closed curves. To each C_i, we attach a disk as follows. Let $A = \{1, 2, \ldots, n\}$ be given the discrete topology, and for each $i \in A$, let $B_i = \mathscr{B}^2$, $S_i = \text{Bd } B_i = \mathscr{S}^1$, and $f_i : S_i \to C_i$ be a homeomorphism. Define $f : \mathscr{S}^1 \times A \to N$ by $f(x,i) = f_i(x)$. Then the adjunction space $\tilde{N} = (\mathscr{B}^2 \times A) \cup_f N$ may be viewed as the manifold N with disks attached to each boundary component. Note that $\tilde{N}$ is a compact connected 2-manifold without boundary. We show that $\tilde{N}$ is simply connected, and it will then follow from (16.B.9), (16.B.10), and (16.B.12) that $\tilde{N}$ is homeomorphic to $\mathscr{S}^2$. Since N is properly contained in $\tilde{N}$, it then follows that N can be embedded in $\mathscr{E}^2$.

Obtain a triangulation of $\tilde{N}$ by first triangulating N (16.A.9) and then taking the cone from the center of each disk that was attached to a boundary component of N. We use (12.C.11) to show that $\tilde{N}$ is simply connected. Let $f : \mathscr{S}^1 \to \tilde{N}$ be a continuous map. We wish to extend f to $\mathscr{B}^2$. Consider $\mathscr{S}^1$ as a triangulated 1-complex and apply (14.C.12) to obtain a map $\hat{f} : \mathscr{S}^1 \to \tilde{N}$ that is simplicial on a subdivision of $\tilde{N}$ and is homotopic to f. We first extend $\hat{f}$ to $\mathscr{B}^2$ and then show how this extension may be used to extend f to $\mathscr{B}^2$.

For each boundary component C_i of N, let K_i be a collar for C_i in N (16.C.1) and let D_i denote the disk attached to C_i. Since $\hat{f}$ is simplicial and each boundary component is collared, it is not difficult to see that $\hat{f}$ is homotopic to a continuous function $\hat{f}$ that maps $\mathscr{S}^1$ into Int N. Thus, we may assume that $\hat{f}$ also has this property. However, Int N is a subset of the simply connected space M, and hence $\hat{f}$ may be extended to a continuous function $\hat{F} : \mathscr{B}^2 \to M$.

Let G be the component of $\mathscr{B}^2 \setminus \bigcup_{i=1}^{n} \hat{F}^{-1}(C_i)$ that contains $\mathscr{S}^1$. Let $\{X_j \mid j \in L\}$ be the set of components of the frontier of G (in $\mathscr{B}^2$). Note that for each j, there is an $i_j \in A$ such that $X_j \subset \hat{F}^{-1}(C_{i_j})$. Furthermore, for each $j \in L$, there is an open set U_j and by (5.C.28) a simple closed curve $\mathscr{S}_j$ such that $X_j \subset I(\hat{S}_j) \subset I(\hat{S}_j) \cup \hat{S}_j \subset U_j \subset \hat{F}^{-1}(K_{i_j})$. We may assume that the $\hat{S}_j$'s are mutually disjoint.

For each $j \in L$, we have that $\hat{F}_{|\hat{S}_j} : \hat{S}_j \to D_{i_j} \cup C_{i_j} \cup K_{i_j}$. Since $D_{i_j} \cup C_{i_j} \cup K_{i_j}$ is a disk, $\hat{F}_{|\hat{S}_j}$ may be extended to a continuous function $F_j : (\hat{S}_j \cup I(\hat{S}_j)) \to D_{i_j} \cup C_j \cup K_{i_j}$. Now define $F : \mathscr{B}^2 \to \tilde{N}$ by

$$F(x) = \begin{cases} \hat{F}(x) & \text{if} \quad x \in \mathscr{B}^2 \setminus \bigcup_{j \in L} I(\hat{S}_j) \\ F_j(x) & \text{if} \quad x \in \hat{S}_j \cup I(\hat{S}_j \cup I(\hat{S}_j) \end{cases}$$

Then F extends $\hat{f}$, and it follows from (4.B.21) and (15.A.1) that f may also be extended to $\mathscr{B}^2$ (also see problem C.13); hence, $\tilde{N}$ is simply connected. Therefore, by (16.B.9) and (16.B.10) $\tilde{N}$ is homeomorphic to $\mathscr{S}^2$, and since N is a proper subset of $\tilde{N}$, N can be embedded in $\mathscr{E}^2$.

(16.C.5) Lemma. Suppose that N_1 and N_2 are compact connected 2-manifolds with boundary and that $N_1 \subset N_2$. Suppose further that $f_1 : N_1 \to$

$\mathscr{S}^2$ and $f_2 : N_2 \to \mathscr{S}^2$ are embeddings. Then there is an embedding $\tilde{f}_2 : N_2 \to \mathscr{S}^2$ such that $\tilde{f}_{2|N_1} = f_1$.

Proof. Let $J_1, J_2, \ldots, J_n$ be the boundary components of N_1. Then by (16.C.1), each J_i is a simple closed curve. For each $i = 1, 2, \ldots, n$, let D_i be the disk in $\mathscr{S}^2$ bounded by $f_2(J_i)$ with the property that $D_i \cap f_2(N_1) = f_2(J_i)$ (why does such a disk exist?). Note that $\mathscr{S}^2 = f_2(N_1) \cup (\bigcup D_i)$. Then $f_1 f_{2|f_2(J_i)}^{-1} : f_2(J_i) \to f_1(J_i)$ is a homeomorphism. Extend $f_1 f_{2|f_2(J_i)}^{-1}$ to a homeomorphism $g_i : D_i \to E_i$, where E_i is the disk in $\mathscr{S}^2$ with boundary $f_1(J_i)$ such that $E_i \cap f_1(N_1) = J_i$.

Define $\tilde{f}_2 : N_2 \to \mathscr{S}^2$ by

$$\tilde{f}_2(x) = \begin{cases} f_1(x) & \text{if} \quad x \in N_1 \\ g_i f_2(x) & \text{if} \quad x \in [f_2^{-1}(D_i \cap f_2(N_2))] \end{cases}$$

Then f_2 is the desired embedding.

(16.C.6) Lemma. Suppose that K is a compact connected subset of a 2-manifold M. Then there is a compact connected 2-manifold with boundary N such that $K \subset \text{Int } N \subset M$.

Proof. This follows immediately from (16.C.2) and the fact that K is compact and connected.

Proof of (16.C.3). Let T be a triangulation of M, and list the closed 2-simplices in T, $s_1, s_2, \ldots$ (why are there only a countable number?). Let $N_1 = s_1$, and let $f_1 : N_1 \to \mathscr{E}^2$ be an embedding. If $N_1 \cap s_2 \neq \varnothing$, let $K_2 = N_1 \cup s_2$. If N_1 and s_2 have empty intersection, let α_1 be an arc in M with one end point in N_1 and the other in s_2. Then $K_2 = N_1 \cup \alpha_1 \cup s_2$ is a connected compact subset of M and by (16.C.6), there is a connected compact 2-manifold with boundary N_2 in M that contains K_2 in its interior. If $s_3 \cap N_2 = \varnothing$, connect these two sets with an arc α_2 as before, to form a compact connected subset $K_3 = s_3 \cup \alpha_2 \cup N_2$ of M (if $s_3 \cap N_2 \neq \varnothing$, let $K_3 = s_3 \cup N_2$). Again, (16.C.16) may be applied to obtain a compact connected 2-manifold with boundary N_3 that contains K_3 in its interior. Thus, there is a sequence of connected compact 2-manifolds with boundary $N_1, N_2, \ldots$ such that for each, i, $N_i \subset \text{Int } N_{i+1}$ and $\bigcup_{i=1}^{\infty} N_i = M$. By (16.C.4), each N_i may be embedded in $\mathscr{E}^2$ via an embedding f_i. Lemma (16.C.5) may be inductively employed to "fix up" the f_i's so as to yield an embedding $F : M \to \mathscr{E}^2$. Clearly, $F(M)$ is embedded as an open simply connected subset of $\mathscr{E}^2$, and hence by (16.A.1), $F(M)$ is homeomorphic to $\mathscr{E}^2$, which completes the proof.

PROBLEMS

Section A

1. Triangulate the torus. [Hint: Triangulate the unit square in a suitable manner and identify edges.]
2. Triangulate the Klein bottle.
3. Triangulate the projective plane.
4. Triangulate the Moebius band.
5. Show that if T is a triangulation of a compact 2-manifold, then T has only finitely many simplices.
6. Suppose that T is a triangulation of a compact 2-manifold. Let

$$v = \text{number of vertices in } T$$

$$e = \text{number of 1-simplices in } T$$

$$t = \text{number of 2-simplices in } T$$

Define $\mathscr{E}(M) = v - e + t$. The number $\mathscr{E}(M)$ is called the *Euler characteristic* of M. It can be shown that $\mathscr{E}(M)$ is a topological invariant (it does not depend on the triangulation used). Assuming this to be the case, find the Euler characteristic of the torus, the Klein bottle, and the projective plane.

7.* Suppose that M is a compact connected 2-manifold and that T is a triangulation of M. With the notation of the preceding problem, show that
   (a) $3t = 2e$;
   (b) $e = 3(v - \mathscr{E}(M))$;
   (c) $v \geq \frac{1}{2}(7 + \sqrt{49 - 24\mathscr{E}(M)})$.

Sections B and C

1. Suppose that M and N are compact 2-manifolds. In each space, remove a small open disk and let h be a homeomorphism from the boundary of the disk removed in M onto the boundary of the one removed in N. Then $M \cup_h N$ is called the *connected sum* of M and N. It can be shown that the connected sum does not depend on the disks removed or on the homeomorphism used. Find the connected sum of two 2-spheres.
2. Suppose that M and N are connected compact 2-manifolds. Let S be the connected sum of M and N. Show that S is a compact 2-manifold.

3. Show that the connected sum of a torus and a projective plane is homeo-morphic to the connected sum of three projective planes.
4. Show that the connected sum of two projective planes is homeomorphic to a Klein bottle.
5. Show that the connected sum of a Klein bottle and a projective plane is homeomorphic to the connected sum of a torus and a projective plane.
6. Show that the connected sum of the Moebius strip and a torus is homeo-morphic to the connected sum of a Moebius strip and a Klein bottle.
7. Find the Euler characteristic of the connected sum of a projective plane and n tori.
8. Show that the Euler characteristic of the connected sum of two 2-manifolds is equal to the sum of their Euler characteristics.
9. What is the universal covering space of $\mathscr{P}^2$?
10. Show that the universal covering space of a compact 2-manifold is either $\mathscr{S}^2$ or $\mathscr{E}^2$.
11. Show that if M is a compact 2-manifold and M is neither $\mathscr{S}^2$ nor the projective plane, then the universal covering space of M is $\mathscr{E}^2$.
12. Let S be the boundary of a 2-simplex. Suppose that f is a continuous map from S into a triangulated 2-manifold with boundary and let $\varepsilon > 0$ be given. Show that there is a simplicial map $g : S \to M$ that is homo-topic to f and such that

    (i) g is a local homeomorphism, and
    (ii) the homotopy between f and g does not move points more than a distance of ε.

13. Suppose that X is a topological space and that $f, f' : \mathscr{S}^1 \to X$ are homotopic. Show that if f' has a continuous extension $g' : \mathscr{B}^2 \to X$, then f may also be extended to $\mathscr{B}^2$. (Use no cannons!)

Chapter 17

AN INTRODUCTION TO n-MANIFOLDS

Investigations of the properties of n-manifolds constitute perhaps the main body of research in topology today. Although the results that we obtain are not trivial (indeed, they include some rather profound theorems) we nevertheless barely scratch the surface of manifold theory.

A. SOME ASSORTED PRELIMINARIES

In Chapter 1, an n-manifold M was defined to be a separable metric space with the property that every point in M has a neighborhood homeomorphic to $\mathscr{E}^n$. This definition eliminates such common spaces as $\mathbf{I}$, half-open intervals, disks, etc. However, these spaces do satisfy the following more general concept.

(17.A.1) Definition. An *n-manifold with boundary* is a separable metric space M with the property that each point in M has a neighborhood homeomorphic with $\mathscr{B}^n$. The *interior of M* (Int M) consists of those points possessing neighborhoods homeomorphic to $\mathscr{B}^n \setminus \mathscr{S}^{n-1}$ (or, equivalently, $\mathscr{E}^n$). The *boundary of M* (Bd M) is defined to be $M \setminus$ Int M.

(17.A.2) Remark. With the foregoing terminology, an n-manifold is simply an n-manifold with boundary whose boundary is empty.

(17.A.3)   Example.   Let $C = \{(x,y,z) \in \mathscr{E}^3 \mid x^2 + y^2 = 1, -1 \leq z \leq 1\}$. Then C is a 2-manifold with boundary where

$$\text{Int } C = \{(x,y,z) \in \mathscr{E}^3 \mid x^2 + y^2 = 1, -1 < z < 1\},$$

$$\text{Bd } C = \{(x,y,z) \in \mathscr{E}^3 \mid x^2 + y^2 = 1, |z| = 1\}$$

Observe that in the preceding example, Int C and Bd C (in the manifold context) do not agree with the interior and frontier of C when C is considered as a subspace of $\mathscr{E}^3$.

The following theorem states that for connected n-manifolds, the separability requirement in (17.A.1) is redundant.

(17.A.4)   Theorem.   If X is a connected metric space such that each point has a neighborhood homeomorphic to $\mathscr{B}^n$, then X is separable.

Proof.   Since X is a metric space, it is paracompact (10.A.8), and hence, by (10.A.18), X may be written as the disjoint union of σ-compact spaces. However, X is connected; consequently, X is itself a σ-compact metric space. By (10.A.17), X is Lindelöf and therefore separable by (2.I.13).

As was the case with connected manifolds, connected manifolds with boundary are arc connected.

(17.A.5)   Theorem.   If M is a connected n-manifold with boundary, then M is arc connected.

Proof.   Suppose that $x \in M$, and that $C = \{y \in M \mid \text{there is an arc } f : \mathbf{I} \to M \text{ with } f(0) = x \text{ and } f(1) = y\}$. It suffices to show that C is both open and closed. To see that C is open, suppose that $z \in C$ and that B is a cell neighborhood of z. Then clearly we have that $B \subset C$, and hence C is open. To see that C is closed, suppose that $z \in M \setminus C$. Then if B is a cell neighborhood of z, it must be the case that $B \cap C = \varnothing$ (for otherwise, an arc from x to z could be constructed that passes through a point in $B \cap C$).

Since M is connected, it follows that $C = M$, which completes the proof.

The simplest n-manifolds with boundary are $\mathscr{E}^n$, $\mathscr{S}^n$, and $\mathscr{B}^n$. Homeomorphic copies of these latter two spaces are of special importance.

(17.A.6)   Definition.   A space C is an *n-cell* if and only if C is homeomorphic to $\mathscr{B}^n$, and C is an *open n-cell* if and only if C is homeomorphic to Int $\mathscr{B}^n$. A space S is an *n-sphere* if and only if S is homeomorphic to $\mathscr{S}^n$.

An apparently innocuous, but nevertheless very useful result is given by the next theorem.

(17.A.7) *Theorem.* If $h : \mathscr{B}^n \to \mathscr{E}^n$ is an embedding, then $h(\mathscr{S}^{n-1})$ is the frontier of $h(\mathscr{B}^n)$.

Proof. First, note that $h(\mathscr{S}^{n-1}) = h(\mathscr{B}^n) \setminus h(\text{Int } \mathscr{B}^n)$. Suppose that $x \in h(\mathscr{S}^{n-1})$ and that U is an open set (in $\mathscr{E}^n$) containing x. If $U \cap (\mathscr{E}^n \setminus h(\mathscr{B}^n)) = \varnothing$, then $U \subset h(\mathscr{B}^n)$, and consequently, $h^{-1}(U)$ is open in $\mathscr{B}^n \subset \mathscr{E}^n$. However, by the invariance of domain (15.B.4), this implies that $h^{-1}(U) \subset \text{Int } \mathscr{B}^n$, a contradiction, since $x \in h(S^{n-1})$. Thus, $h(\mathscr{S}^{n-1}) \subset \text{Fr } h(\mathscr{B}^n)$.

Now suppose that $x \in \text{Fr } h(\mathscr{B}^n)$. Again by (15.B.4), $h(\text{Int } \mathscr{B}^n)$ is open in $\mathscr{E}^n$, and hence $x \notin h(\text{Int } \mathscr{B}^n)$. Therefore, $x \in h(\mathscr{S}^{n-1})$, and consequently, $\text{Fr } h(\mathscr{B}^n) \subset h(\mathscr{S}^{n-1})$.

If C is an n-cell, then by definition it is homeomorphic to $\mathscr{B}^n$. However, it is easy to see that an infinite number of homeomorphisms can be constructed between C and $\mathscr{B}^n$. Although C may have a very peculiar topological frontier (or even none at all), we would nevertheless like to be able to speak of the "boundary of C," which, logically, should be the image of $\mathscr{S}^{n-1}$ under any of the homeomorphisms. However, it is conceivable that not all such images coincide. That this fear is unfounded is the content of the next theorem.

(17.A.8) *Theorem.* Suppose that C is an n-cell. Then there is a unique $(n-1)$-sphere X in C such that $h(\mathscr{S}^{n-1}) = X$ for any homeomorphism $h : \mathscr{B}^n \to C$.

Proof. Let $h : \mathscr{B}^n \to C$ be an arbitrary homeomorphism, and let $X = h(\mathscr{S}^{n-1})$. Suppose that $f : \mathscr{B}^n \to C$ is another homeomorphism. We show that $f(\mathscr{S}^{n-1}) = X$. The map $f^{-1}h : \mathscr{B}^n \to \mathscr{B}^n \subset \mathscr{E}^n$ is a homeomorphism (and an embedding into $\mathscr{E}^n$), and hence we have that $f^{-1}h(\mathscr{S}^{n-1}) = \text{Fr } \mathscr{B}^n = \mathscr{S}^{n-1}$ by the previous theorem. Thus, it must be the case that $f(\mathscr{S}^{n-1}) = X$.

(17.A.9) *Notation.* Henceforth, the X in the previous theorem will be denoted by Bd C.

We have already shown that any compact convex subset of $\mathscr{E}^n$ is an n-cell (3.C.2). The following exercise yields another way that cells may arise.

(17.A.10) *Exercise.* Suppose that C is an n-cell, where $C \subset \mathscr{E}^n \subset \mathscr{E}^{n+1}$, and that $p \in \mathscr{E}^{n+1} \setminus \mathscr{E}^n$. Show that the cone over C from p (the union of the line segments connecting p and points of C) is an $(n+1)$-cell.

(17.A.11) *Exercise.* Show that the suspension of an $(n-1)$-sphere is an n-sphere.

The following extension property of n-cells is frequently useful.

(17.A.12)   *Theorem.*   If C is an n-cell and $h : \text{Bd } C \to \text{Bd } C$ is a homeomorphism, then h may be extended to a homeomorphism from C onto C.

Proof.   Let $g : C \to \mathscr{B}^n$ be a homeomorphism. Then $k = ghg^{-1}_{|\mathscr{S}^{n-1}}$ is a homeomorphism from $\mathscr{S}^{n-1}$ onto itself. Extend k to $\mathscr{B}^n$ by defining

$$
k'(x) = \begin{cases} \|x\| k\left(\dfrac{x}{\|x\|}\right) & \text{if} \quad x \neq \mathbf{O} \\[2mm] \mathbf{O} & \text{if} \quad x = \mathbf{O} \end{cases}
$$

Then $h' = g^{-1}k'g$ is the desired homeomorphism that extends h.

(17.A.13)   *Exercise.*   Suppose that $h : \text{Bd } C \to \text{Bd } C'$ is a homeomorphism, where C and C' are n-cells. Show that h can be extended to a homeomorphism from C onto C'.

(17.A.14)   *Exercise.*   Suppose that F is a closed subset of $\mathscr{B}^n$ and that $F \subset \text{Int } \mathscr{B}^n$. Show that there is a homeomorphism $h : \mathscr{E}^n \to \text{Int } \mathscr{B}^n$ such that $h|_F = id$.

We now show that n-cells are homogeneous (almost), i.e., if $x, y \in \text{Int } C$, then there is a homeomorphism of C onto itself that carries x to y. In fact, we give a slightly stronger result.

(17.A.15)   *Theorem.*   Suppose that $x, y \in \text{Int } C$, where C is an n-cell. Then there is a homeomorphism $h : C \to C$ such that $h(x) = y$ and $h|_{\text{Bd } C} = id$.

Proof.   Before proceeding, the reader should note that it is sufficient to prove the theorem for $C = \mathscr{B}^n$ and $y = \mathbf{O}$. Let $P_x : \mathscr{B}^n \setminus \{x\} \to \mathscr{S}^{n-1}$ be the obvious geometric projection map. Now define $h : \mathscr{B}^n \to \mathscr{B}^n$ by setting

$$
h(z) = \begin{cases} \dfrac{\|z - x\|}{\|P_x(z) - x\|} P_x(z) & \text{if} \quad z \neq x \\[2mm] \mathbf{O} & \text{if} \quad z = x \end{cases}
$$

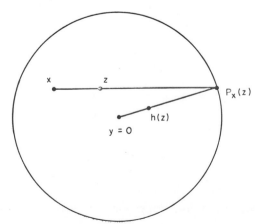

It is easily seen that h is the desired homeomorphism.

Not only are open cells homogeneous, but so are all n-manifolds. This result is contained in the proof of the next theorem.

(17.A.16) *Theorem.* Suppose that M is a connected n-manifold with boundary and that $x,y \in$ Int M. Then there is an n-cell $C \subset M$ such that $x,y \in$ Int C.

Proof. Cover Int M with a collection $\mathscr{C}$ of open n-cells whose closures are n-cells lying in Int M. By (2.F.2), there is a simple chain $C_1, C_2, \ldots, C_n$ in $\mathscr{C}$ such that $x \in C_1$ and $y \in C_n$. Let $x = x_1$, and for $i = 1, 2, \ldots, n - 1$, pick a point $x_{i+1} \in C_i \cap C_{i+1}$. Let $y = x_{n+1}$. For $i = 1, \ldots, n$, there is a homeomorphism $h_i : \bar{C}_i \to \bar{C}_i$ such that $h|_{\mathrm{Bd}\ \bar{C}_i} = id$ and $h(x_i) = x_{i+1}$. Since $h_{i|\mathrm{Bd}\ \bar{C}_i}$ is the identity, h_i may be extended to a homeomorphism $\hat{h}_i : M \to M$. Then $h = \hat{h}_n \hat{h}_{n-1} \cdots \hat{h}_2 \hat{h}_1$ is a homeomorphism from M onto itself such that $h(x) = y$, and $h(\bar{C}_1)$ is an n-cell that contains both x and y.

(17.A.17) *Exercise.* Extend the foregoing result to finite sets of points.

B. n-ANNULI

(17.B.1) *Definition.* A space homeomorphic to $\mathscr{S}^{n-1} \times \mathbf{I}$ is called an *n-annulus*; a space homeomorphic to $\mathscr{S}^{n-1} \times (0,1)$ is an *open n-annulus*; a space homeomorphic to $\mathscr{S}^{n-1} \times (0,1]$ is a *half-open n-annulus*.

(17.B.2) *Notation.* We denote $\{x \in \mathscr{E}^n \mid \|x\| \leq r\}$ by $\mathscr{B}_r^n$.

Clearly, $\overline{\mathscr{B}^n \setminus \mathscr{B}_{1/2}^n}$ is an n-annulus, $\mathscr{B}^n \setminus \mathscr{B}_{1/2}^n$ is a half-open n-annulus, and $\mathrm{Int}(\mathscr{B}^n \setminus \mathscr{B}_{1/2}^n)$ is an open n-annulus.

(17.B.3) *Notation.* Suppose that C and C' are n-cells and that $C' \subset$ Int C. Then $[C',C]$ will denote $\overline{C \setminus C'}$.

It is reasonable to conjecture that $[C',C]$ is an n-annulus. In fact, if C and C' are 2-cells, we have seen (5.E.5) that $[C',C]$ is a 2-annulus. However, for $n > 2$, counterexamples to this conjecture may be constructed. Nevertheless, if certain conditions are placed on the n-cell C', then more positive results are obtainable. Recall that an n-cell lying in $\mathscr{E}^n$ is tame if and only if there is a homeomorphism $h : \mathscr{E}^n \to \mathscr{E}^n$ that carries C onto $\mathscr{B}^n$, and that C is wild otherwise: a similar definition applies to $(n - 1)$-spheres lying in $\mathscr{E}^n$ (15.E.2). The annulus problem may be slightly rephrased to yield one of the most famous conjectures emeriti of topology.

Annulus Conjecture Suppose that C and C' are tame n-cells lying in $\mathscr{E}^n$ such that $C' \subset$ Int C. Then $[C',C]$ is an n-annulus.

In 1969, it was proved that the annulus conjecture is valid if $n \neq 4$ (Kirby, Siebenmann, and Wall [1969]). The status for $n = 4$ is still unknown. The following is a considerably more modest result.

(17.B.4) **Theorem.** Suppose that C and D are tame n-cells in $\mathscr{E}^n$ such that $C \subset \text{Int } D$. Let (C,D) denote $\text{Int } [C,D]$. Then (C,D) is an open n-annulus.

Proof. Clearly, an open n-annulus is homeomorphic with $\mathscr{E}^n \setminus \mathscr{B}^n$, and hence it suffices to show that (C,D) and $\mathscr{E}^n \setminus \mathscr{B}^n$ are homeomorphic. Since D is tame, there is a homeomorphism $h : \mathscr{E}^n \to \mathscr{E}^n$ that maps D onto $\mathscr{B}^n$. Since $h(C)$ is a closed subset of the interior of $\mathscr{B}^n$, by (17.A.14) there is a homeomorphism $f : \mathscr{E}^n \to \text{Int } \mathscr{B}^n$ that is the identity on $h(C)$. Since $h(C) = f^{-1}h(C)$ is a tame n-cell, there is a homeomorphism $g : \mathscr{E}^n \to \mathscr{E}^n$ such that $gf^{-1}h(C) = \mathscr{B}^n$. Then $gf^{-1}h|_{(C,D)}$ is the desired homeomorphism.

Although we are in no position to prove the annulus theorem, we can establish a somewhat weakened version of it.

(17.B.5) **Theorem.** Suppose that for each tame n-cell C in $\mathscr{E}^n$, there is a homeomorphism $h : \mathscr{E}^n \to \mathscr{E}^n$ and a compact set $A \subset \mathscr{E}^n$ such that $h(C) = \mathscr{B}^n$ and $h|_{\mathscr{E}^n \setminus A} = id$. Let C_1 and C_2 be tame n-cells in $\mathscr{E}^n$ such that $C_1 \subset \text{Int } C_2$. Then $[C_1,C_2]$ is an n-annulus.

Proof. Let $g : \mathscr{E}^n \to \mathscr{E}^n$ be a space homeomorphism that carries C_2 onto $\mathscr{B}^n$. For convenience, we shall assume that $\mathbf{O} \in \text{Int } g(C_1)$. Apply (17.A.14) to obtain a homeomorphism $f : \text{Int } \mathscr{B}^n \to \mathscr{E}^n$, which is the identity on $g(C_1)$. Since $fg(C_1) = g(C_1)$ is a tame n-cell, by hypothesis there is a homeomorphism $h : \mathscr{E}^n \to \mathscr{E}^n$ and a compact set A such that

(i) $hfg(C_1) = \mathscr{B}^n$, and
(ii) $h_{\mathscr{E}^n \setminus A} = id$.

Note that we may assume that $\mathscr{S}^{n-1} \subset \text{Int } A$. (One can actually prove that $\mathscr{S}^{n-1}$ must be in the interior of A; see problem B-C.1.)

Now choose ε small enough so that $\mathscr{B}^n_{1+\varepsilon} \subset \text{Int } A$, and let B' be an n-ball with center $\mathbf{O}$ contained in the interior of $fg(C_1) = g(C_1)$. There is a space homeomorphism $k : \mathscr{E}^n \to \mathscr{E}^n$ such that $k(\mathscr{B}^n) = B'$, $k([\mathscr{B}^n,\mathscr{B}^n_{1+\varepsilon}]) = [\mathscr{B}',\mathscr{B}^n_{1+\varepsilon}]$, and k is the identity on $\mathscr{E}^n \setminus \text{Int } \mathscr{B}^n_{1+\varepsilon}$. Let $q = kh$. Note that there is a compact set F, outside of which q is the identity. Finally, set $\phi = f^{-1}qf :$ $\text{Int } \mathscr{B}^n \to \text{Int } \mathscr{B}^n$. Then ϕ is onto, and since $q|_{\mathscr{E}^n \setminus F} = id$, it follows that there is a positive number $r < 1$ such that $\phi(x) = x$ for $x \in \mathscr{B}^n \setminus \mathscr{B}^n_r$. Thus, ϕ may be extended (homeomorphically) to $\hat{\phi} : \mathscr{E}^n \to \mathscr{E}^n$. It is easily checked that $\hat{\phi}g(C_1) = B'$ and $\hat{\phi}g(C_2) = \mathscr{B}^n$, which proves that $[C_1,C_2]$ is an n-annulus.

C. CELLULAR SETS

Cellular subsets of a manifold are defined as follows.

(17.C.1) *Definition.* A subset K of an n-manifold M is *cellular* if and only if there is a sequence of n-cells $C_1, C_2, \ldots$ in M such that $C_i \subset \text{Int } C_{i-1}$ for each i, and $\bigcap_{i=1}^{\infty} C_i = K$.

Cellular sets are compact and connected (4.A.8); however, it is clear from the following examples that they need not be cells.

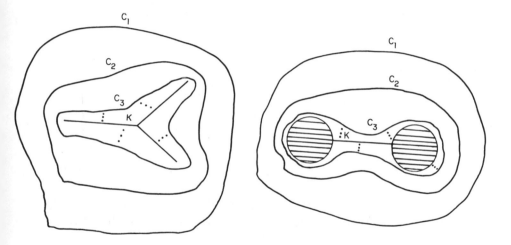

On the other hand, it can be shown that not all cells are cellular. For example, the wild cell constructed previously (15.E.13) is not a cellular set, as we shall prove presently. Tame cells, however, are cellular.

(17.C.2) *Exercise.* Show that if C is a tame n-cell lying in $\mathscr{E}^n$, then C is cellular.

Suppose that K is a cellular subset of $\mathscr{E}^n$. Then there is a decreasing sequence of cells whose intersection in K. Can it be assumed that these cells are tame? The answer is yes; this is an immediate consequence of the following theorem.

(17.C.3) *Theorem.* Suppose that C_1 and C_2 are cells in $\mathscr{E}^n$ such that

$C_1 \subset$ Int C_2. Then there is a tame cell C such that $C_1 \subset$ Int $C \subset C \subset$ Int C_2.

Proof. Let D be a tame n-cell that lies in Int C_1, and let $h : C_2 \to \mathscr{B}^n$ be a homeomorphism. By the invariance of domain, $h(D)$ has nonempty interior. We may assume that $h(D)$ contains $\mathbf{O}$. Choose an $\varepsilon > 0$ such that $\mathscr{B}^n_\varepsilon$ is contained in Int $h(D)$, and select r, $0 < r < 1$, such that $h(C_1) \subset \mathscr{B}^n_r$.

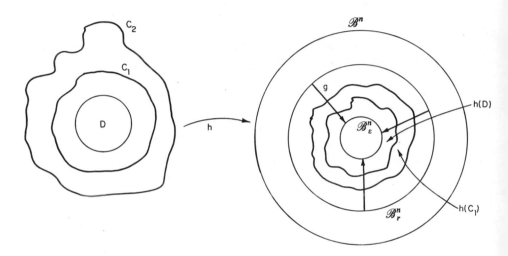

Let $g : \mathscr{B}^n \to \mathscr{B}^n$ be a homeomorphism that leaves Bd $\mathscr{B}^n$ fixed and carries $\mathscr{B}^n_r$ onto $\mathscr{B}^n_\varepsilon$. Define a space homeomorphism $f : \mathscr{E}^n \to \mathscr{E}^n$ by setting

$$f(x) = \begin{cases} h^{-1}gh(x) & \text{if} \quad x \in C_2 \\ x & \text{if} \quad x \in \mathscr{E}^n \setminus C_2 \end{cases}$$

Then $C = f^{-1}(D)$ is tame, and certainly $C \subset$ Int C_2. To see that $C_1 \subset$ Int C, observe that $C_1 \subset h^{-1}(\text{Int } \mathscr{B}^n_r) = h^{-1}g^{-1}(\text{Int } \mathscr{B}^n_\varepsilon) \subset h^{-1}g^{-1}(\text{Int } h(D))$ $= \text{Int } h^{-1}g^{-1}h(D) = \text{Int } f^{-1}(D) = \text{Int } C$.

(17.C.4) Exercise. Show that $[D,C]$ is an n-annulus, where C and D are defined as in the foregoing proof.

(17.C.5) Corollary. Suppose that K is a cellular set in $\mathscr{E}^n$. Then there is a sequence of tame n-cells $C_1, C_2, \ldots$ such that $C_{i+1} \subset$ Int C_i, $K = \bigcap_{i=1}^{\infty} C_i$, and $[C_{i+1}, C_i]$ is an n-annulus.

Next we introduce a concept somewhat more general than cellularity, which can be defined in spaces other than manifolds.

(17.C.6) Definition. A compact, connected subset K of a topological space X is *pointlike* if and only if $X \setminus K$ is homeomorphic to $X \setminus \{p\}$, where p is some point in X.

In practice, one usually restricts the notion of being pointlike to spaces where $X \setminus \{p\}$ is homeomorphic to $X \setminus \{q\}$ for any two points p and q of X.

(17.C.7) Exercise Show that if M is a connected n-manifold and $p,q \in M$, then $M \setminus \{p\}$ is homeomorphic to $M \setminus \{q\}$.

In $\mathscr{E}^n$, subsets are cellular if and only if they are pointlike (problem C.9). That this is not true for n-manifolds follows from the next example that exhibits a pointlike but noncellular subset of a 2-manifold.

(17.C.8) Exercise. (*Christenson and Osborne* [*1968*]). Let M be $\mathscr{E}^3$ minus the integers on the positive x axis and minus 1-spheres of radius $1/4$ centered at the negative integers on the x axis. Then M is a connected 3-manifold. Show that the 1-sphere of radius $1/4$ and center at **O** is pointlike but not cellular.

It is immediate from the next theorem that cellular sets are always pointlike.

(17.C.9) Definition. Suppose that X and Y are topological spaces, $f : X \to Y$, and $y \in Y$. Then $f^{-1}(y)$ is an *inverse set* if and only if $f^{-1}(y)$ contains more than one point.

(17.C.10) Theorem. If C is an n-cell and $K \subset \text{Int } C$ is cellular, then there is a continuous surjection $f : C \to C$ such that

  (i) $f|_{\text{Bd } C} = id$,
  (ii) $f|_{C \setminus K}$ is a homeomorphism, and
  (iii)   the only inverse set under f is K.

Proof. Let $C_1, C_2, \ldots$ be a sequence of n-cells lying in Int C such that $C_{i+1} \subset \text{Int } C_i$ and $K = \bigcap_{i=1}^{\infty} C_i$. We define a sequence of homeomorphisms from C onto C as follows. Let $f_1 : C \to C$ be a homeomorphism such that

  (i) $f_1|_{\text{Bd } C} = id$, and

(ii) $\operatorname{diam} f_1(C_1) < 1/2$.

(Why does such a homeomorphism exist?)

     Suppose that f_m has been defined. Let $f_{m+1} : C \to C$ be a homeomorphism with the property that $f_{m+1}|_{C \setminus C_m} = f_m|_{C \setminus C_m}$ and $\operatorname{diam} f_{m+1}(C_{m+1}) < 1/2^{m+1}$. Then $f(x) = \lim\limits_{m \to \infty} f_m(x)$ is the desired map.

(17.C.11) Corollary. A cellular subset of an n-manifold is pointlike.

     Note that it follows from (17.C.11) the cell C constructed in (15.E.13) is not cellular. If C were cellular, then it would be pointlike and hence the fundamental group of its complement would be trivial.

     It should be noted that in spite of exercise (17.C.8), Christenson and Osborn [1968] established that in a compact n-manifold with boundary, any pointlike subset that lies in an open n-cell is cellular.

     Theorem (17.C.10) may be refined slightly. The easy proof of the next theorem is left to the reader.

(17.C.12) Theorem. If K is a cellular subset of $\mathscr{E}^n$, $p \in K$, and W is an open set in $\mathscr{E}^n$ containing K, then there is a homeomorphism $h : \mathscr{E}^n \setminus K \to \mathscr{E}^n \setminus \{p\}$ such that

    (i) $h(W \setminus K) = W \setminus \{p\}$, and
    (ii) $h(x) = x$ if $x \in E^n \setminus W$.

     We shall use the next sequence of theorems to obtain a generalization of the 2-dimensional Schönflies theorem that was stated and proved in Chapter 5. Recall that if S is an $(n-1)$-sphere in $\mathscr{S}^n$, then a complementary domain is one of the components of $\mathscr{S}^n \setminus S$. Throughout the remainder of the chapter, we shall assume the following theorem. Its proof is beyond our means, and is most elegantly established using homology theory.

(17.C.13) Theorem (Jordan and Brouwer). If S is an $(n-1)$-sphere in $\mathscr{S}^n$, then $\mathscr{S}^n \setminus S$ has exactly two components, and S is the frontier of each.

(17.C.14) Exercise. Suppose that $h : \mathscr{S}^{n-1} \times I \to \mathscr{S}^n$ is an embedding. Show that $\mathscr{S}^n \setminus h(\mathscr{S}^{n-1} \times I)$ has exactly two complementary domains.

     Note that (17.C.13) does not assert that the closure of each complementary domain is an n-cell. If fact, we have seen (15.E.12) that if $n = 3$, then there is a 2-sphere in $\mathscr{S}^3$ with the property that the closures of both complementary domains fail to be 3-cells. However, for $n = 2$ the following result holds.

(17.C.15) **Exercise.** Show that a simple closed curve in $\mathscr{S}^2$ separates $\mathscr{S}^2$ into two complementary domains and that the closure of each is a 2-cell.

We now develop criteria under which $(n-1)$-spheres in $\mathscr{S}^n$ bound n-cells. The results appearing in the next several pages were originally established by Brown [1960b].

(17.C.16) **Theorem.** Suppose that S is an $(n-1)$-sphere in $\mathscr{S}^n$, D is a complementary domain of S, C is an n-cell, and $f : \bar{D} \to C$ is a continuous surjection whose only inverse set is a cellular set in D. Then D is an n-cell.

Proof. Let $K \subset D$ be the unique inverse set under f and let C_0 be an n-cell containing K in its interior.

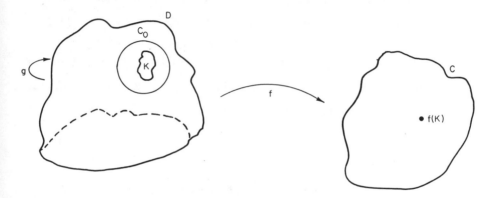

Since K is cellular, by (17.C.10) there is a continuous map $g : \bar{D} \to \bar{D}$ such that $g|_{\bar{D} \setminus C_0} = id$, and K is the unique inverse set of g. Then $fg^{-1} : \bar{D} \to C$ is clearly a bijection. Furthermore, if A is a closed subset of $\bar{D}$, then A is compact and $fg^{-1}(A)$ is a compact subset of C (and thus closed). Consequently, $(fg^{-1})^{-1} : C \to \bar{D}$ is continuous and hence, by (2.G.11), fg^{-1} is a homeomorphism. Therefore, $\bar{D}$ is an n-cell.

In the next theorem, we see that single inverse sets are cellular.

(17.C.17) **Theorem.** Suppose that C is an n-cell and that $f : C \to \mathscr{S}^n$ is a continuous map with a unique inverse set K. If $K \subset \text{Int } C$, then K is cellular.

Before proving this theorem, we first establish two lemmas.

(17.C.18) **Lemma.** Suppose that C is an n-cell and that $f : C \to \mathscr{S}^n$ is a continuous map with a finite number of inverse sets $K_1, \ldots, K_m$, each of

which lies in the interior of C. Then $f(C)$ is the union of $f(\text{Bd } C)$ and one of its complementary domains.

Proof. Since $f|_{\text{Bd } C}$ is an embedding, it follows that $f(\text{Bd } C)$ has two complementary domains D and D' and is the frontier of each (17.C.13). Since $f(C)$ is connected, $f(C)$ lies either in $\bar{D}$ or $\bar{D}'$, say $\bar{D}$. It remains to show that $\bar{D} \subset f(C)$.

We first show that $f(\text{Int } C \setminus \bigcup_{i=1}^{m} K_i)$ is an open subset of $\mathscr{S}^n$. If x is an element of $\text{Int } C \setminus \bigcup_{i=1}^{m} K_i$, then there is an n-cell B such that $x \in B \subset \text{Int } C \setminus \bigcup_{i=1}^{m} K_i$, and $f|_B : B \to \mathscr{S}^n$ is an embedding. Therefore, $f|_{\text{Int } B}$ is an embedding, and by invariance of domain, $f(\text{Int } B)$ is open in $\mathscr{S}^n$ (why can (15.B.4) be applied here?). Consequently, we have that $\text{Fr } f(C) \subset f(\text{Bd } C) \cup (\bigcup_{i=1}^{m} f(K_i))$, and thus $\text{Fr } f(C)$ contains at most m points that do not lie in $f(\text{Bd } C)$.

Now suppose that $x \in \bar{D} \setminus f(C)$. Since $\text{Fr } D = f(\text{Bd } C) \subset f(C)$, it follows that $x \in D \setminus f(C)$. Select a point $z \in f(\text{Int } C \setminus \bigcup_{i=1}^{m} K_i)$, and let $\alpha_1, \ldots, \alpha_m$, α_{m+1} be mutually disjoint (except at the end points) arcs from x to z that lie in D. These arcs must cross $\text{Fr } f(C)$ at $m + 1$ distinct points, none of which are in $\text{Fr } D = f(\text{Bd } C)$. This, however, contradicts the last statement in the preceding paragraph, and completes the proof of the lemma.

(17.C.19) Lemma. Suppose that C is an n-cell and that $f : C \to \mathscr{S}^n$ is a continuous function with only one inverse set K. If $K \subset U \subset \text{Int } C$, where U is open, then $f(U)$ is open in $\mathscr{S}^n$.

Proof. Since K is the unique inverse set and $K \subset U$, it is immediate that $f(C) \setminus f(U) = f(C \setminus U)$. Since $C \setminus U$ is compact, it follows that $f(C \setminus U)$ is closed in $\mathscr{S}^n$. Note that $\mathscr{S}^n \setminus f(U) = f(C \setminus U) \cup \bar{D}$, where D is the complementary domain of $f(\text{Bd } C)$ disjoint from $f(C)$. Thus, the complement of $f(U)$ is closed, and consequently $f(U)$ is open.

Proof of (17.C.17). By (17.C.18), we have that $f(C) = f(\text{Bd } C) \cup D$ for some complementary domain D of $f(\text{Bd } C)$.

Suppose that U is an arbitrary open set in $\text{Int } C$ containing K. We will fit an n-cell inside of U that contains K in its interior. By (17.C.19), $f(U)$ is open. Let V be a "small" neighborhood of $f(K)$ in $f(U)$. Since $\bar{D} \neq \mathscr{S}^n$, there is a homeomorphism $h : \mathscr{S}^n \to \mathscr{S}^n$ that carries $\bar{D}$ into $f(U)$ and is fixed on V. Define $g : C \to U \subset C$ by

$$g(x) = \begin{cases} x & \text{if} \quad x \in K \\ f^{-1}hf(x) & \text{if} \quad x \notin K \end{cases}$$

Then g is an embedding, and hence $g(C)$ is an n-cell contained in U and containing K in its interior. Since U was arbitrary, K must be cellular.

(17.C.20) Exercise. Show that the homeomorphism h exists (by returning to $\mathscr{E}^n$ if necessary) and find a decreasing sequence of cells whose intersection is K.

Each member of a pair of inverse sets is also cellular.

(17.C.21) Theorem. If $f : \mathscr{S}^n \to \mathscr{S}^n$ is a continuous surjection and has precisely two inverse sets K_1 and K_2, then both K_1 and K_2 are cellular in $\mathscr{S}^n$.

Proof. Let S be an $(n-1)$-sphere in $\mathscr{S}^n \setminus (K_1 \cup K_2)$ such that the closures of the complementary domains of S are n-cells. If by some odd happenstance, S separates K_1 and K_2, then (17.C.17) may be applied to conclude the proof. Suppose, then, that there is a complementary domain D of S that contains $K_1 \cup K_2$. Let $k_1 = f(K_1)$, $k_2 = f(K_2)$, and $E = f(\bar{D})$. Let U be an open subset in E that contains k_1 but not k_2. Then, as in the previous theorem, there is a homeomorphism $h : \mathscr{S}^n \to \mathscr{S}^n$ that maps E into U and is fixed on an appropriately small neighborhood V of k_1. Now define $g : \bar{D} \to \mathscr{S}^n$ by

$$g(x) = \begin{cases} x & \text{if} \quad x \in K_1 \\ f^{-1}hf(x) & \text{if} \quad x \in \bar{D} \setminus K_1 \end{cases}$$

Note that K_2 is the only inverse set of g, and hence by (17.C.17) we have that K_2 is cellular. A similar argument shows that K_1 is cellular.

D. THE GENERALIZED SCHÖNFLIES THEOREM

We showed in Chapter 5 that a 1-sphere lying in $\mathscr{S}^2$ (or $\mathscr{E}^2$) always encloses a 2-cell; on the other hand, we constructed (15.E.12) an example of a 2-sphere that fails to bound a 3-cell. What went wrong when the dimension was increased? The basic problem is that for $n \geq 2$, an n-sphere S may be severely "entangled" with itself. Such entanglements are eliminated if a "protective collar" can be built around the sphere. The generalized Schönflies theorem states that if such a protective strip is available, then the sphere S does in fact bound an n-cell (or n-cells). We begin with a definition and a mildly technical lemma.

(17.D.1) Definition. An $(n-1)$-sphere S in an n-manifold with boundary M is *bicollared* if and only if there is an embedding $h : \mathscr{S}^{n-1} \times \mathbf{I} \to M$ such that $h(\mathscr{S}^{n-1} \times \{1/2\}) = S$; S is *collared* if and only if there is an

embedding $h : \mathscr{S}^{n-1} \times I \to M$ such that $h(\mathscr{S}^{n-1} \times \{0\}) = S$. An n-cell $C \subset M$ is *collared* if and only if there is an embedding $h : \mathscr{S}^{n-1} \times I \to M$ such that $h(\mathscr{S}^{n-1} \times \{0\}) = \text{Bd } C$ and $h(\mathscr{S}^{n-1} \times (0,1]) \subset M \setminus C$.

(17.D.2) *Lemma.* Suppose that $h : \mathscr{S}^{n-1} \times I \to \mathscr{S}^n$ is an embedding. Let A and B be the two complementary domains of $h(\mathscr{S}^{n-1} \times I)$. Then there is a continuous surjection $f : \mathscr{S}^n \to \mathscr{S}^n$ such that

  (i) $\bar{A}$ and $\bar{B}$ are the only inverse sets of f, and
  (ii) $f(h(\mathscr{S}^{n-1} \times \{1/2\})$ is the equator of $\mathscr{S}^n$.

  Proof. Recall that $\mathscr{S}^n$ may be considered as the suspension of $\mathscr{S}^{n-1}$ (17.A.11). Let $q : \mathscr{S}^{n-1} \times I \to \mathscr{S}^n$ be the quotient map, where $q(\mathscr{S}^{n-1} \times \{1\}) = a$, $q(\mathscr{S}^{n-1} \times \{0\}) = b$, and $q(\mathscr{S}^{n-1} \times \{1/2\})$ is the equator.

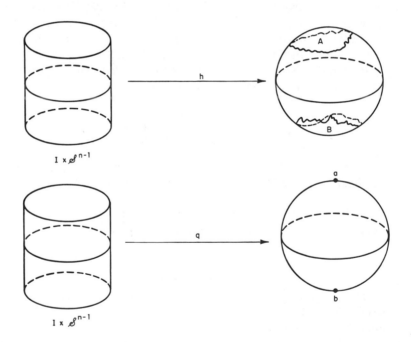

Define $f : \mathscr{S}^n \to \mathscr{S}^n$ by

$$f(x) = \begin{cases} a & \text{if} \quad x \in \bar{A} \\ b & \text{if} \quad x \in \bar{B} \\ qh^{-1}(x) & \text{if} \quad x \in h(\mathscr{S}^{n-1} \times I) \end{cases}$$

($\bar{A}$ and $\bar{B}$ may be relabeled if necessary.)

(17.D.3)   Theorem (Generalized Schonflies Theorem). (*Brown* [*1960b*]). $h : \mathscr{S}^{n-1} \times I \to \mathscr{S}^n$ is an embedding, then the closure of either complementary domain of $h(\mathscr{S}^{n-1} \times \{1/2\})$ is an n-cell.

Proof. Let A and B be the complementary domains of $h(\mathscr{S}^{n-1} \times I)$. By the previous lemma, there is a continuous map $f : \mathscr{S}^n \to \mathscr{S}^n$ whose unique inverse sets are $\bar{A}$ and $\bar{B}$. Hence, it follows from (17.C.21) that $\bar{A}$ and $\bar{B}$ are cellular. Let $\bar{D}$ be a complementary domain of $\mathscr{S}^{n-1} \times \{1/2\}$. Again from the previous lemma, we have that the restriction $f|_{\bar{D}}$ maps $\bar{D}$ onto one of the two hemispheres of $\mathscr{S}^n$. Hence, (17.C.16) may be applied to conclude that $\bar{D}$ is an n-cell.

(17.D.4)   Corollary. Suppose that S is an $(n-1)$-sphere in $\mathscr{E}^n$ and that $h : \mathscr{S}^{n-1} \times (0,1) \to \mathscr{E}^n$ is an embedding. If $h : (\mathscr{S}^{n-1} \times \{1/2\}) = S$, then S bounds a tame n-cell in $\mathscr{E}^n$.

E.   COMPACT n-MANIFOLDS

We saw in Chapter 16 that any compact 2-manifold may be obtained from a 2-cell by appropriate identifications on the boundary. Surprisingly, this result can be generalized to compact n-manifolds.

(17.E.1)   Theorem (*Brown*, [*1962b*]). If M is a compact n-manifold, then there is a continuous surjection $f : \mathscr{B}^n \to M$ such that

(i) $f|_{\text{Int } \mathscr{B}^n}$ is a homeomorphism,
(ii) $f(\text{Bd } \mathscr{B}^n) \cap f(\text{Int } \mathscr{B}^n) = \varnothing$, and
(iii) $\dim f(\text{Bd } \mathscr{B}^n) \leq n - 1$ (i.e., no point of $f(\text{Bd } \mathscr{B}^n)$ has an n-cell neighborhood contained entirely in $f(\text{Bd } \mathscr{B}^n)$).

Thus, just as was the case for 2-manifolds, M may be considered as the quotient space resulting from identifications on the boundary of the standard n-cell $\mathscr{B}^n$. The proof of the theorem will follow readily once we have established Theorem (17.E.3) below. This latter result is of some independent interest, since it is an example of what are often referred to as "engulfing" theorems. In this type of theorem, certain subsets of a space are swallowed up (or engulfed) by other sets (usually via homeomorphisms). Engulfing theorems are amazingly versatile, and are used in a wide variety of situations. The example to be given is one of the more primitive engulfing results, but it does serve to illustrate the concept involved. In the theorem, note how D' sweeps out to engulf both D and X before it becomes satiated.

(17.E.2)   Exercise.   Suppose that E is an open n-cell. Show that

    (i)   any two points in E may be connected by a tame arc, and

    (ii) if α is a tame arc in E and $\varepsilon > 0$, then there is an n-cell $B \subset S_\varepsilon(\alpha)$ such that $\alpha \subset \text{Int } B$.

(17.E.3)   Theorem (Primitive Engulfing Theorem).   Suppose that D is a collared n-cell in an n-manifold M and that ε and δ are positive numbers with $\delta < 1$. Let $\mathscr{E} = \{E_1, E_2, \ldots, E_k\}$ be a finite cover of M by open n-cells each with diameter less than ε, and suppose that $E_i \cap D \neq \varnothing$ for each i. Let $X = \{x_1, x_2, \ldots, x_m\}$ be a finite collection of points in M, and let $h : \mathscr{B}^n \to D$ be a homeomorphism. Then there is a collared n-cell D' and a homeomorphism $h' : \mathscr{B}^n \to D'$ such that

    (i) $D \cup X \subset D'$,

    (ii) $h'|_{\mathscr{B}^n_{1-\delta}} = h$, and

    (iii) $d(h(x),h'(x)) < \varepsilon$ for all $x \in \mathscr{B}^n$.

Proof.   Since D is collared, there is a homeomorphism $\hat{h} : \mathscr{B}^n_2 \to M$ such that $\hat{h}|_{\mathscr{B}^n} = h$. We may assume that δ is small enough so that if $d(x,x') < \delta$, then $d(\hat{h}(x),h(x')) < \varepsilon$. Furthermore, we may assume that X contains only points in $M \setminus D$. For each i, let E_{k_i} denote a member of $\mathscr{E}$ that contains x_i and let y'_i be any point in $E_{k_i} \cap (D \setminus h(\mathscr{B}^n_{1-\delta/2}))$. It follows from (17.E.2) that x_i and y'_i may be joined by a tame arc α_i lying in E_{k_i}. Finally, we can assume that all of these arcs are mutually disjoint (why?). Let α'_i be a subarc of α_i that lies outside of D and joins x_i to some point y_i contained in $h(\mathscr{B}^n_{1+\delta/2})$.

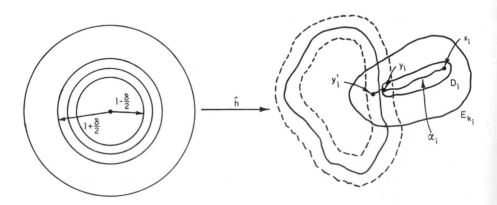

Let $\gamma = \min\{\min d(\alpha_p,\alpha_q), \min d(\alpha'_j,D)\}$, where $1 < p,q \le m$, $p \neq q$, and $1 \le j \le m$. By the previous exercise, each α'_i may be "blown up" to an n-cell

D_i that lies in a $(\gamma/3)$-neighborhood of α'_i, and contains x_i and y_i in its interior. Then the D_i's are disjoint, have diameter less than ε, and do not intersect D.

For each i, let $h_i : M \to M$ be a homeomorphism such that $h_i|_{M \setminus D_i} = id$ and $h_i(y_i) = x_i$. Let g be a homeomorphism of $\mathscr{B}_2^n$ onto itself that is the identity on $\mathscr{B}_{1-\delta/2}^n$ and maps $(\mathscr{B}^n \setminus \text{Int } B_{1-\delta/2}^n)$ to $(\mathscr{B}_{1+\delta}^n \setminus \text{Int } \mathscr{B}_{1-\delta/2}^n)$. Moreover, choose g so that it sends $(\mathscr{B}_2^n \setminus \text{Int } \mathscr{B}^n)$ onto $(\mathscr{B}_2^n \setminus \text{Int } \mathscr{B}_{1+\delta/2}^n)$; then $h' = h_m h_{m-1} \cdots h_1 hg$ is the desired homeomorphism.

Proof of (17.E.1). For each positive k, let $\mathscr{U}_k$ be a finite open cover of M by open sets of diameter less than $1/2^k$. Let X_k be a finite collection of points that contains at least one point from each $U \in \mathscr{U}_k$. By (17.A.17), there is an n-cell D_1 in M that contains X_1; furthermore, we may assume that D_1 is collared (why?). Let $h_1 : \mathscr{B}^n \to D_1$ be any homeomorphism. By the previous theorem, corresponding to $\varepsilon = 1/2^2$ and $\delta = 1/2$, there is an n-cell D_2 and a homeomorphism $h_2 : \mathscr{B}^n \to D_2$ such that

(i) $X_2 \cup D_1 \subset D_2$,
(ii) $h_2|_{\mathscr{B}_1^n - 1/2} = h_1|_{\mathscr{B}_1^n - 1/2}$, and
(iii) $d(h_1(x), h_2(x)) < (1/2)^2$.

The inductive step should now be clear. If not, clarify it, and show that $f : \mathscr{B}^n \to M$ defined by $f(x) = \lim_{n \to \infty} h_n(x)$ is the desired homeomorphism. (Note that f is the limit of a sequence of uniformly continuous functions, and hence is continuous.)

We conclude this chapter with a characterization of $\mathscr{E}^n$. We need the following two lemmas.

(17.E.4) Lemma (*Brown* [1961]). Suppose that S is a bicollared $(n - 1)$-sphere lying in the interior of an n-cell D and that U is an open set in the complementary domain $I(S)$ of S that has compact closure in $\text{int } D$. Let $h : D \to D$ be a homeomorphism such that $h|_U = id$, and suppose that $S \subset I(h(S))$. Then $[S, h(S)]$ is an n-annulus.

Proof. Since S is collared, there is an embedding $f : \mathscr{S}^{n-1} \times I \to I(h(S))$ such that $f(\mathscr{S}^{n-1} \times \{0\}) = S$ and $f(\mathscr{S}^{n-1} \times (0,1])$ lies in $D \setminus I(S)$. Clearly, there is a homeomorphism $g : D \to D$ such that $g|_{h(S)} = id$, $g(S) \subset U$, and $gf(\mathscr{S}^{n-1} \times \{\frac{1}{2}\}) = S$. Then $g^{-1}hgf(\mathscr{S}^{n-1} \times [0,\frac{1}{2}]) = [S, h(S)]$ and hence $[S, h(S)]$ is an n-annulus.

(17.E.5) Lemma. Suppose that D is an n-cell, S is a bicollared $(n - 1)$-sphere in $\text{Int } D$, and K is a compact subset of $\text{Int } D$. Then there is an $(n - 1)$-sphere $\hat{S} \subset \text{Int } D$ such that $K \cup S \subset I(\hat{S})$ and $[S, \hat{S}]$ is an n-annulus.

Proof. We assume that $D = \mathcal{B}^n$. Let p be a point in $I(S)$ and let U be an open neighborhood of p such that $\bar{U} \subset I(S)$. There is a homeomorphism $h : D \to D$ such that $h|_U = id$ and $h(K \cup S) \subset I(S)$. Then by (17.E.4), $\hat{S} = h^{-1}(S)$ is the desired sphere.

(17.E.6) Theorem. If M is a connected n-manifold with the property that every compact subset of M lies in an open n-cell, then M is homeomorphic to $\mathcal{E}^n$.

Proof. It follows from (10.A.18) and (10.A.16) that there is a countable family, $\{K_1, K_2, \ldots\}$ of compact subsets of M such that $M = \bigcup\limits_{i=1}^{\infty} K_i$ and $K_i \subset \text{Int}(K_{i+1})$. By hypothesis, there is an open n-cell C_1, and hence a bicollared closed n-cell D_1 such that $K_1 \subset \text{Int } D_1 \subset \hat{D}_1 \subset C_1$, where $\hat{D}_1$ is the union of D_1 and its collar. Similarly, there is an open n-cell C_2 and by (17.E.5) a collared n-cell D_2 such that $C_1 \cup K_1 \subset \text{Int } D_2 \subset \hat{D}_2 \subset C_2$ and [Bd D_1, Bd D_2] is an n-annulus ($\hat{D}_2$ is the union of D_2 and its collar). Proceeding inductively, we obtain a sequence of n-cells $D_1, D_2, \ldots$ such that $M = \bigcup\limits_{k=1}^{\infty} D_i$ and [Bd D_i, Bd D_{i+1}] is an n-annulus.

A homeomorphism $f : M \to \mathcal{E}^n$ is constructed inductively as follows. Let $f_1 : D_1 \to \mathcal{B}^n$ be an arbitrary homeomorphism and for $i = 2, 3, 4, \ldots$, let $f_i : D_i \to \mathcal{B}_i^n$ be a homeomorphism such that $f_{i|D_{i-1}} = f_{i-1}$. Then $f : M \to \mathcal{E}^n$ defined by $f(x) = f_i(x)$ if $x \in D_i$ is the desired homeomorphism.

(17.E.7) Corollary. Suppose that $E = \bigcup\limits_{i=1}^{\infty} E_i$, where for each i, E_i is open in E, E_i is an open n-cell, and $E_i \subset E_{i+1}$. Then E is an open n-cell.

PROBLEMS

Section A

1. Give another proof of (17.A.5) using simple chains.
2. Show that invariance of domain holds for n-manifolds.
3. Suppose that M and N are manifolds with boundary, and that $h : M \to N$ is a homeomorphism. Show that $h(\text{Bd } M) = \text{Bd } N$.
4. Find a locally Euclidean space that is not T_2.
5. Suppose that M is a compact n-manifold and that $U \subset M$ is homeomorphic to $\mathcal{E}^n$. Identify points in $M \setminus U$ to a single point and show that the resulting quotient space is $\mathcal{S}^n$.

6. Suppose that X is a second countable T_2 space with the property that every point has a neighborhood homeomorphic to $\mathcal{E}^n$. Is X an n-manifold?

7. Suppose that X is paracompact and that every point $x \in X$ has a neighborhood homeomorphic to $\mathcal{E}^n$. Show that X is an n-manifold.

8. Suppose that M is an n-manifold with boundary. Let $x \in M$ and suppose that x has a neighborhood W with the property that there is a homeomorphism $h : W \to \mathcal{B}^n$ such that $h(x) \in \text{Bd } \mathcal{B}^n$. Show that x does not have a neighborhood homeomorphic to Int $\mathcal{B}^n$.

9.* Show that every compact n-manifold can be embedded in a finite product of spheres.

10. Suppose that M is an n-manifold with boundary and that Bd $M \neq \emptyset$. Show that Bd M is an $(n - 1)$-manifold.

11. Show that if M and N are manifolds with boundary, then $M \times N$ is a manifold with boundary and $\text{Bd}(M \times N) = (M \times \text{Bd } N) \cup (\text{Bd } M \times N)$.

12. Show that an n-manifold cannot be homeomorphic to an m-manifold if $m \neq n$.

13. Suppose that $X \times Y$ is a manifold and $(x,y) \in \text{Int}(X \times Y)$. Suppose further that $X \times Z$ is a manifold. Show that there is a point $z \in Z$ such that $(x,z) \in \text{Int }(X \times Z)$.

14.* Let X be a topological space and suppose that $X \times \mathcal{E}^n$ is a manifold. Suppose that M is an m-manifold and $m \geq n$. Show that $X \times M$ is a manifold.

15.* Suppose that X and Y are manifold factors (i.e. there are spaces W_1 and W_2 such that $X \times W_1$ and $Y \times W_2$ are manifolds). Show that $X \times Y$ is a manifold factor and that Bd $(X \times Y) = (X \times \text{Bd } Y) \cup (\text{Bd } X \times Y)$.

16. (a) Exhibit a homeomorphism of the surface of a torus that interchanges circles a and b.

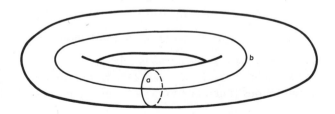

   (b) Let S denote the solid torus. Can the homeomorphism of part (a) be extended to a continuous function $f : S \to S$?

17. Assume the following result of Hanner [1951]. If X is a metric space,

and U_1, U_2, . . . is an open cover of X, where each U_i is an ANR_M, then X is an ANR_M. Show that every n-manifold is an ANR_M.

18. (a) Suppose that M is a compact connected n-manifold, and that $N \subset M$ is a compact n-manifold. Show that $N = M$.

   (b) A topological space Y *dominates* a topological space X if and only if there are continuous maps f and g such that the following diagram is commutative.

Suppose that M and N are compact connected manifolds, and that M dominates N and N dominates M. Show that N and M are homeomorphic.

19. Show by an example that if the hypothesis "manifold" is dropped in problem 18(b), then the conclusion fails.

Sections B and C

1. Show that in (17.B.5), not only may we assume that $\mathscr{S}^{n-1} \subset \text{Int } A$, but that this actually must be the case.

2. Suppose that $C \subset \mathscr{E}^n$ is connected and $\mathscr{E}^n \setminus C$ is homeomorphic to $\mathscr{E}^n \setminus \{p\}$. Show that C is compact.

3. Suppose that C_1, C_2, and C_3 are n-cells and that $C_1 \subset \text{Int } C_2 \subset C_2 \subset \text{Int } C_3$. Show that if $[C_1, C_2]$ and $[C_2, C_3]$ are n-annuli, then there is a homeomorphism $h : C_3 \to \mathscr{B}_3^n$ such that for $i = 1, 2, 3$, $f(C_i) = \mathscr{B}^n$.

4. Suppose that $D \subset \mathscr{E}^n$ is an n-cell lying in $\text{Int } \mathscr{B}^n$ such that $[\text{Bd } D, \text{Bd } \mathscr{B}^n)$ is a half-open annulus. Show that D is tame.

5. Suppose that K_1, K_2, . . . , K_m are pairwise disjoint pointlike subsets of $\mathscr{E}^n$. Show that $\mathscr{E}^n \setminus \bigcup_{i=1}^{m} K_i$ is homeomorphic to $\mathscr{E}^n$ minus m points.

6. Show that the topologist's sine curve is cellular.

7. Let C be an n-cell in $\mathscr{E}^n$ and let $K \subset \text{Int } C$ be a continuum. Suppose that there is a continuous function $f : C \to \mathscr{E}^n$ and a point $p \in \mathscr{E}^n$ such that

   (i) $f_{|C \setminus K}$ is a homeomorphism,
   (ii) $f(K) = p$, and
   (iii) $p \in f(C \setminus K)$.

   Show that $p \in \text{Int } f(K)$.

8. Suppose that K is a cellular subset of $\mathscr{E}^3$ that lies in a plane Q. Show that K is cellular in Q.

9. Show that a subset A of $\mathscr{E}^n$ is cellular if and only if A is pointlike.

10. Suppose that $K_1, K_2, \ldots, K_m$ are cellular subsets of an n-manifold M. Let G be a decomposition of M into points and the sets K_i. Show that M/G is homeomorphic to M.

11. Give an example of a 2-manifold M and a pointlike subset p of M that is not cellular.

12. Modify the example given in (17.C.8) so that all of the 1-spheres are consequtively linked, and display a pointlike subset of the manifold that is not homotopic to a point in the manifold.

Sections D and E

1. Call an n-cell C in $\mathscr{S}^n$ tame if there is a homeomorphism $h : \mathscr{S}^n \to \mathscr{S}^n$ such that $h(C)$ is the upper hemisphere of $\mathscr{S}^n$. Suppose that D is a tame n-cell in $\mathscr{E}^n$. Consider $\mathscr{S}^n$ to be the one-point compactification of $\mathscr{E}^n$. Show that $\mathscr{S}^n \setminus \operatorname{Int} D$ is a tame n-cell in $\mathscr{S}^n$.

2. Suppose that D is an n-cell and that S is an $(n-1)$-sphere contained in $\operatorname{Int} D$. Suppose further that there is an embedding $f : \mathscr{S}^{n-1} \times I \to D$ such that $f(\mathscr{S}^{n-1} \times \{1/2\}) = S$. Let U be an open subset of $I(S)$ and suppose that $h : D \to S$ is a homeomorphism such that $h_{|U} = id$ and $S \subset I(h(S))$. Show that $[S, h(S)]$ is an n-annulus.

3. Suppose that M is a compact manifold and that $(M \setminus U) \subset V$, where U and V are open n-cells. Show that M is an n-sphere.

4. Suppose that K is a compact subset of $\mathscr{E}^n$. Show that K is cellular if and only if $\mathscr{E}^n/G$ is homeomorphic to $\mathscr{E}^n$, where G is the decomposition of $\mathscr{E}^n$ obtained by identifying K to a point.

5. Show that an n-cell is collared if and only if its boundary is bicollared.

6.* A subset B of a space C is locally collared if B may be covered by a collection of open (in B) subsets each of which is collared in X. Show that the boundary of every manifold is locally collared. (The reader may wish to consult Brown [1962a] for a proof of the fact that if a manifold M is contained in a manifold N and M is locally collared in N, then M is collared in N.)

7.* A space is *invertible* if and only if for each $x \in X$ and for each neighborhood U of x, there is a homeomorphism $h : X \to X$ such that $h(X \setminus U) \subset U$. Prove the following result of Hocking and Doyle, using the steps indicated below: If M is an invertible n-manifold, then M is an n-sphere.

(a) Show that M is compact.

(b) Let C be a bicollared n-cell in M and let $D = \overline{M \setminus C}$. Define a homeomorphism $h : M \to M$ with the property that $h(D) \subset \operatorname{Int} C$.

(c) Let $g : C \to \mathscr{S}^n$ be an embedding, consider $gh(\operatorname{Bd} C)$, and then apply (17.D.3).

8.   Suppose that M is an invertible noncompact n-manifold. Show that M and $\mathscr{E}^n$ are homeomorphic.

9.   Suppose that $\{C_i\}$ is a sequence of subsets of a space X such that $X = \bigcup_{i=1}^{\infty} C_i$ and for each i, $C_i \subset C_{i+1}$. Let $\{h_i : X \to Y \mid i \in \mathbf{Z}^+\}$ be a sequence of homeomorphisms such that $h_{i+1|C_i} = h_{i|C_i}$ for all i and $\bigcup_{i=1}^{\infty} h_i(X) = Y$. Show that $h = \lim_{i \to \infty} h_i$ need not be a homeomorphism.

10.   Show that an arc α contained in $\mathscr{E}^n \subset \mathscr{E}^{n+1}$ is tame in $\mathscr{E}^{n+1}$.

Chapter 18

UPPER SEMICONTINUOUS DECOMPOSITIONS

The emphasis in this chapter is on demonstrating to the reader his proximity to the research frontier of at least one rather specialized area of topology. Many theorems are left unproven in order that a more general overview of decomposition theory may be gained. A number of research questions are posed, and it is hoped that the reader will begin to seek for himself answers to these and related problems. The presentation of the material in this chapter has been heavily influenced by the survey article of Armentrout [1966] and by seminars given by Armentrout and Price.

A. GENERALITIES

The reader is advised to first review Section F of Chapter 8. There, an upper semicontinuous decomposition of a topological space was defined as follows.

(18.A.1) Definition. A decomposition $G = \{g_\alpha \mid \alpha \in \Lambda\}$ of a topological space X is *upper semicontinuous* (hereafter denoted by usc) if and only if for each $g_\alpha \in G$ and for each open set U containing g_α there is a saturated open set V such that $g_\alpha \subset V \subset U$. (A subset V of X is *saturated* if and only if $V = P^{-1}(P(V))$, where P is the quotient map from X onto the decomposition space X/G.)

(18.A.2)   Notation.   If G is an usc decomposition of X, then H_G will denote the collection of nondegenerate elements of the decomposition (those elements consisting of more than one point) and H_G^* will denote the union of these elements.

(18.A.3)   Exercise.   Show that a decomposition G of a space X is usc if and only if for each open set U in X, the union of all sets of G contained in U is open.

The following characterization of usc decompositions is frequently useful.

(18.A.4)   Theorem.   A decomposition G of a space X is usc if and only if for each closed subset D of X, the union of members of the decomposition that intersect D is closed.

Proof.   Suppose that G is a usc decomposition of X and that D is closed in X. Let $F = \bigcup \{g \mid g \in G$ and $g \cap D \neq \emptyset\}$. Since $X \setminus D$ is open, it follows from (18.A.3) that $U = \bigcup \{g \mid g \in G$ and $g \subset X \setminus D\}$ is open. Since $F = X \setminus U$, we have that F is closed.

Conversely, suppose that whenever D is closed, then $\bigcup \{g \mid g \in G$ and $g \cap D \neq \emptyset\}$ is closed. Let $g' \in G$ be a subset of U where U is open in X, and let $V = \bigcup \{g \mid g \in G$ and $g \subset U\}$. Then $V = X \setminus \bigcup \{g \mid g \in G$ and $g \cap (X \setminus U) \neq \emptyset\}$, and hence V is open and saturated; therefore, it is immediate from (18.A.1) that G is usc.

Recall from Chapter 8, that geometrically, usc decompositions have the property that sequences of "little" elements of the decomposition may approach big elements, but that the situation illustrated below is proscribed.

(18.A.5)   Exercise.

(a) Suppose that $X = \mathscr{E}^2$, and for each $x \in \mathscr{E}^1$ let $g_x = \{(x,y) \in \mathscr{E}^2 \mid -\infty < y < \infty\}$. Let $G = \{g_x \mid x \in \mathscr{E}^1\}$. Show that G is usc and X/G is homeomorphic to $\mathscr{E}^1$.

(b) Suppose that $X = \mathscr{E}^1$ and that G is the usc decomposition of $\mathscr{E}^1$ whose nondegenerate elements consist of closures of components of the complement in $\mathbf{I}$ of the Cantor set. Show that $\mathscr{E}^1$ is homeomorphic to $\mathscr{E}^1/G$.

(c) Suppose that $X = \mathscr{S}^1$ and that $H_G = \{\{s_1,s_2,s_3\}\}$ where $s_1,s_2,s_3 \in \mathscr{S}^1$ are the vertices of an equilateral triangle. Describe X/G.

(d) Suppose that $X = \mathscr{S}^1$ and that G is the collection of all triples s_1, s_2, s_3 in $\mathscr{S}^1$ that form the vertices of an equilateral triangle. Describe X/G.

(e) Suppose that $X = \mathscr{E}^3$ and that H_G consists of a single element, the unit circle. Are $\mathscr{E}^3$ and X/G homeomorphic?

(f) Suppose that $X = \mathscr{E}^3$ and that G is the collection consisting of $\{\mathbf{O}\}$ and of all 2-spheres of the form $\{(x,y,z) \mid x^2 + y^2 + z^2 = r\}$, where $0 < r < \infty$. Describe X/G.

The notions of lim inf and lim sup are quite useful in the study of usc decompositions.

(18.A.6) Definition. Suppose that $\{A_n\}$ is a sequence of subsets of a space X. Then *lim inf* $\{A_n\}$ is defined to be $\{x \in X \mid$ every neighborhood of x intersects all but a finite number of the $A_n\}$, and *lim sup* $\{A_n\}$ is defined to be $\{x \in X \mid$ every neighborhood of x intersects infinitely many of the $A_n\}$. A set L *is the limit of the* A_n if and only if lim inf $\{A_n\}$ = lim sup $\{A_n\}$ = L.

(18.A.7) Examples.
1. For each $n \in \mathbf{Z}^+$, let $A_n = \{(x,y) \in \mathscr{E}^2 \mid x = 1/n$ and $0 \le y \le 1\}$. Then lim inf $\{A_n\}$ = lim sup $\{A_n\}$ = $\{(x,y) \in \mathscr{E}^2 \mid x = 0, 0 \le y \le 1\}$.
2. For each $n \in \mathbf{Z}^+$, let $A_{2n} = \{(x,y) \in \mathscr{E}^2 \mid x = 1/2n$ and $0 \le y \le 1\}$, and $A_{2n+1} = \{(x,y) \in \mathscr{E}^2 \mid x = 1/(2n + 1)$ and $-1 \le y \le 0\}$. Then lim inf $\{A_n\}$ = $\{(0,0)\}$ and lim sup A_n = $\{(x,y) \in \mathscr{E}^2 \mid x = 0$ and $-1 \le y \le 1\}$.
3. For each $n \in \mathbf{Z}^+$, let $A_n = \{(x,y) \in \mathscr{E}^2 \mid x = 1/n$ and either $0 \le y \le 1/3$ or $2/3 \le y \le 1\}$. Note that neither the lim sup $\{A_n\}$ nor lim inf $\{A_n\}$ is connected.

(18.A.8) Exercise. Suppose that $\{A_n\}$ is a sequence of subsets of a space X. Show that lim inf $\{A_n\} \subset$ lim sup $\{A_n\}$ and that both of these sets are closed.

(18.A.9) Theorem. Suppose that $\{A_n\}$ is a sequence of connected subsets of a compact T_2 space X and lim inf $\{A_n\} \ne \varnothing$. Then lim sup $\{A_n\}$ is connected.

Proof. If T = lim sup $\{A_n\}$ is not connected, then T may be written as the union of disjoint closed subsets D and F. By (18.A.8), D and F are also closed in X. Since X is normal, there are disjoint open sets U and V such that $D \subset U$ and $F \subset V$. Observe that the sequence $\{A_n\}$ is eventually in $U \cup V$, for if not, there would be a sequence of points $\{x_n\}$, $x_{n_i} \in A_{n_i}$ lying outside of $U \cup V$ with a cluster point x in $X \setminus (U \cup V)$. However, it then follows that $x \in T$, an obvious contradiction.

Since $\lim \inf \{A_n\} \neq \varnothing$, there is a point common to both $\lim \inf \{A_n\}$ and either U or V, say U. Then, U intersects all but finitely many of the sets $\{A_n\}$. Note, however, that for sufficiently large n, if $A_n \cap U \neq \varnothing$, then $A_n \cap V = \varnothing$ (the A_n's are connected and eventually lie in $U \cup V$). Since this implies that $V \cap \lim \sup \{A_n\} = \varnothing$, we have reached another contradiction.

The next theorem helps make precise what is meant by saying that small sets may approach large sets in a usc decomposition, but that big sets must keep their distance from a given small set.

(18.A.10)   Theorem.   Let G be a usc decomposition of a T_3 space X into closed sets. Suppose that $\{g_n\}$ is a sequence of elements in G, $g \in G$, and $g \cap \lim \inf \{g_n\} \neq \varnothing$. Then $\lim \sup \{g_n\}$ is a subset of g.

Proof.   Suppose that $x \in g \cap \lim \inf \{g_n\}$ and that $y \in g' \cap \lim \sup \{g_n\}$, where $g' \neq g$ $(g' \in G)$. Since X is regular, there are disjoint open sets U and V in X such that $x \in U$ and $g' \subset V$. We may assume that V is saturated. Since $x \in \lim \inf \{g_n\}$, there is an $N > 0$ such that for each $i > N$, we have that $g_i \cap U \neq \varnothing$. On the other hand, this cannot happen because an infinite number of the g_n lie entirely in V. Thus, $\lim \sup \{g_n\}$ must be contained in g.

In compact metric spaces, the property described in the previous theorem characterizes usc decompositions.

(18.A.11)   Exercise.   Suppose that G is a decomposition of a compact metric space into closed subsets. Show that G is usc if and only if whenever $\{g_n\}$ is a sequence of elements in G with $\lim \inf \{g_n\} \cap g \neq \varnothing$, then $\lim \sup \{g_n\} \subset g$.

A natural way of generating usc decompositions is given next. Let f be any closed continuous map from a space X onto a space Y. Then $\{f^{-1}(y) \mid y \in Y\}$ forms a usc decomposition of X. This is an easy consequence of the following basic result.

(18.A.12)   Theorem.   Suppose that G is a decomposition of a space X. Then G is usc if and only if the quotient map $P : X \to X/G$ is closed.

Proof.   Suppose that G is a usc decomposition of X and that F is a closed subset of X. Then $P^{-1}(P(F)) = \{g \mid g \in G$ and $g \cap F \neq \varnothing\}$. The latter set is closed by (18.A.4), and since X/G has the quotient topology, it follows that $P(F)$ is closed.

Conversely, suppose that P is a closed map and that F is a closed subset of X. We show that $\bigcup \{g \mid g \in G \text{ and } g \cap F \neq \varnothing\}$ is closed. Since P is closed, $P(F)$ is closed. Therefore, $\bigcup \{g \mid g \in G \text{ and } g \cap F \neq \varnothing\} = P^{-1}(P(F))$ is closed, and hence G is usc.

(18.A.13) **Corollary.** Suppose that $f : X \to Y$ is closed, onto, and continuous. Then $G = \{f^{-1}(y) \mid y \in Y\}$ is a usc decomposition of X.

Proof. Consider the following diagram, where $\phi(P(x)) = f(x)$.

By (8.A.6), ϕ is a homeomorphism. If A is a closed subset of X, then $P(A) = \phi^{-1}f(A)$, which is closed in X/G. Hence, P is a closed map and G is usc.

Components of compact metric spaces also yield usc decompositions.

(18.A.14) **Theorem.** Suppose that X is a compact metric space and that $G = \{g \subset X \mid g \text{ is a component of } X\}$. Then G is a usc decomposition of X.

Proof. We apply (18.A.11). Suppose that $\{g_n\}$ is a sequence of elements of G, $g \in G$, and $(\lim \inf\{g_n\}) \cap g \neq \varnothing$. By (18.A.9), $L = \lim \sup\{g_n\}$ is connected, and since L intersects g, it must lie in the component g.

In general, if $f : X \to Y$ is continuous, onto, and is an identification, then the map ϕ described in (18.A.13) is a homeomorphism. Thus, any information gleaned from X/G is immediately applicable to Y, and hence it is natural to inquire which properties of X are inherited by X/G. Of particular interest is the question of whether or not X and X/G are themselves homeomorphic. If G is a usc decomposition of X, then by (18.A.12), X/G will retain those properties of X that are preserved under closed continuous mappings, e.g., connectedness, compactness, local connectedness, separability, etc. Nevertheless, a number of things may go wrong. For instance, not even the T_0 separation property need be maintained.

(18.A.15) **Exercise.** Suppose that $X = (0,\infty)$ is given a topology whose open sets are of the form $(0,b)$, $b \in \mathbf{R}^1$. Let $G = \{A,B\}$, where A is the set of

positive rationals and B is the set of positive irrationals. Show that G is usc, X is T_0, but that X/G is not T_0.

More topological structure is preserved under usc decompositions if members of the decomposition are closed. However, some difficulties still remain.

(18.A.16)   Example.   Suppose that $X = \mathscr{E}^2$ and that G is the usc decomposition of X whose only nondegenerate element is the real line $L = \{(x,0) \in \mathscr{E}^2 \mid -\infty < x < \infty\}$. Then X/G is not even first countable. To see this, suppose that $U_1 \supset U_2 \supset \cdots$ is a countable basis at the point $P(L) \in X/G$. Then $P^{-1}(U_1), P^{-1}(U_2), \ldots$ is a decreasing sequence of open subsets of X containing L. For each $n \in \mathbf{Z}^+$, choose $y_n > 0$ such that $(n,y_n) \in P^{-1}(U_n)$. Then $V = \mathscr{E}^2 \setminus \{(n,y_n) \mid n \in \mathbf{Z}^+\}$ is a saturated open set containing L, but no U_n is contained in $P(V)$.

It follows from the preceding example that first and second countability and in particular metrizability are lost under usc decompositions. However, if additional conditions are imposed on the nondegenerate elements, somewhat greater control is gained over the resulting decomposition space.

(18.A.17)   Theorem.   Suppose that G is a usc decomposition of a space X into compact subsets.

  (i)   If X is T_2, then X/G is T_2.
  (ii)   If X is second countable, then X/G is second countable.
  (iii) If X is metrizable, then X/G is metrizable.

Proof.   (i) This follows easily, since compact subsets of a T_2 space may be enclosed in disjoint open subsets.

(ii)   Let **B** be a countable base for X, and let **B**′ be the family of all finite unions of members of **B**. For each $B' \in \mathbf{B}'$, let $B'' = \bigcup \{g \in G \mid g \subset B'\}$. Then $\{B'' \mid B' \in \mathbf{B}'\}$ is clearly a countable collection of saturated open sets in X. We show that $\{P(B'') \mid B' \in \mathbf{B}'\}$ is a basis for X/G. Suppose that $P(g) \in U$, where $g \in G$ and U is open in X/G. Since g is compact, there is a finite number of members of **B** that cover g and lie in $P^{-1}(U)$. If V is the union of these sets, then $V \in \mathbf{B}'$; hence, the corresponding set V'' contains g, and we have that $P(g) \in P(V'') \subset U$.

(iii)   This is an immediate consequence of Stone's metrization theorem (10.C.7).

(18.A.18)   Question 1.   By the previous theorem, if G is a usc decomposition of X into compact sets, then X/G is metrizable. Is there a reasonable way of actually defining this metric directly from the metric given for X?

Upper semicontinuous decompositions into compact sets are well behaved in the following sense as well.

(18.A.19) Theorem. If G is a usc decomposition of a space X into compact subsets and K is a compact subset of X/G, then $P^{-1}(K)$ is compact.

Proof. We apply (2.H.2). Suppose that $\mathscr{C} = \{C_\alpha \mid \alpha \in \Lambda\}$ is a family of closed subsets in $P^{-1}(K)$ with the finite intersection property. Let $\mathscr{F}$ be the family of all finite intersections of members of $\mathscr{C}$. Then $\mathscr{F}$ also has the finite intersection property, as does the collection $\{P(F) \mid F \in \mathscr{F}\}$. Since K is compact, there is a point $x \in \bigcap \{P(F) \mid F \in \mathscr{F}\}$. Then clearly, $\{P^{-1}(x) \cap C_\alpha \mid C_\alpha \in \mathscr{C}\}$ is a family of closed subsets of the compact set $P^{-1}(x)$ and enjoys the finite intersection property. Hence, the intersection of all of the members of this collection is nonempty. Consequently, $\bigcap \{C_\alpha \mid \alpha \in \Lambda\}$ is nonempty, which shows that $P^{-1}(K)$ is compact.

(18.A.20) Exercise. Suppose that G is a decomposition of a metric space into compact subsets. Show that the following statements are equivalent:

(i) G is usc, and
(ii) if whenever $x_i, y_i \in g_i \in G$ $(i = 1, 2, \ldots)$ and the sequence $\{x_i\}$ converges to a point $x \in g \in G$, then there is a subsequence $\{y_{n_i}\}$ of $\{y_i\}$ that converges to a point $y \in g$.

Connectedness does not fare so well under decompositions into compact subsets. For instance, if $X = [0,1]$ and G is the usc decomposition of X whose only nondegenerate element is the set $g = \{0,1\}$, then $P(g)$ does not pull back to a connected subset. This is corrected in the obvious manner.

(18.A.21) Theorem. Suppose that G is a usc decomposition of a space into connected subsets. If D is a connected subset of X/G, then $P^{-1}(D)$ is connected.

Proof. Suppose that $P^{-1}(D)$ is not connected. Then $P^{-1}(D)$ may be written as the disjoint union of closed (in $P^{-1}(D)$) sets A and B. Since each member of the decomposition is connected, it follows that A and B are saturated. Let $\hat{A}$ and $\hat{B}$ be closed subsets of X such that $\hat{A} \cap P^{-1}(D) = A$ and $\hat{B} \cap P^{-1}(D) = B$. Since P is a closed map, we have that $P(\hat{A}) \cap D$ and $P(\hat{B}) \cap D$ split D into disjoint closed subsets, which contradicts the connectedness of D.

(18.A.22) Definition. Decompositions whose elements are continua are called *monotone decompositions.*

(18.A.23)   Corollary.   If G is a monotone usc decomposition of a space X and C is a continuum in X/G, then $P^{-1}(C)$ is a continuum.

Many additional properties are preserved under monotone usc decompositions; for instance, we have the following theorem due to Bing.

(18.A.24)   Theorem (Bing [1951]).   If G is a monotone usc decomposition of a chainable continuum X, then X/G is chainable.

Proof.   Let ρ be the metric for X, d the metric for X/G, and suppose that $\varepsilon > 0$ is given. Since the quotient map P is uniformly continuous, there is a $\delta > 0$ such that if $\rho(x,y) < \delta$, then $d(P(x),P(y)) < \varepsilon/5$. The continuum X is chainable; hence, there is a δ-chain $\{A_1, A_2, \ldots, A_n\}$ that covers X. It is easy to see that there is an increasing sequence of integers $1 = n_1, n_2, \ldots$, $n_j = n$ such that some element of G intersects A_{n_i} and $A_{n_{i+1}}$ but no member of G intersects A_{n_i} and $A_{n_{(i+1)}+1}$. For integers i and k, let $U_{i,k} = \{P(g) \mid g \in G$ and $g \subset A_i \subset A_{i+1} \subset \cdots \subset A_k\}$. Then $U_{i,k}$ is open and $\{U_{1,5}, U_{n_4,n_8}, U_{n_7,n_{11}}, \ldots, U_{n_{3k+1},n_j}\}(j - 5 \le 3k + 1 < j - 3)$ is an ε-chain that covers X/G.

B.   CELLULAR DECOMPOSITIONS

The following astounding result shatters the notion that monotone usc decompositions might represent a panacea for the problem of inheritability.

(18.B.1)   Theorem (Hurewicz [1930]).   Suppose that K is a compact metric space. Then there is a monotone usc decomposition of $\mathscr{E}^3$ such that K is homeomorphic to a subset of $\mathscr{E}^3/G$.

Proof.   Let T be a standard tetrahedron in $\mathscr{E}^3$ and let ℓ_1 and ℓ_2 be nonintersecting edges of T.

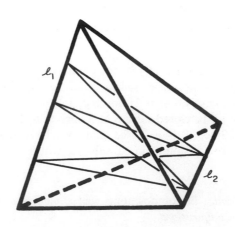

Let C_1 and C_2 be Cantor sets in ℓ_1 and ℓ_2 respectively. By (6.C.12), there are maps $f_1 : C_1 \to K$ and $f_2 : C_2 \to K$ that are continuous and onto. For each $x \in K$, let $A_x = f_1^{-1}(x)$ and $B_x = f_2^{-1}(x)$. Elements of the decomposition are defined as follows. If $x \in K$, let g_x be the subset of $\mathscr{E}^3$ consisting of the union of all possible line segments running from points of A_x to points of B_x. The reader may show with the aid of (18.A.20) that the monotone decomposition of $\mathscr{E}^3$ with nondegenerate elements $\{g_x \mid x \in K\}$ is usc.

We show now that if G is the usc decomposition constructed above, then K may be embedded in $\mathscr{E}^3/G$. Let $j : C_1 \to \mathscr{E}^3$ be the inclusion map and define $H : K \to \mathscr{E}^3/G$ by $H(x) = Pjf_1^{-1}(x)$. It is clear that H is continuous and 1–1. Since K is compact, it follows that H is the desired embedding.

The implications of the preceding result are startling. For example, it follows that the Hilbert cube may be embedded in $\mathscr{E}^3/G$ for some monotone usc decomposition G of $\mathscr{E}^3$. Hence, the decomposition space under a monotone usc decomposition may differ drastically from the original space. Obviously, then, still additional conditions are necessary in order that the decomposition space associated with a space X be homeomorphic to X. R. L. Moore [1925] showed that if the plane is decomposed upper semicontinuously into cellular sets, then the resulting decomposition space is again the plane. An identical result holds for cellular usc decompositions of 2-manifolds. Whyburn [1936] seems to have been the first to pose the obvious question: "If G is cellular decomposition of $\mathscr{E}^3$, is $\mathscr{E}^3/G$ homeomorphic to $\mathscr{E}^3$?" The issue remained unresolved for many years until Bing constructed the dogbone space described in Chapter 8. Armentrout [1970] subsequently described a decomposition of $\mathscr{E}^3$ into straight line segments such that the decomposition space was not homeomorphic to $\mathscr{E}^3$. A number of other examples of such decompositions have also been given. In all these cases, the proof that the decomposition space is distinct from the original space is extremely involved. In fact, one of the major unresolved problems in decomposition theory is that of finding adequate criteria for determining whether or not the decomposition space and the original space are homeomorphic. Perhaps the most applicable test at present is in terms of shrinkability.

(18.B.2) Definition. A usc decomposition G of a metric space X is *shrinkable* if and only if for each $\varepsilon > 0$, for each saturated open cover $\mathscr{U}$ of H_G^*, and for each homeomorphism $h : X \to X$, there is a homeomorphism $f : X \to X$ such that

(i) $f = h$ on $X \setminus \mathscr{U}^*$, where $\mathscr{U}^*$ denotes the union of the members of $\mathscr{U}$, and

(ii) for each $g \in H_G$,
   (a) $\operatorname{diam} f(g) < \varepsilon$, and
   (b) there is a $U \in \mathscr{U}$ such that $h(g) \cup f(g) \subset h(U)$.

The proof of the following theorem is difficult, but should be within the grasp of most readers.

(18.B.3)   Theorem   (McAuley [1962]).   Suppose that G be a monotone usc decomposition of a locally compact metric space X. If G is shrinkable, then X/G is homeomorphic to X.

The converse also holds for cellular usc decompositions of 3-manifolds. This was shown by Voxman [1970] and subsequently generalized to n-manifolds by Siebenmann [1972]. Surprisingly, if G is a cellular usc decomposition of a 3-manifold and M/G and M are homeomorphic, then the shrinking can be realized by a continuous family of homeomorphisms.

(18.B.4)   Definition.   If G is a decomposition of a metric space M, then G *may be realized by a pseudoisotopy* if and only if there is a continuous function $H : M \times [0,1] \to M$ such that

  (i) $H_0(x) = x$ for each $x \in M$,
  (ii)   if $g \in G$, then $H_1(g)$ is a point in M and if $g, g' \in G$, $g \neq g'$, then
         $H_1(g) \neq H_1(g')$, and
  (iii) for $0 \leq t < 1$, H_t is a homeomorphism, and H_1 is onto.

A proof of the following theorem may be found in Voxman [1972].

(18.B.5)   Theorem.   Suppose that G is a cellular usc decomposition of a 3-manifold M and M/G is homeomorphic to M. Then if U is any open set containing H_G^*, there is a pseudoisotopy H that realizes the decomposition such that $H|_{M \setminus U}$ is the identity.

One of the most useful and profound results in decomposition theory is a theorem due to Armentrout (and later generalized to n-manifolds ($n \neq 4$) by Siebenmann [1972]). It states that if a cellular usc decomposition of a 3-manifold yields a 3-manifold, then the decomposition space is homeomorphic to the original space; furthermore, a homeomorphism may be found that approximates the projection map.

(18.B.6)   Theorem   (Armentrout, [1968]).   Suppose that G is a cellular usc decomposition of a 3-manifold M such that M/G is a 3-manifold. Then for each positive number ε, there is a homeomorphism from M onto M/G such that $d(P(x), h(x)) < \varepsilon$ for each $x \in M$.

What other conditions guarantee the existence of a homeomorphism between a given space and an associated decomposition space? The reader should be able to establish easily with the aid of (17.C.10) that if G is a cellular usc decomposition of an n-manifold with only a finite number of non-degenerate elements, then M/G and M are homeomorphic. However, problems arise when one passes to cellular decompositions with a countable number of nondegenerate elements. In fact, Bing [1962] has described a cellular decomposition of $\mathcal{E}^3$ with only countable many nondegenerate elements each of which lies in one of two intersecting planes and with the property that the decomposition space is not $\mathcal{E}^3$. The same author, in addition to having destroyed many fine conjectures with ingenious counter-examples, has contributed the following positive result (Bing, [1957b]).

(18.B.7) Theorem. Suppose that G is a cellular decomposition of $\mathcal{E}^3$ such that H_G is countable and H_G^* is a G_δ subset of $\mathcal{E}^3$. Then $\mathcal{E}^3/G$ and $\mathcal{E}^3$ are homeomorphic.

(18.B.8) Question 2. Is the preceding theorem valid for n-manifolds? What other conditions may be imposed on the nondegenerate elements for this theorem to hold?

Another major unresolved question along these lines is the following.

(18.B.9) Question 3. Suppose that G is a usc decomposition of $\mathcal{E}^3$ with countably many nondegenerate elements each of which is a tame disk (or a tame 3-cell). Is $\mathcal{E}^3/G$ homeomorphic to $\mathcal{E}^3$?

One of Bing's most remarkable discoveries was that if G is the dogbone decomposition, then $(\mathcal{E}^3/G) \times \mathcal{E}^1$ is homeomorphic to $\mathcal{E}^4$ [1959a]. By Armentrout's result (18.B.6), $\mathcal{E}^3/G$ cannot be a manifold, but yet, incredibly, it is a factor of a manifold. Various mathematicians have generalized this to the following extent (e.g., Edwards and Miller [to appear]; Eaton and Pixley [1974]).

(18.B.10) Theorem. Suppose that G is a usc cellular decomposition of $\mathcal{E}^3$ such that $P(H^*)$ is closed and 0-dimensional. Then $(\mathcal{E}^3/G) \times \mathcal{E}^1$ is homeomorphic to $\mathcal{E}^4$.

(18.B.11) Question 4. Must $P(H_G^*)$ be closed and 0-dimensional for the preceding theorem to hold?

More generally, we have the following problem.

(18.B.12) Question 5. If G is cellular decomposition of $\mathcal{E}^n$, is $\mathcal{E}^n/G \times \mathcal{E}^1$ homeomorphic to $\mathcal{E}^{n+1}$?

At present, it is not even known whether cellular decompositions of 3-manifolds preserve dimension. It follows from a result of Dyer [1955] that if G is a cellular decomposition of an n-manifold and dim X/G is finite, then dim $X/G = n$. Zemke [1974] has shown that if G is a cellular decomposition of $\mathscr{E}^3$ into polyhedral sets or if G is decomposition of $\mathscr{E}^3$ into compact convex sets, then dim $\mathscr{E}^3/G \leq 3$, but the following basic question still remains open.

(18.B.13)   Question 6. Suppose that G is a cellular decomposition of an n-manifold M. Is M/G finite dimensional?

(18.B.14)   Question 7. Suppose that G is a decomposition of an n-manifold, such that each nondegenerate element is an arc (or a disk). Is X/G finite dimensional?

(18.B.15)   Question 8. If G is a cellular decomposition of $\mathscr{E}^n$, is $\mathscr{E}^n/G$ embeddable in $\mathscr{E}^{n+1}$? (This is known to be true if $P(H_G^*)$ is closed and 0-dimensional.)

One further question involving cellularity is the following.

(18.B.16)   Question 9. Suppose that G is a monotone decomposition of $\mathscr{E}^3$ such that $\mathscr{E}^3/G$ is homeomorphic to $\mathscr{E}^3$. Under what conditions is G cellular?

C. HOMOTOPIES AND FIXED POINTS

The homotopy relationship between a manifold M and the space resulting from a cellular decomposition of M is especially interesting. In fact, for all known examples of such decompositions, it has turned out to be the case that the original space and the decomposition space are homotopically equivalent. Furthermore, Armentrout and Price [1969] have shown that if G is a cellular decomposition of an n-manifold M, then the induced homeomorphism P_* is actually an isomorphism between $\pi_n(M)$ and $\pi_n(M/G)$. In order that the reader may gain an insight into some of the basic techniques used here, we present in detail a part of the proof of the Armentrout and Price result.

A major disadvantage of cellularity is that it is defined only in the context of manifolds. To escape this restriction, the following notion, which is meaningful in any topological space, is frequently employed.

(18.C.1)   Definition. Suppose that X is a topological space and A is a

subset of X. Then A has property i-UV if and only if for each open set U containing A there is an open set V, such that $A \subset V \subset U$ and if $f : \mathscr{S}^i \to V$, then f is null homotopic in U. The set A has property UV^n if and only if A has property i-UV for each $i \leq n$; A has property UV^ω if and only if A has property i-UV for each i such that $0 \leq i < \infty$.

Clearly, cellular subsets possess property UV^ω. The Armentrout-Price theorem is the following.

(18.C.2) **Theorem.** Suppose that X is a metric space, n is a nonnegative integer, and G is a UV^n decomposition of X (i.e., each member of H_G has property UV^n). Suppose that U is an open subset of X/G and $x \in P^{-1}(U)$. Then for $0 \leq k \leq n$, $P_* : \pi_k(P^{-1}(U),x) \to \pi_k(U,P(x))$ is an isomorphism onto. (For the definition of the higher homotopy groups, see problem 12.D.2.)

We shall examine some of the main ingredients of the proof.

(18.C.3) **Definition.** Suppose that $\mathscr{U}$ is a cover of a space X. Then a collection $\mathscr{V}$ of open subsets of X *star n-homotopy refines* $\mathscr{U}$ if and only if for each $V \in \mathscr{V}$ there is a $U \in \mathscr{U}$ such that

(i) $\operatorname{St}(V,\mathscr{V}) \subset \mathscr{U}$, and
(ii) for each $0 \leq k \leq n$, if $f : \mathscr{S}^k \to \operatorname{St}(V,\mathscr{V})$ is a continuous function, then f may be extended continuously to a map $F : \mathscr{B}^{k+1} \to U$.

The proof of the next lemma is not difficult.

(18.C.4) **Lemma.** Suppose that X is a metric space, $n \in \mathbf{N}$, G is a UV^n decomposition of X, and A is a subset of X/G. If $\mathscr{U}$ is an open cover of A, then there exists an open cover $\mathscr{V}$ of A such that $\{P^{-1}(V) \mid V \in \mathscr{V}\}$ star n-homotopy refines $\{P^{-1}(U) \mid U \in \mathscr{U}\}$.

The key lemma in the proof of Theorem (18.C.2) is the following lifting type result.

(18.C.5) **Lemma.** Suppose that X is a metric space, $n \in \mathbf{N}$, and G is a UV^{n-1} decomposition of X. Suppose further that k is a nonnegative integer such that $k \leq n$, K is a finite simplicial k-dimensional complex, and L is a subcomplex of K. Let $\varepsilon > 0$ be given and suppose that $f : |L| \to X$ and $g : |K| \to X/G$ are continuous maps such that $g_{|L} = Pf$. Then there is a continuous extension F of f to K such that $d(g(x),PF(x)) < \varepsilon$ for each $x \in K$.

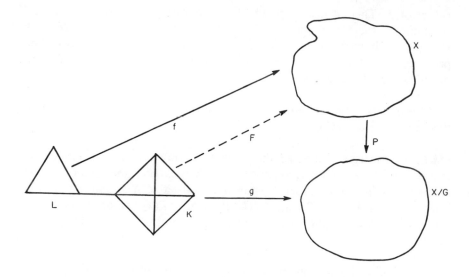

Proof. The basic idea of the proof is to construct F by first defining it on suitable vertices, then on 1-simplices, then on 2-simplices, etc. In this proof all simplicies are assumed to be closed. Let $\mathcal{U}_k$ be a finite open cover of $P^{-1}(g(|K|))$ by sets with the property that diam $P(U) < \varepsilon/3$ for each $U \in \mathcal{U}_k$. The previous lemma is applied k times to obtain finite open covers $\mathcal{U}_{k-1}$, $\mathcal{U}_{k-2}, \ldots, \mathcal{U}_0$ of $P^{-1}g(|K|)$ such that if $0 \le i \le k - 1$, each set of $\mathcal{U}_i$ is an open inverse set and $\mathcal{U}_i$ star $(n - 1)$-homotopy refines $\mathcal{U}_{i+1}$. For $0 \le i \le k$, $\{U_{i_1}, \ldots, U_{i_{m_i}}\}$ will denote the members of $\mathcal{U}_i$.

Let T be a subdivision of K that is fine enough so that if $\sigma \in T$, then $g(\sigma) \subset P(U_{0j})$ for some j. A function F_0 is defined on the 0-skeleton of T as follows. Suppose that v is a vertex of T. If $v \in L$, let $F_0(v) = f(v)$; if $v \notin L$, let $F_0(v)$ be any point in $P^{-1}(g(v))$. Note that if σ is a 1-simplex of T, then for some j, $P^{-1}(g(\sigma')) \subset U_{0j}$, and consequently, $F_0(\sigma') \subset P^{-1}(g(v))$.

We now proceed inductively to define functions F_1, F_2, etc. on the 1-skeleton, 2-skeleton, etc., of T. The map F_{i+1} will always be an extension of F_i. Suppose that $0 \le i < k$ and F_i has been defined on the i-skeleton of T in such a way that (i) if $i > 0$, then F_i extends F_{i-1}, (ii) if x is both in $|L|$ and the carrier of the i-skeleton of T, then $F_i(x) = f(x)$, and (iii) if σ^i is an i-simplex of T, then for some j, $F_i(\sigma^i) \subset U_{ij}$. The function F_i is extended to the carrier of the $(i + 1)$-skeleton as follows.

If σ is an $(i + 1)$-simplex in T that lies in L, then we set $F_{i+1|\sigma} = f_{|\sigma}$. If σ is an $(i + 1)$-simplex of T that is not in $|L|$, then we first work on the i-faces of σ. For each i-face τ of σ, there is a j_τ such that $F_i(\tau) \subset U_{ij_\tau}$. If $i > 0$, then each i-face of $\dot{\sigma}$ must intersect τ, and hence $F_i(\dot{\sigma}) \subset \text{St}(U_{ij_\tau}, \mathcal{U}_i)$. If $i = 0$, then there is a j such that $F_i(\dot{\sigma}) \subset U_{ij}$. Since $\mathscr{U}_i$ star $(n-1)$-homotopy refines $\mathscr{U}_{i+1}$, we have that for some r, $F_i(\dot{\sigma}) \subset U_{(i+1)r}$ and $F_{i|\dot{\sigma}}$ is homotopic to a constant map. Thus, F_i may be extended to F_{i+1} in such a way that (i) F_{i+1} maps the carrier of the $(i+1)$-skeleton T_{i+1} of T onto X, (ii) if $x \in |L| \cap |T_{i+1}|$, then $F_{i+1}(x) = f(x)$, and (iii) if σ^{i+1} is an $(i+1)$-simplex of T, then $F_{i+1}(\sigma^{i+1}) \subset U_{(i+1)j}$ for some j. This completes the induction.

To complete the proof the reader should verify that the map F derived from the F_i's is an "approximate" lifting of g.

Similar techniques eventually lead to a proof of the next lemma.

(18.C.6) Lemma. Suppose that X is a metric space, $n \in \mathbb{N}$, G is a UV^n decomposition of X, $k \leq n$, K is a finite simplicial k-dimensional complex, and L is a subcomplex of K. Suppose that f^0 is a map from $|L|$ into X and g is a map from $|K|$ into X/G such that $g_{||L|} = Pf^0$. Then if ε is a positive number, there is an extension f of f^0 that sends $|K|$ into X and a homotopy $H: |K| \times \mathbf{I} \to X/G$ such that $H_0 = Pf$, $H_1 = g$, and if $t \in \mathbf{I}$, $H_{t||L|} = g_{||L|}$.

Proof of (18.C.2). To see that P_* is 1–1, suppose that $f: (\mathscr{S}^k, s) \to (P^{-1}(U), x)$ and $P_*([f]) = e$. Then Pf can be extended to a map $g: \mathscr{B}^{k+1} \to P^{-1}(U)$. By (18.C.5), there is an extension F of f that maps $\mathscr{B}^{k+1}$ into $P^{-1}(W)$. Therefore, f is homotopic to a constant map in $P^{-1}(\mathscr{U})$, and consequently P_* is 1–1.

That P_* is onto may be established in an equally painless manner. Suppose that $g: (\mathscr{S}^k, s) \to (U, P(x))$. Let $f^0: \{s\} \to P^{-1}(U)$ be defined by $f^0(s) = x$. Then by (18.C.6), there is an extension f of f^0 that maps $\mathscr{S}^k$ into $P^{-1}(U)$ such that Pf and g are homotopic in U. Therefore, $P_*([f]) = [g]$ and P_* is onto.

(18.C.7) Question 10. When is the quotient map P a homotopy equivalence? (If G is a cellular decomposition of an n-manifold M and M/G is finite dimensional, then it is known that P is a homotopy equivalence (Armentrout and Price [1969]).)

A large number of spaces with the fixed point property may be created as a result of the next theorem, whose proof is based on (18.C.5) above.

(18.C.8) Theorem. Suppose that K is an n-dimensional finite simplicial complex whose carrier has the fixed point property and that G is a UV^{n-1} usc decomposition of $|K|$. Then $|K|/G$ has the fixed point property.

Proof. Suppose that h maps $|K|/G$ continuously into itself. A fixed point is found by first showing that for each $\varepsilon > 0$, there is a map $F: |K| \to$

$|K|$ such that $d(PF,hP) < \varepsilon$ (i.e., $d(PF(y),hP(y)) < \varepsilon$ for each $y \in |K|$). Let $x \in |K|$ and choose $x' \in P^{-1}(hP(x))$. Let $f : \{x\} \to \{x'\}$. Since G is a UV^{n-1} decomposition of $|K|$, there is by (18.C.5) an extension F of f to all $|K|$ such that $d(PF,hP) < \varepsilon$. Thus, for each positive integer n, there is a map $F_n : |K| \to |K|$ such that $d(PF_n,hP) < 1/n$. Let x_n be a fixed point of F_n and let z be a limit point of the sequence $\{x_n\}$. Then $P(z)$ is a fixed point of h, since

$$d(hP(z),P(z)) \le d(hP(z),hP(x_n)) + d(hP(x_n),PF_n(x_n)) + d(PF_n(x_n),P(z))$$

and the sum on the right tends toward 0 for increasing values of n.

The following fixed point result may be obtained with the aid of techniques similar to those used in (18.C.2).

(18.C.9)   *Theorem* (*Cobb and Voxman* [*1972*]).   Suppose that G is a UV^{ω} decomposition of a compact metric space X such that X/G is finite dimensional. Then if X has the fixed point property, so does X/G.

(18.C.10)   *Question 11.*   Must X/G be finite dimensional in the preceding theorem?

PROBLEMS

Section A

1. Suppose that G is a decomposition of space X such that G has only finitely many nondegenerate elements each of which is closed in X. Show that G is upper semicontinuous. Is the modifier "closed" necessary?

2. Show that a decomposition G of a space X is usc if and only if for each $g \in G$ and each open set U containing g, there is an open set V such that $g \subset V \subset U$ and with the property that if $g' \in G$ and $g' \cap V \ne \varnothing$, then $g' \subset U$.

3. Show that if G is an usc decomposition of a normal space X into closed subsets, then X/G is T_2.

4. Show that if G is an usc decomposition of a T_0 space such that G has only one nondegenerate element, then X/G is T_0. Can T_0 be replaced by T_1?

5. Show that if G is an usc decomposition of a regular space X into compact sets, then X/G is regular.

6. Show that if G is an usc decomposition of a locally compact T_2 space X into compact subsets, then X/G is locally compact.

7. Suppose that $\{C_i \mid i \in \mathbf{Z}^+\}$ is a family of connected subsets of a T_2

space X and that $\bigcup\limits_{i=1}^{\infty} C_i$ is compact. Show that if $\lim \inf C_i \neq \varnothing$, then $\lim \sup C_i$ is connected.

8. Suppose that G is a monotone usc decomposition of $\mathscr{E}^3$. Show that $P_{|\mathscr{E}^3 \backslash H_G*}$ embeds $\mathscr{E}^3 \backslash H_G^*$ in $\mathscr{E}^3/G$.

Section B

1. Suppose that G is an usc decomposition of $\mathscr{E}^3$ into points and a finite number of compact subsets such that $\mathscr{E}^3/G$ is homeomorphic to $\mathscr{E}^3$. Show that G is a cellular decomposition.

2. Show that if G is a usc cellular decomposition of an n-manifold M with finitely many nondegenerate elements, then M and M/G are homeomorphic.

3. Suppose that G is a monotone usc decomposition of $\mathscr{E}^3$ such that $\mathscr{E}^3/G$ is homeomorphic to $\mathscr{E}^3$. Let B be a 2-sphere in $\mathscr{E}^3/G$ that misses $P(H_G^*)$. Show that $P^{-1}(B)$ is a 2-sphere, and furthermore that $I(P^{-1}(B)) = P^{-1}(I(B))$. (If S is a 2-sphere in $\mathscr{E}^3$, then $I(S)$ denotes the bounded complementary domain of S in $\mathscr{E}^3$.)

4.* A set $F \subset \mathscr{E}^3$ is a *tame Cantor set* if and only if there is a homeomorphism from $\mathscr{E}^3$ onto $\mathscr{E}^3$ that maps F onto the standard Cantor set lying in $[0,1] \subset \mathscr{E}^1 \subset \mathscr{E}^3$. Show that if G is a monotone usc decomposition of $\mathscr{E}^3$ such that $\mathscr{E}^3/G$ is homeomorphic to $\mathscr{E}^3$ and $P(H_G^*)$ is a tame Cantor set, then G is cellular.

5.* Show that if G is a monotone decomposition of $\mathscr{E}^3$ with countably many nondegenerate elements such that $\mathscr{E}^3/G$ is homeomorphic to $\mathscr{E}^3$, then G is a cellular decomposition.

6.* Suppose that G is a monotone decomposition of $\mathscr{E}^3$ such that each nondegenerate element of G lies on a line, L, formed by the intersection of two planes Q_1 and Q_2. Show that there exists a pseudoisotopy $H : \mathscr{E}^3 \times I \to \mathscr{E}^3$ that realizes the decomposition and leaves the planes Q_1 and Q_2 and the line L set-wise fixed.

7. Let M be a metric space, and suppose that K is a collection of mutually disjoint subsets of M. If $g \in K$, then K is said to be *continuous* at g if and only if for each $\varepsilon > 0$, there exists an open set V such that $g \subset V$ and if $g' \in K$ and $g' \cap V \neq \varnothing$, then $g \subset S_\varepsilon(g')$ and $g' \subset S_\varepsilon(g)$. A decomposition G of a metric space is *nondegenerately continuous* in case H_G is continuous at g for each $g \in H_G$. The decomposition G is *continuous* if and only if G is continuous at g for each $g \in G$. Show that the dogbone space is a nondegenerately continuous usc decomposition of $\mathscr{E}^3$.

8. Use (18.B.7) to show that if G is a cellular nondegenerately continuous

decomposition of $\mathscr{E}^3$ with countably many nondegenerate elements, then $\mathscr{E}^3/G$ and $\mathscr{E}^3$ are homeomorphic.

9.   Suppose that G is a continuous decomposition of a metric space X into compact sets. Show that the metric for X/G is equivalent to the Hausdorff metric.

10.   Show that G is a continuous decomposition of a metric space X if and only if the projection map $P : X \to X/G$ is open and closed.

11.*   Show that if G is a shrinkable usc decomposition of an n-manifold M into compact sets, then G is cellular.

Section C

1.*   Use techniques similar to those employed in (18.C.5) to prove the following result of Price [1966]. Suppose that G is a cellular usc decomposition of E^n. If U is a saturated open subset of $\mathscr{E}^n$ such that $P(U)$ is simply connected, then $P^{-1}(U)$ is simply connected.

2.   Suppose that X is a topological space, $p \in X$, and $n \in \mathbf{Z}^+$. Then X is n-LC at p if and only if for each open set U containing p, there is an open set V such that $p \in V \subset U$ and with the property that if $f : \mathscr{S}^n \to V$ is a continuous function, then f is null homotopic in U. The space X is LC^n at p if and only if X is i-LC at p for each $i \le n$, and X is LC^n if and only if X is LC^n at each point $p \in X$. Suppose that X is a metric LC^n space, and G is a usc decomposition of X into compact sets each of which has property UV^n. Show that X/G is n-LC.

3.   Suppose that X is a locally compact, locally connected metric space and that M is a compact subset of X. Show that M has *property* 0-UV if and only if M is connected.

4.   Suppose that G is an usc decomposition of a metric space into compact sets each with property UV^n. Show that if A is a compact subset of X/G with property UV^n, then $P^{-1}(A)$ has property UV^n.

5.   Suppose that X is a topological space and that M is a subset of X with property UV^n. Show that if U is an open set containing M, then there is an open set V such that $M \subset V \subset U$ and with the property that if K is any finite n-complex and $f : K \to V$ is continuous, then f is null homotopic in U.

6.   Suppose that M is a subset of a space X. Then M has *property* CC if and only if for each open set U containing M, there is an open set V such that $M \subset V \subset U$ and with the property that if $f : \mathscr{S}^1 \to V \setminus M$ is continuous, then f is null homotopic in $U \setminus M$. Suppose that G is a usc decomposition of a metric space X into compact subsets each with property UV^1. Show that if A is a compact subset of X/G with property CC, then $P^{-1}(A)$ has property CC.

7. Use (18.C.8) to show that if the Brouwer fixed point theorem is assumed to be true for $k = N$, then it is true for $i \leq N$.
8. Find an example to show that the UV condition in (18.C.8) is a necessary one.

APPENDIX

This appendix includes material that is usually omitted from elementary courses in group theory. It is assumed that the reader is familiar with the basic concepts of group, homomorphism, isomorphism, normal subgroup, quotient group, etc. Much of the material discussed here may be found in Crowell and Fox [1963].

A. FREE GROUPS

(Ap.A.1) **Definition.** Suppose that A is a set. A *syllable on A* is an element of $A \times \mathbf{Z}$. We shall denote (a,n) by a^n. A *word on A* is a finite sequence of syllables. The word with no syllables is denoted by 1.

Let $W(A)$ denote the set of all words on A. Define a binary operation on $W(A)$, i.e., a function $\cdot$, that maps $W(A) \times W(A)$ into $W(A)$, by

$$(a_1^{n_1} a_2^{n_2} \cdots a_p^{n_p}) \cdot (b_1^{m_1} b_2^{m_2} \cdots b_q^{m_q}) = a_1^{n_1} a_2^{n_2} \cdots a_p^{n_p} b_1^{m_1} b_2^{m_2} \cdots b_q^{m_q}.$$

The operation $\cdot$ converts $W(A)$ into a semigroup, i.e., the binary operation $\cdot$ is associative, and $W(A)$ has an identity element, 1.

(Ap.A.2) **Definition.** Let $W(A)$ be the word semigroup on the set A. Suppose that the following hold: $a \in A$; $w_1, w_2 \in W(A)$; $w = w_1 a^0 w_2$; $u = w_1 w_2$; $t = w_1 a^n a^m w_2$; $s = w_1 a^{n+m} w_2$. Then we say that u is obtained from w by an *elementary contraction of type I* and w is obtained from u by an *ele-*

mentary expansion of type I. We say that the element s is obtained from t by an *elementary contraction of type II*, and t is obtained from s by an *elementary expansion of type II.*

Two words $\hat{w}$ and $\hat{u}$ of $W(A)$ are *equivalent*, denoted by $\hat{w} \sim \hat{u}$, if and only if $\hat{w}$ can be obtained from $\hat{u}$ by a finite sequence of elementary expansions and contractions.

The proofs of the following two lemmas are trivial and omitted.

(Ap.A.3)   Lemma.   The relation $\sim$ defined above is an equivalence relation.

(Ap.A.4)   Lemma.   The equivalence relation $\sim$ agrees with the product on $W(A)$, i.e., if $w_1 \sim w_2$ and $w_3 \sim w_4$, then $w_1 \cdot w_3 \sim w_2 \cdot w_4$.

Note that $a^0 \sim 1$, for each $a \in A$.

(Ap.A.5)   Notation.   If A is a set, then $F[A]$ will denote $\{[w] \mid w \in W(A)\}$, where $[w]$ is an equivalence class of w determined by $\sim$.

The proof of the next theorem is straightforward.

(Ap.A.6)   Theorem.   The set $F[A]$ with the binary operation $\cdot$ defined by $[w] \cdot [u] = [w \cdot u]$ is a group.

(Ap.A.7)   Examples.
   1.   If $A = \varnothing$, then $F[A] = [1]$.
   2.   If $A = \{a\}$, then $F[A]$ is infinite cyclic.

(Ap.A.8)   Definition.   Suppose that G is a group and E is a subset of G. Then the *subgroup H generated by E* is defined to be $\cap \{S \mid E \subset S$ and S is a subgroup of $G\}$. If the subgroup generated by E is G, then E is called a *set of generators for G.*

Note that every group has at least one set of generators (namely itself).

(Ap.A.9)   Exercise.   If A is a set, show that $\{[a] \mid a \in A\}$ is a generating set for $F[A]$.

(Ap.A.10)   Notation.   Henceforth, we shall denote the equivalence class $[a]$ by a.

(Ap.A.11)   Definition.   A generating set E of a group G is a *free basis* for G if and only if for each group H and for each set function $\phi : E \to H$,

there is a homomorphism $\hat{\phi} : G \to H$ such that $\hat{\phi}_{|E} = \phi$. It is clear that such an extension is unique.

(Ap.A.12) *Definition.* A group G is called a *free group* if and only if G has a free basis.

(Ap.A.13) *Remark.* The reader should note that a free group with a free basis containing more than one element is not abelian. Hence, free groups should not be confused with free abelian groups.

The next theorem is crucial.

(Ap.A.14) *Theorem.* A group G is free if and only if G is isomorphic to $F[A]$ for some set A.

Proof. We first show that if A is a set, then $F[A]$ is free with free basis A. Suppose that H is a group and $\phi : A \to H$. First, we define a map $\hat{\phi} : W(A) \to H$ by $\hat{\phi}(a^n b^m \cdots) = (\phi(a))^n (\phi(b))^m \cdots$ whenever $a^n b^n \cdots \in W(A)$. Note that $\hat{\phi}(w_1 a^0 w_2) = \hat{\phi}(w_1)(\phi(a))^0 \hat{\phi}(w_2) = \hat{\phi}(w_1)\hat{\phi}(w_2) = \hat{\phi}(w_1 w_2)$, and $\hat{\phi}(w_1 a^n a^m w_2) = \hat{\phi}(w_1)(\phi(a))^n (\phi(a))^m \hat{\phi}(w_2) = \hat{\phi}(w_1)(\phi(a))^{n+m}\hat{\phi}(w_2) = \hat{\phi}(w_1 a^{n+m} w_2)$. Hence, the map $\hat{\phi} : F[A] \to H$ defined by $\hat{\phi}([a^n][b^n] \cdots) = (\phi(a))^n (\phi(b))^m \cdots$ is well defined. Furthermore, it is clear that $\hat{\phi}_{|A} = \phi$.

Suppose that $\psi : G \to F[A]$ is an isomorphism. Then clearly, $\psi^{-1}(A)$ is a free basis for G.

Conversely, if G is free with free basis E, then G is isomorphic with $F[E]$. To see this, let $\theta : E \to F[E]$ be the inclusion map. Then there is a homomorphism $\phi : G \to F[E]$ that extends θ. Since θ is one to one, we have that $\theta^{-1} : E \to G$, and again since E is a free basis for $F[E]$, there is a homomorphism $\psi : F[E] \to G$ that extends θ^{-1}. Thus, $\psi\phi$ and $\phi\psi$ are extensions of $\theta^{-1}\theta = 1_E$ and $\theta\theta^{-1} = 1_E$, respectively, and consequently, by the uniqueness of the extensions (Ap.A.11), we have that $\psi\phi = id_G$ and $\phi\psi = id_{F[E]}$. Hence, ϕ and ψ are isomorphisms.

(Ap.A.15) *Corollary.* Every group is the homomorphic image of a free group.

Proof. Let G be a group. Then clearly, $F[G]$ can be mapped homomorphically onto G.

B. GROUP PRESENTATIONS

(Ap.B.1) *Definition.* A *group presentation* $\{A; R\}$ consists of an arbitrary set A and a set R of elements of $F[A]$. The group determined by the

presentation $\{A;R\}$ is the quotient group $F[A]/\langle R\rangle$, where $\langle R\rangle = \bigcap \{N \mid R \subset N$ and N is a normal subgroup of $F[A]\}$. The set R is called the set of *relations* for the presentation; an element $g \in \langle R\rangle$ is called a *consequence* of R; the set A is called the set of *generators*.

In the text, we frequently abuse the notation by referring to the "group $\{X;R\}$"; of course by this we mean $F[X]/\langle R\rangle$.

(Ap.B.2) Definition. A *presentation* of a group G consists of a group presentation $\{A;R\}$ and an isomorphism ψ from the group determined by the presentation $\{A;R\}$ onto G.

Often, we refer to $\{A;R\}$ as a presentation of a group G without specifically giving the isomorphism ψ.

(Ap.B.3) Exercise. Suppose that G is a group, $\{A;R\}$ is a group presentation, and $\phi : F[A] \rightarrow G$ is a homomorphism from $F[A]$ onto G such that ker $\phi = \langle R\rangle$. Show that ϕ determines a presentation of G (consider the following diagram, where p is the natural projection map and ψ is defined in the obvious manner).

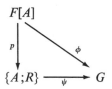

(Ap.B.4) Exercise. Show that every group has a presentation. [Hint: Use (Ap.A.15)].

(Ap.B.5) Examples.
1. $\{x; \}$ and $\{x,y;x\}$ are presentations for the infinite cyclic group.
2. $\{x,y;xyx^{-1}x^{-1}\}$ is a presentation of a free abelian group on two generators (note that this is not a free group).
3. $\{x;x^2\}$ is a presentation for $\mathbf{Z}_2$ (group of integers mod 2).
4. $\{x,y;xyx^{-1}y^{-1},x^2y^3\}$ and $\{x;x^6\}$ are presentation for $\mathbf{Z}_6$ (group of integers mod 6).
5. $\{x,y;x^2,y^2,(xy)^n\}$ is a presentation of the dihedral group of order $2n$.

(Ap.B.6) Remark. Occasionally, we find it convenient to abuse the notation slightly and write a relation in the form $x^2y^3 = 1$, $x^6 = 1$, etc.

(Ap.B.7) Definition. A presentation $\{X; R\}$ is *finitely generated* if and only if X is finite; $\{X;R\}$ is *finitely related* if and only if R is finite. $\{X;R\}$ is *finite* if and only if both X and R are finite. A group G is *finitely generated* if and only if there is a presentation of G that is finitely generated; G is *finitely related* if and only if there is a presentation of G that is finitely related; G is *finitely presented* if and only if there is a presentation of G that is finite.

(Ap.B.8) Remark. Finiteness does not work the way it should. There are examples of finitely presented groups that have subgroups that are not finitely presented. The enterprising reader might seek an example of a non-finitely generated subgroup of $\{x,y; \}$.

(Ap.B.9) Definition. Given two disjoint groups G and H, the *free product of G and H*, denoted by $G * H$, is the group generated by all the elements of G and all the elements of H, subject to the relations $g_1 g_2 g_3^{-1}$ for all $g_1, g_2, g_3 \in G$ such that $g_1 g_2 = g_3$, and $h_1 h_2 h_3^{-1}$ for all $h_1, h_2, h_3 \in H$ such that $h_1 h_2 = h_3$.

(Ap.B.10) Exercise. Show that if $\{x_1, \ldots, x_m; r_1, \ldots, r_n\}$ is a presentation of a group G and $\{y_1, \ldots, y_p; s_1, \ldots, s_q\}$ is a presentation of a group H, then $G * H$ is isomorphic to a group with presentation $\{x_1, \ldots, x_m, y_1, \ldots, y_p; r_1, \ldots, r_n, s_1, \ldots, s_q\}$.

C. HOMOMORPHISMS AND TIETZE'S THEOREM

A major problem that arises in connection with presentations is that of deciding when two presentations determine isomorphic groups.

It can be shown (Rabin [1958]) that there is no general solution to this problem. Partial solutions do exist, however, and in this section we discuss a few such results.

(Ap.C.1) Definition. Suppose that $\{X;R\}$ and $\{Y;S\}$ are presentations. Then a homomorphism $f : F[X] \to F[Y]$ is a *presentation mapping* between the presentations $\{X;R\}$ and $\{Y;S\}$ (denoted by $f : \{X;R\} \to \{Y;S\}$) if and only if for all $r \in R$, $f(r) \in \langle S \rangle$.

(Ap.C.2) Theorem. Every presentation map $f : \{X;R\} \to \{Y;S\}$ determines a unique homomorphism $f_* : F[X]/\langle R \rangle \to F[Y]/\langle S \rangle$ such that the following diagram is commutative. (p_x and p_y are the natural projections.)

$$F[X] \xrightarrow{\quad f \quad} F[Y]$$

$$\downarrow p_x \qquad\qquad p_y \downarrow$$

$$F[X]/\langle R\rangle \xrightarrow{\quad f_* \quad} F[Y]/\langle S\rangle$$

Proof. Note that $f_* = p_y f p_x^{-1}$ is a well defined map and is the desired homomorphism.

(Ap.C.3) Exercise. Let $\{X;R\}$ be a group presentation and G be a group. Show that a homomorphism from $f : F[X]/\langle R\rangle \to G$ is determined by mapping generators of $F[X]$ to elements of G in such a way that if $r = x_1^{n_1} x_2^{n_2} \cdots x_p^{n_p} \in R$ and $f(x_i^{n_i}) = g_i^{n_i}$, then $g_1^{n_1} g_2^{n_2} \cdots g_p^{n_p} = 1_G$.

(A.C.4) Definition. Presentation maps $f_1, f_2 : \{X;R\} \to \{Y;S\}$ are *homotopic* (denoted by $f_1 \simeq f_2$) if and only if for all $x \in X$ we have that $f_1(x)f_2(x^{-1}) \in \langle S\rangle$.

(Ap.C.5) Exercise. Suppose that $f, \hat{f} : \{X;R\} \to \{Y;S\}$ and $g, \hat{g} : \{Y;S\} \to \{Z;T\}$ are presentation maps. Show that

(i) $f \simeq \hat{f}$ if and only if $f_* = \hat{f}_*$;
(ii) if $f \simeq \hat{f}$ and $g \simeq \hat{g}$, then $gf \simeq \hat{g}\hat{f}$;
(iii) for each homomorphism $\phi : F[X]/\langle R\rangle \to F[Y]/\langle S\rangle$, there is a presentation map $f : \{Y;R\} \to \{Y;S\}$ such that $f_* = \phi$, and furthermore, any two such presentation maps are homotopic;
(iv) $(gf)_* = g_* f_*$;
(v)   if *id*: $\{X;R\} \to \{X;R\}$ is a presentation map, then $id_* = id_{F[X]/\langle R\rangle}$.

(Ap.C.6) Definition. Presentations $\{X;R\}$ and $\{Y;S\}$ are of the same *type* if and only if there are presentation maps $f : \{X;R\} \to \{Y;S\}$ and $g : \{Y;S\} \to \{X;R\}$ such that $gf \simeq id$ and $fg \simeq id$. In this case, f (or g) is called a *presentation or homotopy equivalence*.

(Ap.C.7) Theorem. Presentations $\{X;R\}$ and $\{Y;S\}$ are of the same type if and only if $F[X]/\langle R\rangle$ is isomorphic to $F[Y]/\langle S\rangle$.

Proof. If $\phi : F[X]/\langle R\rangle \to F[Y]/\langle S\rangle$ is an isomorphism, then so is ϕ^{-1}. Use part (iii) of (Ap.C.5) to obtain presentation maps f and g. Since $\phi^{-1}\phi$ and $\phi\phi^{-1}$ are identities, we have that $gf \simeq id$ and $fg \simeq id$.

Conversely, suppose that f and g are presentation maps such that $gf \simeq id$ and $fg \simeq id$. Then by parts (i), (iv), and (v) of (Ap.C.5), we have that $g_* f_* = id_*$ and $f_* g_* = id_*$. Hence, f_* and g_* are isomorphic.

(Ap.C.8) Definition. Suppose that $\{X;R\}$ is a presentation. Let $s \in \langle R \rangle$ and suppose that $\{Y;S\}$ is a presentation with $Y = X$ and $S = R \cup \{s\}$. Then the identity map $I : F[X] \to F[Y]$ is a presentation mapping $I : \{X;R\} \to \{Y;S\}$, and similarly $I' : F[Y] \to F[X]$ is a presentation map $I' : \{Y;S\} \to \{X;R\}$. The mappings I and I' constitute *Tietze equivalences of type I and I'*, respectively.

(Ap.C.9) Theorem. The mappings I and I' are presentation equivalences, and hence $F[X]/\langle R \rangle$ is isomorphic to $F[Y]/\langle S \rangle$.

   Proof. Consider the following diagram.

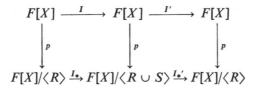

Here, I and I' are identity maps. It is easy to see that they are both presentation maps. Furthermore, $I'I$ is the identity presentation map from $\{X;R\}$ onto $\{X;R\}$. Therefore, by (Ap.C.5, parts (iv) and (v)), we have that $I'_* I_* = id|_{F[X]/\langle R \rangle}$. Similarly, $I_* I'_* = id|_{F[X]/\langle R \cup \{s\} \rangle}$.

(Ap.C.10) Remark. The import of (Ap.C.9) is that given a group presentation, one can add a relation that is a consequence of the others, or delete a relation that is a consequence of the others, without significantly changing the group of the presentation.

(Ap.C.11) Example. The group $\{x;x^3\}$ is isomorphic to $\{x;x^3,x^6\}$.

(Ap.C.12) Definition. Let $\{X;R\}$ be a group presentation. Suppose that y is an element of some set and that $y \notin X$. Let $\xi \in F[X]$, and let $\{Y;S\}$ be a presentation, where $Y = X \cup \{y\}$ and $S = R \cup \{y\xi^{-1}\}$. Then $II : F[X] \to F[Y]$, defined by $II(x) = x$, yields a presentation map $II : \{X;R\} \to \{Y;S\}$, and $II' : F[Y] \to F[X]$, defined by

$$II'(z) = \begin{cases} \xi & \text{if} \quad z = y \\ z & \text{if} \quad z \in X \end{cases}$$

yields a presentation map $II' : \{Y;S\} \to \{X;R\}$. The mappings II and II' constitute *Tietze equivalences of types II and II'*, respectively.

(Ap.C.13)   *Theorem.*   The mappings *II* and *II'* are presentation equivalences, and hence $F[X]/\langle R\rangle$ and $F[Y]/\langle S\rangle$ are isomorphic.

Proof.   Note that $II'II(x) = id(x)$. Also we have that

$$II\,II'(z) = \begin{cases} z & \text{if} & z \in X \\ \xi & \text{if} & z = y \end{cases}$$

and since $\xi y^{-1} = (y\xi^{-1})^{-1} \in \langle S\rangle$, it follows that $II\,II' \simeq id$.

(Ap.C.14)   *Remark.*   The import of (Ap.C.13) is that a new generator may be added to a group presentation, provided that a new relation is added that expresses the new generator as a word in the old ones. Similarly, one can delete a generator if it is expressible as a word in other generators.

(Ap.C.15)   *Example.*   We have that

$$\{x,y;xyx^{-1}y^{-1},x^2y^3\}$$

$$\overset{\text{notation}}{=} \quad \{x,y;xy = yx, x^2 = 1, y^3 = 1\}$$

$$\overset{II}{\simeq} \{x,y,z;xy = yx, x^2 = 1, y^3 = 1, z = xy\}$$

$$\overset{\text{I and I'}}{\simeq} \{x,y,z;z = yx, x = x^{-1}, y = y^{-2}, z = xy\}.$$

Now since $z = yx = y^{-2}x^{-1} = y^{-1}(y^{-1}x^{-1}) = y^{-1}z^{-1}$, it follows that

$$\{x,y,z;z = yx, x = x^{-1}, y = y^{-2}, z = xy\}$$

$$\simeq \{x,y,z;z^{-2} = y, x = x^{-1}, y = y^{-2}, z = xy\}$$

$$\overset{II'}{\simeq} \{x,z;x = x^{-1}, z^{-2} = z^4, z = xz^{-2}\}$$

$$= \{x,z;x = x^{-1}, z^6 = 1, z^3 = x\}$$

$$\overset{II'}{\simeq} \{z;z^3 = z^3, z^6 = 1\} \overset{I'}{\simeq} \{z;z^6 = 1\}.$$

Although we do not use the following theorem, it is of fundamental interest in presentation theory. A proof be found in Magnus, Karass, and Solitar [1966].

(Ap.C.16)   *Theorem (Tietze Theorem).*   Suppose that $\{X;R\}$ and $\{Y;S\}$ are finite presentations for isomorphic groups. Then there is a finite sequence of Tietze Equivalences *I, I', II, II'* that change one presentation into the other.

D.   DIRECT LIMITS OF GROUPS

(Ap.D.1)   Definition.   A *direct system of groups*, $(G_\alpha, f_{\alpha\beta}, D)$, consists of a directed set D, a family of groups $\{G_\alpha\}_{\alpha \in D}$ and a family of homomorphisms $\{f_{\alpha\beta} \mid \alpha,\beta \in D$ and $\alpha \le \beta\}$ such that

  (i) $f_{\alpha\beta} : G_\alpha \to G_\beta$,
  (ii)   if $\alpha \le \beta \le \gamma$, then $f_{\alpha\gamma} = f_{\beta\gamma}f_{\alpha\beta}$, and
  (iii) $f_{\alpha\alpha} = id$ for each α.

Let $\underset{\alpha \in D}{\times}\, G_\alpha$ be the *direct product* of the groups G_α, i.e., $\underset{\alpha \in D}{\times}\, G_\alpha = \{\{x\}_{\alpha \in D} \in \underset{\alpha \in D}{\prod}\, G_\alpha \mid x_\alpha \in G_\alpha$ and $x_\alpha = 1_{G_\alpha}$ for all but a finite number of $\alpha\}$, where the group operation is defined coordinatewise.

(Ap.D.2)   Exercise.   Show that for each group H and for each family of homomorphisms $\{h_\alpha : G_\alpha \to H \mid \alpha \in D\}$ such that $h_\alpha(g_\alpha)h_\beta(g_\beta) = h_\beta(g_\beta)h_\alpha(g_\alpha)$ for all $\alpha,\beta \in D$ and for all $g_\alpha \in G_\alpha$ and $g_\beta \in G_\beta$, there is a unique homomorphism $h : \underset{\alpha \in D}{\times}\, G_\alpha \to H$ such that for each $\alpha \in D$, $hi_\alpha = h_\alpha$, where i_α is the natural injection from G_α into the direct product.

(Ap.D.3)   Definition.   Let K be the smallest normal subgroup of $\underset{\alpha \in D}{\times}\, G_\alpha$ generated by $\{i_\alpha(g)^{-1}i_\beta f_{\alpha\beta}(g) \mid \alpha,\beta \in D, \alpha \le \beta, g \in G_\alpha\}$. The *direct limit of the system* $(G_\alpha, f_{\alpha\beta}, D)$ is $(\underset{\alpha \in D}{\times}\, G_\alpha)/K$ and is denoted by $\varinjlim G_\alpha$.

(Ap.D.4)   Theorem.   Suppose that $(G_\alpha, f_{\alpha\beta}, D)$ is a direct system of groups, and let $\varinjlim G_\alpha$ be the corresponding direct limit. For each α, define a homomorphism $f_\alpha : G_\alpha \to \varinjlim G_\alpha$ by $f_\alpha = pi_\alpha$, where p is the natural homomorphism (we denote $p(g)$ by $[g]$ or $[g]_K$) from $\underset{\alpha \in D}{\times}\, G_\alpha$ onto $\varinjlim G_\alpha = (\underset{\alpha \in D}{\times}\, G_\alpha)/K$, and where K is defined as above. Then whenever $\alpha < \beta$ the following diagram is commutative, and $\varinjlim G_\alpha$ is generated by $\underset{\alpha \in D}{\bigcup}\, \mathrm{Im}\, F_\alpha$.

Proof. Suppose that $x_\alpha \in G_\alpha$. By the definition of K, we have that $i_\alpha(x_\alpha)^{-1} i_\beta f_{\alpha\beta}(x_\alpha) \in K$. Hence, $[i_\alpha(x_\alpha)] = [i_\beta f_{\alpha\beta}(x_\alpha)]$, which implies that $f_\alpha(x_\alpha) = f_\beta(f_{\alpha\beta}(x_\alpha))$.

It is clear that $\varinjlim G_\alpha$ is generated by $\bigcup_{\alpha \in D} \operatorname{Im} f_\alpha$, since $\underset{\alpha \in D}{\times} G_\alpha$ is generated by $\bigcup_{v \in D} \operatorname{Im} i_\alpha$.

(Ap.D.5)   Examples.
   1.   Let $D = \mathbf{N}$. For each $n \in \mathbf{N}$, let $G_n = \mathbf{Z}$ and set $f_{ni}(z) = z$ for all $n \leq i$. Then $\varinjlim G_n$ is isomorphic to $\mathbf{Z}$.
   2.   Let $D = \mathbf{N}$. For each $n \in \mathbf{N}$, let $G_n = \{m2^n \mid m \in \mathbf{Z}\}$ with group operation $+$, and set $f_{ni}(z) = 2^{i-n}z$ for $n < i$. Then $\varinjlim G_n$ is isomorphic to $\mathbf{Z}$.
   3.   Let $D = \mathbf{N}$. For each $n \in \mathbf{N}$, let $G_n = \mathbf{Z}$ and set $f_{ni}(z) = 2^{i-n}z$ for $n < i$. Find $\varinjlim G_n$.

(Ap.D.6)   Exercise. If A, B, and C are groups, and $f : A \to B$ and $g : A \to C$ are homomorphisms such that f is surjective, then there is a homomorphism $h : B \to C$ such that $hf = g$ if and only if $\operatorname{Ker} f \subset \operatorname{Ker} g$.

(Ap.D.7)   Theorem. Let $(G_\alpha, f_{\alpha\beta}, D)$ be a direct system of groups and $\varinjlim G_\alpha$ be the corresponding direct limit. Suppose that H is a group and that $\{h_\alpha : G_\alpha \to H \mid \alpha \in D\}$ is a family of homomorphisms such that for each $\alpha \leq \beta$, the following diagram is commutative. Suppose further that H is generated by $\bigcup_{\alpha \in D} \operatorname{Im} h_\alpha$.

Then there is a unique homomorphism h from $\varinjlim G_\alpha$ onto H such that for each $\alpha \leq \beta$, the following tetrahedral diagram is commutative.

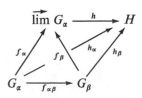

Proof. Since $\varinjlim G_\alpha$ is generated by $\bigcup\limits_{\alpha \in D} \operatorname{Im} f_\alpha$, it suffices to define h

for the generating set $\{f_\alpha(g) \mid \alpha \in D,\ g \in G_\alpha\}$. Let $f_\alpha(g)$ be a generator of $\varinjlim G_\alpha$. Since the following diagram is to be commutative, it follows that h must be defined by $h(f_\alpha(g)) = h_\alpha(g)$.

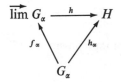

We show that if $k \in K$, then $h(k) = 1_H$, and hence h may be extended homomorphically to all of $\varinjlim G_\alpha$. Let k be a generator of K, i.e., $k = i_\alpha(g_\alpha)^{-1} i_\beta f_{\alpha\beta}(g_\alpha)$. Then $h(k) = 1_H$ if and only if $h(f_\alpha(g_\alpha)) = h(f_\beta f_{\alpha\beta}(g_\alpha))$. By the definition of h, we have that $h(f_\alpha(g_\alpha)) = h_\alpha(g_\alpha)$, and $h(f_\beta f_{\alpha\beta}(g_\alpha)) = h_\beta(f_{\alpha\beta}(g_\alpha))$. It follows from the commutativity of the following diagram that $h_\beta(f_{\alpha\beta}(g_\alpha)) = h_\alpha(g_\alpha)$, and hence h may be extended to $\varinjlim G_\alpha$.

The remaining properties of h are easily established.

(Ap.D.8)   **Theorem.**   Suppose that $(G_\alpha, f_{\alpha\beta}, D)$ is a direct system of groups and that $\varinjlim G_\alpha$ is the corresponding direct limit. Suppose that H is a group and $\{h_\alpha : G_\alpha \to H \mid \alpha \in D\}$ is a family of homomorphisms such that H is generated by $\bigcup\limits_{\alpha \in D} \operatorname{Im} h_\alpha$ and that for all $\alpha \le \beta$ the diagram below is commutative.

Suppose further that whenever $\hat{G}$ is a group and $\{\hat{h}_\alpha : G_\alpha \to \hat{G} \mid \alpha \in D\}$ is a family of homomorphisms such that the following diagram is commutative whenever $\alpha < \beta$.

Then there is a unique homomorphism $h : H \to \hat{G}$ such that the following diagram is commutative.

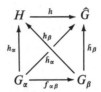

Then H is isomorphic with $\varinjlim G_\alpha$.

Proof. The proof is almost identical to that of (12.F.8).

(Ap.D.9) Theorem. Suppose that for each $n \in \mathbf{Z}^+$, $G_n = \{x_1, x_2, \ldots, x_{m_n};$ $r_1, r_2, \ldots, r_{p_n}\}$ and that for each n, $m_n \leq m_{n+1}$ and $p_n \leq p_{n+1}$. Suppose further that $f_{n+1,n} : G_n \to G_{n+1}$ defined by $f_{n+1,n}(x_i) = x_i$ for $1 \leq i \leq m_n$ is a 1–1 homomorphism for each n. Let $G = \{x_1, x_2, \ldots ; r_1, r_2, \ldots\}$. Then $\varinjlim G_n = G$.

Proof. The theorem follows easily from (Ap.D.8).

BIBLIOGRAPHY

Aleksandrov, P. (1927). Simpliziale Approximationen in der allgemeinen Topologie, *Math. Ann. 96*, 489–511.

Aleksandrov, P. (1939). Bikompakte Erweiterungen topologische Räume, *Mat. Sb. 5*, 403–423.

Aleksandrov, P., and Uryson, P. (1923). Une condition nécessaire et suffisante pour qu'une classe (L) soit une classe (D), *C. R. Acad Sci. Paris 177*, 1274–1277.

Aleksandrov, P., and Uryson, P. (1929). Memoire sur les espaces topologiques compactes, *Verh. Ak. Wet. Amst. 14*, 1–96.

Alexander, J. W. (1924). On the subdivision of 3-space by a polyhedron, *Proc. Nat. Acad. Sci. U.S.A. 10*, 6–8.

Alexander, J. W. (1924). An example of a simply connected surface bounding a region which is not simply connected, *Proc. Nat. Acad. Sci. U.S. 10*, 8–10.

Anderson, R. D. (1958). One-dimensional continuous curves and a homogeneity theorem. *Ann. of Math. 68*, 1–16.

Anderson, R. D. (1966). Hilbert Space is Homeomorphic to the Countable Infinite Product of Lines, *Bull. Amer. Math. Soc. 72*, 515–519.

Anderson, R. D., and Bing, R. H. (1968). A complete elementary proof that Hilbert space is homeomorphic to the countable mfinite product of lines. *Bull. Am. Math. Soc. 74*, 771–792.

Anderson R. D., and Choquet, G. (1959). A plane continuum no two of whose nondegenerate subcontinua are homeomorphic: an application of inverse limits, *Proc. Am. Math. Soc. 10*, 347–353.

Antoine, L. (1921). Sur l'homeomorphie de deux figures et de leurs voisinages, *J. Math. pures et appl. (8) 4*, 221–325.

502 Bibliography

Arens, R., and Dugundji, J. (1951). Topologies for function spaces, *Pac. J. Math. 1*, 5–31.
Armentrout, S. (1966). Monotone decompositions of E^3, pp. 1–27, Topology Seminar, Wisconsin, 1965. Annals of Mathematical Studies No. 60, Princeton University Press, Princeton.
Armentrout, S. (1967). Topology Conference, Arizona State Univ., Tempe, Arizona. 22–35.
Armentrout, S. (1968). Concerning cellular decompositions of 3-manifolds that yield 3-manifolds, *Trans. Am. Math. Soc. 133*, 307–332.
Armentrout, S. (1970). A decompostition of E^3 into straight arcs and singletons, *Diss. Math. Rozp. Mat. 68*, 46.
Armentrout, S., and Price, T. (1969). Decompositions into compact sets with UV-properties, *Trans. Am. Math. Soc. 141*, 433–442.
Bellamy, D. (1971). Mappings of indecomposable continua, *Proc. Am. Math. Soc. 30*, 179–180.
Bing, R. H. (1946). The Kline sphere characterization problem, *Bull. Am. Math. Soc. 52*, 644–653.
Bing, R. H. (1947). Extending a metric, *Duke Math. J. 14*, 511–519.
Bing, R. H. (1948). A homogeneous indecomposable plane continuum, *Duke Math J. 15*, 729–742.
Bing, R. H. (1949), A convex metric for a locally connected continuum, *Bull. Amer. Math. Soc. 55*, 812–819.
Bing, R. H. (1951a). Snake-like continua, *Duke Math. J. 18*, 653–663.
Bing, R. H. (1951b). Concerning hereditarity indecomposable continua, *Pac. J. Math. 1*, 43–51.
Bing, R. H. (1954). Locally tame sets are tame, *Ann. of Math. 59*, 145–158.
Bing, R. H. (1957a). A decomposition of E^3 into points and tame arcs such that the decomposition space is topologically different from E^3, *Ann. of Math. 65*, 484–500.
Bing, R. H. (1957b). Upper semi-continuous decompositions of E^3, *Ann. of Math. 65*, 363–374.
Bing, R. H. (1959a). The Cartesian product of a certain non-manifold and a line is E^4, *Ann. of Math. 70*, 399–412.
Bing, R. H. (1959b). An alternative proof that 3-manifolds can be triangulated, *Ann. of Math. 69*, 37–65.
Bing, R. H. (1962). Pointlike decompositions of E^3, *Fund. Math. 50*, 431–453.
Bing, R. H. (1963). Approximating surfaces from the side, *Ann, of Math. 77*, 145–192.
Bing, R. H. (1965). A translation of the normal Moore space conjecture. *Proc. Am. Math. Soc. 16*, 612–619.
Birkhoff, G. (1937). Moore-Smith convergence in general topology, *Ann. of Math. 38*, 39–56.
Borsuk, K. (1931). Sur les retracts, *Fund. Math. 17*, 152–170.
Borsuk, K. (1931). Quelques théorèmes sur les ensembles unicohérents, *Fund. Math.* 171–209.

Borsuk, K. (1932). Uber eine Klasse von lokal zusammenhängenden Räumen, *Fund. Math. 19*, 220–242.

Borsuk, K. (1933a). Drei Sätze über die n-dimensionale Euklidische Sphäre, *Fund. Math. 20*, 177–190.

Borsuk, K. (1933b). Über die Abbildungen der metrischen kompaktam Räume auf die kreislinie, *Fund. Math. 20*, 224–231.

Brahana, H. R. (1922). Systems of Circuits on two-dimensional manifolds, *Ann. of Math. 23*, 144–168.

Brody, E. J. (1960). The topological classification of the lens spaces, *Ann. of Math. 71*, 163–184.

Brouwer, L. E. J. (1910). Beweis des Jordanschen Kurvensatz, *Math. Ann. 69*, 169–175.

Brouwer, L. E. J. (1912). Beweis des Jordanschen Sätze für den n-dimensionalen Räum, *Math, Ann. 71*, 314–319.

Brouwer, L. E. J. (1912). Beweis des Invarianz des n-dimemsionalen Gebietes, *Math. Ann. 71*, 55–56.

Brown, M. (1960a). Some applications of an approximation theorem for inverse limits, *Proc. Am. Math. Soc. 11*, 478–483.

Brown, M. (1960b). A proof of the generalized Schoenflies theorem, *Bull. Am. Math. Soc. 66*, 74–76.

Brown, M. (1961). The monotone union of open n-cells is an open n-cell, *Proc. Am. Math. Soc. 12*, 812–814.

Brown, M. (1962a). Locally flat embeddings of topological manifolds, *Topology of 3-manifolds and Related Topics*. Englewood Cliffs, N.J.: Prentice-Hall, pp. 83–91.

Brown, M. (1962b). A mapping theorem for untriangulated manifold, *Topology of 3-manifolds and Related Topics*. Englewood Cliffs, N.J.: Prentice-Hall, 92–94.

Brown, R. F. (1974). Elementary consequences of the noncontractibility of the circle, *Am. Math. Monthly 81*, 247–252.

Burgess, C. E. (1959). Chainable continua and indecomposibility, *Pac. J. Math. 9*, 653–659.

Burgess, C. E. (1961). Homogeneous continua which are almost chainable, *Can. J. Math 13*, 519–528.

Burgess, C. E., and Cannon, J. W. (1971). Embeddings of surfaces in E^3, *Rocky Mt. J. Math. 1*, 259–344.

Cantor, G. (1883). Uber unendliche lineare Punktmannigfaltigkeiten, *Math. Ann. 21*, 545–591.

Cartan, H. (1937a). Théorie des Filtres, *C. R. Acad. Sci. Paris 205*, 595–598.

Cartan, H. (1937b). Filtres et Ultrafiltres, *C. R. Acad. Sci. Paris 205*, 777–779.

Casler, B. G. (1965). An imbedding theorem for connected 3-manifolds with boundary, *Proc. Am. Math. Soc. 16*, 559–566.

Catlin, D. E. (1968). A short proof of the nest characterization of compactness, *Am. Math. Monthly 75*, 751.

Chittenden, E. W. (1927). On the metrization problem and related problems in the theory of abstract sets, *Bull. Am. Math. Soc. 33*, 13–34.

Christenson, C. O. Short note on a paper of Porter (*to appear*).

Christenson, C. O., and Osborne, R. P. (1968). Pointlike subsets of a manifold, *Pac. J. Math. 24*, 431–435.

Cobb, J. I., and Voxman, W. L. (1972). Some fixed point results for UV decompositions of compact metric spaces, *Proc. Am. Math. Soc. 33*, 156–160.

Cohen, D. E. (1954). Spaces with weak topology, *Quart. J. Math. Oxford 5*, 77–80.

Comfort, W. W. (1969). A short proof of Marczewski's Separability theorem. *Am. Math. Monthly 76*, 1041–1042.

Crowell, R. H., and Fox, R. H. (1963). *Introduction to Knot Theorey*, Boston: Ginn and Company.

Dieudonné, J. (1944). Une généralization des espaces compacts, *J. Math. pures et appl. 23*, 65–76.

Dold, A. (1972). *Lectures on Algebraic Topology*, Berlin: Springer Verlag.

Dold, A. (1963). Partitions of unity in the theory of fibrations, *Ann. of Math. 78*, 223–255.

Dugundji, J. (1966). *Topology*, Boston: Allyn and Bacon.

Dyer, E. (1956). A fixed point theorem, *Proc. Am. Math. Soc. 7*, 662–672.

Dyer, E. (1956). Certain transformations which lower dimension, *Ann. of Math. 63*, 15–19.

Eaton, W. T., and Pixley, C. (1975). S^1 cross and UV^∞ decomposition of S^3 yields $S^1 \times S^3$. Geometric Topology, 1974. Berlin: Springer–Verlag *Lecture Notes 438*, 166–194.

Edwards R. D., and Miller R. T. (to appear). Cell-like closed 0-dimensional decompositions of R^3 are R^4 factors.

Fox, R. H. (1943). On homotopy type and deformation retracts, *Ann. of Math. 44*, 40–50.

Fox, R. H. (1945). On topologies for function spaces, *Bull. Am. Math. Soc. 51*, 429–432.

Fox, R. H., and Artin, E. (1948). Some wild cells and spheres in three-dimensional space, *Ann. of Math. 49*, 979–990.

Fraser, R. B. Jr. (1972). A new characterization of Peano continua, *Prace Mat. 16*, 247–248.

Frink, A. H. (1937). Distance functions and the metrization problem, *Bull. Am. Math. Soc. 43*, 133–142.

Fugate, J. B. (1965). *Chainable Continua*, Topology Seminar, Univ. of Wisconsin. Princeton, N.J.: Princeton Univ. Press. 129–135.

Furch, R. (1930). Polyedrale Gebilde verschiedenes Metrik, *Math. Zeit. 32*, 512–544.

Greever, J. (1967). *Theory and Examples of Point-set Topology*, Belmont, California: Brooks/Cole.

Hahn, H. (1914). Mengentheoretische Charakterisierung der stetigen Kurve, *Akad. Wiss. Wien, Math. -Nat. Klasse 123*, 1–57.

Halmos, P. R. (1960). *Naive Set Theory*, New York: Van Nostrand.

Hamilton, O. H. (1951). A fixed point theorem for pseudo-arcs and certain other metric continua. *Proc. Am. Math. Soc. 2*, 173–174.

Hanner, O. (1951). Some theorems on absolute neighborhood retracts, *Ark. Mat. 1*, 389–408.

Hausdorff, F. (1930). Erweiterung einer Homoömorphie, *Fund. Math. 16*, 353–360.

Hawkins, T. (1970). *Lebesgue's Theory of Integration*, Madison: Univ. of Wisconsin Press, p. 37.

Heath, R. W. (1964). Screenability, pointwise paracompactness and metrization of Moore-spaces. *Can. J. Math. 16*, 763–770.

Henderson, G. W. (1964). The pseudo-arc as an inverse limit with one binding map, *Duke Math. J. 31*, 421–425.

Hewitt, E. (1960). The role of compactness in analysis, *Am. Math. Monthly 67*, 499–516.

Hilton, P. J., and Wylie, S. *Homology Theory*, London: Cambridge Univ. Press.

Hirsch, M. W., and Zeeman, E. C. (1966). Engulfing, *Bull. Am. Math. Soc. 72*, 113–115.

Hu, S. (1959). *Homotopy Theory*, New York: Academic Press.

Hudson, J. F. P. (1969). *Piecewise Linear Topology*, New York: Benjamin.

Hurewicz, W. (1930). Über oberhalb-stetige Zerlegungen von Punktmengen in Kontinua, *Fund. Math. 15*, 57–60.

Hurewicz, W. (1935; 1936). Beiträge zur Topologie des Deformationen I-IV, *Proc. Ak. Wet. Amst. 38*, (1935) 112–119, 521–528; and *39* (1936) 117–126, 215–224.

Jolley, R. F., and Rogers, J. T. Jr. (1970). Inverse limit spaces defined by only finitely many distinct bonding maps, *Fund. Math. 68*, 117–120.

Jones, F. B. (1937). Concerning normal and completely normal spaces, *Bull. Am. Math. Soc. 43*, 671–677.

Jones, F. B. (1951). Certain homogeneous unicoherent indecomposible continua, *Proc. Am. Math. Soc, 2*, 855–859.

Kampen, E. R. van (1933). On the connection between the fundamental group of some related spaces, *Am. J. Math. 55*, 261–267.

Kelley, L. J. (1950a). Convergence in Topology, *Duke Math. J. 17*, 277–283.

Kelley, J. L. (1950b). The Tychonoff product theorem implies the axiom of choice, *Fund. Math. 37*, 75–76.

Kelley, J. L. (1955). *General Topology*, New York: Van Nostrand

Kirby, R. C., and Siebenmann, L. C. (1969). On the triangulation of manifolds and the Hauptvermutung, *Bull. Am. Math. Soc. 75*, 742–749.

Kirby, R. C., Seibenmann, L. C., and Wall C. T. C. (1969). The annulus conjecture and triangulation, *Not. Am. Math. Soc. 16*, 432.

Knaster, B., and Kuratowski, C. (1921). Sur les ensembles connexes, *Fund. Math. 2*, 206–255.

Knight, C. J. (1964). Box topologies, *Ouart. J. Math. Oxford 15*, 41–54.

Kresimar, D., and Mardesic, S. (1968). A necessary and sufficient condition for the n-dimensionality of inverse limits, Proceedings of the International Symposium of Topology and its Applications, Belgrade, pp. 124–129.

Levine, N. (1960). Remarks on uniform continuity in metric spaces, *Am. Math. Monthly 67*, 562–563.

Magnus, W., Karrass, A., and Solitar, D. (1966). *Combinatorial Group Theory*, New York: Wiley.

Markov, A. A. (1960). Unsolvability of the homeomorphism problem, Proceedings of the International Congress of Mathematicians, 1958, Cambridge Univ. Press, p. 300.

Massey, W. S. (1967). *Algebraic Topology: An Introduction*, New York: Harcourt, Brace, and World.

Maunder, C. R. F. (1970). *Algebraic Topology*, London: Van Nostrand Reinhold.

Mazurkiewicz, S. (1920). Sur les lignes de Jordan, *Fund. Math. 1*, 166–209.

McAuley, L. F. (1962). Upper semicontinuous decompositions of E^3 into E^3 and generalizations to metric spaces, *Topology of 3-manifolds and Related Topics*, Englewood Cliffs, New Jersey: Prentice-Hall, pp. 21–26.

Milnor, J. (1961). Two complexes which are homeomorphic but combinatorially distinct, *Ann. of Math. 74*, 575–590.

Moise, E. E. (1948). An indecomposable plane continuum which is homeomorphic to each of its nondegenerate subcontinua, *Trans. Am. Math. Soc. 63*, 581–594.

Moise, E. E. (1949). Grille decomposition and convexification theorems for compact metric locally connected continua, *Bull. Am. Math. Soc. 55*, 1111–1121.

Moise, E. E. (1952a). Affine structures in 3-manifolds II. Positional properties of 2-spheres, *Ann. of Math. 55*, 172–176.

Moise, E. E. (1952b). Affine structures in 3-manifolds V. The triangulation theorem and Hauptvermutung, *Ann. of Math. 56*, 96–114.

Moise, E. E. (1954). Affine structures in 3-manifolds VIII; invariance of knot types, local tame embeddings, *Ann. of Math. 59*. 159–170.

Monk, J. D. (1969). *Introduction to Set Theory*, New York: McGraw-Hill.

Moore, R. L. (1925). Concerning upper semi-continuous collections of continua, *Trans. Am. Math. Soc. 27*, 416–428.

Moore, E. H., and Smith H. L. (1922). A general theory of limits, *Am. J. Math. 44*, 102–121.

Munkres, J. R. (1975). *A First Course in Topology*. Prentice-Hall, Englewood Cliffs, N. J.

Nadler, S. B. Jr. (1970). A note on inverse limits of finite spaces, *Can. Math. Bull. 13*, 69–70.

Nadler, S. B. Jr. (1973). The indecomposibility of the dyadic solenoid, *Am. Math. Monthly 80*, 677–679.

Nagami, K. (1970). *Dimension Theory*, New York: Academic Press.

Nagata, J. (1950). On a necessary and sufficient condition of metrizability, *J. Inst. Poly. Osaka City Univ. 1*, 93–100.

Newman, M. H. A. (1964). *Elements of the topology of plane sets of points*, London: Cambridge Univ. Press.

Newman, M. H. A. (1966). The engulfing theorem for topological manifolds, *Ann. of Math. 84*, 555–571.

Olum, P. (1958). Nonabelian cohomology and van Kampen's theorem, *Ann. of Math. 68*, 658–668.

Papakyriakopoulos, C. D. (1943). A new proof of the invariance of the homology groups of a complex, *Bull. Soc. Math. Gréce 22*, 1–54.

Pittman, C. R. (1970), An elementary proof of the triod theorem, *Proc. Am. Math. Soc. 25*, 919.

Price, T. M. (1966). A necessary condition that a cellular upper semicontinuous decomposition of E^n yield E^n, *Trans. Am. Math. Soc. 122*, 427–435.

Price. T. (1969). Mimeographed notes, Iowa City: Univ. of Iowa.

Rabin, M. O. (1958). Recursive unsolvability of group theortic problems, *Ann. of Math. 67*, 172–194.

Radó, T. (1925). Über den Begriff des Riemannschen Fläche, *Acta Lit. Sci. Szegad 2*, 101–121.

Reidemeister, K. (1928). Über Knotengruppen, *Abh. Math. Sem. Univ. Hamburg 6*, 56–64.

Reidemeister, K. (1935). Homotopieringe und Linsenräume, *Abh. Math. Sem. Univ. Hamburg 11*, 102–109.

Rolfson, D. (1970). Characterizing the 3-cell by its metric, *Fund. Math. 68*, 215–223.

Rudin, M. E. (1969). A new proof that metric spaces are paracompact, *Proc. Am. Math. Soc. 20*, 603–605.

Rushing, T. B. (1973). *Topological Embeddings*, New York: Academic Press.

Schori, R. (1965). A universal snake-like continuum, *Proc. Am. Math. Soc. 16*, 1313–1316.

Schori, R., and West, J. E. (1972). 2^I is homeomorphic to the Hilbert cube, *Bull. Am. Math. Soc. 78*, 402–406.

Schubert, H. (1964). *Topologie*, Stuttgart: B. G. Teubner.

Seifert, H. (1931). Konstrucktion dreidimendionaler geschlossener Räume, *Ber. Sachs. Ak. Wiss. 83*, 26–66.

Seifert, H., and Threlfall, W. (1934). *Lehrbuch der Topologie*, Leipzing: B.G.Teubner [New York: Chelsea (1945)].

Siebenmann, L. (1972). Approximating cellular maps by homeomorphisms, *Topology 11*, 271–294.

Sierpinski, W. (1916). Sur une courbe cantorienne qui contient une image biunevoque et continuede toute courbe donnée, *C. R. Sci. Paris 162*, 629.

Singer, I. M., and Thorpe, J. A. (1967). *Lecture Note on Elementary Topology and Geometry*, Glenview, Illinois: Scott Foresman.

Smirnov, Yu. M. (1953) On metrization of topological spaces, *Amer. Math. Soc. Transl. Ser. 1*, 91.

Spanier, E. H. (1966). *Algebraic Topology*, New York: McGraw-Hill.

Steen, A. L., and Seebach, J. A. Jr. (1970). *Counterexamples in Topology*, New York: Holt, Rinehart and Winston.

Stone, A. H. (1948). Paracompactness and product spaces, *Bull. Am. Math. Soc. 54*, 977–982.

Stone, A. H. (1949). Incidence relations in unicoherent spaces, *Trans. Am. Math. Soc. 65*, 427–447.

Stone, A. H. (1956). Metrizability of decompostion spaces, *Proc. Am. Math. Soc. 7*, 690–700.

Stone, A. H. (1959). Metrizability of unions of spaces, *Proc. Am. Math. Soc. 10*, 361–366.

Tietze, H. (1915). Über Funktionen die auf einer abgeschlossene Menge stetig sind, *J. R. Ang. Math. 145*, 9–14.

Tihonov, A. (1935). Über einen Funtionenräum, *Math. Ann. 111*, 762–766.

Urysohn, P. (1925). Zum Metrisation Problem, *Math. Ann. 94*, 309–315.

Veblen, O. (1905). Theory of plane curves in non-metrical analysis situs, *Trans. Am. Math. Soc. 6*, 83–98.

Voxman, W. L. (1970). On the shrinkability of decompostions of 3-manifolds, *Trans. Am. Math. Soc. 150*, 27–39.

Voxman, W. L. (1972). Decompositions of 3-manifolds ånd pseudoisotopies, *Trans. Am. Math. Soc. 164*, 503–508.

Whitehead, J. H. C. (1939). Simplicial spaces, nuclei, and m-groups, *London Math. Soc. Proc. 45*, 243–327.

Whitehead, J. H. C. (1949). Combinatorial homotopy I, *Bull. Am. Math. Soc. 55*, 312–245.

Whyburn, G. T. (1936). On the structure of continua, *Bull. Am. Math. Soc. 42*, 49–73.

Wilder, R. L. (1929). *Topology of Manifolds*, Am. Math. Soc. Colloq. Publ. 32.

Young, G. S. (1964). A condition for the absolute homotopy extension property, *Am. Math. Monthly 71*, 896–897.

Zeeman, E. C. (1963). *Seminar on Combinatorial Topology*, Inst. des Hautes Etudes Sci., Paris.

Zeeman, E. C. (1964). Relative simplicial approximation, *Proc. Camb. Phil. Soc. 60*, 39–43.

Zemke, C. (1974). *Dimension, Decompositions, and Pseudo-isotopies*, Thesis, Univ. of Idaho, Moscow.

INDEX